Chemistry:
A Unified Approach

Chemistry:
A Unified Approach

J. W. BUTTLE, B.Sc., M.A.
University of London, Goldsmiths' College

D. J. DANIELS, B.Sc., Ph.D., F.R.S.C.
University College of Arts, Science and Education, Bahrain

P. J. BECKETT, B.Sc., Ph.D., A.R.S.C.
North Kesteven School, Lincoln

Butterworths
London Boston Sydney Wellington Durban Toronto

First published 1966
Second edition 1970
Reprinted 1973
Third edition 1974
Fourth edition 1981

© Butterworth & Co (Publishers) Ltd 1981

British Library Cataloguing in Publication Data

Buttle, J. W.
  Chemistry – 4th ed.
  1. Chemistry
  I. Title   II. *Daniels, D. J.*
  III. *Beckett, P. J.*
  540   QD33

  ISBN 0-408-70938-3

Filmset by Mid-County Press, London SW15
Printed and bound by Robert Hartnoll Ltd, Bodmin, Cornwall

Preface to the Fourth Edition

This Fourth Edition represents a response to comments from previous readers and to the changing climate of Science Education (for example, as revealed in the recent Association for Science Education Consultative document *Alternatives for Science Education*). Thus, the historical section has been rewritten to take account of current philosophical and sociological debates. The final chapter on industrial chemistry has been extended to deal with the social implications and the current economic crisis in the Western World. Chapter 6, Energetics and Kinetics, has also been rewritten with the intention of adding greater clarity and giving a more detailed consideration of energy changes. A glossary has been added to the collection of Appendixes. Questions from recent examination papers are included at the end of the appropriate chapters; we thank the various examination boards for their permission to use the questions.

Preface to the First Edition

We believe that, for too long, the teaching of Chemistry has been handicapped by a forbidding mixture of fact, recipe and mysticism. This has been due, in large measure, to the content of the examination syllabus, with its emphasis on descriptive aspects and uncomprehending memorizing; the teacher, aware of his responsibilities to his students, has thus felt obliged to follow this pattern and provide his pupils with 'all the facts'. There is, however, no need for the textbook to emulate the teacher: rather, we think, it should extend, stimulate and provoke. Perhaps then the Science student will be encouraged to make fuller use of his critical faculties and be more prepared to participate in discussion.

This particular book is based upon our own A level Chemistry courses, but we think it should also prove of value to Open Scholarship candidates and for University students up to the level of a General degree. Much of the material is also relevant to the industrially-biased O.N.C. courses.

It has been our intention to provide the student with a book that is complementary to his laboratory notes. We have therefore omitted the practical detail which could obscure the main issues involved and render the book unwieldy, although a chapter has been devoted to experimental techniques, to indicate the type of evidence upon which chemical theories may be based and reaction mechanisms formulated. Principles rather than uncorrelated fact have been emphasized throughout the book, in the belief that they are not always so difficult to understand as their belated entry into present examination syllabuses might suggest. At the same time, we have tried to introduce modern knowledge and ideas in a manner compatible with present examination requirements without, we hope, being too precise in our approach and terminology. It would be unwise to suggest that a chemical formula always gives a wholly accurate impression or that a chemical equation conveys a full account of what actually takes place. We have endeavoured to give those representations that seem most relevant to the context. The teacher can, in fact, play a very useful role by discussing, from time to time, the limitations inherent in the symbolism of the subject. The student should not be encouraged to take a definition too precisely or to believe, for example, that chemical bonding can, at the present time, be completely elucidated in terms of electrons or electron clouds. (The electron itself illustrates perfectly the dilemma that awaits the chemist who insists on the 'cut and dried' approach.) In short, the student should be made fully aware that Chemistry, like the other sciences, is imperfectly understood and is in a constant state of flux, with theories being rendered obsolete, new techniques being devised and the frontiers of the subject being continually extended.

The book is really in three sections. In the first Part, the physical aspect of the subject is developed in order to make the rest of the book more meaningful. We make no apology for discussing free energy and entropy, nor for introducing such things as s, p, d and f orbitals. The second Part is a comparative review of the chemical elements from the point of view of the Periodic Table. Because so much chemistry is the chemistry of the non-metals and anions, we have progressed through the Periodic Classification from right to left. We have found it convenient to deal with the elements of Groups 1, 2 and 3 as a whole and to regard the Transitional Metals as members of the d block of elements rather than as members of undefined sub-groups of the representative elements.

The last Part of the book deals with Organic Chemistry. Here the emphasis is on the functional group and the characteristic property. We have not found it necessary to perpetuate the division between aliphatic and aromatic compounds; instead, the aromatic nucleus has been regarded as a modifying influence upon the functional group. Because we feel that they are a rationalizing influence in Chemistry, we have used reaction mechanisms widely.

We have tried to provide a broad view of Chemistry, in the hope that the student will get things better in perspective. For instance, we have discussed topics such as the chemistry of inheritance, respiration and photosynthesis, the lanthanoids and actinoids and various geological aspects of the distribution of the elements, as well as covering the more important industrial processes. Throughout the book the student has been introduced to systematic nomenclature, and at the end of each Chapter are included selections of questions which we hope will prove helpful to him. We are very grateful to the publishers for their advice and assistance.

J. W. Buttle   D. J. Daniels   P. J. Beckett

Contents

1 The development of chemistry

Science and chemistry • Development of atomic and molecular theories • The classification of the elements • Deeper theories of the atom • Notes

Science and chemistry

We have attempted to provide in the following chapters a coherent account of a present-day body of knowledge conventionally known as 'Chemistry'. As you come to grips with the contents of this book, you should, however, keep in mind the following:

(1) We have made our own selection of material from an enormous quantity of fact and theory, bearing examination requirements in mind.
(2) Many of these 'facts' and theories are the subject of controversy and argument.
(3) 'Chemistry' is a man-made category and like any other social construct is liable to change with the passage of time and under the pressure of historical events.

Given, then, that Chemistry is a label for a particular body of knowledge, and the activities carried out by chemists in amassing that knowledge, what might it involve? Putting the question differently, what might be distinctive about 'Chemistry'? The *Concise Oxford Dictionary* defines Chemistry as 'the science of the elements and compounds & their laws of combination & behaviour under various conditions'.

This definition, of course, relies for its meaning upon the interpretation given to the concept 'Science', so, as with most dictionary definitions, we need to look further afield. Fortunately, philosophers through the ages have devoted a considerable amount of thought to just such a question as:

What marks off Science from other human activities?

Philosophers in the past were inclined to regard Science as an activity whereby truth was achieved by an *inductive* process, that is, by arguing from the particular to the general and arriving at a universal law of nature. Induction, as a basis for scientific method, has however come under increasing attack because:

(1) There is always the possibility of an exception to the rule turning up.
(2) It implies an *empiricist* stance, with the facts 'speaking for themselves' to an openminded observer free from preconceptions.

So, Science, of late, has come to be associated with probability rather than certainty and with a *hypothetico-deductive* approach. According to this alternative conception, the scientist generates a hypothesis as a candidate for explanation, deduces testable consequences from it and proceeds to test the credibility of the hypothesis by experiment and observation. Controversy among the supporters of this hypothetico-deductive model arises in terms of whether the emphasis is upon confirmation or falsification of the hypothesis[2]. Supporters of the latter view argue that the 'confirmation' approach suffers from the same weakness as the inductive process described above, in that an exception might one day turn up. On this view the status of the truth arrived at by falsification is higher and 'purer', even if it is only the truth of what is *not* the case.

So the debate goes on. All very confusing to students who may have been brought up to believe that Science represents one of the few areas of objective certainty in an uncertain world. Perhaps one of the sources of confusion is the tendency for philosophers to deal with abstractions; in this case, with the 'ideal' scientist. Perhaps it will prove more constructive to look at the work of *real* scientists in their social and historical context. So, for the rest of this Chapter we will explore the work of some scientists engaged in problems concerning basic chemical concepts such as 'element', 'compound' and 'atom'. You should read this section again after reading the rest of the book.

Development of atomic and molecular theories

We start towards the end of the eighteenth century, when several scientists were working in the field of what we would nowadays call chemical composition, noting, for example, the loss in weight on heating certain substances as well as the nature of the various 'airs' given off. They had inherited from previous ages and thinkers certain ideas which they could use to explain their observations[3]. One of these was 'elective affinity' which asserted that the elementary atoms of substances were held together by forces of mutual affinity. Not only was this force invoked for what we now recognize as chemical compounds but also for solutions. Thus no distinction was drawn between metal dissolving in acid and salt dissolving in water. Indeed, the only substances regarded as physical mixtures were those that could be visibly distinguished as such.

Nevertheless, at the turn of the century, Prout put forward the Law of Constant Composition as a result of his experiments on certain substances (which we would now recognize as pure compounds). What was really happening was that Proust was challenging the idea of elective affinity, as it was then applied, and was carrying out his research in this critical spirit:

> The attraction which makes sugar dissolve in water may or may not be the same as that which makes a definite quantity of charcoal and hydrogen dissolve in another quantity of oxygen, to form the sugar of our plants; but we can see clearly that the two sorts of attraction are so different in their results that it is impossible to confound them.

He was naturally challenged by orthodoxy, notably in the shape of Berthollet, who claimed that, unless one constituent crystallized out or distilled over, elements could combine in variable proportions. Since they were arguing from different perspectives the conflict remained unresolved, for the time being.

The person who was to provide the decisive push into a different mode of interpretation of chemical changes was John Dalton. Through his interests in meteorology he was investigating the extent of dissolution of gases in water and the effect of mixing gases upon the final pressure. He had shown from his writings that he was out of sympathy with Berthollet's notion of variable chemical composition:

> The opinions I more particularly allude to are those of Berthollet on the Laws of chemical affinity; such that chemical agency is proportional to mass, and that in all chemical unions there exist insensible gradations in the proportions of the constituent principles. The inconsistence of these opinions, both with reason and observation, cannot, I think, fail to strike everyone who takes a proper view of the phenomena.

He now sought to explain the diffusion of gases into a vacuum or into each other through repulsive forces between similar particles (exerting their partial pressure on the container). He also went on to postulate that these similar particles were identical by calling upon what in retrospect appears unremarkable, even commonplace, evidence:

> If some of the particles of water were heavier than others, if a parcel of the liquid were constituted principally of these heavier particles, it must be supposed to affect the specific gravity of the mass, a circumstance not known. Similar observations may be made on other substances. Therefore we may conclude that *the ultimate particles of all homogeneous bodies are perfectly alike in weight, figure etc.* In other words, every particle of water is like every other particle of water; every particle of hydrogen is like every other particle of hydrogen etc.

Dalton did not stop there: if he had, the impact of his hypothesis would have probably been much less. Instead, he spelt out some of the practical implications (or, alternatively, testable consequences):

> In all chemical investigations, it has justly been considered an important object to ascertain the relative *weights* of the simples which constitute a compound. But unfortunately the inquiry has terminated here; whereas from the relative weights in the mass, the relative weights of the ultimate particles or atoms of the bodies might have been inferred, from which their number and weight in various other compounds would appear, in order to assist and to guide other future investigations, and to correct their results. Now it is one great object of this work, to shew the importance and advantage of ascertaining the relative weights of the ultimate particles, both of simple and compound bodies, the number of simple elementary particles which constitute one compound particle, and the number of less compound particles which enter into the formation of one more compound article.
>
> If there are two bodies, A and B, which are disposed to combine, the following is the order in which the combinations may take place, beginning with the most simple, namely:
>
> 1 atom of A + 1 atom of B   = 1 atom of C, binary
> 1 atom of A + 2 atoms of B = 1 atom of D, ternary
> etc. etc.

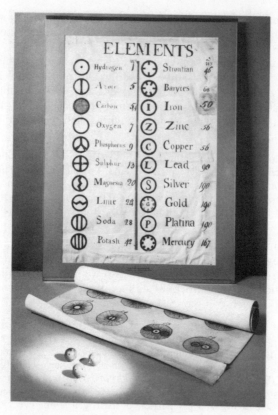

The following general rules may be adopted as guides in all our
investigations respecting chemical synthesis.
(1) When only one combination of two bodies can be obtained, it must be
    presumed to be a *binary* one, unless some cause appear to the contrary.
(2) When two combinations are observed, they must be presumed to be a
    binary and a *ternary* etc. etc.

Dalton, then, besides redefining such concepts as atom, compound atom,
and element, was proposing new rules for chemical investigation. These rules,
however, did not necessarily give the correct formulae for compounds;
because Dalton was aware of only one compound of hydrogen and oxygen,
he assumed that water was a binary compound HO. More evidence was
needed, which would make Dalton's rules redundant.

In 1808, the year in which Dalton's 'A New System of Chemical
Philosophy' was published (from which the above extracts are taken) Gay-
Lussac publicly formulated the law which still bears his name:

Not only, however, do gases combine in very simple proportions... but the
apparent contraction of volume which they experience on combination has
also a simple relation to the volume of the gases, or at least to that of one of
them.

This was just the generalization needed by chemists searching for the
chemical formulae of compounds. Despite the fact that it was perfectly in

Figure 1b John Dalton
(Crown copyright)

keeping with his atomic theory, Dalton was unable to accommodate it to his world-view and dismissed it as spurious:

> The truth is, I believe, that gases do not unite in equal or exact measures; when they appear to do so, it is owing to the inaccuracy of our experiments.

Gay-Lussac himself reconciled the laws of both fixed and variable proportions (and the conflicting positions of Proust and Berthollet) as follows:

> We must first of all admit, with M Berthollet, that chemical action is exercised indefinitely in a continuous manner between the molecules of substances, whatever their number and ratio may be, and that in general we can obtain compounds with very variable proportions. But then we must admit at the same time that — apart from insolubility, cohesion and elasticity, which tend to produce compounds in fixed proportions — chemical action is exerted more powerfully when the elements are in simple ratios... In this way we reconcile two opinions and maintain the great chemical law.

So Gay-Lussac, like Berthollet, did not see the need to take up Dalton's notion of 'compound atom' and hence distinguish between what *we* would now call solutions and compounds.

It was Avogadro who, in 1811, brought a consistent perspective to bear. Discussing the significance of Gay-Lussac's Law, he wrote:

> ...the first hypothesis to present itself...is the supposition that the number of integral molecules in any gas is always the same for equal volumes.'

Avogadro realized that if his hypothesis were true, it would be possible to determine the relative masses of molecules and in fact he himself calculated that the molecule of oxygen was about 15 times heavier than a molecule of hydrogen. (Embedded in Avogadro's hypothesis is the idea of the *mole*[4]. A given volume of *any* gas, at a particular temperature and pressure, contains a fixed number of molecules. In fact 22.4 dm^3 of a gas at standard temperature and pressure (*see* p. 67) contains the relative molecular mass and, by Avogadro's hypothesis, a constant number of molecules = the mole.) But, although the atomic theory as formulated by Dalton directed attention to the determination of relative atomic (and molecular) masses, no-one at that time seemed prepared to apply Avogadro's hypothesis. Instead, because there was at that time no way of determining the precise formula of a compound, chemists when attempting to determine the relative atomic mass usually inadvertently arrived at the *equivalent*; that is, the mass of element combined with one gram of hydrogen. For example, eight grams of oxygen are found to combine with one gram of hydrogen to form water; if one assumes (as Dalton did) that water is a binary compound, then one also assumes that the relative atomic mass of oxygen is 8.

Almost half a century went by before Cannizzaro revived Avogadro's hypothesis as a means of resolving this dilemma. By measuring the masses of equal volumes of the gaseous element as well as various gaseous compounds of it (and therefore, by Avogadro's hypothesis, equal numbers of molecules) and determining the amount of the element in each compound, he arrived at the kind of information shown in *Table 1.1* for the case of hydrogen and its compounds.

Table 1.1 Diatomic nature of hydrogen (after Cannizzaro)

Name of substance	Mass of one molecule	Mass of hydrogen present in one molecule
Hydrogen	2x	2x
Hydrogen chloride	36.5x	1x
Methane	16x	4x
Ammonia	17x	3x
Hydrogen sulphide	34x	2x

Since the smallest amount of hydrogen in a compound was half the hydrogen molecule, it was assumed that the hydrogen molecule is diatomic. In Cannizzaro's own words:

> Compare the various quantities of the same element contained in the molecule of the free substance and in those of all its different compounds, and you will not be able to escape the following law: The different quantities of the same element contained in different molecules are all whole multiples of one and the same quantity which, always being entire, has the right to be called an atom.

In this way, then, Cannizzaro made the abstract concepts of atomic mass and atomicity more real.

By means of similar experiments and reasoning, the molecular formula for certain compounds could be ascertained. For example, water could be shown to be a ternary and not a binary compound (as Dalton had thought) by establishing the combining volumes of the various substances:

2 volumes of hydrogen + 1 volume of oxygen → 2 volumes of steam

By Avogadro's hypothesis:

2 molecules of hydrogen + 1 molecule of oxygen → 2 molecules of steam

Since both hydrogen and oxygen are diatomic (established by Cannizzaro's method above),

1 molecule of steam contains 2 atoms of hydrogen and 1 atom of oxygen.

That is, the formula is H_2O.

The relative molecular mass of comparatively volatile compounds can also be determined by measuring the vapour density[5] and applying Avogadro's hypothesis and the knowledge that the hydrogen molecule is diatomic.

Cannizzaro's method of determining relative atomic masses was applicable only to gaseous elements. In the case of solid elements, one possible approach was to determine the equivalent mass; that is, the mass of element displacing one gram of hydrogen. Since the valency, V, is the number of atoms of hydrogen displaced by one atom of element, it follows that[6]

Atomic mass = Equivalent mass × valency

$A = E \times V$

It remained to determine the valency. This could possibly be calculated by recourse to two laws established in 1819:
(1) *Mitscherlich's Law of Isomorphism:* 'Substances which have similar crystalline form (isomorphous substances) have similar formulae.'[7]
(2) *Dulong and Petit's Law:* 'The atoms of all simple substances have the same capacity for heat.'[8]

The scientific revolution[9] launched by John Dalton at the beginning of the nineteenth century had clearly, by the middle of the century, changed both the course of chemical research and the conceptual framework by which sense was made of experimental data. The basic concepts of atom, molecule, atomic mass, formulae and valency had taken on new meaning and an impressive and relatively coherent body of knowledge was being built up along the lines indicated above. *It is probably not too much to claim that the first half of the century was decisive in marking out new territory distinct from physics and putting chemistry on a respectable footing.*

The classification of the elements

This new way of looking at the physical world, allied to the abandonment of the Greek concept of 'element'[3], as well as the development of new techniques, resulted in a rapid increase in the discovery of chemical elements. As the list grew, chemists sought to introduce order and system by classifying these elements into groups or families. The major achievement in this respect was that of the Russian chemist Mendeleev, who published his Periodic Table, as well as his reasoning, in the 1869 issue of the *Journal of the Russian Chemical Society*, from which the following extracts are taken.

First of all, Mendeleev reviews the overall position with regard to the criteria used up to then by chemists in their attempt to arrange the elements systematically:

> The most frequent classification of elements into metals and non-metals is based upon physical differences...ever since it became known that an element such as phosphorus could appear in non-metallic as well as in metallic form, it became impossible to found a classification on physical differences...
>
> Those systems which are based upon the behaviour of elements with respect to hydrogen and oxygen likewise show many uncertainties and separate elements which, without doubt, exhibit great similarity...
>
> The ordering of the elements according to their electrochemical sequence is considered by the history of chemistry to be as unfortunate an attempt as is the ordering according to their relative resemblance...
>
> In recent times the majority of chemists is inclined to achieve a correct ordering of the elements on the basis of their valency. Elements such as vanadium, molybdenum, tungsten...and the elements of the platinum group form compounds with different valencies which are so characteristic that, at least for the present, it is impossible to think that we can make use of the strict concept of valency in order to understand the compounds of these elements...
>
> Thus, there does not exist a single principle which can withstand criticism that would permit their arrangement in a more or less strict system...
>
> However, everybody does understand that in all changes of properties of elements, something remains unchanged. In this regard only a numerical value is known and this is the atomic weight appropriate to the element... For this reason I have endeavoured to found the system upon the quantity of the atomic weight.

T a b e l l e I.

Typische Elemente								*der chemischen Elemente.*
			K = 39	Rb = 85	Cs = 133	—	—	
			Ca = 40	Sr = 87	Ba = 137	—	—	
			—	?Yt = 88?	?Di = 138?	Er = 178?	—	
			Ti = 48?	Zr = 90	Ce = 140?	?La = 180?	Th = 231	
			V = 51	Nb = 94	—	Ta = 182	—	
			Cr = 52	Mo = 96	—	W = 184	U = 240	
			Mn = 55	—	—	—	—	
			Fe = 56	Ru = 104	—	Os = 195?	—	
			Co = 59	Rh = 104	—	Ir = 197	—	
			Ni = 59	Pd = 106	—	Pt = 198?	—	
H = 1	Li = 7	Na = 23	Cu = 63	Ag = 108	—	Au = 199?	—	
	Be = 9,4	Mg = 24	Zn = 65	Cd = 112	—	Hg = 200	—	
	B = 11	Al = 27,3	—	In = 113	—	Tl = 204	—	
	C = 12	Si = 28	—	Sn = 118	—	Pb = 207	—	
	N = 14	P = 31	As = 75	Sb = 122	—	Bi = 208	—	
	O = 16	S = 32	Se = 78	Te = 125?	—	—	—	
	F = 19	Cl = 35,5	Br = 80	J = 127	—	—	—	

Figure 1.2 Mendeléev's Periodic Table (Crown copyright)

Mendeleev then considers various arrangements before arriving at his Periodic Table (*see Figure 1.2*). He summarizes his discussion thus:

(1) The elements, if arranged according to their atomic weights, show a distinct periodicity of their properties.
(2) Elements exhibiting similarities in their chemical behaviour have atomic weights which are approximately equal (as in the case of Pt, Ir, Os) or they possess atomic weights which increase in a uniform manner (as in the case of K, Rb, Cs).
(3) The arrangement of elements or of groups of elements according to their atomic weights corresponds to their so-called valencies.
(4) The bodies most abundantly found in nature possess a small atomic weight.
(5) The magnitude of the atomic weight determines the character of the element.
(6) The discovery of numerous unknown elements is still to be expected.
(7) The atomic weight of an element will have to be corrected eventually when its analogues become known.

Mendeleev's was a notable achievement. He had the vision and boldness to discount anomalous atomic masses and to leave gaps in his Table. He even predicted the properties of the element to be discovered, on the basis of the place awaiting it in the Table. It is worth bearing in mind, however, that we are all more or less creatures of our time, working with contemporary ideas and techniques. As is often the case in the history of Science, although subsequent recognition immortalizes the name as well as the achievement, the impression thus given of an individual pushing back the frontiers of knowledge alone and with the rest of humanity lost from sight is quite false. In this case, a contemporary of Mendeleev, Lothar Meyer, by plotting *atomic volume* (that is, atomic mass ÷ density) against *atomic number* (the number of the element in the series) revealed the periodicity of the elements in a striking and elegant manner (*see* page 40).

These twin achievements led to the development of new concepts in the chemist's vocabulary. Apart from atomic number, 'group' and 'period' entered the language in a new way. But there was still no understanding of why elements should show family relationships and why families should reveal progressive differences. Satisfactory explanation had to await a more sophisticated model of the atom itself.

Deeper theories of the atom (see *Table 2.10*)

Although Michael Faraday put forward his laws of electrolysis in 1834, it was almost fifty years later that Helmholtz, in 1881, drew attention to a significant implication:

Now the most startling result of Faraday's laws is perhaps this. If we accept the hypothesis that the elementary substances are composed of atoms, we cannot avoid the conclusion that electricity also ... is divided into definite elementary portions which behave like atoms of electricity.

Before the century was out, J. J. Thomson, investigating the rays emanating from the cathode of a discharge tube, had discovered these 'atoms

Figure 1.3 Mendeléev (Ann Ronan Picture Library)

of electricity' or *electrons*. At about the same time, Becquerel discovered the phenomenon of *radioactivity*.[10] The nature of the radioactivity:

> Alpha rays, with the mass of the helium atom and two positive charges;
> Beta rays, consisting of rapidly-moving electrons; and
> Gamma rays, made up of electromagnetic waves of very high frequency,

suggested a heterogeneous and electrical atom. The immediate response was to picture the atom as a mass of positively charged particles (protons) embedded in a 'sea' of electrons. In order to maintain the neutral structure of the atom, the number of protons was equal to the number of electrons.

In 1909, Rutherford, Geiger and Marsden used alpha particles to bombard thin gold foil and found that only a very small proportion were reflected back (*Figure 1.4*). Evidently the latter had come under the influence of the positvely-charged part of the atom and so this must form only a small part of the total volume. The model which Rutherford put forward to explain his observations consisted of a very small nucleus containing the positive charge and most of the mass around which the electrons rotated — rather like a solar system in miniature. The analogy, however, is not entirely appropriate. A revolving electron emits energy as radiation and would therefore gradually collapse into the nucleus. To overcome this objection, Niels Bohr in 1913 postulated the existence of 'closed' orbits, incorporating the relatively new quantum theory[11]: provided the electrons remained in their orbits, no energy was emitted. But movement from one orbit to another was accompanied by a quantum of energy — absorbed if moving from an inner to an outer orbit and emitted if the reverse. Impressive support comes from the line spectra of elements, each line being associated with a particular quantum of energy (*Figure 1.5*). This model, with various modifications, holds considerable appeal to chemists, since it provides a convincing explanation for both periodicity (*see* Chapter 2) and valency (*see* Chapter 3).

It remains to provide some explanation for relative atomic masses being fractional (instead of the simple whole numbers originally suggested by

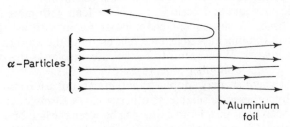

α-Particles

Aluminium foil

Figure 1.4 Geiger, Marsden and Rutherford's experiment

Prout) and for the anomalous atomic masses which perplexed Mendeleev. The explanation came with the discovery by Chadwick (in 1932) of the *neutron*. This is a particle having the mass of the proton but no electrical charge — hence the name neutron. The neutron could now be invoked as a means of increasing the mass of the nucleus without increasing the positive charge, so that the number of electrons outside the nucleus remained unchanged. (For example, the nucleus of the oxygen atom would contain eight neutrons as well as eight protons, giving the appropriate relative atomic mass of 16, whilst still only having the 'correct' number of eight electrons in

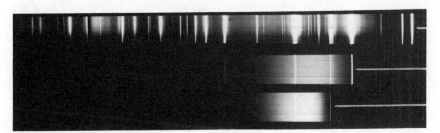

Figure 1.5 Spectrogram of uv light sources (P. W. Allen & Co.)

orbits outside.) Since by now it was realized that the distinctive chemical character of an element was essentially a function of its electronic structure, it would be possible theoretically for different atoms of the *same* element to have different numbers of neutrons and hence different relative atomic masses, whilst retaining identical chemistry. Here was an explanation of the existence of *isotopes* (Soddy, 1913) and of fractional atomic masses, since these as customarily measured would be made up of the arithmetical mean of the relevant isotopes. For instance, in the case of the element chlorine, two isotopes are believed to exist, of mass 35 and 37, in the proportions 75.4 and 24.6 per cent respectively, giving as the average relative atomic mass 35.457. Here, at the same time, is an explanation of anomalous atomic masses, since there could be widely varying numbers of neutrons in the various nuclei of the element, without in any way changing the chemistry. In fact, despite its utilization by Mendeleev, the relative atomic mass is irrelevant to the position of the element in the Periodic Table (as Mendeleev seemed at times to suspect). The atomic number becomes the key concept; apart from being the number of the element in the Periodic Table, it is also the number of extranuclear electrons in the various isotopes of the element and hence a measure of the chemistry of that particular element.

Atomic structure and periodicity will be dealt with more fully, and on a 'factual' basis, in subsequent chapters. It remains to ask here what the historical development sketched out tells us about the way chemists might go about their work. What seems to be the case, insofar as this development is typical, includes a number of features:

(1) Whilst scientists may adopt a hypothetical stance it by no means follows that hypotheses, falsifiable or otherwise, are formulated explicitly.

(2) There can be a considerable time-lag between the first break-through to a new way of interpreting evidence and the absorption of the new ideas into a consistent and coherent perspective. Witness, for example, the contributions of Gay-Lussac and Avogadro which 'went begging' for almost half a century because their significance to the new movement was not appreciated. This transitional period is, according to T. S. Kuhn, typical of 'science in crisis' and is a time of pronounced professional insecurity.

(3) Much of the time when 'normal science'[9] prevails is a time of fact-gathering, development of new techniques and the generation of quantitative laws. In our example, it was a period of intense activity in the determination of 'atomic weights' and the search for new elements. It would be strange to deny these research workers the title of 'scientist' because they did not proceed on a formal hypothetico-deductive basis. They were, in fact, asking certain questions of nature at a point in historical time. The answers obtained either fitted into the scientific wisdom of the day or they raised further questions by not fitting in. Such would appear to be the stuff of which scientific progress is made. It is not the direct approach to objective, certain and permanent truth; rather is it a matter of trying to make increasing sense of natural phenomena.

Notes

1.   A prominent advocate of this inductive approach was Francis Bacon (1561–1626). His advocacy is to be found in, for example, his *First and Second Books of Aphorisms.*

2.   Karl Popper is the leading protagonist. He maintains
   (1) Those hypotheses that are most refutable but have not yet been refuted are always to be preferred to those that are less refutable.
   (2) Those hypotheses that are most refutable are those which are most simple.
   (3) Refutability is a sufficient and necessary condition for separating science from non-science (*see* his *The Logic of Scientific Discovery*).

3.   Several Greek philosophers speculated upon the ultimate nature of matter, in both qualitative and quantitative terms. Thus, *Empedocles* reduced all matter qualitatively to the four Elements, Fire, Earth, Water and Air, which in various combinations gave the properties of hot, cold, dry and wet. *Democritus* argued that all matter was made up of absolutely solid and

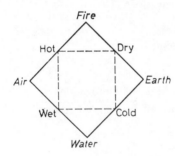

Figure 1.6 The 'four elements'

indestructible atoms and that existing space involved an alternation of atoms and 'vacuity', since it was neither completely full nor completely empty. But nothing experimental followed from either, since their statements were assertive rather than hypothetical. If they were scientists, then their 'science' would not now be recognized as such. But their ideas entered the consciousness of Western man and doubtless had an important effect on how they saw the natural world.

4.   That the number of molecules in a *mole* of compound (or the number of atoms in a mole of element) is constant can be shown as follows:

$$\text{Number of molecules in 1 mole} = \frac{\text{Relative molecular mass, } M}{\text{Mass of 1 mole substance}}$$

But

The mass of 1 molecule of substance $= M \times$ mass of 1 atom of hydrogen
Therefore

$$\text{Number of molecules in 1 mole} = \frac{M}{M \times \text{mass of hydrogen atom}}$$

$$= \frac{1}{\text{Mass of hydrogen atom}}$$

$$= \text{Constant (about } 6.023 \times 10^{23})$$

A *molar* (M) solution contains one mole of substance per dm^3 of solution; a decimolar (0.1M) a tenth of a mole, and so on. Hence, given volumes of solutions of the same molarity contain the same number of molecules, ions etc.

5.   The relationship between vapour density and relative molecular mass can be established as follows:

Vapour density =

$$= \frac{\text{Mass of vapour}}{\text{Mass of equal volume hydrogen (at same temp. and pressure)}}$$

$$= \frac{\text{Mass of } n \text{ molecules of vapour}}{\text{Mass of } n \text{ molecules hydrogen}} \text{ (by Avogadro's principle)}$$

$$= \frac{\text{Mass of 1 molecule of vapour}}{\text{Mass of 1 molecule hydrogen}}$$

$$= \frac{\text{Mass of 1 molecule of gas}}{\text{Mass of 2 atoms hydrogen}} \tag{1}$$

But

$$\text{Relative molecular mass} = \frac{\text{Mass of 1 molecule}}{\text{Mass of 1 atom hydrogen}} \tag{2}$$

$$\frac{(2)}{(1)} = 2$$

that is,

Relative molecular mass = 2 × vapour density.

6.   That $A = E \times V$ can be shown as follows:

If the valency of the element is V, then

1 atom combines with V atoms of hydrogen

If the mass of 1 atom of hydrogen is x grams, then

the mass of 1 atom of element is Ax grams

and

Ax grams combine with or displace Vx grams of hydrogen

that is

$\frac{A}{V}$ grams combine with or displace 1 gram of hydrogen

But this is the equivalent mass E

Therefore

$$\frac{A}{V} = E \quad \text{or} \quad A = E \times V$$

7. The law of Isomorphism has its limitations, but it did permit the valency of several elements to be evaluated by comparison with isomorphous compounds of known formulae. For example, in the case of the alums — a series of double sulphates — if the original aluminium ion can be replaced by another ion, the latter will have the same valency as aluminium.

8. Dulong and Petit's Law can be written:

The atoms of all simple substances have the same capacity for heat.

Because there is the same number of atoms in a mole of any element (the Avogadro constant) it follows that the '*atomic heat*' (i.e. the product of relative atomic mass and specific heat capacity) should also be constant. With very few exceptions this is true approximately for solids at ordinary temperatures, as shown in *Table 1.2*. A knowledge of the specific heat capacity of the element can therefore be instrumental in determining relative atomic masses.

Table 1.2 Some atomic heats (for units, *see* Appendix 1)

Element	Atomic heat/J mol^{-1} K^{-1}	Element	Atomic heat/J mol^{-1} K^{-1}
Magnesium	25.6	Gold	26.0
Zinc	25.6	Chromium	26.4
Silver	25.6	Bismuth	26.4
Phosphorus	26.0	Iron	26.8

For example, the specific heat capacity of nickel is 0.46 J g^{-1} K^{-1}. The relative atomic mass is therefore, from Dulong and Petit's Law

$\frac{26.4}{0.46}$ or approximately 57.

Now the equivalent mass, by experiment, is 29.35, so the valency ($= A/E$) is approximately $57 \div 29.35$ which, to the nearest integer, is 2. The *accurate* relative atomic mass ($E \times V$) is therefore

$29.35 \times 2 = 58.70$.

9. At least that is how Thomas Kuhn in his book *The Structure of Scientific Revolutions* describes it. It was an example, still in Kuhn's terms, of *science-in-crisis* where an alternative theory is competing with the established theory as a more successful resolver of growing anomalies. When a theory (or better a *paradigm*) holds unchallenged sway, *normal science* prevails and scientists busy themselves in solving puzzles suggested by the theory. (Note that Karl Popper would probably not wish to call these puzzle-solvers scientists at all.)

10. A classic example of *serendipity*, a happy chance occurrence. The story

goes that, quite by accident, a key was lying on some photographic plates in a drawer, together with what we now know to be a radioactive substance. When Becquerel later developed the top plate, he discovered a key-shaped shadow. He turned his attention to the third occupant of the drawer and subsequently discovered the phenomenon which we now call radioactivity.

11. To explain certain anomalies in his researches into radiation, Max Planck, in 1900, proposed that energy was not continuous but existed as discrete 'packets' which he called *quanta*. These quanta were fixed for particular frequencies and varied directly with the frequency. That is:

Quantum of energy $= h \times$ frequency

where

$h =$ Planck's constant, 6.6×10^{-34} J s.

Questions

(1) Correlate the four laws of chemical combination with the atomic theory. Can Berthollides be reconciled with any aspects of these laws and theory?

(2) Comment on the statement: 'The structure of an atom represents a miniature solar system'.

(3) A metal chloride has a vapour density of 95.0 and the percentage by mass of chlorine in the compound is 74.6. Calculate the relative atomic mass of the metal.

(4) An element has a specific heat capacity of 0.113 J g^{-1} K^{-1} and an equivalent mass of 39.67. Find the relative atomic mass of the element and the valency state in the compound considered.

(5) The following values were found for five compounds each containing silicon:

| Vapour density | 16.1 | 31.1 | 46.2 | 85.0 | 267.8 |
| per cent Si | 87.5 | 90.5 | 91.4 | 16.6 | 10.4 |

Calculate the relative atomic mass of silicon from these figures.

(6) Modify Dalton's Atomic Theory so that it is in accord with modern views.

(7) Who, of the following, do you think has had the greatest impact on the course of chemical development, and why? Dalton, Mendeleev, Cannizzaro, Lavoisier, Boyle.

2 The atom

Size and composition

An atom can be roughly described as a very small, dense nucleus, around which electrons circulate in a comparatively large volume, as expounded by Rutherford.

The volume of the atomic nucleus is about 10^{-44} m^3 and it consists of positively charged *protons* and electrically neutral particles known as *neutrons* which have a mass almost identical with that of the protons. These are collectively known as *nucleons* and their sum is numerically equal to the *mass number*, A. The number of protons themselves represents the *atomic number*, Z, of the atom.

The electrons are sufficient in number to balance the positive charge on the nucleus, each electron having a negative charge numerically equal to the positive charge of a proton, although the mass of an electron is only about $\frac{1}{1840}$ of that of a nucleon. The total volume of an atom is very large compared with the volume occupied by the nucleons, being of the order of 10^{-30} m^3, so that the volume of the nucleus is only about 10^{-14} that of an atom.

Some of the properties of these fundamental particles are shown in *Table 2.1* and *2.2*, while *Table 2.3* illustrates the particulate composition of some representative elements.

The four quantities, atomic number, atomic mass, charge and atomicity are used to designate an atomic species, for example, a singly ionized dinitrogen molecule consisting of atoms of mass number 14 can be shown as: $^{14}_{7}N_2^+$, for which $Z = 7$.

The electronic structure of the atom

Planck had suggested in 1900 that energy (like matter) is discrete or quantized, that is, energy changes occur in 'packets' or *quanta* and not continuously. In 1913, Bohr utilized this quantum theory and proposed that an electron revolved in a closed orbit around the nucleus and that there was a clearly defined energy associated with each orbit. Any energy changes then occurred as discrete quanta when an electron was transferred from one orbit to another of different energy. The energy change involved, ΔE, is related to

Table 2.1 The atom

Component	Radius/m
Electron	2×10^{-15}
Hydrogen nucleus	1.5×10^{-15}
Uranium nucleus	9×10^{-15}
An atom	1×10^{-10} (approx.)

Table 2.2 Fundamental particles

	Electron	Proton	Neutron
Rest mass/g	9.1×10^{-28}	1.67×10^{-24}	1.67×10^{-24}
Absolute charge/C	-1.6×10^{-19}	$+1.6 \times 10^{-19}$	
Mass relative to H = 1	0.00054	1.0076	1.0089
Relative charge	-1	$+1$	0
Designation	$_{-1}^{0}e$, β	$_{1}^{1}H$ or $_{1}^{1}p$	$_{0}^{1}n$

Table 2.3 Composition of some representative atoms (the most common isotopes (p. 12) have been selected)

Element	Atomic number	Mass number	protons	Number of neutrons	electrons
H	1	1	1	0	1
He	2	4	2	2	2
Na	11	23	11	12	11
U	92	238	92	146	92

the frequency, v, of the spectral line associated with the transition, by the equation:

$$\text{quantum} = \Delta E = hv$$

where h is Planck's constant (6.625×10^{-34} J s).

As spectra are manifestations of the changes in energy of electrons, their observation can be made to yield much useful information regarding the *energy levels* of the *atomic orbitals*. The *emission spectra* of all the elements consist of lines occurring in different parts of the electromagnetic spectrum (p. 572) and may be observed by passing an electric discharge through the elements in their gaseous state, or by using the more complicated *arc spectra*, produced by striking an electric arc between electrodes of the metal. In the spectra so obtained, the lines represent electrons of the element which, having been excited to higher energy levels, are returning to lower energy states. The return of an electron from different, higher energy levels or shells to a certain lower energy level gives rise to a series of spectral lines. The energy picture for

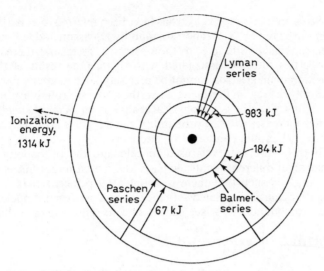

Figure 2.1 Energy picture of the hydrogen atom
(diagrammatic)

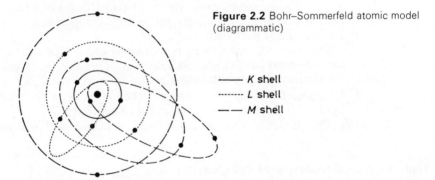

Figure 2.2 Bohr–Sommerfeld atomic model
(diagrammatic)

———— *K* shell
-------- *L* shell
— — *M* shell

different transitions of the single excited electron of the hydrogen atom is
given in *Figure 2.1*. The energy difference between consecutive shells
decreases as the distance from the nucleus increases and ultimately
coalescence of the orbits occurs, so that any electrons reaching this level can
then escape from the influence of the nucleus, and they are therefore said to
have become ionized.

In Bohr's original theory, successive closed shells were shown to be
capable of accommodating 2, 8, 18 and 32 electrons respectively.
However in 1923 Pauli propounded his *Exclusion Principle*, according to
which each electron in an atom is unique, so that two electrons of equal
energy must differ by spinning in opposite directions.

As a result of this principle and an examination of spectra, it became
necessary to split the shells of Bohr's atom into sub-shells (orbits) of different
eccentricities. Because the energy of the electrons in a given orbit is constant,
the maximum number per orbit is two, the electrons spinning in opposite
directions on their own axes (*see Figure 2.2*).

So far in this Chapter electrons have been referred to as particles, but it was demonstrated by G. P. Thomson and by Davisson and Germer in 1927 that electrons can be diffracted by the atoms in a crystal and, therefore, some wave properties must be associated with them. One result of this is that the position of an electron cannot be precisely located at any particular instant, but only the region in which there is the greatest probability of finding it. The probability can be given physical reality as an *electron cloud*, the density of which increases with the probability. Such regions are called *orbitals*, each of which, when full, contains two electrons.

If this wave theory is used to calculate the probability of finding an electron at different points in space, and a continuous line is drawn around the nucleus such that there is about a 99 per cent chance of finding the negative charge associated with the electron between the nucleus and the line, then the boundary surface can be said to represent the orbital.

Types of orbital

The first, K, shell contains one orbital which is spherically symmetrical and is referred to as the *1s orbital*.

The second, L, shell contains one *s* orbital (the *2s*) and three others, known as *2p orbitals*, which each consist of two lobes and which are in three mutually perpendicular directions; the total complement of electrons in this second shell is thus eight.

The third, M, shell, as well as containing a *3s orbital* and three *3p orbitals* can accommodate five orbitals known as *3d orbitals*, thus allowing a further ten electrons to enter this shell, giving 18 in all.

The next, N, shell, can hold a further 14 electrons in a set of seven *4f orbitals*, giving a total of 32 electrons (*Table 2.4*).

The shapes of some of these orbitals are illustrated in *Figure 2.3*.

Table 2.4 Accommodation of electrons

Electron shell	Principal quantum number, n	Angular quantum number, l	Electron type	Magnetic quantum number, m	Spin quantum number, s	Number of electrons accommodated
1st (K)	1	0	s	0	$\pm\frac{1}{2}$	2
2nd (L)	2	0	s	0	$\pm\frac{1}{2}$	2 } 8
		1	p	$+1, 0, -1$	$\pm\frac{1}{2}$	6
3rd (M)	3	0	s	0	$\pm\frac{1}{2}$	2
		1	p	$+1, 0, -1$	$\pm\frac{1}{2}$	6 } 18
		2	d	$+2, +1, 0, -1, -2$	$\pm\frac{1}{2}$	10
4th (N)	4	0	s	0	$\pm\frac{1}{2}$	2
		1	p	$+1, 0, -1$	$\pm\frac{1}{2}$	6
		2	d	$+2, +1, 0, -1, -2$	$\pm\frac{1}{2}$	10 } 32
		3	f	$+3, +2, +1, 0, -1, -2, -3$	$\pm\frac{1}{2}$	14

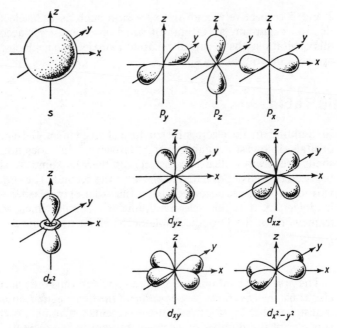

Figure 2.3 Atomic orbitals

The four quantum numbers

The arrangement of electrons shown in *Table 2.4*, and therefore the shape of the Periodic Table, is governed by *four quantum numbers*. The first three arise from solutions of Schrödinger's wave equation (1926) for an electron in the field of an electric charge (i.e., in an atom), and these three numbers characterize an orbital.

The *principal* quantum number n gives the number of the shell and is mainly concerned with energy. The subsidiary or azimuthal or *angular* quantum number l is concerned with the shape or eccentricity of the orbital. The *magnetic* quantum number m takes account of the behaviour of the electron in an external magnetic field.

n can have any integral value from 1 upwards. l can have any integral value from zero to $(n-1)$. m can have any integral value, including zero, between $-l$ and $+l$.

The fourth quantum number s can be $+\frac{1}{2}$ or $-\frac{1}{2}$ and is related to the direction in which an electron spins, and arises from experiments of Stern and Gerlach (1921).

Pauli's exclusion principle postulates that no two electrons in the same atom can have exactly the same set of quantum numbers. If this is applied to the above rules, we arrive at the result that the first shell can hold two s electrons, the second shell two s and six p electrons, the third shell two s, six p and ten d electrons, and the fourth shell two s, six p, ten d, and fourteen f electrons.

Note that the letters s, p, d, and f are shorthand forms of specifying $l = 0, 1,$

2 and 3 respectively, and an expression such as a $3d$ electron means an electron occupying an orbital for which $n=3$ and $l=2$; according to Pauli this orbital may be occupied by up to two electrons, which must then have opposed or antiparallel spins.

The Periodic Table

In 'building up' the electronic structure of an element, the electrons enter the orbitals of lowest available energy. However, this does not mean that the shells are always filled successively, as there is some overlap in energies between orbitals of different shells after the second. The order in which the various orbitals are occupied, that is the order of increasing energy, is shown in *Table 2.5*; it is this selective filling of orbitals which accounts for the framework of the Periodic Table of the elements (*Table 2.6*).

Period one

The first shell contains one s orbital and can thus only accommodate two electrons; the *electronic configuration* of the first electron entering the orbital is designated as $1s^1$ (the first figure represents the number of the shell and the superscript the number of electrons present in the orbital). This electronic configuration is that of the hydrogen atom in the *ground state*, i.e., that of lowest energy. With two electrons, the configuration is $1s^2$. It represents the electronic structure of the helium atom; the electrons must be of opposite spin and these two configurations are often shown as

$$
\begin{array}{cc}
\mathbf{1} & \mathbf{2} \\
\mathbf{H}\ \boxed{\uparrow} & \mathbf{He}\ \boxed{\uparrow\downarrow}
\end{array}
$$

This $1s$ orbital, and hence the first shell, is now full and addition of a further electron must take place in the shell of next higher energy. For a useful mnemonic to remember the order of orbital filling, *see Figure 2.7* (p. 33).

Period two

The third element, lithium, thus has the configuration $1s^22s^1$ and belongs to Group 1 of the Periodic Table. Beryllium, of atomic number four, is represented $1s^22s^2$ or $K2s^2$, and belongs to Group 2.

Unlike the first shell, the second shell has three p orbitals also available and the next electron must enter one of these. Thus boron, of atomic number five, has the configuration $K2s^22p_x^1$, and is the first member of Group 3. The placing of the next electron is indicated by *Hund's rule* which states that, under normal circumstances, no available orbital of any one type will take up a second electron until all the available orbitals of that type have acquired one electron each, e.g., the p orbitals will each receive one electron before any pairing occurs. The next element, carbon (Group 4) will accordingly have the configuration $K2s^22p_x^12p_y^1$; nitrogen, atomic number seven, will be

$K2s^22p_x^12p_y^12p_z^1$ (Group 5), oxygen (in Group 6) $K2s^22p_x^22p_y^12p_z^1$, fluorine (Group 7) $K2s^22p_x^22p_y^22p_z^1$ and neon (Group 0) $K2s^22p_x^22p_y^22p_z^2$.

3	4	5		6		7		8		9		10	
Li	Be	B		C		N		O		F		Ne	
2s	2s	2s	2p	2s	2p	2s	2p	2s	2p	2s	2p	2s	2p

For most purposes it is not necessary to distinguish the various *p* orbitals from each other; for example, nitrogen may be quoted as $K2s^22p^3$.

Period three

The configurations of the next group of eight elements, sodium to argon, follow an analogous pattern, except that it is the third shell which is in the process of being filled (although not completely at this stage).

11	12	13		14		15		16		17		18	
Na	Mg	Al		Si		P		S		Cl		Ar	
3s	3s	3s	3p	3s	3p	3s	3p	3s	3p	3s	3p	3s	3p

Period four

The two elements following, potassium and calcium, have electrons in the 4*s* orbital, but reference to *Table 2.6* shows that after this is full, the next available set of orbitals is not the 4*p* but the 3*d*. These will therefore be the next to be occupied, giving rise in all to ten elements called 'transitional elements', with atomic numbers from 21 to 30. After this the 4*p* orbitals fill up, thus accounting for elements of atomic numbers 31 to 36.

19	20	21	......	25	......	30			31	
K	Ca	Sc		Mn		Zn			Ga	
4s	4s	4s	3d	4s	3d	4s	3d		4s	4p
								KLM		etc.

Period five

This *s*, *p*, *d* sequence is now repeated with elements 37 to 54.

Periods six and seven

For periods six and seven this sequence is modified to include the seven 4*f* orbitals in the former and the seven 5*f* orbitals in the latter but the basic principles remain the same, until ultimately the largest certainly known atom, that of the element with an atomic number of 104 is reached.

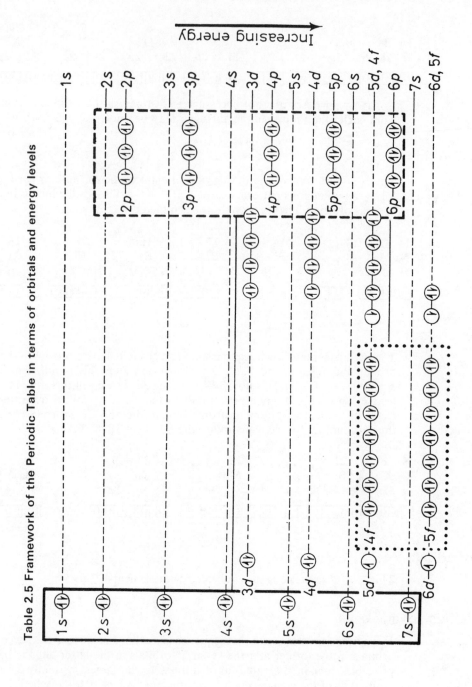

Table 2.5 Framework of the Periodic Table in terms of orbitals and energy levels

Table 2.6 The Periodic Table — Long Form

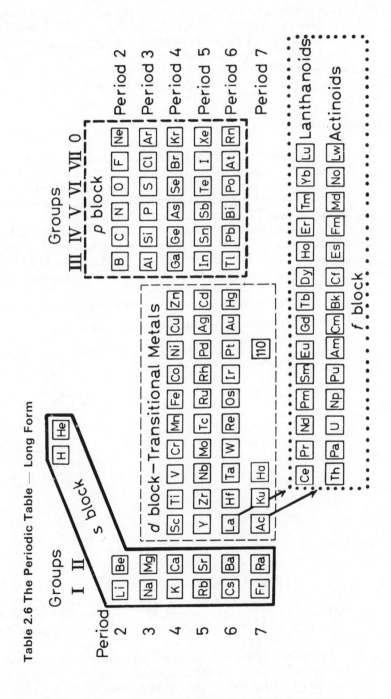

It can be seen that all the elements of Group 1 have one electron in the outer shell, those of Group 2 have two electrons in the outer shell, and similarly for the other Groups. As the chemistry of elements is to a large extent dictated by the outer electronic configurations, the elements in any one Group have great chemical similarity and there is therefore a certain *periodicity* in traversing the Table. Some meaning has now been given to the ideas of Mendeleev expressed in the previous Chapter.

Nuclear stability

Since the number of electrons in an atom is not directly dependent upon the number of neutrons present, it is possible for atoms to have the same number of electrons and protons, and therefore the same chemical properties, but yet contain different numbers of neutrons and hence possess different relative atomic masses. These different atoms of the same element are called *isotopes*. The relative atomic mass of an element, as used chemically, is then an average value representative of the different isotopes present and their relative proportions. For example, the percentage abundance of the two isotopes of chlorine of mass numbers 35 and 37 are 75.4 and 24.6, respectively, giving the relative atomic mass of chlorine as 35·46.

The stable isotopes of some representative elements are given in *Table 2.7*.

Table 2.7 Isotopic composition of some elements

Element	Atomic number	Stable isotopes (*in order of abundance*)	Element	Atomic number	Stable isotopes (*in order of abundance*)
H	1	1, 2	V	23	51
C	6	12, 13	Fe	26	56, 54, 57, 58
O	8	16, 18, 17	Sn	50	120, 118, 116, 119, 117, 124, 122, 112, 114, 115
Na	11	23	I	53	127
S	16	32, 34, 33	Hg	80	202, 200, 199, 201, 198, 204, 196

If the mass number of an element is small, then the relative differences in the isotopic masses can be appreciable, and the isotopes can show some marked divergencies in properties. This point is well illustrated by hydrogen (mass number 1) and its main isotope, deuterium, D (mass number 2), as shown in *Table 2.8*.

The nature of the forces, millions of times more powerful than those between a proton and an extranuclear electron, which operate in the nucleus and bind together the protons and neutrons, is still not understood. It has been suggested that the interconversion neutron + proton → proton + neutron

provides the basis for the force of attraction; certainly a neutron can decay into a proton and an electron, with the emission of energy. Furthermore, many stable atoms possess equal (often even) numbers of protons and neutrons, although the elements of high relative atomic mass possess more neutrons than protons.

Table 2.8 Isotopes of hydrogen

	H_2	HD	D_2
Melting temperature/K	13.8	16.5	18.5
Boiling temperature/K	20.3		23.5
Specific latent heat of fusion/J kg^{-1}	0.118	0.155	0.191
Specific latent heat of vaporization/J kg^{-1}	0.93	1.11	1.27

Notwithstanding the nature of the nuclear force, the stability of the nucleus is revealed by the fact that the mass of the nucleus is not exactly equal to the total of the masses of the individual nucleons. Mass (equivalent to energy in terms of the Einstein equation, $E = mc^2$, where c is the velocity of light) is lost as the nucleons so interact as to produce a more stable situation. The energy required to separate the nucleons of the nucleus is known as the *binding energy* of the nucleus. *Figure 2.4* shows the dependence of binding energy per nucleon upon the mass number, and it is clear from this that the most stable elements tend to be of intermediate mass number and that conversion of elements at the extremes into these more stable elements would result in the release of energy. This is the basis for the nuclear processes known as fusion and fission (pp. 30, 32).

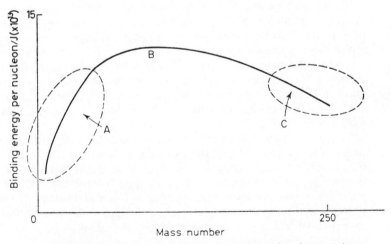

Figure 2.4 Stability of atomic nuclei. Nuclei in region *A* can undergo fusion processes to give larger nuclei with liberation of energy. *B* is the region of nuclear stability. In region *C*, fission processes produce smaller atoms of greater stability with liberation of energy

Many of the elements of very high mass number (all the natural isotopes of elements 84 to 92) are spontaneously unstable and undergo *radioactive decay*, that is, they emit various particles of very high energy as they form a more stable nucleus (*Figure 2.5*).

For example, the loss of a helium nucleus (an α-particle) leaves a nuclear residue of mass number 4 units and atomic number 2 units less than the original atom:

$$^{226}_{88}Ra \rightarrow ^{222}_{86}Rn + ^{4}_{2}He$$

The loss of an electron (a β-particle) from the nucleus involves the conversion of a neutron into a proton, leaving the mass virtually unchanged but giving a new element with atomic number one unit more than the original:

$$^{228}_{88}Ra \rightarrow ^{228}_{89}Ac + ^{0}_{-1}e$$

The disintegration of an unstable isotope is not influenced by any external conditions, and so the rate of decay is proportional to the number of unstable atoms present; i.e. it corresponds to a first-order reaction (p. 140). Clearly, the rate will decline as the number of unstable atoms remaining becomes less, and so it will take an infinitely long time for decay to become complete. Hence we refer to the *half-life* of a radioactive element, or the time taken for half of the atoms to decay.

A knowledge of the half-lives of radioactive elements can be useful in calculating the effect of 'fall-out' from nuclear explosions and in dating rocks and fossils. For example, radioactive carbon-14 (^{14}C) is produced in the atmosphere by the absorption of a neutron (activated by cosmic radiation) into an atom of nitrogen:

$$^{14}_{7}N + ^{1}_{0}n \rightarrow ^{14}_{6}C + ^{1}_{1}p$$

Plants, whilst alive, constantly absorb this radioactive carbon into their system by photosynthesis; death, however, puts a stop to this and so the proportion of this isotope decreases as it undergoes decay and is not replenished. A knowledge of the proportion and the half-life (5700 years) for this isotope permits calculations of the approximate time of death to be carried out—values that can be of great importance to geologists and anthropologists. Similar calculations, based on the proportion and half-lives of various uranium isotopes, suggest that the earth was formed about 5000 million years ago.

Artificially radioactive isotopes can be made by bombardment of atoms with atomic particles, e.g., in the atomic pile. Because of the radiation which these emit on decay, they act as 'tracers' in the elucidation of reaction mechanisms; for example, carbon dioxide containing ^{14}C is used in photosynthesis—compounds subsequently found to contain this particular isotope can then be assumed to have been formed, directly or indirectly, from carbon dioxide. In this way, progressively more detail is obtained about the reactions taking place.

A further very important use of artificial radioisotopes is in medicine, where they are used for both diagnostic and therapeutic purposes. In this context it is important to use isotopes of short half-life so as to minimize the

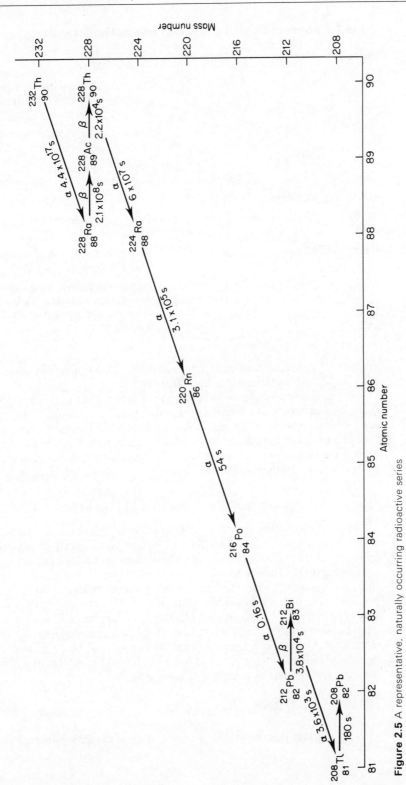

Figure 2.5 A representative, naturally occurring radioactive series

Table 2.9 Some radioactive isotopes of medical interest

Isotope	Half-life	Emission	Uses include
^{131}I	8.0 days	$\beta + \gamma$	Diagnosis and treatment of thyroid conditions, scans of adrenal glands, liver and gallbladder
^{99m}Tc	6 hours	γ	Scans of brain, thyroid, liver, bone-marrow, lungs, renal organs and the pericardium
^{60}Co	5.3 years	$\beta + \gamma$	Cancer therapy
^{32}P	14.3 days	β	Treatment of bone metastases
^{198}Au	2.7 days	$\beta + \gamma$	Treatment of pleural and peritoneal metastases
^{59}Fe	45 days	$\beta + \gamma$	Measurement of iron absorption from the digestive tract
^{239}Pu	24 400 years	α	Batteries in cardiac pacemakers; in this case the radioactivity is used as a source of energy rather than of ionizing radiation

amount of unintended cellular damage. *Table 2.9* summarizes some medically important radioisotopes and their uses.

The best-known isotope which undergoes fission, that is, splits into two or more smaller atoms of roughly comparable mass, is uranium-235 (^{235}U). In each fission process, more neutrons are produced than are used, so that further fission is induced and the process, once initiated, can become a chain reaction with consequent explosion, provided that the material exceeds the critical size, i.e., retains sufficient neutrons and allows sufficient heat build-up, e.g.,

$$^{235}_{92}\text{U} + {}^{1}_{0}\text{n} \rightarrow {}^{236}_{92}\text{U} \rightarrow {}^{143}_{56}\text{Ba} + {}^{90}_{36}\text{Kr} + 3{}^{1}_{0}\text{n} + 28 \text{ pJ fission}^{-1}$$

This is the situation for pure ^{235}U; natural uranium contains about 140 times as much ^{238}U as ^{235}U. As the former is an excellent absorber of neutrons, there is little prospect of a chain reaction developing, and natural uranium is therefore far safer than it might have been thought at first to be the case. By using a 'moderator' for slowing down neutrons, however, it is possible to ensure a favourable balance, as ^{238}U does not readily capture slow neutrons; ^{235}U therefore still undergoes appreciable fission. This is what happens in the slow reactor, where the heat generated is converted into electricity. At the same time, the capture of neutrons by ^{238}U leads to the 'breeding' of fissile material in the form of plutonium, which, like ^{235}U, has military and industrial potential (*Figure 2.6*):

$$^{238}_{92}\text{U} + {}^{1}_{0}\text{n} \xrightarrow{-\beta} {}^{239}_{93}\text{Np} \xrightarrow{-\beta} {}^{239}_{94}\text{Pu}$$

With the fast reactor, ^{235}U is enriched over the 238 isotope by taking

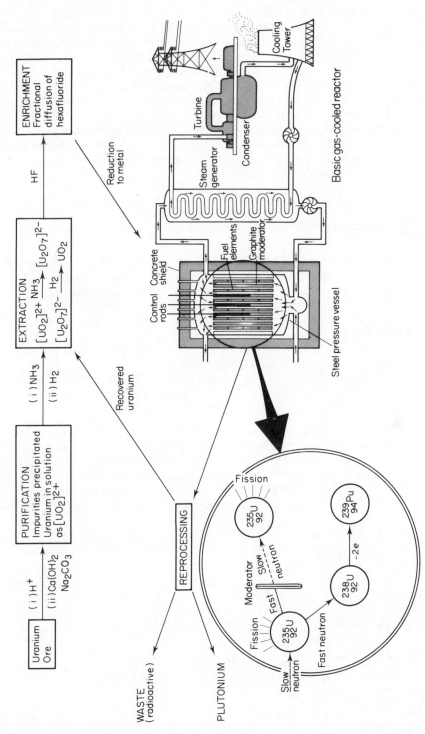

Figure 2.6 Atomic fission

advantage of the slight difference in their masses, commonly by gas centrifugation, which aids the fractional diffusion of the volatile compounds $^{238}UF_6$ and $^{235}UF_6$. Since plutonium is different chemically from uranium, rods of uranium which have been used in atomic reactors can be 'reprocessed' chemically to extract the fissile plutonium. Apart from its military potential, this plutonium can be used in the same way as can ^{235}U to produce electricity. Naturally great care and remote control must be used in handling these very radioactive substances.

At very high temperatures, such as those of stellar atmospheres and thermonuclear devices, *fusion* of hydrogen atoms or their isotopes can take place, with the formation of helium. The binding energies (*Figure 2.4*) are such that considerable energy is evolved, e.g.

$$\underbrace{{}^2_1H + {}^3_1H}_{5.032} \longrightarrow \underbrace{{}^4_2He + {}^1_0n}_{5.012} + 2.8 \text{ pJ}$$

2.015 3.017   4.003 1.009   relative masses

i.e., 0.020 atomic mass units are converted into 2.8 pJ fission^{-1} of energy. Using this information a check can be made on Einstein's equation $E = mc^2$; to convert atomic mass units into mass requires the fact that 1.008 g of hydrogen contain 6.023×10^{23} atoms, also $c = 3 \times 10^8$ m s^{-1}, and therefore the energy released by the loss of mass can be calculated. How can the discrepancy be resolved?

Summary

The development of ideas of atomic structure is shown in *Table 2.10*. The electronic arrangement of an atom is governed by the permissible values taken up by the four quantum numbers, which represent the availability of suitable orbitals. The order of filling these orbitals is given by following the arrows in *Figure 2.7*. The types of outer electronic configurations shown in *Table 2.11* result, giving the characteristic structure of the Periodic Table.

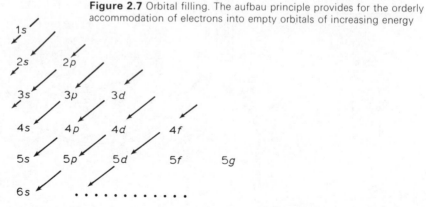

Figure 2.7 Orbital filling. The aufbau principle provides for the orderly accommodation of electrons into empty orbitals of increasing energy

Table 2.10 Development of ideas of atomic structure

Originators	Pictorial representation	Accountability	Limitations
Democritus, Dalton	 solid sphere	Indestructibility of matter Combining ratio in molecules as whole numbers	Radioactivity and disintegration of atoms
Rutherford	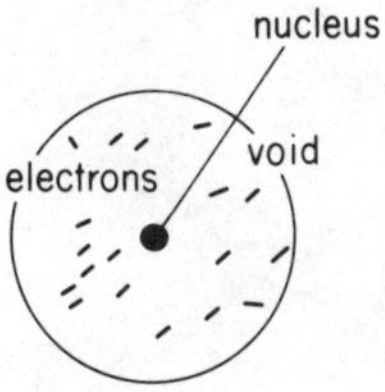 electrons, nucleus, void	Electrons, nucleons and void (Geiger–Marsden experiment, p. 11)	Collapse of electrons into nucleus; continual radiation of energy as electrons accelerate to nucleus
Bohr	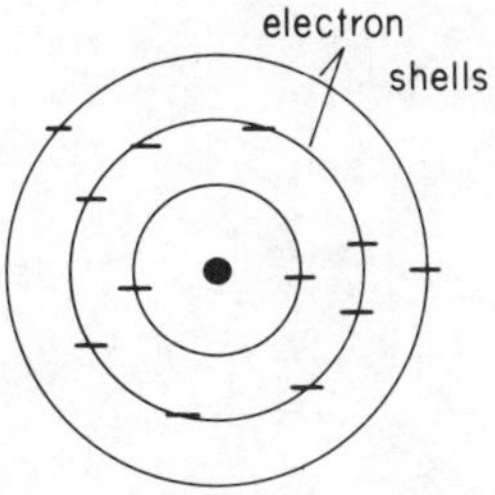 electron shells	Radiation of energy only occurs when electrons jump from one orbit to another	'Makeshift' combination of mechanical and quantum ideas. Violation of Pauli's principle (p. 19)
Bohr–Sommerfeld	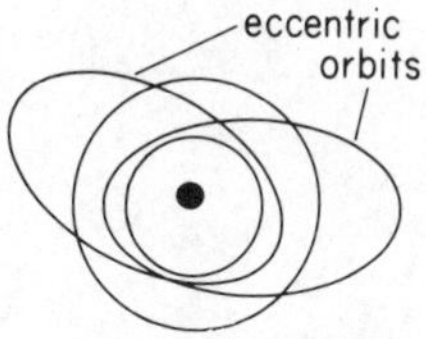 eccentric orbits	Sub-shells (orbits) overcomes Pauli objection. 'Explains' structure of Periodic Table	Unwieldy in 3-D; corpuscular — avoids wave properties of electron
Heisenberg, Schrödinger	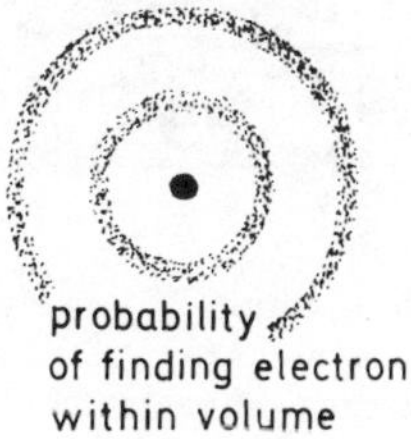 probability of finding electron within volume	Wave-properties of electron, i.e., probability, not certainty, of locating electrons	Solution of complex differential equations

Table 2.11 The structure of the Periodic Table

Outer electron configuration	Elements
$ns^{1\ or\ 2}$	s-Block (representative or normal)
$ns^2np^{1\ to\ 6}$	p-Block (representative or normal)
$(n-1)d^{1\ to\ 10}ns^2$	d-Block (transitional)
$(n-2)f^{1\ to\ 14}(n-1)d^1ns^2$	f-Block (inner transitional)

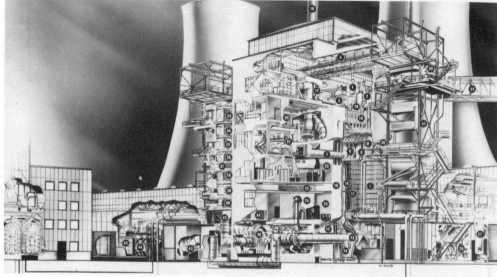

Figure 2.8 A first-generation atomic power station: Calder Hall, Cumbria, opened in 1956. Neutrons, from nuclear fission, are slowed down by graphite moderators, prior to capture by uranium-235 and further fission.
1 Graphite moderator. 2 Charge machine. 3 Discharge machine. 4 Fuel basket. 5 Overhead electric travelling crane. 6 Sampling outlet tubes. 7 Gas outlet duct Number 4. 8 Charging chute. 9 Control rod. 10 Charge tubes. 11 Fuel elements. 12 Viewing tube. 13 Gas inlet duct from Number 4 boiler. 14 Pressure vessel support frame. 15 Thermal shield. 16 Biological shield. 17 Shield cooling fans. 18 Shield cooling stack. 19 Control rod actuators. 20 Fuel element preparation room. 21 Gas sampling plant room. 22 Control room. 23 Battery room. 24 Switch room. 25 Cable gallery. 26 Main circulator Number 4. 27 Blow-down tank. 28 Gas cyclone and filter room. 29 High pressure steam drum. 30 Low pressure steam drum. 31 Boiler Number 1. 32 Duct to boiler Number 2. 33 HP superheater bank, boiler Number 4. 34 HP evaporator bank (upper). 35 HP evaporator bank (lower). 36 HP HT economizer bank. 37 LP superheater bank. 38 LP evaporator bank (upper). 39 LP evaporator bank (lower). 40 HP and LP mixed economizer bank. 41 Gas outlet to circulator Number 4. 42 Motor alternator room. 43 Number 4 Ward – Leonard set. 44 Standby diesel generators. 45 Diesel control panel.

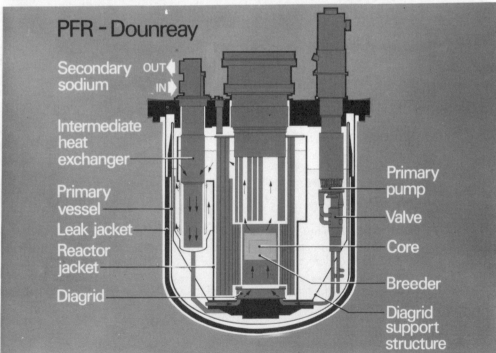

Figure 2.9 (a) Aerial view of Prototype Fast Reactor (PFR) and Dounreay Fast Reactor, in Caithness, Scotland; and (b) simplified cross-section of Core Tank.

   The use of enriched uranium-235 obviates the need to slow down neutrons. Capture of some of these by uranium-238 leads to the 'breeding' of further fissile material in the shape of plutonium. (Copyright United Kingdom Atomic Energy Authority)

Some atomic nuclei are unstable and undergo radioactive decay, at a rate indicated by the half-life characteristic of the nuclei in question. Absorption of radiation, e.g., alpha and beta particles, or neutrons, can lead, by subsequent disintegration, to artificial radioactivity.

The spontaneous conversion of less stable into more stable atoms results in the production of large quantities of energy; in fission processes, large atoms rearrange as smaller ones, while in fusion, small atoms like hydrogen are converted into larger atoms such as helium in the plasma state, a highly ionized, high-temperature state.

Questions

(1)   Calculate the abundance of the two isotopes of bromine of mass numbers 79 and 81 from its relative atomic mass of 79.916.

(2)   Dr Prout in 1815 suggested that all relative atomic masses were simple multiples of that of hydrogen. Discuss this hypothesis in the light of modern knowledge.

(3)   Discuss the reasons why some atoms are unstable.

(4)   The count rate (s^{-1}) of a radioactive isotope fell from 1000 to 250 in an hour. Find the half-life of the isotope.

(5)   In one year one gram of radium produces 0.043 cm^3 of helium formed from the emission of 11.6×10^{17} alpha particles. Calculate a value for the Avogadro constant.

(6)   Calculate the energy which would be released by the reaction:

$$_1^1H + _1^3H \rightarrow _2^4He$$

(7)   In the Periodic Table (p. 24), hydrogen and helium are placed apart from the main table. Comment on this and discuss alternative ways of accommodating these two elements.

(8)   The relative atomic mass of lead found in two different minerals is 206.01 and 207.9. Suggest a reason for this discrepancy.

(9)   Write an essay on the use of radioisotopes in industry and medicine.

10)   Rewrite and complete in the form of equations, the radioactive disintegrations of the following naturally occurring nuclides:

[e.g., $^{227}Ac(\beta^-, \alpha)$ is $_{89}^{227}Ac + _{-1}^0e \rightarrow _{86}^{223}Rn + _2^4He$]

$^{10}B(\alpha, n)$, $^{14}C(\beta^-)$, $^3H(\beta^-)$, $^{223}Fr(\beta^-, \gamma, \alpha)$, $^{231}Pa(\alpha,\gamma)$, $^{226}Ra(\alpha,\gamma)$, $^{222}Rn(\alpha)$, $^{228}Th(\beta^-)$, $^{238}U(\alpha,\gamma)$, $^{27}Al(p,\alpha)$, $^7Li(p,\gamma)$.

(11)   Calculate the potential difference required to accelerate an electron from rest to a velocity of 10^6 m s^{-1}.

(12)   Using the principles of conservation of energy and momentum, show that if an alpha particle strikes a hydrogen atom centrally, the velocity of the latter is 1.6 times greater after collision.

(13)   Niels Bohr (1913) assumed that the angular momentum (mvr) of an electron mass m, travelling at a speed v in a circular orbit of radius r, is quantized in units of $h/2\pi$, i.e., $mvr = nh/2\pi$, where n, the principal quantum number, is an integer. The energy, E, of such an electron is given by:

$$E = \tfrac{1}{2}mv^2 - Ze^2/4\pi\varepsilon_0 r$$

where the first term is the contribution of the kinetic energy and the second term is the potential energy (attractive); Z is the atomic number of the atom, e the electronic charge and ε_0 the permittivity of free space. The electron is prevented from falling into the nucleus by the centrifugal force, i.e.,

$$mv^2/r = Ze^2/4\pi\varepsilon_0 r^2 \qquad \text{(cf. Coulomb's law, p. 45)}$$

Writing the Planck equation as $E = hv$, and using the speed of light, c, and the relationship $c = \lambda v$, where λ is the wavelength and v the frequency, show that:

$$1/\lambda = -Z^2 R/n^2$$

where R, the Rydberg constant, is equal to $e^4 m/8h^3 c\varepsilon_0^2$ and has the dimensions of (length)$^{-1}$.

3 The chemical bond

Stability of noble gases • The electrovalent (ionic) bond • The covalent bond • Dative covalency • Multiple bonds • The hydrogen bond • The metallic bond • Oxidation state

Attention must now be turned to the ways in which atoms can combine to form molecules.

Stability of noble gases

As seen in Chapter 2, the noble gases have completely full inner shells and an outer octet of electrons (except helium, which has only two electrons in its electron shell), each electron in every orbital being paired with another of opposite spin. The noble gases are monatomic and form only a very limited number of compounds. A reasonable deduction is therefore that a fairly stable state for any atom to exist in is that in which its electronic configuration has been modified to correspond to that of the nearest noble gas. This is typical of all the elements except, to some extent, those of the transition series.

An atom can attain such a configuration in one of three ways: by the acceptance or removal of electrons or by sharing electrons with other atoms. These possibilities give rise to the two basic types of valence—electrovalence and covalence.

Although it is customary to refer to compounds as being of one type or the other, it must be realized that few compounds are ever completely covalent or electrovalent: the majority of compounds have bonding of an intermediate character.

The electrovalent (ionic) bond

An electrovalent bond involves a transference of one or more electrons from one atom to another, with the resultant formation of charged particles or ions.

Simple electronic transfer to produce an electrovalent bond is illustrated by sodium chloride:

$$\text{Na} \xrightarrow{\hspace{2cm}} \text{Na}^+ + e^-$$
$$1s^2 2s^2 2p^6 3s^1 \qquad\qquad 1s^2 2s^2 2p^6 \quad \text{(now with core of noble gas, neon)}$$

$$\text{Cl} + e^- \xrightarrow{\hspace{2cm}} \text{Cl}^-$$
$$1s^2 2s^2 2p^6 3s^2 3p^5 \qquad 1s^2 2s^2 2p^6 3s^2 3p^6 \text{(now with core of noble gas, argon)}$$

Ionization energy

The removal of an electron from an element requires a certain amount of energy, expressed in terms of the ionization energy (*Table 3.1*). The two main factors which influence the size of this energy are:

(1) *The distance of the electron from the nucleus*   The larger the atom, the greater will be the distance between the positively charged nucleus and the negative electron which is to be removed, and hence, since electrostatic attraction varies inversely as the square of the distance separating opposite charges, the lower will be the ionization energy.
(2) *The charge on the nucleus*   The greater the positive charge on the nucleus, the greater will be the attractive forces between it and the outer electrons, and hence the greater will be the ionization energy. The situation is, however, complicated by the varying shielding powers of different types of electron separating the nucleus from the valency electrons; unpaired inner electrons of the transition elements do not screen the nucleus as effectively as do paired electrons.

Since the number of atoms in a mole of any element is a constant (Avogadro's constant, 6.023×10^{23}), relative atomic mass/density will be a function of the volume of an atom. This ratio, called the atomic volume, is the volume (in cm^3) occupied by one mole of the element. The periodic variation in atomic volume with atomic number reflects the modern classification of the elements (compare *Table 2.3* and *Figure 3.1*). The connection between the size of atoms and their ionization energies is well illustrated by the elements of the first and second Periods (*Figures 3.1* and *3.2*).

Table 3.1 Some representative first ionization energies

Element	Ionization energy/kJ mol^{-1}	Element	Ionization energy/kJ mol^{-1}
H	1314	Li	519
He	2372	Na	489
Ne	2080	K	418
Ar	1522	Cu	745
Xe	1163	Ag	732
Cl	1255	Tl	590

As further electrons are removed from an atom, the ionization energy increases, since there is then a net positive charge on the atom. The removal of electrons from a completely filled shell generally requires far more energy

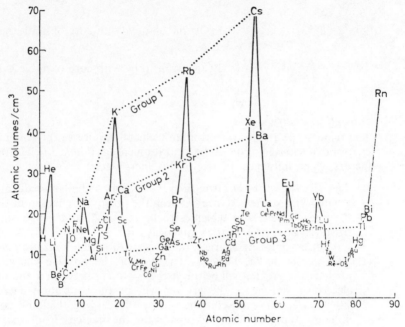

Figure 3.1 Atomic volumes

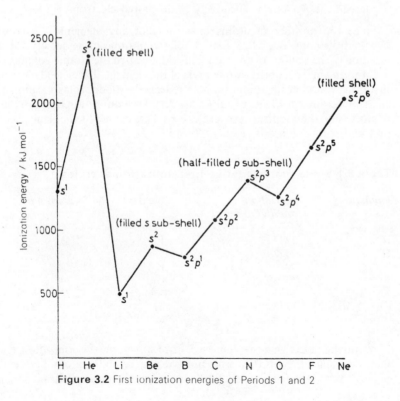

Figure 3.2 First ionization energies of Periods 1 and 2

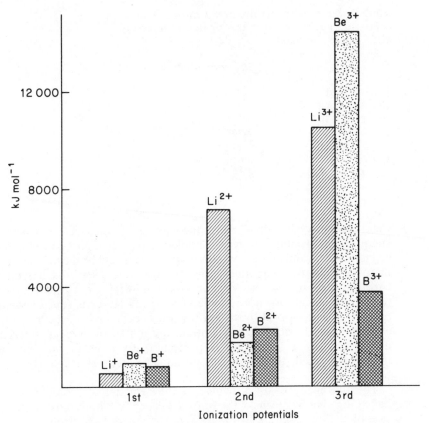

Figure 3.3 Comparison of ionization potentials of lithium, beryllium and boron. In Li^{2+}, an electron has been removed from a complete inner shell and in Be^{3+}, a third electron has been removed from a complete inner shell

than is normally associated with chemical changes, so that atoms and ions possessing only completely filled shells are generally stable (but *see* Chapter 17). *Figure 3.3* shows not only the increase of successive ionization energies but the excessive energy required to remove an electron from a complete shell.

Electron affinity
When electrons are added to an atom, there will be an absorption or release of energy. This energy is, somewhat misleadingly, referred to as the electron affinity. The only normal anions to be formed with release of energy are those of the halogens. The energy required to add a second electron to an ion already negatively charged results in the electron affinity having positive values (*see Table 3.2*).

Lattice and solvation energies
It would seem reasonable to suppose that an ionic compound would be formed when the ionization energy of the one element was balanced by the

Table 3.2 Electron affinities/kJ mol^{-1}. The value is exothermic for the process shown if the sign is negative

Process	Electron affinity/kJ mol^{-1}
$F + e^- \rightarrow F^-$	-335
$Cl + e^- \rightarrow Cl^-$	-365
$Br + e^- \rightarrow Br^-$	-331
$I + e^- \rightarrow I^-$	-305
$O + e^- \rightarrow O^-$	-142
$O + 2e^- \rightarrow O^{2-}$	$+640$
$S + 2e^- \rightarrow S^{2-}$	$+397$
$Se + 2e^- \rightarrow Se^{2-}$	$+423$

electron affinity of the other. As the halogens are the only elements to liberate energy when forming ions with a noble-gas type configuration, this would suggest that the only ionic compounds possible would be the halides. In fact, many other ionic compounds exist; the previous discussion has omitted considerations such as the energy released in forming a crystal lattice (that required to break down one mole of a crystal lattice is called the *lattice energy*) or that evolved by an ion when it becomes solvated (the *solvation energy*). Thus the above considerations are not by themselves sufficient to decide whether a compound is predominantly ionic or not but, when intelligently used, they can provide an indication of the likely type of bonding.

The total energy changes involved in the formation of a compound can be linked together by means of the Born–Haber cycle. This is illustrated for the case of sodium chloride in *Figure 3.4* (*see also* p. 127). From this energy diagram we can show the overall cyclic changes by *Figure 3.5*, and summation of these energy changes by Hess's law (p. 124) gives

$$-410 \text{ kJ mol}^{-1} + L = 354 \text{ kJ mol}^{-1}$$

Therefore

$$L = 764 \text{ kJ mol}^{-1}$$

Electronegativity

A measure of the tendency for a *combined* atom to pull electrons towards itself can be determined approximately from the average of the ionization energy and electron affinity of the uncombined atom, since both ionization energy and electron affinity are a measure of the electron-attracting capacity of an atom. Originally, electronegativity values were based on the difference between experimental and calculated bond energies (p. 126) because any difference between these two values was taken to represent the difference between a polarized and a pure covalent bond (*Table 3.4*). A direct theoretical assessment of electronegativity can now be made, using Coulomb's law (p. 45), but, nevertheless, the application of electronegativity is largely in qualitative terms. The elements on the right-hand side of the Periodic Table have the greatest electronegativities, and in any one Group the elec-

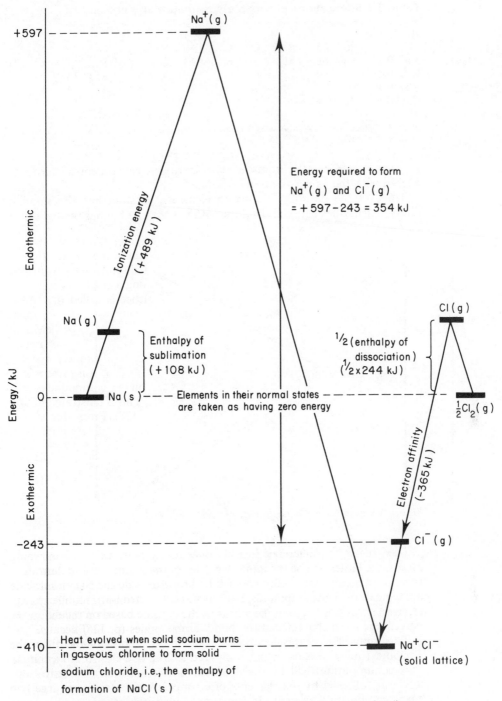

Figure 3.4 Energy changes accompanying the formation of sodium chloride (Born–Haber diagram)

Table 3.3 Some representative electronegativity values

H	2.1												
Li	1.0	Be	1.5	B	2.0	C	2.5	N	3.0	O	3.5	F	3.9
Na	0.9	Mg	1.2	Al	1.5	Si	1.8	P	2.1	S	2.5	Cl	3.0
K	0.8	Ca	1.0			Ge	1.8	As	2.0	Se	2.4	Br	2.8
Rb	0.8	Sr	1.0			Sn	1.8	Sb	1.9	Te	2.1	I	2.5
Cs	0.7	Ba	0.9										

(From E. Cartmell and G. W. A. Fowles, *Valency and Molecular Structure*, 3rd ed., p. 140, London, Butterworths, 1970)

tronegativity falls as the atomic number increases. This general pattern is in accord with the values of atomic volumes.

Thus, of the naturally occurring elements, caesium has the smallest electronegativities between the atoms at the ends of a bond is related to the *3.3*).

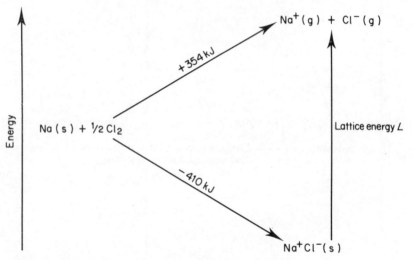

Figure 3.5 Estimation of lattice energy of sodium chloride

The greater the difference in electronegativity between two combining atoms, the larger will be the ionic character of the bond; caesium fluoride is therefore a highly ionic compound, while carbon dioxide and silicon sulphide are typical covalent compounds. Pauling has calculated how the difference in electronegativities between the atoms at the ends of a bond is related to the percentage ionic character of the bond. Some values are in *Table 3.4*.

To summarize, it can safely be said that the elements of Group 1 of the Periodic Table (i.e., those which only need to lose one electron to attain an s^2p^6 configuration) will be those which most readily form positive ions (*cations*), followed by the elements of Group 2, which require to lose two electrons. Further, as each Group is descended, the atomic radii become larger and loss of electrons becomes easier. In Groups 3 and 4 there will be a

Table 3.4 Electronegativity difference and ionic character of bonds

Difference in electronegativity	0.2	0.5	0.8	1.0	1.3	1.6	1.7	2.0	2.4	2.9	3.2
Ionic character/%	1	6	15	22	34	47	51	63	76	88	92

progressively smaller tendency to form ions with s^2p^6 configurations because this would entail the loss of three and four electrons, respectively.

In the same way it may be expected that the elements which would most readily form negative ions (*anions*) by a gain of electrons would be those of Group 7 (the halogens), followed by those of Groups 6 and 5, in that order.

It must be stressed that the conclusions drawn from this discussion are of a very general nature and exclude considerations of the transition elements. The changes in the Transition Periods are neither as great nor as regular as for the non-transitional elements, since inner quantum levels are being filled with electrons. This aspect will be dealt with in more detail in Chapter 16.

Properties of electrovalent compounds

The ionic bond confers very distinct properties, but it must be remembered that very few compounds are completely ionic, and so deviations from these properties must be expected.

An important feature of ionic solids is that the forces holding the ions together are of an electrostatic nature and have no preferential direction. They are said to be coulombic, as they obey Coulomb's law:

$$F \propto \frac{z_1 e \times z_2 e}{r^2}$$

where

e	= electronic charge
z_1, z_2	= the two valencies of the ions
r	= distance between the charges
F	= a force of attraction if the charges are unlike and of repulsion if alike.

The solid therefore will consist of ions regularly disposed with regard to packing and charge considerations (*see* Chapter 4). No discrete molecules exist in the lattice and the structure is firmly held together in all directions; it will therefore be hard. Because of the large electrostatic forces which have to be overcome to separate the constituent particles, ionic compounds have a high melting point. For similar reasons the liquid produced has a high boiling point. The larger the distance between the ions, the smaller the force of attraction, in accordance with Coulomb's law, and generally the lower the m.p. and b.p. Thus, whereas lithium fluoride boils at 1680 °C, sodium chloride boils at 1440 °C and potassium iodide at 1320 °C. Similarly, ions of multiple charge experience correspondingly larger forces and have higher boiling points, e.g., magnesium oxide, $Mg^{2+}O^{2-}$, does not even melt until 2800 °C while calcium chloride, $Ca^{2+}Cl_2^-$, boils at 1600 °C.

A further feature is that, when fused, an ionic compound is capable of carrying an electric current. This is transported by the ions present in the liquid, the positively charged cations migrating to the cathode and the negatively charged anions to the anode.

Ionic compounds only dissolve in solvents which themselves are of a polar nature (i.e., compounds such as water and liquid ammonia, in which there is an uneven distribution of charge or dipole; *see* p. 49). Briefly, the explanation is that, for the solid to dissolve, the ions must somehow acquire a corresponding stability to counteract the potential energy of the crystal lattice which is destroyed. This they can do if the solvent molecules can solvate some or all of the ions, thus giving them a solvation energy. Polar solvents are able to interact in this way, and hence the solid will tend to dissolve. Differences in solubility between compounds can thus be partly attributed to differences in lattice and solvation energies. A further consideration is that high dielectric solvents (i.e., insulators) such as water can insulate any ion which has dissolved from the resultant oppositely charged solid, thus hindering its return to the lattice. Such solvents will therefore assist the dissolution of ionic compounds.

When in solution, the reactions of ionic compounds are the reactions of their constituent ions; thus, all bromides will instantaneously react with silver nitrate solution to precipitate silver bromide:

$$Br^- + Ag^+ \rightarrow AgBr\downarrow$$

The covalent bond

Simple covalency

This involves the sharing of electrons of opposite spin from different atoms. The sharing normally continues until no unpaired electrons remain in the outer orbitals. It has been shown that in ionic compounds the atoms usually acquire a noble-gas-type configuration; by the process of sharing electrons this is often the case also with covalently bound atoms, although here the atom does not exert complete control over all the electrons. It is also possible for elements in Period 3 and above to acquire more than eight electrons in their outer shells by this process of sharing, because of the availability of d and sometimes f as well as the usual s and p orbitals; it is this fact that is responsible for many marked differences between the elements of Periods 2 and 3.

Covalent bond formation is best explained by some examples. Consider first the molecule of hydrogen. The hydrogen atom has the configuration $1s^1$, so that by sharing this electron with a second hydrogen atom, both atoms acquire the stable s^2 configuration. This process can be represented in two ways:

$$H^{\boldsymbol{\cdot}} + H^{\boldsymbol{\cdot}} \rightarrow H\!:\!H \text{ (or } H\!-\!H)$$
$$s^1 \quad s^1 \quad s^2$$

or, in terms of orbitals

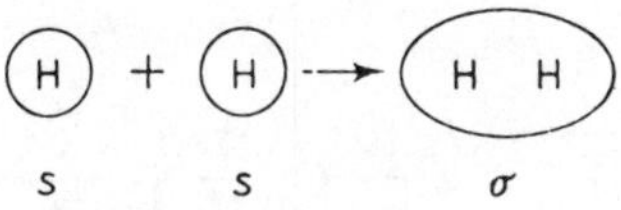

$$s \qquad\qquad s \qquad\qquad \sigma$$

i.e., *the two s orbitals form a σ (sigma) bond.*

Similarly, for the case of chlorine, the electronic configuration of the atoms is $KL3s^23p^5$, which can be alternatively represented as

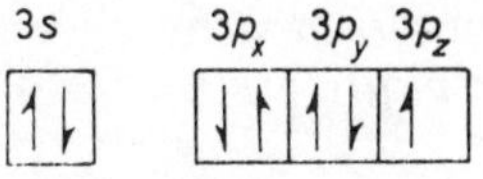

One of the p electrons is unpaired so that, by sharing another p electron from a second chlorine atom, a stable configuration is obtained:

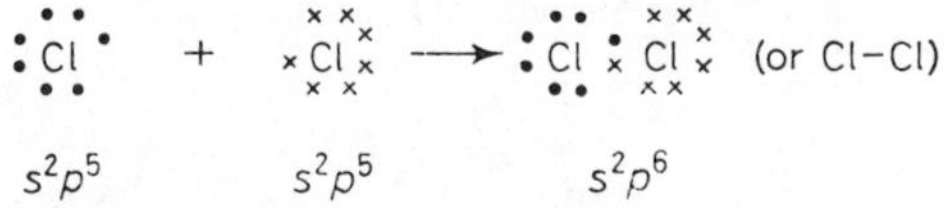

$$s^2p^5 \qquad\qquad s^2p^5 \qquad\qquad s^2p^6$$

(Only the electrons in the valency shell are shown, since these are the ones mainly affected by bond formation. The electrons are represented as ˙ and ×, although after chemical union the electrons of equal energy from the different atoms are indistinguishable from one another.)

In terms of orbitals, this bond formed by the p orbitals can be represented as

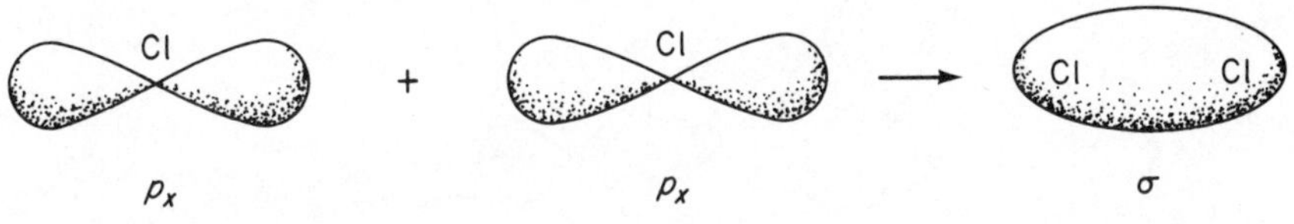

$$p_x \qquad\qquad p_x \qquad\qquad \sigma$$

i.e., *the p orbitals form a σ (sigma) bond.*

As a further example, consider tetrachloromethane. Carbon initially has the configuration $K2s^22p^2$, and to complete its octet it requires a share in one electron from each of four chlorine atoms:

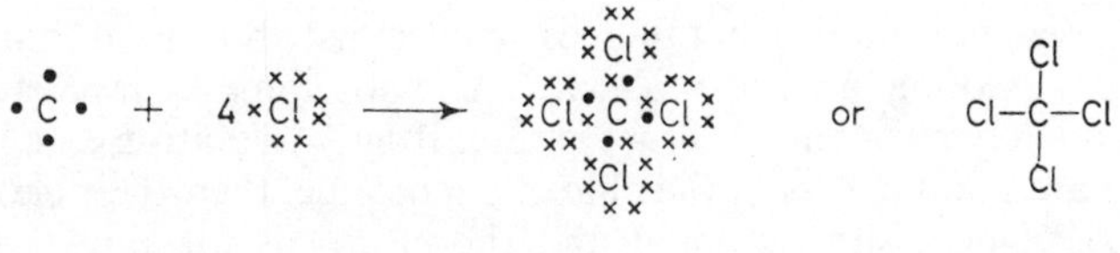

two shared electrons being equivalent to a single chemical bond. In its state of lowest energy, the carbon atom has only two unpaired electrons (the p electrons) and hence, before it can form four chemical bonds, one of the s electrons must be promoted to the vacant p orbital, thus giving four unpaired electrons and allowing a valency of four:

C

Ground state

Excited state

The description of the tetrachloromethane molecule in terms of orbitals is more complicated than those so far discussed. The four orbitals required of the carbon atom can be described as formed by a combination ('*hybridization*') of the one *s* and the three *p* orbitals:

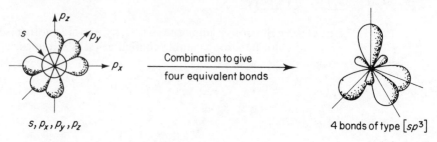

s, p_x, p_y, p_z Combination to give four equivalent bonds   4 bonds of type $[sp^3]$

The bonds are then formed by the overlap of these orbitals with the appropriate *p* orbitals of the chlorine atoms:

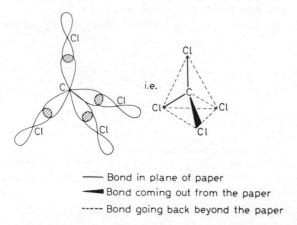

—— Bond in plane of paper

◀ Bond coming out from the paper

----- Bond going back beyond the paper

Covalency implies that there is an even distribution of charge between the atoms making up the molecule. As with ionic compounds, however, a completely covalent structure is rare. If only two atoms are linked together and are identical, as in the chlorine molecule, then their electronegativities must of necessity be equal and the electrons forming the bond will be symmetrically placed between the atoms, with the result that the molecule will be non-polar. However, if one of the atoms has a greater electronegativity than the other, then it will attract the electrons towards its end of the bond and the other atom will become relatively positive; there will thus be a tendency towards an ionic nature, although in this case the electrons have not been completely transferred from the one atom to the other, and the

Table 3.5 Electronegativity and ionic character and relative polarizability of the anions X^-

HX	Electronegativity of X	Percentage ionic character	Relative polarizability of X^-
HF	3.9	43	0.96
HCl	3.0	17	3.57
HBr	2.8	11	4.99
HI	2.5	5	7.57

tendency will be greater the greater the difference in electronegativity. This polar nature is very common and is well represented by the hydrogen halides, $H^{\delta+} \longrightarrow X^{\delta-}$ (*Table 3.5*). The displacement of charge gives rise to a *dipole moment* (by analogy with the magnetic moment of a magnet, the dipole moment of a bond is equal to the displaced charge multiplied by the length of the bond; for a molecule, the vector sum of the bond dipole moments gives the dipole moment of the molecule; *see* p. 575). The values for the hydrogen halides are shown in *Table 3.6*.

The formation of a covalent bond is favoured when the elements concerned are of similar electronegativity and in those compounds where a small cation of high charge would be the alternative structure. This high charge would confer on the cation such a great attractive force for the electrons of the anion—an effect known as *polarizing power*—that the latter would be pulled towards the cation, resulting in a return to a covalent structure (as the anion is polarized). This is the basis of *Fajans' rules*:

(1) The larger the charge on the cation or anion, and
(2) the smaller the size of the cation and/or the larger that of the anion, the higher will be the percentage of covalent bonding involved (*Figure 3.6*).

Table 3.6 Dipole moments

HX	$10^{30} \times$ Dipole moment/C m
HF	6.3
HCl	3.3
HBr	2.6
HI	1.3

$$\overset{\delta+}{\text{H}}\text{———}\overset{\delta-}{\text{X}}$$

$$\longleftarrow l \longrightarrow$$

Dipole moment $= l \,.\, \delta+$

Dative covalency

Covalency has been defined as the sharing of electrons between two atoms, one electron being supplied from each atom. There is, however, a further possibility, namely that one of the atoms supplies both of the electrons for the bond. This type of bond is often referred to as a coordinate bond or as dative covalency, but there is no fundamental difference between this and ordinary

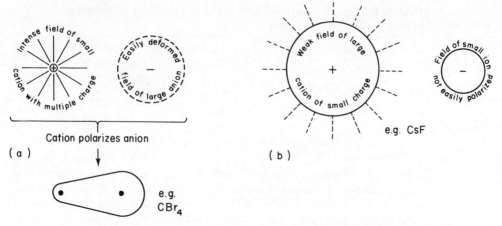

Cation polarizes anion

(a)

e.g. CBr_4

(b)

e.g. CsF

Figure 3.6 Variation of polarization by cation. (a) Considerable polarization: structure largely covalent; (b) negligible polarization: structure largely ionic. For relative polarizability of anions, *see Table 3.5*

covalency. The conditions necessary for the formation of such a bond are that the *donor atom* shall have a pair of electrons (a *lone pair*) not already involved in bond formation and that the *acceptor atom* requires two electrons to attain a stable configuration. As its name implies, this type of bond is particularly common in coordination compounds (p. 352), but a suitable illustration is provided by aluminium chloride. If we consider the molecule $AlCl_3$, in which the constituent atoms are held together by covalent bonds, then the aluminium atom has only six electrons in its outer orbitals:

The chlorine atoms, however, each have three pairs of electrons not involved in bonding. If one of these atoms were to donate a pair of electrons to another aluminium atom in a second $AlCl_3$ unit, then this second aluminium atom would be surrounded by eight electrons. Similarly, one of the chlorine atoms attached to this aluminium atom can donate a pair of electrons to the first aluminium atom, completing its octet and conferring stability on the resulting dimeric molecule, Al_2Cl_6:

(Aluminium chloride exists mainly as this dimer at temperatures just above sublimation point but increasingly as the monomer at temperatures above 400 °C.)

A coordinate bond may be represented by an arrow pointing from the donor to the acceptor atom:

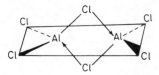

Properties of covalent compounds

Covalent molecules are usually discrete individuals held together by strong bonds in definite directions. As the atoms which form a bond do so by sharing electrons, it may be supposed that they will approach each other in such a way that the relevant electronic orbitals will overlap as much as possible. It was seen in Chapter 2 that p, d and f orbitals are directional and it is therefore to be expected that covalent molecules will have definite fixed shapes.

Consider methane, CH_4. There will be eight electrons in the outer valency orbitals of the carbon atom (four from the carbon and one from each of the four hydrogen atoms), divided into four pairs. As electrons are all negatively charged, these four pairs will arrange themselves as far apart as possible. This is equivalent to saying that they will be tetrahedrally arranged:

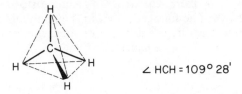

$\angle$ HCH $= 109°\ 28'$

This is in agreement with the shape of the tetrachloromethane molecule which has earlier (p. 47) been shown in terms of hybridization of the s and three p orbitals of the valency electrons of the carbon atom.

The Sidgwick–Powell theory states that the number of pairs of electrons in the combined valency shell determines the overall geometry of a covalent compound. An extension of this theory involves the assumption that the electrostatic repulsion between electron pairs decreases in the order:

lone pair–lone pair > lone pair–bond pair > bond pair–bond pair

From these considerations an understanding of spatial arrangements in covalent compounds is possible.

In the case of the ammonia molecule, NH_3, there will still be eight electrons in the valency orbitals of the central atom (five from the nitrogen and one from each of the hydrogen atoms), but of the four pairs only three will be involved in bond formation, and the fourth will exist as a lone pair. As there are still four pairs altogether, they must be orientated towards the corners of a tetrahedron; the lone pair however, will reside closer to the nitrogen atom than the four pairs in the N–H bonds and will exert a greater repulsive effect on these bonds, making the bond angles contract somewhat. Thus the ammonia molecule is pyramidal, with $H\hat{N}H = 106°45'$ (as compared with the tetrahedral angle of $109°28'$)

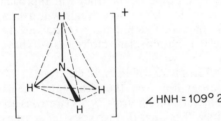   ∠ HNH = 106°45'

The lone pair on the nitrogen atom can be donated to a proton, thus forming the tetrahedral ammonium ion (it must be stressed that in the ammonium ion, all four bonds are equivalent and no one bond can be identified as the coordinate bond):

Similarly, in Group 6, oxygen exerts a valency of two, but in its hydride, water, there are two lone pairs. These exert a greater distorting effect on the bonding pairs than the one lone pair does in the ammonia molecule, and although the overall structure is still based on a tetrahedron, the molecule will be V-shaped, with an even smaller bond angle:

$H\hat{O}H = 104°\,31'$

Although the bonds in covalent molecules are themselves very strong, there are no great forces between individual molecules. In the crystalline state, the various molecules are conveniently packed but the only forces holding the molecules together are weak forces (*van der Waals' forces*, p. 69) such as that produced when two adjacent molecules distort the electronic clouds of each other and induce a weak electrostatic attraction between them, that is, form an *induced dipole*.

As a result, crystals of most covalent compounds are easily broken, soft, and have low melting and boiling points. There are, however, some structures which are bonded throughout by covalent links; these will be very hard and have high melting points. A notable example of such a structure is that of diamond, which may be thought of as a *giant molecule* (*see* p. 83). In some crystals the properties are also appreciably modified by the presence of the weak bond, known as the hydrogen bond (p. 57).

The contrast between ionic and covalent compounds is well illustrated by the change in melting points of the chlorides of the elements of the Second Period in proceeding from left to right, i.e., from a predominantly ionic to a predominantly covalent molecular compound, as shown in *Table 3.7.* Unlike ionic compounds, fused covalent compounds will not conduct an electric current because of the absence of any charged particles to act as transferring agents (*see* Chapter 7): molar conductivities are shown in *Table 3.7.*

Table 3.7 Melting points and molar conductivities of chlorides in the Second Period

Chloride	NaCl	$MgCl_2$	Al_2Cl_6	$SiCl_4$	PCl_3	SCl_2	Cl_2
M.p./°C	800	712*	193	-70	-112	-78	-102
$\Lambda/\Omega^{-1}\,m^2\,mol^{-1}$	0.0133	0.0029	1.5×10^{-10}	0	—	—	—

* At 226 kPa

Also, in contrast to electrovalent compounds, covalent substances generally dissolve in non-polar solvents but not in polar solvents, although in water hydrolysis may occur; e.g., silicon tetrachloride is hydrolysed to an indefinite hydrate of silica, $SiO_2 . nH_2O$. Their reactions with other substances in solution usually take place slowly because actual chemical bonds have to be broken and reformed.

Multiple bonds

Sometimes two atoms combine together by sharing not two but four or even six electrons, giving rise to multiple bonds. Nitrogen normally exists as a diatomic molecule; it can readily be seen that for each nitrogen atom to have an octet of electrons in its outer orbitals, each atom must share three electrons with the other:

$$\overset{\text{xx}}{\underset{\text{x}}{N}}{}^{\text{x}}_{\text{x}} \;+\; \overset{\text{oo}}{\underset{\text{o}}{}}\overset{}{N}{}^{\text{o}}_{\text{o}} \longrightarrow \overset{\text{x}}{\underset{\text{x}}{}}N{\overset{\text{o}}{\underset{\text{o}}{\overset{\text{x}}{\underset{\text{x}}{}}}}}N{}^{\text{o}}_{\text{o}} \quad \text{or} \quad N{\equiv}N$$

From this it might be thought that all these bonds are equivalent and that the bond strength is three times that of a single bond, but this is not the case. The electronic configuration of the nitrogen atom is $K2s^2 2p^3$, i.e.,

1s   2s $2p_x\ 2p_y\ 2p_z$

↑↓		↑↓		↑	↑	↑

and the bonding electrons are the three *p* electrons, which give rise to two types of bond.

It has been shown above (p. 47) that when the *p* orbitals overlap in an end-on manner, they form an orbital which embraces the two atoms in the same way as when two *s* orbitals overlap; such bonds, whatever types of orbital

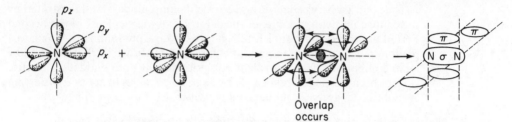

Figure 3.7 The nitrogen molecule

they are formed from, are known as σ (*sigma bonds*. A second possible mode
of interaction between p orbitals is by sideways overlap; in this case the
resultant orbital consists of two negative charge clouds—one above and one
below the line joining the centres of the two atoms. The bond formed by this
process is called a π (*pi*) *bond*.

Both types of bond are exhibited in the nitrogen molecule (*Figure 3.7*).

In view of its importance in organic chemistry, a suitable element to
consider more fully is carbon. It has been shown that the four bonds in
methane are equivalent and pointing towards the corners of an imaginary
regular tetrahedron, and that the s and three p orbitals do not retain their
individual identities but instead they give rise to four equal orbitals ([sp^3]
hybrids) separated by the tetrahedral angle (p. 51).

This accounts very satisfactorily for compounds in which carbon is
exerting a valency of four, but other carbon compounds exist in which, for
example, each carbon atom is only joined to three other atoms, e.g., ethene,
C_2H_4. In this case hybridization of the s orbital and of two of the p orbitals of
carbon occurs, to produce three [sp^2] hybrid bonds, all in one plane and
directed towards the corners of an imaginary equilateral triangle. The p
orbital remaining is at right angles to the plane of the triangle:

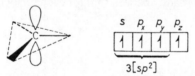

This lone p orbital can interact with one in the same plane on an adjacent
atom to yield a π bond, as in the nitrogen molecule. The molecule of ethene
can thus be pictured as shown in *Figure 3.8*.

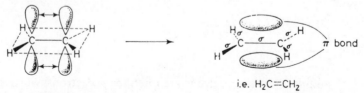

Figure 3.8 The ethene molecule

It is also possible for carbon to use the s orbital and one p orbital to form
two [sp] hybrid orbitals.

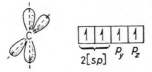

This will produce two linear σ bonds and there will be two p orbitals remaining which will be at right angles both mutually and also to the direction of the σ bonds. These p orbitals can interact to give two π bonds, as for example in ethyne, C_2H_2 (*Figure 3.9*).

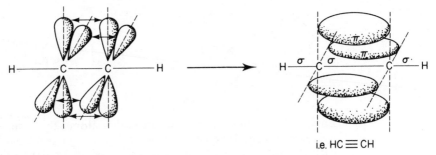

i.e. HC≡CH

Figure 3.9 The ethyne molecule

This molecule can therefore be described as being surrounded by a cylindrical cloud of negative charge, equal in magnitude to that of four electrons (cf. N≡N).

A further example is afforded by buta-1,3-diene, CH_2:CH·CH:CH_2. The formation of the π bonds in this molecule can be represented as

but there is obviously a probability that the p orbital on the second carbon atom will interact with that on the third:

This is equivalent to saying that the complete molecule will be enveloped in a *delocalized* π-type charge cloud and that all three of the C—C bonds will exhibit some double bond character (a phenomenon often called *resonance*). That this is so is shown by the intermediate lengths of the bonds in the molecule, as compared with normal single and double bonds between carbon atoms:

$$C—C$$
154 pm (1·54Å)

$$H_2C{=}CH—CH{=}CH_2$$
135 pm   144 pm   135 pm
(1·35Å)   (1·44Å)   (1·35Å)

$$C{=}C$$
133 pm (1·33Å)

One other compound is worthy of consideration, namely benzene, C_6H_6. This is often formulated as

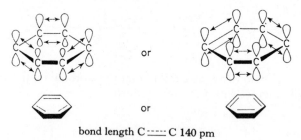

or more simply as

Each carbon atom is joined to three other atoms; this is an example of sp^2 hybridization, and the benzene molecule is a regular, planar hexagon, each carbon atom having one spare p orbital perpendicular to the plane of the ring. These six orbitals can interact together in either of the two ways shown in *Figure 3.10*.

As there is an equal probability of the two forms occurring, the actual structure is intermediate between the two, and the individual π orbitals are fused into one large delocalized π cloud (*see* above). Such a view accounts very satisfactorily both for the observed equality in length of all the C—C bonds in the molecule and for the chemical properties of this compound (p. 423).

or

or

bond length C ----- C 140 pm

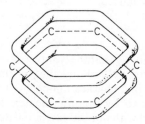

Figure 3.10 The benzene ring

The hydrogen bond

A hydrogen atom is so small that it very nearly represents a 'point' charge, as a result of which, although it has only a valency of one, another suitable atom can be held by an attraction of mainly electrostatic character. Because of the electrostatic nature of the force, only atoms which have a large electronegativity will form such hydrogen bonds; indeed this phenomenon appears to be confined to fluorine, oxygen, nitrogen and possibly sulphur. Chlorine, although highly electronegative, does not appear to form hydrogen bonds (but *see* chloral hydrate, p. 507), presumably because it is too large to approach sufficiently closely. A hydrogen bond may be represented by:

$$\overset{\delta-}{-X} - \overset{\delta+}{H} \cdots \overset{\delta-}{Y} -$$

H bond

Reference to *Table 3.8* will show that the energy associated with a hydrogen bond is very small in comparison with the bond energies of normal chemical bonds, but nevertheless the presence of hydrogen bonding between molecules can produce very marked effects, as examples in later Chapters will show.

Table 3.8 Bond energies (i.e., energy needed to *break* the bond)

Hydrogen bond	Bond energy $kJ\ mol^{-1}$	Covalent bond	Bond energy $kJ\ mol^{-1}$
$H\cdots F$	41.8	$H{-}F$	433
$H\cdots O$	29.4	$O{-}H$	461
$H\cdots N$	8.4	$N{-}H$	391

The metallic bond

About 70 per cent of all the elements are metals, and a discussion of valency would be incomplete if it did not include some mention of the way in which metallic atoms are held together.

Normally in metals there is simply a close packing of identical atoms, each atom having eight or twelve equidistant neighbours, i.e. it is said to have a *coordination number* of eight or twelve (*see* Chapter 4). Such a structure confers the typical properties of metals such as high melting point, high density, malleability, ductility and opacity.

Covalent bonding in the normal sense is precluded, not only because a large number of electrons would be required, but also because such a large number of suitable orbitals is not available. The only way in which a covalent structure could exist with such a large coordination number would be by having a *resonating* structure.

Resonance implies that the real structure has an intermediate position between the various possible structures which can be drawn conventionally. For example, for a coordination number of six and a valency of three for all the atoms in a structure, the following possibilities can be drawn:

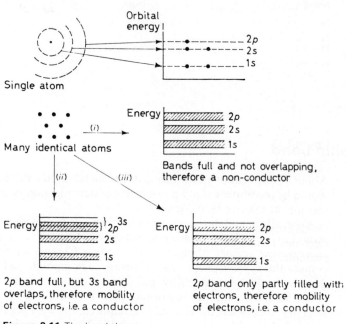

This would logically lead to a situation in which the valency electrons would be mobile. The picture thus emerges of identical cations in a 'sea' of mobile electrons; these electrons are the valency electrons of the original atoms, now no longer connected to any one atom but instead acting as the 'cement' binding the cations together. This idea of mobile electrons immediately explains the good electrical and thermal conductivities of metals generally. In ionic compounds, the ions carry the current whereas, in metals, electrons perform this service.

An extension of this model can be used to explain the different conducting powers of metals and also the phenomenon of *semiconduction* in certain elements, e.g., silicon, germanium, grey tin and tellurium. The energy of each electron in the 'sea' of electrons is clearly not the same because the electrons may not all have come from the same type of orbital. Further, when many atoms come together to form a metal there are many electrons present which

Figure 3.11 The band theory

are in identical quantum states. By Pauli's principle (p. 21) this situation is not permissible, so the energy levels 'expand' into energy bands (*Figure 3.11*). If (*a*) the bands do not overlap and (*b*) each readily available band is completely filled with electrons, the element is a non-conductor at low temperatures. At high temperatures, enough energy may be supplied to promote some of the electrons to empty energy bands of greater energy, thus making conduction possible. The nearer an empty band is to the highest filled band, the lower will be the temperature at which conduction begins, and as the temperature is increased above this point, so the conductivity will increase. This behaviour characterizes a semiconductor. The presence of impurities can alter the semiconductor properties considerably, either by providing fresh empty energy bands or by altering the position of existing bands relative to each other (*Figure 3.12*). Semiconductors find application in *transistors* and as *photoelectric devices*, e.g., in exposure meters.

The higher the atomic number of an element, the closer are the energy levels; the energy bands also get progressively nearer to each other as the quantum numbers become larger. The availability of conduction bands in metals is increased by the overlapping of the energy levels with those of higher quantum numbers. This is illustrated by comparing the electrical conductivities of potassium and copper:

Element	K	Cu
Atomic conductance/(Ω^{-1} m^{-1} ÷ atomic volume in cm^3)	4×10^{-7}	8.4×10^{-6}

Copper is a better conductor than potassium, even though they both possess only one unpaired electron: in copper, however, there is a greater overlapping of bands, as the unpaired electron is in a higher quantum shell.

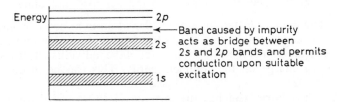

Figure 3.12 Effect of an impurity on a non-conductor

Oxidation state

Oxidation can be defined as a reaction involving partial or complete loss of electrons (and reduction, as a reaction involving a gain of electrons). For example, in the reaction

$$Fe^{2+} + \tfrac{1}{2}Cl_2 \rightarrow Fe^{3+} + Cl^-$$

iron is oxidized and chlorine reduced.

Many reactions involve oxidation and reduction, although their names do not suggest this:

Displacement such as $Cu^{2+} + Zn \rightarrow Cu + Zn^{2+}$ involves oxidation of the dissolving metal and reduction of the metal being displaced from solution.

Combination between metal and non-metal involves oxidation of the metal and reduction of the non-metal: $2Mg + O_2 \rightarrow 2Mg^{2+} O^{2-}$.

If a metal loses one electron per atom, it is said to have gained an oxidation number or state of $+1$; a non-metal gaining one electron per atom is similarly said to acquire an oxidation number of -1. In the case of the formation of potassium chloride from the elements:

$$K + \tfrac{1}{2}Cl_2 \rightarrow K^+ \; Cl^-$$

Oxidation number   0   0 $+1$ -1

Because the usual covalent bond involves elements of different electronegativity, with the less electronegative element acquiring a slight positive charge at the expense of the other, the above concept of oxidation state still holds, and the oxidation number can be regarded as the charge which each atom would carry in the hypothetically ionized state. Thus, in the compound $H^{\delta +} - Cl^{\delta -}$, hydrogen is in an oxidation state of $+1$ and chlorine of -1.

Rules for determining oxidation numbers

The application of simple rules permits the oxidation state of an atom in any given compound to be worked out:

(1) Atoms in the elemental form are in zero oxidation state.
(2) Hydrogen in compounds is in the $+1$ oxidation state except in the ionic hydrides (-1).
(3) Oxygen in compounds is in the -2 oxidation state, except in peroxides (-1) and oxygen difluoride $(+2)$.
(4) The halogens in halides are in the -1 oxidation state.

The oxidation state of a particular atom in an ion or compound can be determined by breaking the latter down into its component atoms. The algebraic sum of the oxidation states of the individual atoms is then equated to the charge on the complete entity. This can be made clear by some examples:

(a) Phosphoric acid has the formula H_3PO_4; each hydrogen atom is in oxidation state $+1$, each oxygen atom is in oxidation state -2, giving a total -5 for the hydrogen and oxygen atoms together. As the charge on the molecule is zero, the atom of phosphorus must be in oxidation state $+5$.

(b) Potassium dichromate, $K_2Cr_2O_7$. If the oxidation number of each chromium atom is x, we can write $(+2) + (2x) + (-14) = 0$, from which the sum of the oxidation states of the two chromium atoms is 12 and the oxidation number of the chromium is therefore $+6$.

Summary

Orbitals are, on the whole, more stable when either empty or containing their full complement of two electrons of opposite spin. Chemical reaction, leading to the formation of chemical bonds, involves the vacating or filling of orbitals.

Metals tend to lose electrons in the course of chemical reaction to give positive ions and are thus described as 'electropositive'. In the presence of non-metals, valency electrons are transferred to the non-metallic atoms and negative ions are formed, so that non-metals are said to be 'electronegative'. The chemical bond so formed is called *electrovalent* or *ionic*. In the absence of non-metals, the valency electrons form a 'cement' binding together the tightly-packed metallic ions. Both this *metallic bond* and the ionic bond give substances capable of conducting electricity: in the case of the metal itself, the valency electrons are free to move under an applied potential, even in the solid state. With ionic compounds, however, electrical conduction is the result of migration of ions and not electrons; these are free to move only in the liquid state, and so ionic compounds conduct electricity only after melting or dissolving has broken down the ionic lattice.

Reaction occurs between atoms of non-metals by sharing electrons. For single bonds, overlapping of orbitals takes place along the line joining the centres of the atoms to give σ- (sigma) bonds, e.g.,

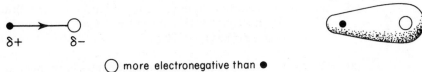

p-orbital p-orbital   σ-bond

Overlapping of any remaining p-orbitals at right angles to the σ-bond produces a π-(pi) bond with electron clouds above and below the single bond: this is the situation for the formation of multiple bonds

π-bond

p-orbitals

Because there are no charged particles available in these covalent compounds, they are incapable of conducting electricity, even in the liquid state. If the covalent link takes place between atoms of differing electronegativity, the bonding electrons will not be shared equally (i.e., the electron cloud will not be uniform), and there will be some separation of charge. The molecule will, in fact, be a dipole:

δ+   δ−

○ more electronegative than ●

When hydrogen is attached to the extremely electronegative elements fluorine, oxygen or nitrogen, there is sufficient separation of charge to give a detectable electrostatic attraction between different parts of adjacent molecules. This attraction is the basis of the *hydrogen bond*.

Electron clouds of adjacent atoms can interact with each other to give electrostatic induction and attraction (rather like the induction and attraction which result from a magnet being brought close to a bar of iron). This is one aspect of the *van de Waals forces* operating between molecules.

There is a wide variation of bond energy associated with these various chemical bonds. Whereas a metallic, ionic or covalent bond can have a bond energy exceeding 400 kJ mol^{-1} a hydrogen bond is usually less than 40 kJ mol^{-1} and a van der Waals bond seldom greater than 8 kJ mol^{-1}.

Questions

(1) Using dots and crosses to indicate valency shell electrons of different atoms, construct structures for the following compounds:

CaO; H_2S; HNO_3; $Cu(NH_3)_4SO_4.H_2O$; $K_3Fe(CN)_6$; CO; CH_3OCH_3

Explain what types of bond are present in each of the compounds.

(2) What factors affect the formation of an ionic bond? Why is a C^{4+} ion unknown in any compound?

(3) Xenon is known to form a tetrafluoride. Discuss the possible nature of the bonds in this compound.

(4) Using the following information, construct a Born–Haber cycle and from it deduce the sublimation energy of iodine:

Heat of formation of sodium iodide   = -291 kJ mol^{-1}
Heat of sublimation of sodium   = 108 kJ mol^{-1}
1st ionization energy of sodium   = 489 kJ mol^{-1}
Heat of dissociation of iodine   = 110 kJ mol^{-1}
Electron affinity of iodine   = -305 kJ mol^{-1}
Lattice energy of sodium iodide   = 647 kJ mol^{-1}

(5) Suggest the shapes of BF_4^-, N_2H_4, H_3O^+, H_2O_2, BF_3NH_3.

(6) How would you expect the shapes of the molecules to change in the series H_2O, H_2S, H_2Se?

(7) What is the oxidation state of the named atom in the following: sulphur in SO_3, nitrogen in NH_2OH, chlorine in $HClO_4$, vanadium in $[V_5O_{14}]^{3-}$, zirconium in $[ZrO(SO_4)_2]^{2-}$?

(8) Discuss the various approaches that chemists have made to the concept of electronegativity.

(9) Write an account of the hydrogen bond in nature, with particular respect to water and the transmission of information inside the living cell.

(10) Write an account of semiconductors.

(11) Comment on *Figure 3.13*.

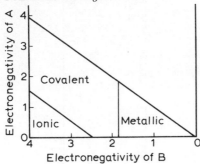

Figure 3.13 Bond type of the compound AB

(12) What conclusions can you draw from the information in *Table 3.9*?

Table 3.9

Substance	B.p./K	Molar heat of evaporation/kJ mol^{-1}
Benzene	353	34.8
Caesium chloride	1573	160
Chlorine	240	20.4
Hydrogen	20	0.92
Potassium	1033	91.6
Rubidium	952	75.6
Silver chloride	1837	178
Sodium	1155	103
Sodium chloride	1738	170

(13) Draw all the possible electronic structures for the molecules shown in *Figure 10.5* (*see* p. 231) and then suggest the reason for the actual structures. What would you expect the structures of $[ICl_4]^-$ and XeF_4 to be?

(14) (*a*) Discuss the essential features and the theoretical implications of the atomic spectrum of hydrogen. Include in your answer an explanation of why the spectrum involves many lines which can be divided into a number of groups, and why the spacing of the lines varies.

(*b*) Give the equations which are involved when measuring the first and second ionization energies of magnesium.

(*c*) State how and why the first ionization energies change within Group 1 (Li to Cs), and across Period 3 (Na to Ar). [J.M.B.]

4 States of matter

Molecular interpretations • The gaseous state • Relative molecular masses and formulae of gases • The liquid state • The solid state • Changes of state

Introduction: molecular interpretations

Matter exists in three common, easily recognized states—the solid, liquid and gaseous states. (A fourth state, plasma, exists at extremely high temperatures; in this, the energy is such that electrons and nucleons have a largely independent existence.) The most obvious difference between the solid, liquid and gaseous states is in the capacity to maintain a fixed shape and volume. At a given temperature, a solid is rigid and of a definite volume, a liquid occupies a definite volume but it can flow to occupy a vessel of any shape, and a gas has neither shape nor a fixed volume but will expand to fill completely any container into which it is introduced.

These differences in behaviour can be explained in terms of a kinetic (Greek *kinesis* = motion) theory of matter. This postulates that at all temperatures above absolute zero, 0 K or $-273\,°C$, atoms and molecules are in motion. In a solid this motion is very small and probably consists essentially of oscillations of the particles about a mean position. As the temperature is raised (and the kinetic energy increased), this motion becomes more vigorous, until the particles collide with each other and are no longer restricted to particular positions; this represents fusion, and the resultant liquid consists of mobile particles. The interatomic or intermolecular forces, however, are still sufficiently strong to cause the volume at a fixed temperature to remain constant. If still more heat energy is applied, a stage is ultimately reached where the kinetic energy imparted to the molecules is sufficiently great to overcome the forces between the molecules, and the result is a gas which can occupy any space available.

For pure substances, melting and boiling occur at definite temperatures; how then can evaporation of liquids be accounted for? At a given temperature it is the average kinetic energy of the system which is constant: it does not follow that all the individual particles will be endowed with this particular kinetic energy. In fact, the kinetic energies of the molecules will be distributed in accordance with normal probability laws (*Figure 4.1*), and there will be a certain proportion with greater kinetic energy, moving more quickly than the remainder.

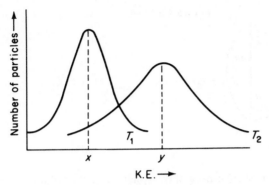

Figure 4.1 Distribution of kinetic energies in a system of molecules. Temperature $T_2 > T_1$ and the average kinetic energy $y > x$. The area under each curve is the same

It is the faster-moving particles which, when at the surface of the liquid, will be able to overcome the attractive forces exerted by the molecules in the liquid. They will accordingly escape into the atmosphere, and the liquid is said to evaporate. As the molecules which escape are those with greatest energy content, it follows that the average kinetic energy, and hence the temperature, of the liquid will tend to fall. Eventually, if the liquid is in a closed space, the vapour will become saturated and particles which have lost energy by collision will re-enter the liquid at the same rate as fast particles leave; a state of *dynamic* equilibrium will then be obtained.

The gaseous state

Deduction of the gas laws

Consider a gas enclosed in a hollow sphere of radius r. The gas will exert a pressure on the walls of the sphere by virtue of the change in momentum of the molecules when they strike the sides.

Assume that collisions between the molecules of both the gas and the wall of the vessel are perfectly elastic, so that there is no loss of momentum. Now consider 1 molecule, of mass m, moving with a speed u; then, if θ is the angle of incidence on collision with the wall, the change in momentum is $2mu \cos \theta$ (*Figure 4.2*). The distance, AB, between two impacts with the wall is $2r \cos \theta$, and hence the number of impacts per second is $u/(2r \cos \theta)$. Therefore, the rate of change of momentum is

$$2mu \cos \theta \times \frac{u}{2r \cos \theta} = \frac{mu^2}{r}$$

Now, the rate of change of momentum is equal to the force (Newton's second law of motion), and pressure is equal to force per unit area. Hence the pressure caused by one molecule moving inside the sphere is given by

$$p = \frac{mu^2}{r} \times \frac{1}{4\pi r^2} = \frac{mu^2}{4\pi r^3}$$

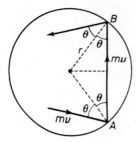

Figure 4.2 Motion of a molecule in a spherical container

where $4\pi r^2$ is the surface area of the sphere. But the volume, V, of a sphere is

$$V = \frac{4}{3}\pi r^3$$

Therefore

$$pV = \frac{1}{3}mu^2$$

and for n molecules moving with speed u

$$pV = \frac{1}{3}nmu^2$$

However, as explained earlier, the molecules in a gas are not all moving with the same speed, and therefore u^2 is the mean of the square velocity of the molecules, i.e., $u^2 = (u_1^2 + u_2^2 + \ldots u_n^2)/n$, and u is the 'root mean square velocity' and is designated $c_{\text{r.m.s.}}$

The kinetic theory assumes that the average kinetic energy of the molecules, $\frac{1}{2}nmc_{\text{r.m.s.}}^2$, is proportional to the absolute temperature; thus at a given temperature the kinetic energy is constant. It follows that $pV = \frac{1}{3}nmc_{\text{r.m.s.}}^2$ is constant for a fixed mass of gas at constant temperature and under these conditions the pressure is inversely proportional to the volume, which is a statement of *Boyle's law*.

Rewriting the above equation:

$$pV = \tfrac{2}{3}(\tfrac{1}{2}nmc_{\text{r.m.s.}}^2)$$

i.e.,

$$pV = \tfrac{2}{3}(\text{average kinetic energy})$$

Therefore $pV \propto T$ (where T is the temperature on the absolute scale) so that, for a mass of gas at constant pressure: $V \propto T$, which is a statement of *Charles' law*.

The above equation shows that $pV \propto T$ and is a combination of Boyle's and Charles' laws, so that pV/T must be a constant. For one mole of an ideal gas, i.e., one which obeys these laws, the constant is called the universal gas constant, R, and for n mole of gas $pV/T = nR$ or

$$pV = nRt$$

which is the *general gas equation*. One mole of a perfect gas occupies 22.414 dm³ (litres) at 101.325 kPa (1 atm) and 273.16 K, thus

$$R = \frac{pV}{nT} = \frac{101.325 \text{ kN m}^{-2} \times 22.414 \times 10^{-3} \text{ m}^3}{1 \text{ mol} \times 273.16 \text{ K}} = 8.314 \text{ N m K}^{-1} \text{ mol}^{-1}$$

$$= 8.314 \text{ J K}^{-1} \text{ mol}^{-1}$$

where 1 Pa (pascal) = 1 N m^{-2} and 1 N m = 1 J (joule).

If there are equal volumes of two gases at the same temperature and under equal pressure, then

$$\tfrac{1}{3} n_1 m_1 c_{1\text{r.m.s.}}^2 = \tfrac{1}{3} n_2 m_2 c_{2\text{r.m.s.}}^2$$

and because the average kinetic energies are equal,

$$\tfrac{1}{2} m_1 c_{1\text{ r.m.s.}}^2 = \tfrac{1}{2} m_2 c_{2\text{ r.m.s.}}^2$$

Thus

$$n_1 = n_2$$

i.e., equal volumes of gases at the same temperature and pressure contain equal numbers of molecules (Avogadro's law).

It follows that, at s.t.p., i.e., at 273 K (0 °C) (the standard temperature) and 101.3 kPa (1 atm or 760 mmHg) (the standard pressure) the volume occupied by one mole of all perfect gases, the Gram Molecular Volume, is constant. It has been found experimentally to be 22.4 dm³.

For example, the density of hydrogen at s.t.p. is 0.09 g dm^{-3}. Its relative molecular mass is 2.016, giving a value for the gram molecular volume (G.M.V.) at s.t.p. of 2.016/0.09 = 22.4 dm³.

Further

$$c_{\text{r.m.s.}}^2 = \frac{3pV}{nm}$$

But nm/V is the density, ρ, of the gas and therefore

$$c_{\text{r.m.s.}}^2 = \frac{3p}{\rho}$$

or, at constant pressure,

$$c_{\text{r.m.s.}}^2 \propto \frac{1}{\rho}$$

or

$$c_{\text{r.m.s.}} \propto \sqrt{\left(\frac{1}{\rho}\right)}$$

This is equivalent to Graham's law of diffusion which can thus be derived from a kinetic theory of matter.

The law of partial pressures also follows from the kinetic theory; since the molecules are independent of each other, their number alone (and not their nature) is relevant. In a mixture of gases, therefore, the *pressure that a particular gas exerts (the partial pressure) is that which it would exert if it alone occupied the volume.* The total pressure is then merely the sum of the partial pressures.

An interesting example of the effect of varying partial pressures is afforded by pulmonary and tissue respiration. Air contains about 21 per cent oxygen and therefore the partial pressure of oxygen in the lungs (at s.t.p) is (21/100) × 760 mmHg = 160 mmHg. This is in contact with the alveoli, in which the partial pressure of oxygen is only 100 mmHg and therefore it flows into the alveoli and from thence into the venous blood, where the oxygen partial pressure is only 40 mmHg. This inflow raises the partial pressure to 90 mmHg and converts the blood into arterial blood, which then releases the oxygen to tissues, in which the partial pressure is 30 mmHg.

Conversely, the carbon dioxide generated in the tissues has a partial pressure of 50 mmHg and it flows into the arterial blood where the partial pressure is 40 mmHg; this process, linked with the release of oxygen to the tissues, converts the blood to venous blood ($P_{CO_2} = 46$ mmHg) which liberates carbon dioxide into the alveoli ($P_{CO_2} = 40$ mmHg) and thence to the lungs ($P_{CO_2} = 0.3$ mmHg) and subsequent exhalation.

The kinetic theory, and hence the gas laws deduced from it, has been derived on the basis of several assumptions which are known to be oversimplifications. Hence it is not surprising that no real gas perfectly obeys these laws. The variations shown by some common gases are shown in *Figure 4.3.* (Assuming Boyle's law to be correct, a plot of pV against p or V should be a straight line parallel to the axis.)

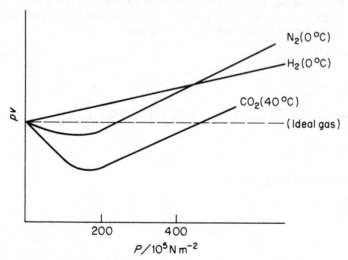

Figure 4.3 Deviations from Boyle's law for some gases

There are two main reasons for the approximate nature of these laws. The derivation of the kinetic theory assumed that the molecules themselves occupied a negligible volume in comparison with that of the gas as a whole. This becomes increasingly incorrect as the pressure on the gas is increased and the molecules are brought more closely together. Thus there will be considerable deviations from the laws at high pressures. Secondly, it has been assumed that there are no forces of attraction between the molecules. But such forces exist and will add to the pressures exerted upon the gas so that the measured, externally applied, pressure will be lower than the true value. This explains the decrease in the product, pV, at low pressures for most gases (*Figure 4.3*). In fact, at sufficiently low temperatures all gases exhibit this phenomenon. Also, because of the molecular attraction, work has to be done by an expanding gas in overcoming this interaction, so that a cooling effect results which is of considerable practical importance in the liquefaction of gases (p. 85).

In an effort to make allowance for these deviations from ideal behaviour, several modifications have been made to the general gas equation, $pV = nRT$. One of the simplest is that due to van der Waals, who suggested the equation

$$\left(p + \frac{a}{V^2}\right)(V - b) = RT$$

where a and b are constants. The former provides for the extra pressure caused by the mutual attraction of the molecules, while the latter makes allowance for the molecular volumes and can be shown to be equal to four times the actual volume of the molecules. As a result of this equation, the intermolecular forces are now referred to as *van der Waals forces*. They are very weak compared with those associated with conventional bonds, seldom exceeding 8 kJ mol^{-1}, whereas the covalent bond can exceed 400 kJ mol^{-1}.

Relative molecular masses of gases

Since the relative molecular mass of a gas is equal to twice its vapour density (p. 14), measurements of the latter afford a convenient method for the determination of the relative molecular mass of any gas or of any volatile substance which can easily be converted into the gaseous state, subject to the inevitable slight inaccuracy caused by the nonideality of the gas.

The most direct method of relative molecular mass determination will accordingly be by comparison of the mass of a known volume of the gas under consideration with the experimentally observed mass of an equal volume of hydrogen measured under the same conditions. This is the basis of Regnault's method. In theory it is very simple, but in practice, owing to the very small differences in mass involved, accurate results are difficult to obtain. The gas which is being weighed must be perfectly dry, since small amounts of water vapour will cause considerable errors; the globe containing the gas must be repeatedly filled and emptied to ensure that no traces of air remain; precautions must be taken to ensure that the gases are both at atmospheric pressure, and also the decrease in weight caused by the buoyancy of the globe must be balanced by counterpoising a similar globe on the other arm of the balance.

A more convenient method for determining vapour densities is that due to Victor Meyer. A known mass of volatile liquid is made to vaporize and displace its own volume of air. This air is then collected over water and measured at atmospheric pressure and temperature. A conventional ap-

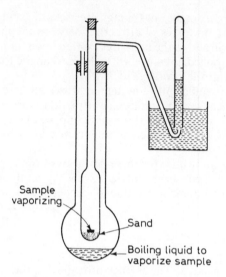

Figure 4.4 Victor Meyer's apparatus

paratus is shown in *Figure 4.4*. The outer tube contains a liquid whose temperature can be maintained at least 20 kelvins above the boiling point of the volatile liquid. After this temperature has been maintained for some time, so that equilibrium has been attained, a small sample of known mass of the substance whose vapour density is required is quickly introduced into the inner tube, usually in a small stoppered vessel or 'Hofmann bottle', and the main stopper replaced. The substance immediately volatilizes and expels an amount of air equal to its own volume into the graduated collecting tube. The volume occupied now will of course be that which the vapour would have occupied if it had been at the temperature and pressure of the receiver. This can be measured, adjusted for the vapour pressure of the water in the graduated tube and converted to s.t.p. Thus the volume occupied at s.t.p. by a known mass of the volatile substance has been determined and hence the vapour density can be deduced from the knowledge that at s.t.p. the density of hydrogen is 0.09 g dm^{-3}. Alternatively, the need to rely upon the vapour density relationship can be avoided by utilizing the fact that the gram molecular mass of any gas occupies the gram molecular volume, i.e., 22.4 dm^3 at s.t.p.

Example. 0.083 g of bromoethane when volatilized in a Victor Meyer apparatus displaced 18.5 cm^3 of air (adjusted to atmospheric pressure), collected over water at 16 °C. The atmospheric pressure was 100.5 kPa; the vapour pressure of water at 16 °C is 1.8 kPa. Calculate the relative molecular mass of bromoethane.

The pressure exerted by the bromoethane itself was $(100.5 - 1.8) = 98.7$ kPa; the volume occupied by 0.083 g of bromoethane at s.t.p. is therefore:

$$\frac{18.5 \text{ cm}^3 \times 98.7 \text{ kPa} \times 273 \text{ K}}{101.3 \text{ kPa} \times 289 \text{ K}}$$

and the mass which would occyuy 22.4 dm^3 (22 400 cm^3) is given by

$$\frac{22\,400 \text{ cm}^3 \times 0.083 \text{ g} \times 101.3 \text{ kPa} \times 289 \text{ K}}{18.5 \text{ cm}^3 \times 98.7 \text{ kPa} \times 273 \text{ K}} = 109 \text{ g}$$

Hence, the relative molecular mass of bromoethane is 109.

The success of this method depends upon strict adherence to certain precautions. The quantity of substance used must be such that none of its vapour escapes from the vessel in which it is volatilized, but only air which has been displaced. If the former occurs, the vapour will condense in the receiver and invalidate the result. It is also important to maintain a constant temperature in the apparatus and to avoid direct contact between the inner tube and the heating liquid; otherwise, superheating might occur and give rise to irregular expansion and contraction of the air in the tube.

Neither of the methods mentioned is suitable for compounds—particularly organic ones—which decompose at their boiling point. For such substances, it is necessary to volatilize them at pressures considerably below atmospheric; by this means, their boiling point will be much lower than usual. A suitable apparatus for such determinations is that of Hofmann. The material is introduced into the vacuum above a column of mercury. The tube containing the mercury can be surrounded by a jacket containing the vapour of a compound which boils at a temperature higher than that being investigated. The substance vaporizes and the pressure is given by the difference produced in the height of the mercury column.

Example. The mercury in a barometer tube of length 98 cm and cross-sectional area 0.8 cm^2 fell from a height of 75.9 cm to 30.8 cm after the introduction of 0.077 g of ethoxyethane at 100 °C. Calculate the relative molecular mass of the ether.

The volume of ether at 100 °C (373 K) and a pressure equivalent to that which would support $(75.9 - 30.8)$ cm of mercury is $(98.0 - 30.8) \times 0.8$ cm^3. Therefore, the volume corrected to s.t.p. is

$$\frac{(98.0 - 30.8) \times 0.8 \text{ cm}^3 \times (75.9 - 30.8) \text{ cmHg} \times 273 \text{ K}}{373 \text{ K} \times 76 \text{ cmHg}}$$

22.4 dm^3 at s.t.p. then has a mass of

$$\frac{0.077 \text{ g} \times 373 \text{ K} \times 76 \text{ cmHg} \times 22.4 \times 10^3 \text{ cm}^3}{(98.0 - 30.8) \times 0.8 \text{ cm}^3 \times (75.9 - 30.8) \text{ cmHg} \times 273 \text{ K}} = 74 \text{ g}$$

Assuming the vapour behaves as an ideal gas at 100 °C, 74 is the relative molecular mass of the ether, since it is that which would occupy 22.4 dm^3 (litre) at s.t.p.

A source of error in all relative molecular mass determinations based on gas densities is the non-ideality of the system. *Figure 4.3* shows that at very low pressures the deviation from ideality is much less; therefore, by referring gases and vapours to conditions of low pressure, more accurate use can be made of the gas laws.

The limiting density is defined as

$$m/p_0 V_0$$

where $p_0 V_0$ is the value extrapolated to zero pressure, an imaginary state (*see*

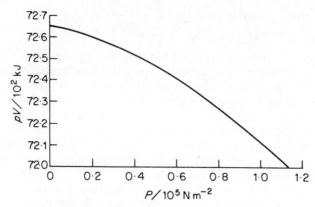

Figure 4.5 Variation of pV with p for hydrogen chloride

Figure 4.5), and m is the mass of the gas.

The normal density is

$$m/p_1 V_1$$

where $p_1 = 101$ kPa (1 atm) and V_1 is the corresponding volume. Therefore

Limiting density = normal density $\times (p_1 V_1 / p_0 V_0)$

The limiting density can then be used to give a more accurate relative molecular mass for the gas or vapour by using the expression

22.4 dm^3 $\times$ limiting density g dm^{-3} = molar mass g

A further method for the determination of the relative molecular mass of a gas or vapour is to determine its rate of diffusion relative to that of a substance of known relative molecular mass and then to make use of Graham's law of diffusion. In practice, it is difficult to determine rates of diffusion and it is more common to study the *effusion* of gases through a small orifice.

The formulae of gases

Avogadro's law permits the formulae of many gases to be determined. For the elucidation of formulae it is often vital to know the atomicity of hydrogen and other elements. These have been determined by Cannizzaro's method (*see* Chapter 1). His reasoning for the diatomicity of hydrogen has since been confirmed by a determination of the specific heat capacity of the element at constant pressure and at constant volume. (The ratio of the specific heat capacity of a gas at constant pressure to that at constant volume, c_p/c_v, is a constant, the value of which depends upon the atomicity of the gas. For the monatomic gases its value is 1.67, for the diatomic gases, 1.4 and for triatomic gases, 1.33. The determination of this ratio is usually made indirectly by measuring the velocity of sound in the gas, for details of which a physics textbook should be consulted.)

In general, there are two methods of approach to the problem of determining the formula of a gaseous molecule, namely by synthesis or by decomposition and measurement of the volumes of gases produced.

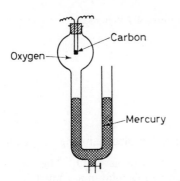

Figure 4.6 Volume composition of carbon dioxide

The first method can be suitably illustrated by the synthesis of hydrogen chloride.

It is found that

1 volume of hydrogen + 1 volume of chlorine yields 2 volumes of hydrogen chloride (temperature and pressure being kept constant)

Hence, by Avogadro's law:

1 molecule of hydrogen + 1 molecule of chlorine yields 2 molecules of hydrogen chloride

Hydrogen and chlorine are both diatomic and therefore

1 atom of hydrogen + 1 atom of chlorine yields 1 molecule of hydrogen chloride suggesting that the formula of hydrogen chloride is HCl.

Synthesis can also be applied to gases such as carbon dioxide: carbon is heated electrically in a closed tube with oxygen. After adjusting the temperature and pressure to those at the beginning of the experiment, it is found that there is no volume change (*Figure 4.6*):

Carbon + 1 volume oxygen yields 1 volume carbon dioxide

xC + 1 molecule oxygen yields 1 molecule carbon dioxide

and so the formula of carbon dioxide is C_xO_2. As is always the case when not all the substances involved are gases, the vapour density is also required. This indicates a relative molecular mass of 44, hence

$12x + 32 = 44$

and therefore

$x = (44 - 32)/12 = 1$

i.e., the formula is CO_2

Synthesis is the method preferred if the gas in question is very stable. Less stable gases are usually decomposed into their elements. For example, if dinitrogen oxide is heated in a closed tube with an iron wire (*Figure 4.7*), decomposition takes place with the formation of oxide on the wire, the final volume of gas, nitrogen, being the same as that of the original oxide, i.e.,

1 volume dinitrogen oxide yields 1 volume nitrogen

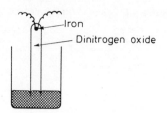

Figure 4.7 Volume composition of dinitrogen oxide

1 molecule dinitrogen oxide yields 1 molecule nitrogen = 2 atoms nitrogen

Hence the formula of dinitrogen oxide is N_2O_x. The vapour density is approximately 22, so the relative molecular mass must be near 44; since $N_2 = 28$, $x = 1$ and the formula is N_2O.

The liquid state

The relative compactness of the liquid state means that there is considerable resistance to flow; this can be regarded as internal friction or viscosity. It is measured in terms of the force required to produce a flow of the liquid. The cause of viscosity is not completely clear, but it appears that liquids consisting of molecules of symmetrical shapes are less viscous (or more mobile) than others and also that viscosity depends upon molecular size. But perhaps the major cause of high viscosity is molecular interaction. A liquid can be regarded in one sense as a disordered solid, the disorder being to a large extent a dynamic one. That is, if the system is 'frozen' in time, a liquid, especially if there are large intermolecular attractions, may reveal many regions of an ordered nature. Hydrogen bonding, as for example in ethane-1,2-diol and propane-1,2,3-triol, is a major cause of high viscosity, and it is doubtless significant that it is also a contributory factor in the ability to attain a high level of order.

Table 4.1 Some typical viscosities, η, at 25 °C

Compound	$\eta/10^{-3}$ kg m^{-1} s^{-1}	Compound	$\eta/10^{-3}$ kg m^{-1} s^{-1}
Ethoxyethane	0.2	Ethanol	1.1
Propanone	0.3	Mercury	1.5
Methanol	0.5	Nitrobenzene	2.0
Water	0.9	Propane-1,2,3-triol	1×10^3

Liquids, unlike gases, possess definite boundary surfaces and the molecules in the surface layer are in a special position in that some of their chemical affinity is unsaturated and the interaction with molecules in the interior is far greater than any attraction towards the few molecules present in the vapour phase (*Figure 4.8*). Consequently, the surface of a liquid tends to shrink and is said to be in a state of *surface tension*. If gravitational forces are in any way overcome, it assumes a spherical shape, because this has a minimum surface area for a given mass. The greater the force of molecular or

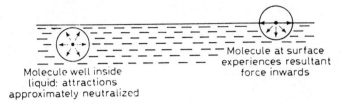

Molecule at surface
experiences resultant
force inwards

Molecule well inside
liquid: attractions
approximately neutralized

Figure 4.8 The cause of surface tension; molecules well inside the liquid as at A experience attractions approximately neutralized, whereas those at the surface as at B experience resultant inward forces

Table 4.2 Typical values of surface tension, σ, at 25 °C

Liquid	$\sigma/10^2$ N m^{-1}	Liquid	$\sigma/10^2$ N m^{-1}
Water	7.3	Benzene	2.9
Soap solution	ca. 2.8	Ethanol	2.2
Ethoxyethane	1.7	Mercury	47.0

atomic attraction within the liquid, the greater will be the surface tension (*Table 4.2*).

The solid state

Broadly speaking, solid substances appear to be either crystalline, i.e., have a definite geometrical form, or amorphous. Substances in the latter category are, however, often microcrystalline and, on the atomic scale, contain regular repeating units. Disorder may still be achieved by having flexible covalent links (as for example in elastomers, which are normally amorphous but on stretching exhibit some crystallinity) or by the distribution of ions, atoms or molecules in a random fashion throughout an irregular crystal structure (e.g., glass).

For crystalline substances, there are seven, and only seven, external shapes which can be fitted together to give a solid structure without voids; accordingly these are the only crystal classes possible (*Figure 4.9*).

Metallic lattices

The atoms of an element can be superficially regarded as identical spheres, and the structures of metals result from their close packing. *Figure 4.10* shows two layers of such a structure. A third layer of closely-packed atoms

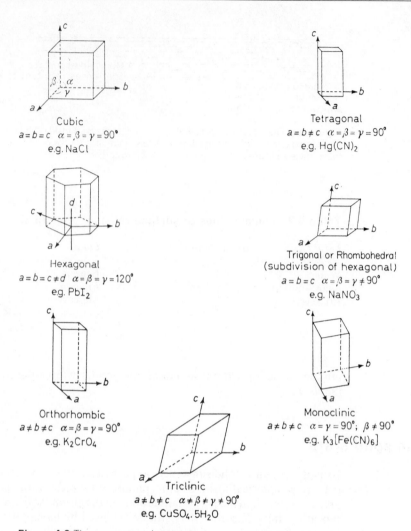

Cubic
$a = b = c \quad \alpha = \beta = \gamma = 90°$
e.g. NaCl

Tetragonal
$a = b \neq c \quad \alpha = \beta = \gamma = 90°$
e.g. $Hg(CN)_2$

Hexagonal
$a = b = c \neq d \quad \alpha = \beta = \gamma = 120°$
e.g. PbI_2

Trigonal or Rhombohedral
(subdivision of hexagonal)
$a = b = c \quad \alpha = \beta = \gamma \neq 90°$
e.g. $NaNO_3$

Orthorhombic
$a \neq b \neq c \quad \alpha = \beta = \gamma = 90°$
e.g. K_2CrO_4

Monoclinic
$a \neq b \neq c \quad \alpha = \gamma = 90°; \quad \beta \neq 90°$
e.g. $K_3[Fe(CN)_6]$

Triclinic
$a \neq b \neq c \quad \alpha \neq \beta \neq \gamma \neq 90°$
e.g. $CuSO_4.5H_2O$

Figure 4.9 The seven crystal systems

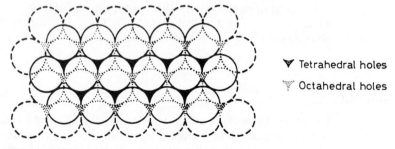

▼ Tetrahedral holes

▽ Octahedral holes

Figure 4.10 Close packing

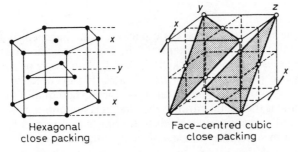

Hexagonal
close packing

Face–centred cubic
close packing

Figure 4.11 Hexagonal and face-centred cubic lattices

may be placed in either of two ways: either directly over the centres of the first layer or over the voids not covered by the previous layers. In the former case, the repeat unit is of the type $XYXYXY...$ and is called hexagonal close packing, whilst in the latter it is $XYZXYZXYZ...$ called cubic close packing. In side elevation the latter looks like a face-centered cube (*Figure 4.11*).

In both hexagonal and cubic close packing, each sphere is in contact with 12 others (six in the same plane and three in each plane above and below) and is said to have a *coordination number* of 12.

Some metals form a less well packed structure in which the coordination number of each atom is only eight; this is equivalent to a body-centred cubic structure (*Figure 4.12*).

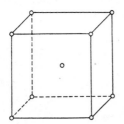

Figure 4.12 Body-centred cube

Metals possessing cubic close packing of their atoms are more malleable and ductile than those with hexagonal close packing. These two properties depend upon the ease with which planes of atoms can glide over each other; such movement can most easily occur between close-packed planes, of which there are more in the cubic than in the hexagonal structure.

A few metals have structures other than those described above; for example, indium has a tetragonal structure.

Some metals exist in more than one crystal form and are then said to exhibit polymorphism or allotropy. For example, both iron and chromium at normal temperatures exist as body-centred cubic structures but at elevated temperatures are converted into face-centred cubic crystals.

Ionic lattices

In simple ionic structures, close packing is maintained as far as possible, consistent with the maintenance of the correct stoichiometry. When equal-sized spheres are closely packed, there will of necessity be some free space (a hole) between them. *Figure 4.10* indicates that in cubic close packing there are two types of such hole. In one type, the hole is equally surrounded by six atoms and is thus octahedral; in the other it is surrounded by four equidistant atoms and is therefore tetrahedral. In ionic structures, anions are normally arranged in some form of close packing and cations, which are often much smaller than the anions, are placed in the octahedral or in the smaller tetrahedral holes. For this to produce a stable structure, the cations must be sufficiently large to hold the anions in their positions without their coming into too close contact with each other. The structure thus depends upon the ratio cation radius: anion radius. The limiting situation for 3-coordination is shown in *Figure 4.13*.

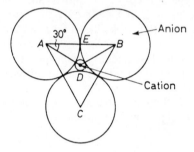

Figure 4.13 The limiting situation for three-coordination

Let the radius of the cation be r_c and that of the anion r_a, then $AE = r_a$ and $AD = r_c + r_a$. But $AE/AD = \cos 30°$, therefore

$$\frac{r_a}{r_c + r_a} = \frac{\sqrt{3}}{2}$$

$$2r_a = \sqrt{3}r_c + \sqrt{3}r_a$$

$$2 = \sqrt{3}r_c/r_a + \sqrt{3}$$

$$\frac{r_c}{r_a} = \frac{0.268}{1.732} = 0.155$$

Similar calculations can be performed for the other possible types of coordination. The results are summarized in *Table 4.3*, whilst the ionic radii of some of the elements are shown in *Table 4.4* and displayed pictorially in *Figure 4.14*.

A very large number of ionic compounds have crystals based on one of seven different repeating units or *unit cells*. These common structures are described on the following pages.

(a) The cubic sodium chloride and caesium chloride lattices

The radius ratio for sodium chloride is 0.52 and thus octahedral coordination should be exhibited; that this is so is shown in *Figure 4.15*. Each

Table 4.3 Radius ratio limits for different coordination numbers

Coordination number	Shape	Radius ratio limits
3	plane triangular	0.155 — 0.225
4	tetrahedral	0.225 — 0.414
6	octahedral	0.414 — 0.732
8	cubic	>0.732

Table 4.4 Some representative ionic radii/pm

Group		The transitional elements					Group				
1	2						3	4	5	6	7
Li^+	Be^{2+}						B^{3+}		N^{3-}	O^{2-}	F^-
60	31						20		171	140	136
Na^+	Mg^{2+}						Al^{3+}			S^{2-}	Cl^-
97	65						50			184	181
K^+	Ca^{2+}	Ti^{4+}	Cr^{3+}	Fe^{3+}	Ni^{2+}	Zn^{2+}	Ga^{3+}				Br^-
133	99	68	69	64	72	74	62				195
Rb^+	Sr^{2+}							Sn^{4+}			I^-
148	113							71			216
Cs^+	Ba^{2+}							Pb^{4+}			
169	135							84			

sodium ion occupies an octahedral hole in a face-centred cubic array of chloride ions and vice versa; such coordination is referred to as 6:6.

In this structure, each corner ion is shared between eight unit cells, each ion on a face of the cell by two cells, each ion on an edge by four cells, while the ion inside the cell belongs only to that particular one. Thus there are

$$8 \text{ corner } Na^+ \text{ ions} = \tfrac{8}{8} = 1 \; Na^+ \text{ per cube}$$
$$6 \text{ face } Na^+ \text{ ions} \;\; = \tfrac{6}{2} = 3 \; Na^+ \text{ per cube}$$
$$\left. \right\} \; 4 \; Na^+$$

$$12 \text{ edge } Cl^- \text{ ions} = \tfrac{12}{4} = 3 \; Cl^- \text{ per cube}$$
$$1 \text{ centre } Cl^- \text{ ion} \; = \tfrac{1}{1} = 1 \; Cl^- \text{ per cube}$$
$$\left. \right\} \; 4 \; Cl^-$$

in agreement with the empirical formula NaCl.

All the alkali halides, except those of caesium, crystallize with the sodium chloride structure. The radius ratios of the caesium halides are too great for octahedral coordination (e.g., $r_{Cs^+}/r_{Cl^-} = 0.93$) and accordingly, the caesium chloride structure (and also that of CsF, CsBr, CsI) has 8:8 coordination (*Figure 4.16*); (in this and subsequent structures, the lines indicated by — show the coordination number of the central atom or ion).

(b) The fluorite lattice
In order to maintain the correct stoichiometry in compounds of type MX_2, it is necessary for the coordination number of the cations to be twice that of the

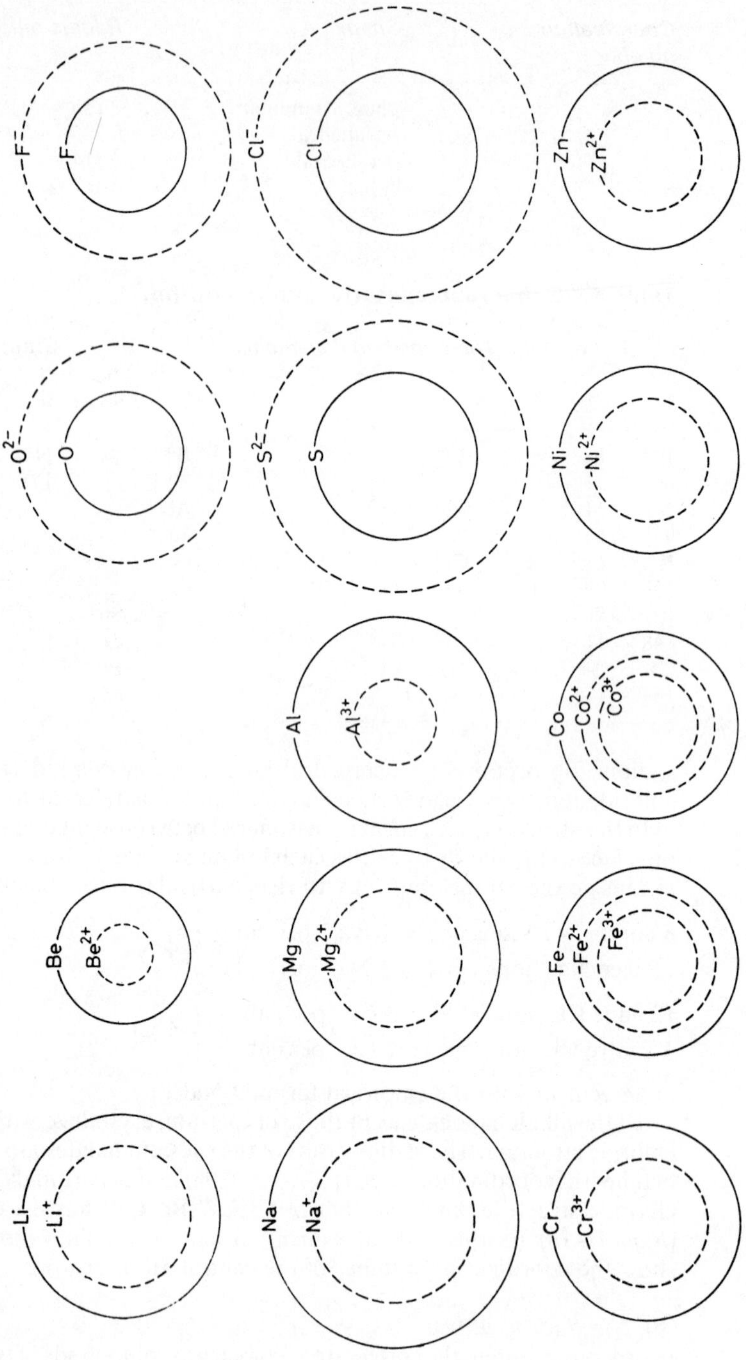

Figure 4.14 Comparison of atomic and ionic sizes

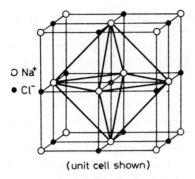

Figure 4.15 The sodium chloride lattice

○ Na⁺
● Cl⁻

(unit cell shown)

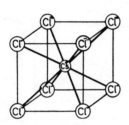

Figure 4.16 The caesium chloride lattice

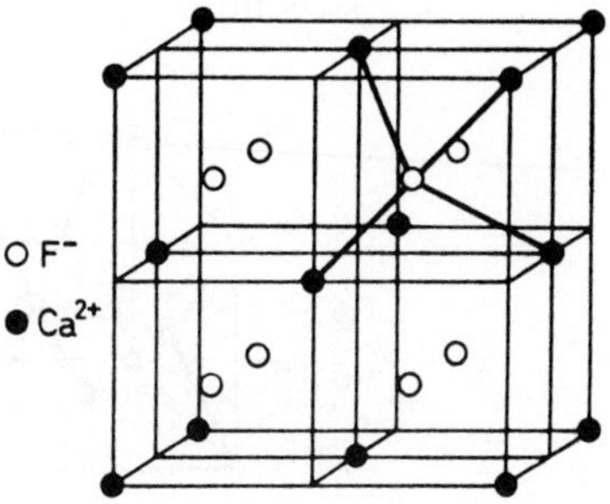

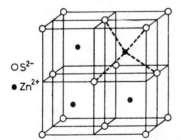

○ F⁻
● Ca²⁺

Figure 4.17 The fluorite lattice

Figure 4.18 The zinc blende lattice

○ S²⁻
● Zn²⁺

anions. An arrangement which satisfies this condition is the fluorite lattice, characteristic of calcium fluoride and the fluorides of several other bivalent metals. In this, the fluoride ions occupy the tetrahedral holes in a face-centred cubic array of calcium ions, thus giving rise to 8:4 coordination (*Figure 4.17*). When the positions of the positive and negative ions are reversed, the structure is referred to as an anti-fluorite lattice, as for example in lithium oxide, Li_2O.

(c) The zinc blende lattice

In the fluorite lattice, the anion occupies *all* the tetrahedral holes of a cubic close-packed array of cations. In the zinc blende lattice (*Figure 4.18*), on the other hand, for compounds of the type MX, the cation occupies *half* the tetrahedral holes of a cubic arrangement of anions. The sulphides of zinc, cadmium and mercury and the halides of copper, all crystallize in this manner. Several of their compounds do not have the radius ratio 0.225 to 0.414 required theoretically for tetrahedral coordination because there is a considerable degree of covalency, and the simple derivation of limiting radius ratio does not hold.

(d) The rutile lattice

A less efficient method of packing than those discussed so far is based on a tetragonal prism and 6:3 coordination. Such a structure is that of rutile, TiO_2, in which the titanium ions are surrounded octahedrally by oxide ions, whilst each oxide ion is only surrounded by three titanium ions (*Figure 4.19*). Many compounds in which the radius ratio is about 0.55 crystallize with this structure.

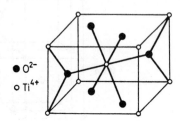

Figure 4.19 The rutile lattice

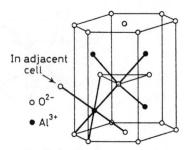

Figure 4.20 The corundum lattice

(e) The corundum lattice

An example of a hexagonal lattice is that of corundum (aluminium oxide). Aluminium ions occupy two out of every three octahedral holes in a closely-packed hexagonal lattice of oxide ions, thus giving rise to 6:4 coordination. This type of lattice is adopted by many metal oxides of formula M_2O_3 (*Figure 4.20*).

(f) The cadmium iodide structure

As a result of the large polarizability of the iodide ions, the cadmium iodide structure is intermediate in character between an ionic and a covalent lattice. The arrangement of ions is hexagonal close packed but the structure forms well-defined layers, each cation being surrounded octahedrally by six iodide ions whilst the three nearest cation neighbours of each anion are all on one side of it (*Figure 4.21*).

Covalent lattices

Covalent structures fall into two types: those in which small individual molecules are held together in a crystal lattice by weak forces such as van der Waals forces or hydrogen bonds, and those in which the covalent bonding continues indefinitely in two or three dimensions. The former are *molecular crystals* whilst the latter are referred to as *macromolecular crystals*. Examples of each type are shown in *Figure 4.22*.

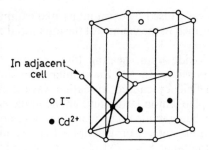

Figure 4.21 The cadmium iodide structure

In adjacent cell

○ I⁻
● Cd²⁺

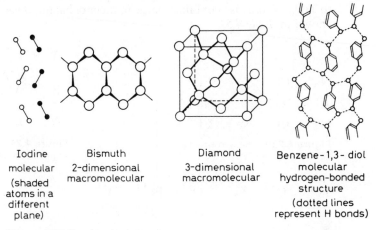

Iodine	Bismuth	Diamond	Benzene-1,3-diol
molecular	2-dimensional	3-dimensional	molecular
(shaded	macromolecular	macromolecular	hydrogen-bonded
atoms in a			structure
different			(dotted lines
plane)			represent H bonds)

Figure 4.22 Covalent lattices

Non-stoichiometry

In ionic lattices there are sometimes marked departures from the 'ideal' formulae. Electrical neutrality of such structures is maintained either by the presence of extra electrons held within the lattices or by atoms of the same element exhibiting different valencies, as for example in $Fe^{II}O$ which is deficient in iron and hence has a proportion of Fe^{III} ions present.

Even in ionic lattices possessing 'ideal' formulae, some of the ions may be displaced from their lattice sites and occupy interstitial holes instead, or may indeed be completely absent.

The different types of non-stoichiometric compounds occur when:

(1)   There is a metal excess caused by anion vacancies. In this case, electrical balance is maintained by electrons (which were originally associated with anions) remaining trapped. This situation is found, for example, when excess of sodium is burnt in chlorine. The sodium chloride formed is yellow because of electrons trapped in what have become known as '*F* centres' [*F* = Farbe (German) = colour] (*Figure 4.23*).

(2) There is a metal excess caused by extra interstitial cations. In this case, free electrons are trapped in the vicinity of the interstitial cations (*Figure 4.24*).

For example, zinc oxide upon heating loses oxygen (and becomes yellow), producing possibly Zn^+ or Zn at interstitial sites.

(3) There is a metal deficiency caused by cation vacancies. The crystal structure has a complete anion lattice and an incomplete cation lattice, the cation deficiency being made up by oxidation of some of the cations to a higher oxidation state. For example, FeO is metastable but compounds within the range $Fe_{0.91}O$ to $Fe_{0.95}O$ are stable and possess both Fe^{2+} and Fe^{3+} ions, as shown in *Figure 4.25*.

A particularly noteworthy compound is $Ti^{II}O$, which exhibits both anionic and cationic deficiencies (ranging in composition from $Ti_{0.75}O$ through TiO to $TiO_{0.69}$) and also contains a large number of balanced vacant sites, e.g., as shown in *Table 4.5*.

$$
\begin{array}{cccc}
Na^+ & Cl^- & Na^+ & Cl^- \\
Cl^- & Na^+ & e^- & Na^+ \\
Na^+ & Cl^- & Na^+ & Cl^-
\end{array}
$$

Figure 4.23 *F* centre in sodium chloride lattice

$$
\begin{array}{ccc}
A^+ & B^- & A^+ \\
 & e^-A^+ & \\
B^- & A^+ & B^- \\
A^+ & B^- & A^+
\end{array}
$$

Figure 4.24 Free electrons trapped near interstitial cations

$$
\begin{array}{cccccc}
Fe^{2+} & O^{2-} & Fe^{2+} & O^{2-} & Fe^{2+} & O^{2-} \\
O^{2-} & Fe^{3+} & O^{2-} & Fe^{2+} & O^{2-} & Fe^{2+} \\
Fe^{2+} & O^{2-} & & O^{2-} & Fe^{3+} & O^{2-} \\
O^{2-} & Fe^{2+} & O^{2-} & Fe^{2+} & O^{2-} & Fe^{2+}
\end{array}
$$

Figure 4.25 Cation deficiency in iron oxide

Table 4.5 Non-stoichiometry in titanium(II) oxide

Composition	Per cent Ti sites occupied	Per cent O sites occupied
$Ti_{0.75}O$	74	98
TiO	85	85
$TiO_{0.69}$	96	66

In non-stoichiometric compounds there is a small but measurable mobility of trapped electrons and ions; as a result, such compounds are semi-conductors (p. 59). Trapped electrons can sometimes be excited by light energy, the resultant conductivity then being known as *photoconductivity*. The energy acquired from the incident light is sometimes emitted as radiation of somewhat longer wavelength, giving rise to *luminescence*. This property is utilized in zinc sulphide screens for the detection of X-rays and electrons and, when mixed with small amounts of an alpha-particle emitter as a permanent source of energy, for the luminous hands and numbers of some clocks and watches.

A further consequence of the availability of electrons and of positive holes (i.e., vacant sites where positive ions could be placed) on the surfaces of many non-stoichiometric compounds, particularly oxides, is that they possess valuable catalytic properties.

Changes of state

Liquefaction of gases

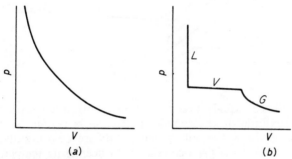

Figure 4.26 Variation of pressure with volume for a gas. Graph (a) is for a high temperature and shows the rectangular hyperbola in accordance with Boyle's law. Graph (b) is for a lower temperature and is in three sections: along the line G the substance behaves as gas, showing a marked change of volume with pressure; along V the gas is changing into liquid, and there are large volume changes for slight variation in pressure, and at L the substance is a liquid and the effect of pressure on volume is almost negligible

It was shown in the previous Section that gases obey Boyle's law only to a limited extent; *Figure 4.26* indicates that the deviation from the law is smaller the higher the temperature. This is generally the case; an increase in temperature causes molecules to move faster, so that the mutual attraction of one for the other becomes relatively less important. On the other hand, if the temperature is progressively lowered, the attractive forces assume a greater importance, until eventually the gas condenses into a liquid. It was Andrews who, in 1869, carried out a thorough investigation into the conditions under which gases could be converted into liquids. He plotted values of volume against the corresponding pressures at various fixed temperatures. The *isothermals* obtained are shown for carbon dioxide in *Figure 4.27*.

What conclusions can be drawn from these isothermals? At 48.1 °C, carbon dioxide follows Boyle's law closely, but at 35.5 °C pronounced departures from ideal behaviour are apparent. At 21.5 °C, when the pressure on gaseous carbon dioxide is increased to about 6 MPa, liquid carbon dioxide appears and the pressure remains constant until all the gas has been

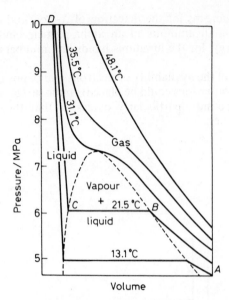

Figure 4.27 Isotherms of carbon dioxide

liquefied. Since liquids are almost incompressible, further increase of pressure has little effect. A similar pattern of behaviour is exhibited at all lower temperatures, the only significant difference being that the pressure needed to liquefy the gas decreases with decreasing temperature. In fact, carbon dioxide can be liquefied by the application of pressure at all temperatures below but not above 31.1 °C. This temperature is known as the 'critical temperature' of carbon dioxide and the corresponding pressure as the 'critical pressure'.

At the critical temperature, the kinetic energy tending to separate molecules equalizes the cohesive forces which maintain the liquid state. If, therefore, a liquid is heated in a closed system (so that the critical pressure can be attained), at the critical temperature the surface tension of the liquid vanishes and with it the meniscus. It is impossible to distinguish liquid from gas and we can say that we have 'continuity of state'.

Carbon dioxide is typical of other gases in its response to pressure at different temperatures, and *Table 4.6* gives the critical temperatures and pressures of other compounds. (Strictly speaking, a substance is only regarded as a gas when above its critical temperature and there is no possibility of liquefying it; below its critical temperature, when it can be liquefied by the application of pressure, it is known as a vapour.)

It can be seen that ammonia and carbon dioxide can be liquefied at room temperature merely by the application of pressure. Other gases need to be cooled considerably before there is any possibility of liquefaction. The production of low temperatures is therefore extremely important. There are three fundamental ways of doing this:

(1) Cooling by bringing into contact with cold substances; the effect can be multiplied by using materials at progressively lower temperatures.
(2) By allowing the gas under high pressure to expand into a region of lower

Table 4.6 Critical values

Substance	Critical temperature/°C	Critical pressure/MPa*
Helium	−268	0.23
Hydrogen	−234	1.28
Nitrogen	−146	3.35
Oxygen	−118	4·97
Methane	−83	4.56
Carbon dioxide	31	7.50
Ammonia	133	11.20
Water	374	21.85

* MPa is approximately 10 atm.

pressure. Because intermolecular attraction exists between the molecules of a gas, internal work is done, as the molecules move apart, in overcoming this attraction; the energy required to perform this work can only be obtained from the kinetic energy of the molecules themselves and so the temperature falls, provided the gas is below a certain temperature, called the *inversion temperature*. For most gases, this temperature is so high (for example, for oxygen it is 700 °C) that it presents no problem. For gases like hydrogen and helium, however, the inversion temperatures are low, and so initial cooling must be applied before this *Joule–Thomson effect* can be utilized (below −80 °C for hydrogen and below −240 °C for helium).

(3) By causing the gas to perform external work, for example, moving a piston and driving an engine.

Work is done when a force moves through a distance.

Now force = pressure × area, and so work done = pressure × area × distance
= pressure × volume change (*Figure 4.28*)

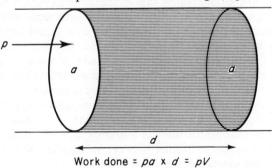

Work done = $pa \times d = pV$

Figure 4.28 Work done by a gas expanding at constant pressure

For an ideal gas

$$p_1 V_1 = p_2 V_2$$

and no work is done when the gas undergoes a change in volume. For a real gas moving from high to low pressure, work is either done by or on the gas, depending upon the way in which the product pV varies with p. In *Figure 4.29(a)*, which applies at lower temperatures, $p_1 V_1 < p_2 V_2$, and so the gas does external work in expanding and the cooling due to this reinforces that due to the Joule–Thomson effect. In *Figure 4.29(b)*, however, which operates at higher temperatures, $p_1 V_1 > p_2 V_2$, and external work is performed *on* the gas as it expands, increasing the kinetic energy and thus minimizing or even reversing the Joule–Thomson effect. (It must be emphasized that all these changes are effected adiabatically, that is, in isolation from the environment.)

The most important gas liquefaction process in industry today is the production of liquid air (*Figure 4.30*), followed by fractional distillation into its components. Air can be liquefied by making use of all the above principles. It is compressed isothermally to about 15 MPa (150 atm) and cooled to $-30\,°C$ by the first heat exchanger. About three-quarters of the air is then

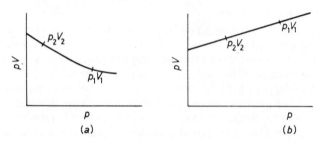

Figure 4.29 Expansion of real gases: in (a) is shown the situation at low temperatures where work is done by the gas as it expands, and (b) corresponds to higher temperatures where work is done on the gas as it expands

expanded to 400 kPa by doing external work as it drives an engine: the temperature falls to about $-160\,°C$. It then goes to the bottom of the fractionator, where it assists vaporization. The remaining air is liquefied by a second heat exchanger and passed to the fractionator at X where, as it descends, it 'washes' ascending vapour free from oxygen (b.p. $-183\,°C$). The more volatile nitrogen (b.p. $-196\,°C$) continues to vaporize but is condensed where it meets the condenser, cooled by cold liquid oxygen. The liquid nitrogen is then expanded through a valve to 100 kPa (1 atm) and enters the top half of the fractionator Y where, as it 'scrubs' the ascending vapour free from oxygen, it is vaporized itself and passes out from the plant through heat exchangers which utilize its low temperature in cooling incoming air. At the same time, liquid air from the base of the fractionator expands to 100 kPa by passage through a valve and enters the top half of the fractionator at Z; as it falls, nitrogen evaporates, leaving liquid oxygen which can be removed as required. The intimate contact between descending liquid and ascending vapour ensures satisfactory fractionation.

The liquid oxygen contains the denser noble gases, notably argon, whilst helium and neon are present in the nitrogen fraction. These gases can be extracted, if required, by further fractionation.

Boiling point and melting point
Although under certain circumstances (at the critical temperature and pressure) the gas and liquid phases merge into one and there is continuity of

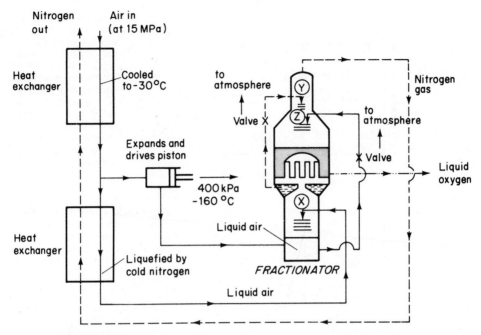

Figure 4.30 Liquefaction of air.

state, usually the two phases are quite distinct. The more energetic liquid molecules are able to escape from the liquid and become vapour, so giving rise to a vapour pressure. The higher the temperature, the more likely are molecules to possess the necessary kinetic energy to enable them to escape from the liquid phase, so that the vapour pressure is then greater. Eventually, at a sufficiently high temperature, it equals the external pressure and all the liquid becomes vapour, without further increase in temperature. The liquid is said to *boil* and the heat provided, instead of producing an increase in temperature, produces an increase in disorder (or entropy, p. 130), measured as the latent heat of vaporization. It follows that the boiling point is dependent upon the value of the external pressure, but it is constant at constant pressure. Indeed, a constant boiling point is usually taken as a criterion of purity.

Similarly, when a solid is heated, there is an increase in kinetic energy of the particles, and the temperature rises. As with the liquid, some particles have energy so much in excess of the average that they escape into the vapour phase, so that the solid, too, exerts a vapour pressure, which also increases with temperature. Eventually, vibrations of the ions, atoms or molecules become so extreme that collisions between adjacent particles occur and the crystal lattice breaks down. The melting point at which this occurs is constant at constant pressure for a pure substance, so that this, like the boiling point, is indicative of purity. At the melting point, the heat taken in, instead of producing an increase in temperature, causes an increase in disorder: because the order in the liquid is far higher than in the gas, the latent heat of fusion is less than that of vaporization (*see Figure 4.31*).

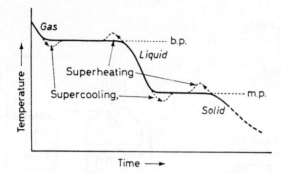

Figure 4.31 Effect of continuous cooling of a gas, or heating of a solid. The thick line shows the ideal situation. At the boiling point (b.p.) liquid and gas are in equilibrium and at the melting point (m.p.) solid and liquid are in equilibrium. However, supercooling and superheating often occur owing to delay in establishing these equilibria

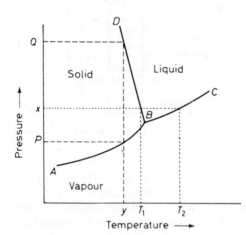

Figure 4.32 Phase diagram for a pure substance with one solid form only. If the pressure is maintained constant at x while the temperature is steadily increased, solid becomes liquid at T_1 and liquid becomes gas at T_2 (at 101 kPa, T_2 represents the normal b.p.). If the temperature is kept constant at y while pressure is steadily increased, vapour is converted into solid at P and the solid melts at Q The triple point for water is at 590 Pa and 0.007 °C

Phase diagrams

Many of the ideas that have been discussed in this Section can be conveniently summarized in graphical form. In many cases, for pure substances, if vapour pressure is plotted against temperature, two distinct curves, *AB* and *AC*, are obtained for solid and liquid, respectively (*Figure 4.32*). The point *B* is known as the triple point and is the temperature where solid, liquid and vapour co-exist. Line *BD* represents the effect of pressure on the melting point and, in the case shown, indicates that an increase in pressure lowers it. Le Chatelier's theorem (p. 144) enables us to predict that in this particular case, when solid melts, there is a diminution in volume.

Polymorphism (Allotropy)

If the solid is capable of existing in different forms (polymorphs or allotropes), *Figure 4.32* requires modification. This depends upon whether the allotropes are each stable under differing conditions or whether one allotrope only is stable. The former, reversible, type of allotropy is called *enantiotropy* and is exemplified by sulphur (*Figure 4.33*) whilst the other variety is termed *monotropy*. Phosphorus is an example of a monotropic element (*Figure 4.34*).

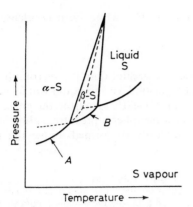

Figure 4.33 Phase diagram for sulphur, showing typical enantiotropy. At A, α-sulphur has a lower vapour pressure than the β form and is therefore the stable polymorph, but at B, β-sulphur has the lower vapour pressure and is the more stable polymorph in this region. Broken lines show unstable equilibria

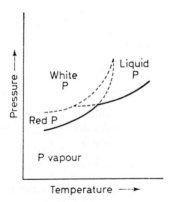

Figure 4.34 Phase diagram for phosphorus, showing typical monotropy. Red phosphorus has consistently lower vapour pressure than white phosphorus and is therefore always the stable solid form. Broken lines show unstable equilibria

In principle, the slow cooling of molten sulphur (or its vapour) leads to the formation of β-sulphur (the monoclinic allotrope) and eventually to α-sulphur (the rhombic form). Heating of rhombic sulphur reverses the transitions.

In the case of phosphorus, the vapour pressure of the white allotrope is always greater than that of the red form, and therefore the former is unstable (i.e., has a higher energy) in comparison with the latter. Because of this, the direct transition of the red into the white allotrope is not possible whereas the reverse change is easily produced. The white form condenses from the vapour.

Sublimation
Provided the triple-point pressure is greater than the surrounding pressure, sublimation will occur on heating a solid, i.e., no liquid will form during the transition to the vapour state. This is clearly seen from the phase diagram of carbon dioxide (*Figure 4.35*); at atmospheric pressure, sublimation occurs while liquid carbon dioxide can only be formed above the triple-point pressure.

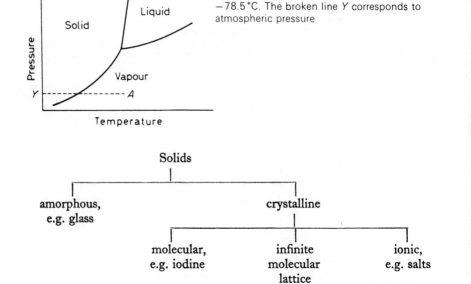

Figure 4.35 Phase diagram for carbon dioxide; the triple point is at 501 kPa and $-78.5\,°C$. The broken line Y corresponds to atmospheric pressure

Figure 4.36 The structural types of solids

Summary

Matter exists between two extremes: that of the perfect gas, where molecules are completely independent of each other and where kinetic energy is high, and the perfect solid at absolute zero, where interaction is at a maximum and kinetic energy zero. The degree of randomness of atoms, molecules and ions consequently decreases from the gaseous state through the liquid to the solid state, where the highest order is attained. Physical constants such as density, melting and boiling points, critical temperature and pressure, surface tension and viscosity are dependent upon the intermolecular and interionic forces as well as the possible geometric arrangement.

The structure of solids may be classified in terms of *Figure 4.36*.

Questions

(1) What properties of the liquid state show that it is intermediate between the solid and gaseous states?

(2) What effect has cubic and hexagonal close packing on the density of a metallic element?

Figure 4.37

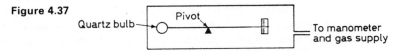

(3) *Figure 4.37* shows a microbalance, in which the pressures of different gases required to float the quartz bulb are determined. If the beam is balanced in oxygen at a pressure of 5.920 kPa and in argon at a pressure of 4.740 kPa, calculate the relative atomic mass of argon, given that it is monatomic and the vapour density of oxygen is 16.

(4) Explain what precautions are required in using the gram molecular volume to calculate relative molecular masses.

If the normal density of hydrogen chloride is 1.64 g dm^{-3}, calculate the relative molecular mass of the gas, using the values given in *Figure 4.5*.

(5) In an experiment, the rate of diffusion of mercury vapour relative to that of radon was 1.05. Knowing the relative atomic mass of mercury and the fact that mercury vapour and radon are monatomic, deduce the relative atomic mass of the latter.

(6) Using the Table of ionic radii (*Table 4.4*), suggest what structures you would expect for rubidium bromide and calcium oxide.

(7) How would you determine the relative atomic masses of argon and bromine?

(8) Write an essay on the 'defect solid state'.

(9) 'Theories are concerned with ideal, non-existent, systems and so are of little use to the practising scientist'. Discuss this statement, with particular reference to gases.

(10) Suggest ways of determining the formulae of nitrogen dioxide, sulphur dioxide and ammonia.

(11) Write notes on (*a*) thixotropy, (*b*) liquid crystals, and (*c*) glasses.

(12)   The *Phase Rule* can enable a chemist to make some predictions about the possible outcome of changing conditions. It is expressed as $P + F = C + 2$ where P is the number of phases, C is the number of components (i.e., the minimum number of substances which are chemically sufficient to define the system), and F is the number of degrees of freedom (i.e., the number of factors which must be fixed to fix the system). It follows that the greater the number of components there are in a system, the greater will be the number of degrees of freedom than in a system with one component. Similarly, the greater the number of phases present, the smaller will be the number of degrees of freedom.

Look at the phase diagrams in this book and interpret them in the light of the Phase Rule.

(13)   A solution of a long-chain acid (relative molecular mass $= 300$) was made by dissolving 0.009 g in $100 \, cm^3$ of an organic solvent. A surface film was formed by placing $0.1 \, cm^3$ of this solution on water and allowing the solvent to evaporate. This film occupied an area of $80 \, cm^3$. Calculate L, the Avogadro constant, assuming that the molecules are closely packed and occupy an area of $50 \, Å^2$ each [Ångström unit (Å) $= 10^{-10}$ m].

[O.]

5 Solutions

The colloidal state • Factors affecting the formation of a
solution • Gas–liquid equilibria • Gas–solid equilibria • Liquid–liquid
equilibria • Solid–liquid equilibria • Colligative properties of dilute
solutions

A 'true' solution implies a breakdown at the molecular level; that is to say, the dissolving substance, or *solute*, is separated into individual molecules (or ions) throughout the *solvent*. Such a solution is said to be a *homogeneous* phase, with the composition being uniform and constant throughout. (This description, however, is hardly more than a convenience: a little thought will make it clear that whether such a system appears homogeneous or heterogeneous depends very much upon the scale of dimensions used, i.e., whether atomic or much larger.)

The colloidal state

If the solute is macromolecular or the size of its aggregates is too great, the average diameter of the solute 'particle' will be much greater than the usual loose solvent aggregates. The mixture then shows deviations from a true solution, and the system breaks down into a solution consisting of a definite disperse phase suspended in a dispersion medium—the continuous phase (*Figure 5.1*). If the particles of the disperse phase are of the order of about 1–1000 nm and remain dispersed, the solution is said to be *colloidal*; if larger and rapidly settling out, the disperse phase is regarded as a coarse suspension. But there are no sharp divisions between the various systems, and it is believed that any solute can be obtained in the form of a colloid (*see Table 5.1*).

Table 5.1 Types of dispersion

	Particle size/nm	Optical characteristics
True solution	<1 (diam.)	Clear
Colloids and fine suspensions	1–1000 (limit of colloidal range)	Scatter light, but individual particles invisible
Precipitate (coarse suspension)	>1000 settle out quickly	Opaque: individual particles visible

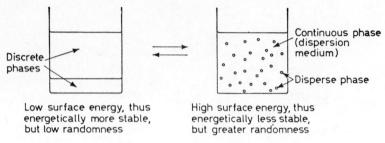

Low surface energy, thus   High surface energy, thus
energetically more stable,   energetically less stable,
but low randomness   but greater randomness

Figure 5.1 Colloidal dispersions

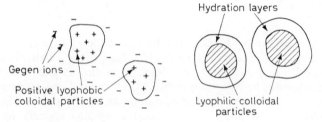

Figure 5.2 Colloidal particles

Colloidal solutions are either *lyophobic* ('solvent-hating') or *lyophilic* ('solvent-loving'). Broadly speaking, the former owe their stability to the repulsion between similarly-charged aggregates, whilst the latter are prevented from flocculating by having large solvation layers. Many lyophilic colloids, however, are also charged and some, for example amino acids and proteins, possess both negative and positive charges: the stability of these is then lowest when the algebraic sum of the fixed charges is zero

Since the colloidal systems as a whole are electrically neutral, the charged particles are surrounded by mobile counter- (or gegen-)ions (*see Figure 5.2*).

The presence and type of charge on colloidal particles may be shown by the migration produced under an applied electrical potential, a procedure known as *electrophoresis* (*Figure 5.3*).

The stability of a lyophobic colloid can be reduced by adding ions of opposite charge which are more strongly adsorbed than the counter-ions; the net charge may then be so reduced that flocculation occurs. In general, the larger the charge on the ions, the more readily are they adsorbed and the greater their neutralizing power. For negative colloids, the flocculating power of ions is of the order $Th^{4+} > Al^{3+} > Mg^{2+} > Na^+$, and for positive

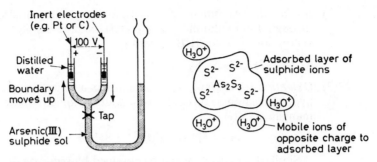

Figure 5.3 Electrophoresis

colloids, $Fe(CN)_6^{4-} > Fe(CN)_6^{3-} > SO_4^{2-} > Cl^-$. Flocculation can also be brought about by mixing colloids of opposite charge.

The stability of lyophilic colloids is not greatly affected by the addition of electrolytes, except in sufficient concentration to compete for the solvent within the solvation layers, when 'salting-out' of the colloid occurs. Very important naturally occurring lyophilic colloids include starch, gelatine, egg white albumin and the proteins of living cell cytoplasm.

The concentration of lyophilic systems can be much higher than that of lyophobic; the taking up of large quantities of solvent often results in the viscosity of lyophilic dispersions differing considerably from that of the dispersion medium (for example, gelatine). Again, because of the solvation layers, lyophilic colloidal particles generally have almost identical refractive indices to the dispersion media, whereas lyophobic particles differ sufficiently to act as scattering surfaces for light waves. This allows a means of observing the latter as a cone of scattered light (the Tyndall cone) when a lyophobic colloid is viewed through the ultramicroscope at right angles to a converging beam of light (*Figure 5.4*).

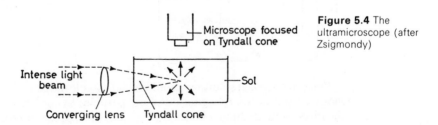

Figure 5.4 The ultramicroscope (after Zsigmondy)

Since the colloidal state represents only an intermediate range of particle size, the method of preparation depends either upon the aggregation of ions in solution *up* to, or dispersion of the solid material *down* to, colloidal dimensions.

Examples of methods involving aggregation are:

(1) *Hydrolysis*, for example by boiling a very dilute solution of iron(III) chloride in distilled water. A red sol of hydrated iron(III) oxide results:

$$Fe^{3+} + 3H_2O \rightarrow 3H^+ + Fe(OH)_3 \rightarrow Fe_2O_3 . xH_2O$$

(2) *Double decomposition*, for example by passing hydrogen sulphide through a hot solution of arsenic(III) oxide in distilled water, giving a yellow sol of arsenic(III) sulphide:

$$As_2O_3 + 3H_2S \rightarrow As_2S_3 + 3H_2O$$

(3) *Oxidation*, for example the production of a white sulphur sol by passing hydrogen sulphide through a very dilute, acidified solution of potassium manganate(VII):

$$2MnO_4^- + 6H^+ + 5H_2S \rightarrow 2Mn^{2+} + 8H_2O + 5S$$

Similarly, a silver sol can be obtained by reducing diamminesilver(I) hydroxide with dilute methanal (formaldehyde):

$$2Ag(NH_3)_2^+ + 2OH^- + H.CHO \rightarrow 2Ag + HCOOH + 4NH_3 + H_2O$$

(4) *Exchange of solvent:* the addition of water to a solution of phosphorus in ethanol results in the formation of a phosphorus suspension in aqueous ethanol.

Examples of dispersion methods include, for lyophilic colloids, shaking with the dispersion medium ('peptization') or grinding the solid in a ball-mill, perhaps in the presence of dispersing (deflocculating) agents. For lyophobic colloids, electrical methods using spark discharge between electrodes of the metal in the solvent or dispersion with ultrasonic waves can be employed.

In order that dispersions, particularly of lyophobic colloids, remain stable, free electrolytes must be removed, usually by *dialysis*. A dialyser, for example, parchment paper or cellophane, allows the simple ions to pass through but retains the larger colloidal particles (*Figure 5.5*). An important natural dialysing membrane is the peritoneal membrane.

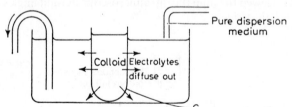

Figure 5.5 Dialysis: S is a semipermeable thimble through which colloid particles cannot pass but electrolytes can

Purified colloidal solutions should remain stable indefinitely, as Brownian motion—the continual agitation of the particles by molecular impact with the dispersion medium—keeps the particles in random motion, preventing their sedimentation under gravity.

Osmotic pressure measurements (p. 114) can be made on the purified colloid and they often reveal, in the case of lyophilic colloids, that the particles are very large (macro) molecules.

Factors affecting the formation of a solution

A useful guide to the possibility of dissolution is the axiom that 'like dissolves like'. Thus ionic or very polar substances tend to be soluble in polar liquids

such as water, whilst covalent organic molecules are usually soluble in
organic solvents of low polarity. What lies behind this generalization?

Two natural tendencies are an increase in the randomness and a decrease
in energy of a system (*see* p. 130). The first factor always favours dissolution,
provided that the solvent has no structure of its own, since there must be an
increase in randomness when two distinct phases (the solute and the solvent)
are mixed. The second factor involves the lattice energy, which for ionic
solids is considerable, that is to say ionic lattices are very stable. Further
stability can therefore only come about if there is a marked interaction with
the solvent, which in turn, means that the solvent molecules must have a
dipole moment to permit solvation of the ions. However this implies that the
solvent has some internal structure which therefore clouds the issue with
respect to the first factor. Indeed the situation in aqueous solutions is far from
simple.

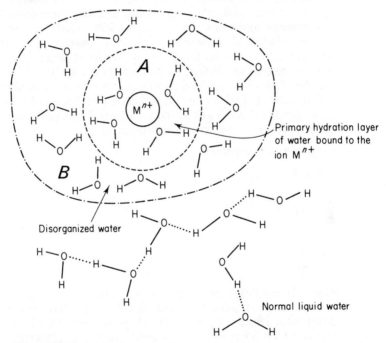

Figure 5.6 Structure of water and aqueous solutions

In aqueous solution, the innermost region of water surrounding an ion
consists of polarized water molecules immobilized by the ions as shown in
Figure 5.6. The outermost region of water has the normal liquid structure
associated with hydrogen bonding, whilst the middle region contains water
molecules more randomly arranged than in normal water because of the
influence of the innermost region. The relative volumes of the regions A and
B (*Figure 5.6*) determine the overall ordering effect produced by the added
ions. If the free energy decrease (p. 131) accompanying this change exceeds
the lattice energy of the solid, then the solid will dissolve. There is no
possibility of this sort of interaction between a non-polar covalent molecule

and water and so the aggregation of water molecules is unaffected, i.e., the non-polar substance is insoluble.

Electrovalent substances do not generally dissolve in non-polar solvents because there is no prospect of interaction to provide the energy necessary to disrupt the ionic lattice. The fact that there is little lattice energy in a substance of low polarity increases the possibility of this dissolving in a non-polar solvent where there is no molecular interaction and aggregation, especially as dissolution increases the randomness of the system.

Gas–liquid equilibria

Because gases have no boundary surfaces and there is little molecular interaction, energy changes accompanying mixing are usually negligible. The move to greater randomness results in gases being freely miscible and the fact that gaseous molecules are moving rapidly means that solution will be rapidly effected. Furthermore, the rapid movement of the molecules generally offsets any tendency for the denser constituent to settle out. In contrast, equilibrium between a gas and a liquid is reached only slowly and a number of possible equilibria may in fact exist if the gas combines chemically with the solvent. Only in the simplest cases does *Henry's law* apply: the mass, m, of a gas dissolved by a fixed amount of liquid is directly proportional to the gas pressure, p, at constant temperature, i.e.,

$$m \propto p \text{ or } m/p = \text{constant} \tag{1}$$

For a perfect gas

$$pV = \frac{m}{M} \cdot RT$$

i.e.,

$$m = \frac{pVM}{RT}$$

where M = relative molecular mass. Substitution in (1) gives VM/RT = constant, and since the term M/RT is constant, the volume of gas dissolved at a fixed temperature by a fixed amount of liquid is independent of pressure. For this reason, gas solubilities are generally expressed in volume rather than mass units; the *solubility coefficient* is the volume of gas dissolved in 1 cm^3 of liquid.

For example, the solubility coefficients of two gases, A and B, in water are 0.03 and 0.01, respectively, at 10 °C. Therefore, if 1 dm^3 of water saturated at 10 °C by being in contact at 100 kPa with an equimolecular mixture of the two gases is boiled, the volume of each gas (measured at 10 °C and 50 kPa, the partial pressure) expelled will be $0.03 \times 1000 = 30$ cm^3 of A and $0.01 \times 1000 = 10$ cm^3 of B.

At s.t.p. these volumes become

$$\frac{30 \text{ cm}^3 \times 50 \text{ kPa} \times 273 \text{ K}}{100 \text{ kPa} \times 283 \text{ K}} = 14.4 \text{ cm}^3 \text{ of } A$$

$$\frac{10 \text{ cm}^3 \times 50 \text{ kPa} \times 273 \text{ K}}{100 \text{ kPa} \times 283 \text{ K}} = 4.8 \text{ cm}^3 \text{ of } B$$

Those gases which dissolve exothermically have, in accordance with Le Chatelier's principle (p. 144), negative temperature coefficients; that is, they are less soluble in hot than cold solvent; this is the case for most gases in water.

Departures from Henry's law are caused by:

(1) the non-ideal nature of the gas over the pressure range employed, and
(2) the chemical solution of the gas, e.g.

$$NH_3(g) \overset{H_2O}{\rightleftharpoons} NH_3(aq) \rightleftharpoons NH_3 \cdot H_2O \rightleftharpoons NH_4OH \rightleftharpoons NH_4^+ + OH^-$$

The more soluble gases in fact are those which react with the solvent.

In the colloidal range, gases dispersed in liquids constitute *foams*, whilst liquids dispersed in gases are generally termed *aerosols*. Foams are stabilized by lowering the surface tension of the liquid and providing some mechanical strength (for example, by increasing the viscosity) by the addition of foaming agents such as proteins and soaps. Aerosols are inherently unstable and are generally produced by the rapid escape of gas, the aerosol propellant (often an inert fluorohydrocarbon), through a dispersed mixture.

Gas–solid equilibria

The vapour pressure of solvated molecules is determined by the strength of the bonding of the coordinated solvent molecules. For hydrates, the terms 'deliquescent', 'efflorescent' and 'hygroscopic' are only relative and depend on the vapour pressure of the water in the air. A substance is hygroscopic and takes in water only if the vapour pressure of the hydrate is less than that of water in the environment. A substance is efflorescent and gives out water vapour only if the vapour pressure of its hydrates exceeds that of the surroundings. The vapour pressure increases, of course, with the extent of hydration. *Figure 5.7* shows the relationship between the vapour pressure and the degree of hydration of copper(II) sulphate.

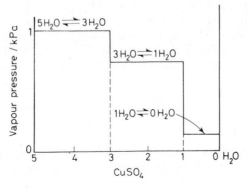

Figure 5.7 Vapour pressures of hydrates of copper sulphate

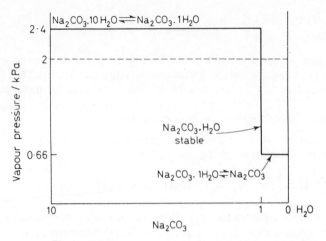

Figure 5.8 Vapour pressures of hydrates of sodium carbonate

The effect of hydration on the behaviour of a substance is well illustrated by sodium carbonate; anhydrous sodium carbonate is hygroscopic, whereas the decahydrate is efflorescent—the stable form is the monohydrate, *under normal atmospheric conditions*, where the vapour pressure is about 2 kPa (*Figure 5.8*)

$$Na_2CO_3 . 10H_2O \underset{2.4\,kPa}{\rightleftharpoons} Na_2CO_3 . H_2O \underset{660\,Pa}{\rightleftharpoons} Na_2CO_3$$

Because of the unsaturated nature of surfaces, due to the presence of residual bonds, adsorption of gases readily takes place, particularly at low temperatures; the adsorbed molecules saturate the residual bonds of the solid and are removed only with some difficulty, for instance in the metallurgical operation of degassing a solid. Solids containing large internal surface areas (that is, having a large porous capillary system) can adsorb very large volumes of gases: a common adsorbent is activated charcoal. In the colloidal range, entrapped gases cause otherwise transparent minerals (by internal reflection) to become opaque, for example milk quartz. Collectively, such gases dispersed in solids are called *solid foams*; they are now commonplace in the form of foam rubbers.

Solids dispersed in gases are referred to as *smokes*. They usually represent a nuisance, both domestically and industrially. It is normally necessary to remove smoke particles from industrial gases such as sulphur dioxide by electrostatic precipitation. The impure gas is passed between insulated metal plates maintained at a high electrical potential difference: the solid particles become charged under such conditions and are attracted to the plate of opposite charge.

Liquid–liquid equilibria

Liquids of similar polarity are completely miscible, on account of the mutual interaction that is possible between the different molecules (*see* previous

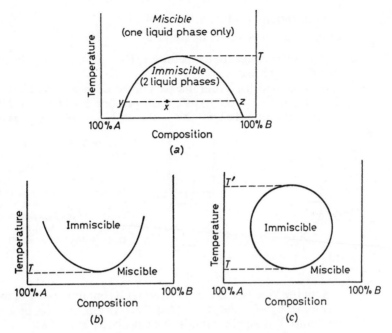

Figure 5.9 Mutual solubility diagrams. (*a*) The point x represents two liquid phases present in the proportions

$$\frac{\text{Amount of } y}{\text{Amount of } z} = \frac{\text{Length } xz}{\text{Length } xy}$$

discussion). Thus, ethanoic (acetic) acid is completely miscible with water and methylbenzene with benzene, while water and benzene themselves are immiscible. Between these extremes can be found liquids which are partially miscible, where the two liquid phases consist of solutions of one in the other. The compositions of these liquid phases are often susceptible to temperature change, as shown in *Figure 5.9*. In some cases, mutual solubilities increase with temperature, as shown in *Figure 5.9(a)*, for example phenol and water. *Figure 5.9(b)* shows mutual solubility decreasing with temperature, as found with triethylamine and water. In some other cases (e.g., nicotine and water) both types of behaviour are found (*Figure 5.9(c)*).

The *consolute* or *critical solution temperatures*, *T*, are very sensitive to impurity, and so measurements of miscibilities can often afford information about the condition of binary and more complex mixtures.

The distribution law

If a solute is capable of simple solution in two immiscible liquids, then it distributes itself between them in accordance with the respective solubilities, so that the ratio

Concentration of solute in liquid A

Concentration of solute in liquid B

is constant at constant temperature. This ratio is called the *distribution* or *partition coefficient*.

If a system with a favourable partition coefficient can be found for a solute, then it can be largely transferred from one liquid to the other, possibly leaving impurities behind: this is the basis for *extraction* (*Figure 5.10*).

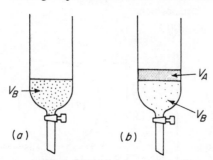

Figure 5.10 Distribution of a solute between two immiscible liquids. Funnel (a) contains a volume V_B containing m_1 g of dissolved solute in liquid B, and funnel (b) contains, at equilibrium, a volume V_A containing m_2 g of dissolved solute in liquid A lying above a volume V_B now containing $(m_1$-$m_2)$ g of dissolved solute in liquid B

Suppose V_B to be the volume of liquid B containing m_1 g of solute to which a volume, V_A, of liquid A is added. Then the mass, m_2 g, of solute extracted by liquid A at equilibrium can be calculated as follows:

$$\frac{\text{Concentration in } A}{\text{Concentration in } B} = \frac{m_2/V_A}{(m_1 - m_2)/V_B} = K \text{ (the partition coefficient)}$$

$$m_2 = \frac{m_1 K V_A}{(V_B + K V_A)}$$

Example   The solubility of iodine in tetrachloromethane and water at $20\,°C$ is 15.8 and $0.29\ \mathrm{g\,dm^{-3}}$, respectively. Find the volume of tetrachloromethane required to extract 90 per cent of the iodine from $3\ \mathrm{dm^3}$ of water at $20\,°C$.

The partition coefficient may be taken to a sufficient approximation as the ratio of the solubilities, if the solutions are dilute:

$$\frac{I_2 \text{ in } CCl_4}{I_2 \text{ in } H_2O} = \frac{15.8}{0.29}$$

If $V\ \mathrm{dm^3}$ is the volume of tetrachloromethane required, then

$$\text{Partition coefficient} = \frac{15.8}{0.29} = \frac{90/V}{10/3}$$

Therefore

$$V = \frac{0.29 \times 90 \times 3}{15.8 \times 10} = 0.50\ \mathrm{dm^3}$$

Three points should, however, be noted:

(1) The solute must be in the same physical condition in both solvents; for example, if association or dissociation of solute molecules occurs in one liquid, then the partition coefficient will not remain constant (in fact,

departure from constancy can provide significant evidence about such things as ionization);

(2) As partition is a manifestation of differential solubility, neither liquid should be saturated with respect to the solute;

(3) More efficient extraction is achieved by shaking separate small samples of solvent with the solution and combining the extracts than by using all the solvent in a single extraction.

Applications of distribution

Ethereal extraction—Ethoxyethane is only slightly soluble in water; it dissolves most organic substances, is chemically inert and, being volatile, easily removed from the liquid phase. It is consequently widely used for extracting organic substances from aqueous systems, the more hydrophilic impurities remaining behind.

Desilverization of lead-Zinc, when just above its m.p. (419 °C), is largely insoluble in molten lead (m.p. 327 °C). The distribution coefficient for silver between these two liquids at this temperature is

$$\frac{\text{Concentration in zinc}}{\text{Concentration in lead}} \sim 3000$$

Zinc is therefore used to desilverize lead. Relatively small quantities of zinc added to molten argentiferous lead remove most of the silver present; it rises to the surface, where it is skimmed off. The zinc is finally removed from the silver by distillation.

Distillation of liquid mixtures

Liquids always have a definite vapour pressure associated with them, the result of certain molecules with excess energy escaping from the liquid phase. At equilibrium, when the 'saturation vapour pressure' prevails, the rate at which liquid molecules leave the liquid equals that at which vapour molecules return (*Figure 5.11*). A liquid of high volatility has a high vapour pressure because its molecules are, on the whole, moving rapidly and are therefore more able to escape into the vapour phase. In the case of an ideal mixture of liquids, where the interactions between molecules are equal and each constituent retains its independent volatility, the total vapour pressure is the sum of the separate vapour pressures (although these are proportional to the relative number of molecules present). It follows that the more volatile component, if of sufficient concentration, contributes the larger proportion to the total. In *Figure 5.12(a)* the total vapour pressure, RX, of the mixture of composition X is made up of XQ, from the more volatile constituent A, and of XP from component B.

As the temperature is increased, so the vapour pressure increases until it equals the external pressure and the liquid boils. It follows from what has been said that the vapour given off is richer in the more volatile constituent than is the liquid left behind. It can be condensed and analysed and a graph such as that in *Figure 5.12(b)* constructed. A mixture of composition x boils at

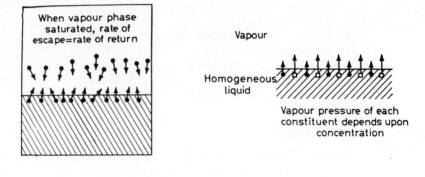

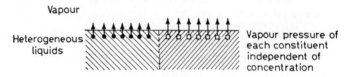

Figure 5.11 Vapour pressure of liquids

temperature T' and its vapour has the composition y. If it is now condensed, it gives a liquid that boils at T'' giving off vapour of composition z. If this process is continued, virtually pure A can be obtained. This is what happens in *fractional distillation*: a temperature gradient is established in a tall vertical column, sufficient to permit the above procedure to take place continuously, the vapour that is formed condensing, falling, warming and reboiling to give off a vapour still richer in the more volatile constituent.

Most liquids are non-ideal in the sense that unequal interaction takes place between molecules of the different components. If loose compound formation occurs, for example in hydration, a minimum value for the total vapour pressure results (*Figure 5.13(a)*). The mixture with this minimum

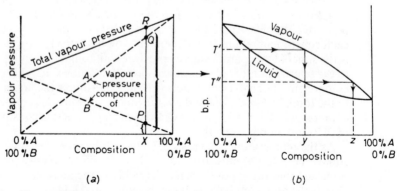

(a)   (b)

Figure 5.12 Vapour pressure and boiling point curves for two ideally miscible liquids

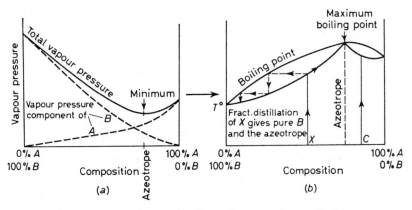

Figure 5.13 Vapour pressure and boiling point curves for two liquids showing negative deviation from Raoult's law

vapour pressure is called an *azeotrope*. It has the highest boiling point of the system; the vapour and liquid curves coincide at this point, so the vapour and liquid have the same composition and in this respect resemble a pure substance (*Figure 5.13(b)*). Fractional distillation of any mixture other than the azeotrope gives a liquid of the azeotropic composition and a vapour of pure *A* or *B* (e.g., hydrochloric acid and water).

The existence of repulsive forces between the molecules of different liquids (or, at least, the non-existence of attractive forces) results in the total vapour pressure being greater than the sum of the separate ones. This once again leads to the existence of an azeotrope, this time with maximum vapour pressure and therefore minimum boiling point. Such a system (e.g. propanone–tetrachloromethane or ethanol–water) is shown in *Figure 5.14*. Distillation of the azeotrope gives a vapour with the same composition as the liquid, so that purification by distillation is impossible. Fractional distillation of any mixture other than the azeotrope gives a vapour with the azeotropic composition and residual liquid of *A* or *B*.

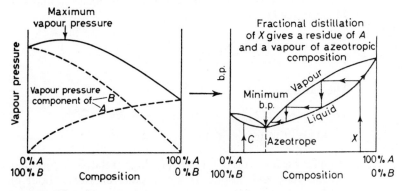

Figure 5.14 Vapour pressure and boiling point curves for two liquids showing positive deviation from Raoult's law

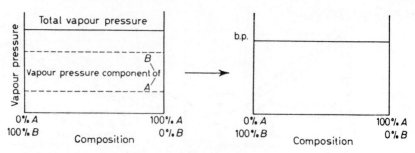

Figure 5.15 Theory of distillation of immiscible liquids, e.g., steam distillation

Repulsive forces between the different molecules result, in the extreme case, in the liquids being immiscible. Now each liquid phase can exert its own true vapour pressure (*Figure 5.11*) independent of the overall composition, so that the boiling point remains constant whilst the two liquids are present, and is not affected by the variation in composition (*Figure 5.15*). Furthermore, because each liquid exerts its normal vapour pressure, the external pressure is reached below the boiling point of either. This fact is utilized in the *steam distillation* of organic substances insoluble in water. Steam is passed in from an external vessel and, by providing a nucleus for the formation of bubbles of vapour, minimizes 'bumping' and superheating.

In the case of the steam distillation of nitrobenzene, the mixture boils at 99 °C, where the vapour pressure of water is 97.7 kPa (733 mmHg) and of nitrobenzene 3.6 kPa (27 mmHg), so that the total vapour pressure is 101.3 kPa (760 mmHg) or atmospheric pressure. Since the vapour pressure (partial pressure) is proportional to the number of moles present ($pV = nRT$, where V, R and T are constant), we can write

$$\frac{\text{Vapour pressure of nitrobenzene}}{\text{Vapour pressure of water}} = \frac{\text{Moles of nitrobenzene}}{\text{Moles of water}}$$

i.e., since the relative molecular masses of nitrobenzene and water are 123 and 18 respectively

$$\frac{3.6}{97.7} = \frac{\text{mass of nitrobenzene}/123}{\text{mass of water}/18}$$

Therefore

$$\frac{\text{Mass of nitrobenzene}}{\text{Mass of water}} = \frac{123 \times 3.6}{18 \times 97.7} \approx \frac{1}{4}$$

Thus, because of the relatively high relative molecular mass of the nitrobenzene and despite its low vapour pressure, about one-fifth of the distillate consists of nitrobenzene, and it is distilling at more than 100 kelvins below its normal boiling point, so that there is far less risk of decomposition.

Solid–liquid equilibria

Solutions of solids in liquids provide the most common type of solute/solvent interaction. With water as the solvent, the characteristic solute is an electrovalent compound. Dissolution results in the ions of the crystal lattice being further separated; work is done when ions of opposite charge are moved apart, and unless the energy required is completely forthcoming in the form of hydration energy of the ions, the dissolution will be endothermic. By Le Chatelier's theorem (p. 144), the substance will then be more soluble in hot than cold water and a positive temperature coefficient will result. *Figure 5.16* illustrates the different types of solubility curves possible for saturated aqueous solutions.

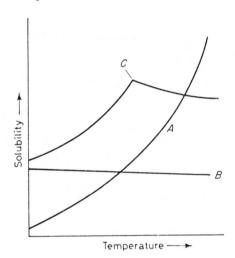

Figure 5.16 Solubility of solids in water. Curve *A* shows how the solubility of most salts increases with temperature, although the solubility of some solids decreases (as for gases), as shown by curve *B*. A break in a solubility curve, as at *C*, indicates a transition between two forms of the same substance

The addition of solid to liquid usually depresses the freezing point of the latter. Two fundamental phase diagrams are possible, depending upon whether the solid deposited on cooling consists of a solution of one component in the other or whether, because of mutual insolubility, two solid phases result. In the case of cobalt and nickel, for example [*Figure 5.17(a)*], progressive substitution of atoms occurs without significant change in the shape of the crystal lattice, and a solid solution is formed. If a liquid mixture of composition X is cooled, solid first appears at T^0, being of the composition represented by Y. The remaining liquid consequently becomes richer in component A and the freezing point progressively falls. The solid crystallizing out also becomes richer in A but, of course, it is never as rich in A as the liquid with which it is in contact. (Such a phase diagram is very similar to that for distillation of an ideal liquid mixture.) It should be noted that only liquid exists above the liquidus curve and only solid below it, whilst between the two curves solid and liquid coexist.

When the atoms of the two components are insufficiently compatible (for example in size) to allow the formation of solid solutions, a phase diagram such as that of *Figure 5.17(b)* operates. If a liquid of composition Y is cooled, the pure component B begins to settle out when temperature T^0 is reached.

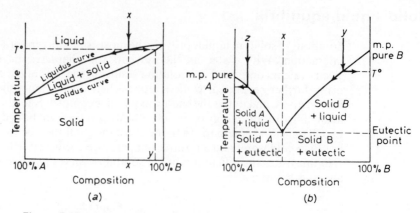

Figure 5.17 (a) Solids form solid solution (e.g., copper and nickel, or naphthalen-1-ol); (b) solids present as separate phases (e.g., ice and salt, or naphthalene and naphthalen-2-ol)

The liquid becomes richer in A as the temperature falls and pure B continually separates out, until the *eutectic* temperature is reached, when the remaining liquid crystallizes out without further change in temperature. The solid so formed, containing fine-grained crystals of both components, is called the eutectic, and embedded in it will be pure B. A similar situation prevails with a mixture of composition Z, but this time it is pure A that settles out until the eutectic is reached. If a liquid mixture of the eutectic composition X is cooled, it freezes at the constant eutectic temperature, giving a solid with the same composition as the liquid (in this respect it resembles a pure substance).

Colligative properties of dilute solutions

For solutions of non-volatile solids in liquids, several relationships can be derived which depend only on the ratio of the number of solute 'particles' to solvent 'particles', provided that the solutions are dilute. Such relationships which depend only on the number ratio and not on the type of 'particles' are termed *colligative*. (It is most important that the units employed are proportional to the number of particles present; the mole, containing the Avogadro constant of molecules can be used, but dissociation or ionization will lead to anomalous results.)

The number ratio is usually expressed in terms of the molar fraction. If n = number of solute molecules and n_0 = number of solvent molecules, then $n+n_0$ = total number of molecules, and the mole fractions of solute and solvent are

$$x = \frac{n}{n_0+n} \quad \text{and} \quad x_0 = \frac{n_0}{n_0+n} \quad (x+x_0 = 1) \tag{1}$$

Lowering of vapour pressure

If a non-volatile solute is added to a liquid solvent, the 'escaping tendency' of the latter is reduced by the presence in the surface of non-volatile molecules:

in other words, the vapour pressure of the liquid is reduced. Clearly, the effect is proportional to the relative numbers of molecules present. If

p = vapour pressure of the solution

p_0 = vapour pressure of the pure solvent

then

$p \propto x_0$ or $p = kx_0$

where k is a constant.

When $x_0 = 1$, i.e., for pure solvent, $p = p_0$, i.e., $p_0 = k$

Therefore

$$p = p_0 x_0 \quad \text{or} \quad \frac{p}{p_0} = x_0$$

and

$$x = 1 - x_0 = \frac{p_0 - p}{p_0} = \frac{n}{n_0 + n} \quad (\sim n/n_0 \text{ for a dilute solution}) \; [\text{from (1)}]$$

That is, *the relative lowering of vapour pressure is equal to the molar fraction of the solute.* This statement is known as *Raoult's law*, and it can obviously be used for determining relative molecular masses.

A dynamic method of measuring the vapour pressure lowering is due to Walker and Ostwald. Air is drawn slowly through tubes containing the solution and then through tubes of pure solvent, all of known mass. The loss in mass caused by solvent being taken up by the air from both sets of tubes is determined. Assuming that the air is saturated as it leaves the last of each set of tubes,

Loss in mass from solution $\propto p$

Loss in mass from pure solvent $\propto p_0 - p$

(since the air is already saturated to the value of p).

Example.   The loss in mass of a set of bulbs containing 5.04 g of non-volatile solute, S, in 30.0 g of ethanol after a current of air had been drawn through was 0.607 g; the further loss from a second set of bulbs, containing pure ethanol, when the same current of air was passed through, was 0.079 g

$p \propto 0.607$

$p_0 - p \propto 0.079$ g

Therefore

$p_0 \propto 0.079 + 0.607 \propto 0.686$ g

$$\frac{p_0 - p}{p_0} = \frac{0.079}{0.686} = x = \text{mole fraction of solute} = \frac{n}{n_0} \sim \frac{m_1/M_{r1}}{m_2/M_{r2}}$$

where

m_1 = mass of solute = 5.04 g

M_{r1} = rel. mol. mass of solute

m_2 = mass of solvent = 30.0 g

M_{r2} = rel. mol. mass of solvent = 46

Therefore

$$\frac{5.04/M_{r1}}{30/46} \sim \frac{0.079}{0.686}$$

i.e.,

$$M_{r1} \sim \frac{5.04 \times 0.686 \times 46}{0.079 \times 30} \sim 68$$

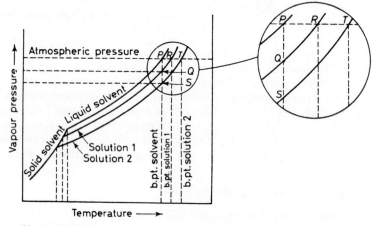

Figure 5.18 Vapour pressure of solvent and dilute solutions

Elevation of boiling point and depression of freezing point

Since a solution boils when its vapour pressure is equal to the surrounding atmospheric pressure, a lowering of vapour pressure is equivalent to an elevation of boiling point. This effect is illustrated by *Figure 5.18*. For dilute solutions the curves are close together and virtually parallel.

Therefore

$$\frac{PR}{PQ} = \frac{PT}{PS} = \text{constant}$$

i.e.,

$$\frac{\text{Elevation of boiling point}}{\text{Depression of vapour pressure}} = \text{constant}$$

It can be seen that the relative lowering of vapour pressure is proportional to the relative elevation of boiling point, i.e.,

$$\frac{p_0 - p}{p_0} = \text{constant} \times \frac{T - T_0}{T_0}$$

$$\frac{T - T_0}{T_0} = \text{constant} \times x$$

$$\sim k \frac{m_1/M_{r1}}{m_2/M_{r2}} \quad \text{(for very dilute solutions)}$$

Consequently, if the relative molecular mass of the solvent is known, together with the constant k and the mass of solute and solvent producing the measured elevation, then the relative molecular mass of the solute can be calculated. It is more usual, however, to introduce the term 'molecular or molal elevation constant' (that is, the elevation produced by one mole of solute in, usually, 1 kg of solvent) and to calculate the relative molecular mass by proportion.

Example. The elevation constant for benzene is 2.7 K mol^{-1} kg. If 1 g of a non-volatile solute in 40 g of benzene produces an elevation of 0.20 kelvin, calculate the relative molecular mass of the solute.

1 g solute in 40 g benzene gives an elevation of 0.20 kelvin

1 g in 1 kg gives an elevation of $0.20 \times 40/1000$ kelvin

$\dfrac{1 \times 1000 \times 2.7}{0.2 \times 40}$ g solute in 1 kg gives one of 2.7 kelvin

Therefore

$$\text{Relative molecular mass of solute} = \frac{1000 \times 2.7}{0.2 \times 40} = 337.5$$

It is important that the solutions used be always homogeneous and dilute and that the solute be non-volatile and neither associated nor dissociated if the correct relative molecular mass is to be obtained. It follows that the elevations measured are inevitably small, so that very sensitive thermometers must be used and precautions taken to minimize superheating and fluctuations of temperature.

Reference to *Figure 5.18* will show that similar considerations apply to the effect of non-volatile solutes on the freezing point of solutions. By use of the term 'molecular (or molal) depression constant', therefore, the relative molecular mass of a solute can be calculated, provided that it behaves normally and that it is pure solvent which separates on freezing. Various solvents are known which have extremely large depression constants, so that the use of very accurate thermometers is obviated; foremost among these is camphor, which is used in the Rast method. As its depression constant is variable, it is usual to calibrate it against a solute of known relative molecular mass, for example naphthalene.

Example. The melting point of camphor was found to be 178 °C. When 0.019 1 g naphthalene was dissolved in 0.401 9 g camphor, the melting point was depressed to 159 °C. The melting point of a solution of 0.014 8 g of an unknown substance in 0.408 5 g camphor was 162 °C. Calculate its relative molecular mass.

Relative molecular mass of naphthalene, $C_{10}H_8$, is 128, i.e., one mole of $C_{10}H_8$ has a mass of 128 g.

If M = relative molecular mass of unknown substance, then $0.0148/M$ = mole of unknown substance.

0.0191 g naphthalene in 0.4019 g camphor causes depression of 19 kelvins

0.0191 g naphthalene in 1000 g camphor causes depression of
$$\frac{19 \text{ K} \times 0.4019 \text{ g}}{1000 \text{ g}}$$

128 g (1 mole) naphthalene in 1000 g camphor causes depression of
$$\frac{19 \text{ K} \times 0.4019 \text{ g} \times 128 \text{ g}}{1000 \text{ g} \times 0.0191 \text{ g}}$$

$0.0148/M$ mole substance in 1000 g causes depression of
$$\frac{19 \text{ K} \times 0.4019 \text{ g} \times 128 \text{ g} \times 0.0148}{1000 \text{ g} \times 0.0191 \text{ g} \times M}$$

$0.0148/M$ mole substance in 0.4085 g causes depression of
$$\frac{19 \text{ K} \times 0.4019 \text{ g} \times 128 \text{ g} \times 0.0148 \times 10^3}{10^3 \times 0.0191 \text{ g} \times M \times 0.4085 \text{ g}}$$

and this calculated depression should equal the observed depression of $(178 - 162)$ kelvin, giving a value for M of 116.

Osmosis

The mixing (diffusion) of a solute and a solvent, until the concentration is uniform throughout, results in an increase in randomness (ΔS is positive, see p. 130) and is therefore a spontaneous process. If the diffusion is restricted by a *semipermeable* membrane, so that only solvent molecules can pass through, the process is termed *osmosis* (*Figure 5.19*). Osmosis will continue as long as there is a difference in concentration on both sides of the membrane or until a hydrostatic pressure is set up, called the *osmotic pressure*, which counteracts the diffusion pressure. Looked at in this way, osmotic pressure can be regarded as negative diffusion pressure.

Some methods of examining osmosis are shown in *Figure 5.19*. The simplest is (1) but better results are obtained by depositing a membrane A of $Cu_2[Fe(CN)_6]$ in the walls of a porous pot by allowing diffusion of solutions of $CuSO_4$ and $K_4[Fe(CN)_6]$ in opposite directions (the membrane is more uniform if diffusion is assisted by electrical migration), as in Pfeffer's method (2). In (3), Berkeley and Hartley applied an external pressure just to prevent osmosis (the manometer level remains constant). This has the advantage that no unknown dilution of the solution occurs. In biological cells (de Vries), if no diffusion of water into or out of the cell sap occurs (4), the solution and the cell sap have the same osmotic pressure and the two are said to be *isotonic*.

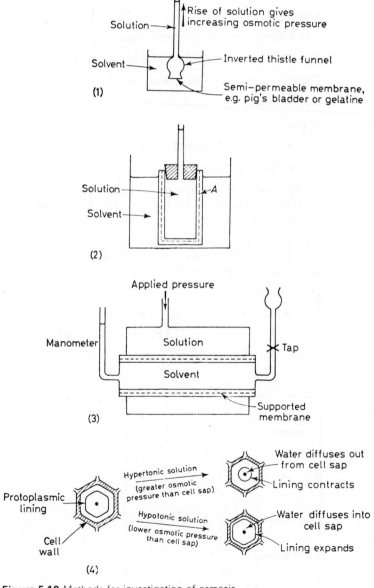

Figure 5.19 Methods for investigation of osmosis

Like the other phenomena discussed in this Section, osmosis is a function of the mol fraction of the solute and affords another means of calculating relative molecular masses. In fact, osmotic pressure, Π, is found experimentally to obey the laws which hold for gases:

$\Pi \propto c$ (or $\Pi \propto 1/V$) at constant temperature

and also

$\Pi \propto T$ at constant concentration

where

V = volume containing a fixed mass of solute

c = concentration and T is the absolute temperature

Therefore the general gas equation holds

$\Pi V = (m/M)RT$

where

m = mass of solute present

M = molar mass of solute.

The constant R is even found to be numerically equal to the universal gas constant. (Just as the gas laws are obeyed only by gases at low pressure, so the solutions used must be dilute.)

Example. The osmotic pressure of a solution of 1.82 g of mannitol in 0.1×10^{-3} m^3 (0.1 litres) water is 232 kPa at 10 °C. Calculate the relative molecular mass of mannitol.

$\Pi V = (m/M)RT$

$$M = \frac{mRT}{\Pi V} = \frac{1.82 \times 10^{-3} \text{ kg} \times 8.31 \text{ J K}^{-1} \text{ mol}^{-1} \times 283 \text{ K}}{232 \text{ kPa} \times 0.1 \times 10^{-3} \text{ m}^3} = 0.180 \text{ kg mol}^{-1}$$

so that the relative molecular mass is 180

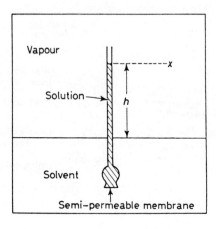

Figure 5.20 Osmotic pressure and vapour pressure

Osmotic pressure and vapour pressure

The relationship between the osmotic pressure of a solution and the relative lowering of vapour pressure can be deduced by considering the equilibrium in the closed system shown in *Figure 5.20*.

Let p_0 = the vapour pressure of the pure solvent at the surface of the liquid. The vapour pressure, p_0, will decrease with height, h, until at the level x it equals p. Since the system is at equilibrium, this must also be the vapour pressure of the solution at the same height.

Now

Pressure = height × density

$$p_0 - p = \rho_v h$$
$$\Pi = \rho h$$

where

ρ_v = mean density of vapour

ρ = density of solution

Substituting for h

$$p_0 - p = \frac{\rho_v}{\rho} . \Pi$$

But

$$\rho_v = \frac{M}{V_m} = \frac{M}{RT} . p_0$$

where

M = molar mass of solvent

V_m = molar volume

Therefore

$$\frac{p_0 - p}{p_0} = \frac{M}{\rho RT} . \Pi = \text{constant} \times \Pi$$

That is, the relative lowering of vapour pressure is proportional to the osmotic pressure, and this ties together the four colligative properties.

Ionization and association

It has been pointed out above that the solute must be neither ionized nor associated if the correct relative molecular mass is to be determined. This is quite obvious when it is realized that the effects described in this Section are functions of only the number but not the nature of the particles. Thus, if a substance ionizes *completely*, each molecule giving rise to two ions, the effect will be doubled; similarly, if a substance associates completely into double molecules, the effect will be halved.

Allowance is made for abnormality by introducing the *van't Hoff factor, i.* If the calculated values for osmotic pressure, vapour-pressure lowering, boiling-point elevation and freezing-point depression (assuming no abnormality) are referred to as c_{calc}, and c_{obs} is the observed colligative value, then

$$i = \frac{c_{obs}}{c_{calc}}$$

A knowledge of i permits the degree of dissociation or association to be evaluated. Thus, for a binary electrolyte, $AB \rightleftarrows A^+ + B^-$; if α is the degree of ionization, one molecule gives rise, after ionization, to $(1-\alpha)$ molecules of undissociated AB (in a statistical sense), and α ions each of A^+ and B^-. That is, the number of particles formed from one particle of AB is $1-\alpha+2\alpha = 1+\alpha$, and since c_{calc} and c_{obs} depend only on the number of particles present

$$i = \frac{\text{number of particles due to ionization}}{\text{number of particles assuming no ionization}}$$

$$= \frac{1+\alpha}{1} \text{ for a binary electrolyte}$$

A similar relationship can be deduced for association. Let substance A associate into double molecules

$$2A \rightleftharpoons (A)_2$$
$$2(1-\alpha) \quad \alpha$$

If α = degree of association, 2 molecules of A give rise to $(2-2\alpha+\alpha)$ particles

Therefore

$$i = \frac{2-\alpha}{2} = 1 - \frac{\alpha}{2}$$

Example. A solution of a monobasic organic acid contains 0.1 mole in 1 kg of water and depresses the freezing point of the water by 0.223 kelvin. A solution of the same molarity in benzene depresses the freezing point of benzene by 0.26 kelvin. What can be deduced? (Freezing-point depression for water, 1.86 K mol^{-1} kg; for benzene, 5.1 K mol^{-1} kg).

Assuming no dissociation 1.0 mole of acid in 1 kg water would produce a depression of 1.86 kelvin, and 0.1 mole a depression of 0.186 kelvin

Hence

$$c_{calc} = 0.186$$

Therefore

$$i = \frac{c_{obs}}{c_{calc}} \frac{0.223}{0.186} = 1.2 = 1 + \alpha_1 \text{ (binary electrolyte)}$$

$$\alpha_1 = i - 1 = 0.2.$$

Thus the acid is 20 per cent dissociated in water.

Assuming no association 1.0 mole of acid in 1 kg benzene would produce a depression of 5.1 kelvin and so, 0.1 mole, a depression of 0.51 kelvin.

Therefore

$$i = \frac{0.26}{0.51} = \frac{2 - \alpha_2}{2}$$

$\alpha_2 = 0.98$. Thus the acid is 98 per cent associated in benzene.

Questions

(1) Give reasons for the following observations:
  (a) Iron(III) oxide sol prepared by boiling a few drops of iron(III) chloride in distilled water is positively charged whilst addition of sodium hydroxide solution under carefully controlled conditions produces a negatively charged colloid.
  (b) Hydrogen ions are more effective than other univalent ions in causing the flocculation of most negatively charged colloids.
  (c) Arsenic(III) sulphide sol is negatively charged.
  (d) Dilute solutions must be used in preparing lyophobic colloids.
  (e) At an oil–water interface, soap molecules orientate in a specific manner.

(2) Calculate the mass of naphthalene, $C_{10}H_8$, required to produce a vapour pressure lowering of 264 Pa (2 mmHg) when dissolved in 50 g methylbenzene at 50 °C. The vapour pressure of methylbenzene at 50 °C is 12.2 kPa (92.6 mmHg).

(3) Derive an expression relating the van't Hoff factor, i, with the degree of dissociation of a molecule producing n ions.

(4) A solution of 6.10 g of a monobasic organic acid in 50 g of benzene has a boiling point 1.5 kelvin higher than pure benzene. At 17 °C, 0.40 g of the acid in 1 dm³ of water has an osmotic pressure of 8.16 kPa (61.4 mmHg). Also 0.50 g of the same acid requires 41.0 cm³ of 0.1 M NaOH for neutralization. Given that the molecular elevation constant for benzene is 2.53 K mol⁻¹ kg, explain the above data.

(5) Describe the precautions necessary in the use of the standard types of apparatus for obtaining accurate colligative values and discuss critically the various types of design that you have met.

(6) Calculate the proportion by mass of chlorobenzene in the distillate during steam distillation. The mixture boils at 90 °C at a pressure of 97.8 kPa (740 mmHg). The vapour pressure of water at 90 °C is 70.0 kPa (530 mmHg).

(7) Find the quantity of solute extracted per dm³ from an aqueous solution containing 50 g dm⁻³ when shaken with (a) one 200 cm³ portion of ether, (b) two 100 cm³ portions, if the partition coefficient for the solute between ether and water is 8:1.

Develop a formula for the extraction of the same solute using ten 20 cm³ portions of ether starting from the same aqueous solution. (Let x_1 be the amount of solute remaining in the aqueous solution at the end of the first extraction, and continue the calculation for the second extraction on this basis, when it will be found that a general equation results.)

(8) Discuss how Le Chatelier's principle can be applied to the effect of pressure on the solubility of a gas in water and to the effect of temperature on the solubility of a solid in water.

Assuming that air contains 80 per cent of nitrogen and 20 per cent of oxygen by volume and that the solubility coefficients of nitrogen and oxygen in water are 0.02 and 0.04 respectively at 20 °C, calculate the volume of each gas expelled by boiling 1 dm³ of water saturated with air at 20 °C.

(9) Trace the changes taking place when mixtures of composition C in *Figures 5.13* and *5.14* are heated.

(10) Comment on the flocculation values in *Table 5.2* (arbitrary scale) for silver iodide and aluminium oxide sols:

Table 5.2

| | Silver iodide | | Aluminium oxide | |
Electrolyte	Flocc. value	Electrolyte	Flocc. value
$NaNO_3$	140	KCl	46.0
$Mg(NO_3)_2$	2.6	K_2SO_4	0.3
$Al(NO_3)_3$	0.07	$K_3[Fe(CN)_6]$	0.08

(11) Write an essay dealing with the use of colloids in industry.

(12) Interpret the phase diagram, *Figure 5.21*, for steel

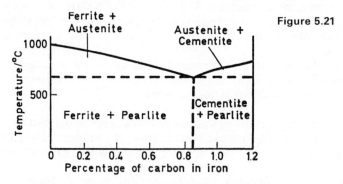

Figure 5.21

(13) Comment on the following:
   (a) Although sodium sulphate is soluble in water barium sulphate is insoluble, whilst barium chloride, like sodium chloride, is freely soluble;

(b) Glucose $C_6H_6(OH)_6$ is soluble in water but insoluble in petrol, whereas hexanol, $C_6H_{13}OH$, is insoluble in water but soluble in petrol.

(14) (a) State
   (i) Henry's Law,
   (ii) Dalton's Law of partial pressures,
   (iii) Raoult's Law.
(b) Give explanations of the following statements:
   (i) When oxygen is dissolved in water Henry's Law is obeyed whereas it is not obeyed when ammonia is dissolved in water under the same conditions.
   (ii) Addition of glucose to water lowers the freezing point of water.
   (iii) A system containing bromobenzene and water boils at 98 °C at normal atmospheric pressure, 101 kPa (760 mmHg) [the vapour pressures of bromobenzene and water at 100 °C are 20 kPa (150 mmHg) and 100 kPa (752 mmHg) respectively].
(c) The saturation vapour pressures of a liquid X and a liquid Y at the same temperature are respectively 66 kPa (497 mmHg) and 80 kPa (602 mmHg). A mixture of X and Y in which the mol fraction of X is 0·4, at the same temperature, has a vapour pressure of 92 kPa (692 mmHg). Does this mixture obey Raoult's law? Explain how you reach your conclusion.

[A.E.B.]

(15) (a) (i) Describe the preparation of a sol of iron(III) hydroxide from iron(III) chloride.
   (ii) The iron(III) hydroxide sol is put in a cellophane bag which in turn is placed in a beaker of water. Chloride ions are soon detectable in the water but none of the red colloid appears on the water side of the cellophane. The process is known as dialysis. How do you account for the observed effects?
(b) Distinguish between the terms lyophobic and lyophilic as applied to colloids. Give two examples of each type.
(c) Name two properties which a sol of iron(III) hydroxide has in common with a gelatin sol which justify calling both of them colloids.
(d) Explain what is meant by electrophoresis. Describe an experiment you would carry out with an iron(III) hydroxide sol to demonstrate electrophoresis.

(16) When a substance is added to a two-phase liquid system it is generally distributed with different equilibrium concentrations in the two phases. Why is this so?
   A certain amount of iodine was shaken with CS_2 and an aqueous solution containing 0.3 mol dm^{-3} of potassium iodide. By titrating with thiosulphate solution, it was found that the CS_2 phase contained 32.3 g dm^{-3} and the aqueous phase 1.14 g dm^{-3} of iodine. The distribution coefficient for iodine distributed between CS_2 and water is 585. Calculate the equilibrium constant for the reaction

$I_2 + I^- \rightleftharpoons I_3^-$ at the prevailing temperature.   [C]

6 Energetics and kinetics

The first law of thermodynamics • Thermochemistry • Work, heat and entropy; the second law of thermodynamics • Chemical kinetics • Opposing reactions: chemical equilibrium

The First Law of Thermodynamics

Energy changes produced in chemical reactions occur either through alterations in the vibrations, rotations, and translations of molecules and ions, or by the breaking and making of chemical bonds. Such changes take place within a *system* but their effect on the *surroundings* must also be considered. When a system does work, w, as in an engine or cell, and at the same time absorbs heat q, from the surroundings, the total energy change of the system, or the *internal energy change*, ΔU, is related to w and q by the equation:

$$\Delta U = q - w$$

This is a statement of the First Law of Thermodynamics, which simply illustrates the conservation of energy, i.e., energy can be converted from one form to another, but cannot be created or destroyed. *Figure 6.1* represents this diagramatically. The change in internal energy of the system is given by

$$\Delta U = U_f - U_i = q - w$$

Thermochemistry

If an increase in volume occurs when a reaction is carried out in an open vessel exposed to the atmosphere, the only work done by the system will be against the constant atmospheric pressure, p, as expansion, i.e., $w = p\Delta V$ (*Figure 4.28*), where ΔV is the change in volume. The First Law then becomes:

$$\Delta U = q_p - p\Delta V$$

or

$$q_p = \Delta U + p\Delta V = \Delta H$$

where q_p is the change in thermal energy at constant pressure and is denoted by ΔH, the *enthalpy change* at constant pressure. Note that we are interested in the energy changes of the system and a positive value for ΔH means that

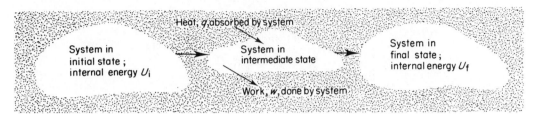

Figure 6.1 Internal energy

the system has absorbed heat and the reaction is therefore *endothermic*; for an *exothermic* reaction, ΔH is negative.

It is customary in thermochemical equations, i.e., those which take account of the energy changes involved, for the reactants and products to be in their *standard* (normal) states. For example, the equation:

$$C(s) + O_2(g) \rightarrow CO_2(g) \qquad \Delta H = -394 \text{ kJ mol}^{-1}$$

where (s) and (g) are used to denote the solid and gaseous states respectively, means that when one mole (12 g) of graphite (the standard, most stable, form of carbon) is reacted with one mole (32 g) of gaseous oxygen to produce gaseous carbon dioxide, the enthalpy change (or heat of the reaction) is -394 kilojoule, and since the value of ΔH is negative, the reaction is exothermic. It is usual to quote thermodynamic data at a temperature of 25 °C, i.e., 298 K.

Enthalpy changes can be redefined in terms of a few specific reactions.

The *enthalpy of formation* (heat of formation) is the enthalpy change when one mole of a compound is formed from its elements. For example the enthalpy of formation of carbon dioxide is given as $\Delta H_f = -394 \text{ kJ mol}^{-1}$.

The *enthalpy of combustion* (heat of combustion) of a substance is the enthalpy change occurring when one mole of it is completely oxidized, and again using the same example, the enthalpy of combustion of graphite is given by: $\Delta H = -394 \text{ kJ mol}^{-1}$.

The *enthalpy of neutralization* (heat of neutralization) of an acid or a base is the enthalpy change occurring when one mole of H^+ or OH^- is neutralized. This value is roughly constant ($\Delta H = -57.5 \text{ kJ mol}^{-1}$) for a strong acid reacting with a strong base (p. 173).

In measuring the enthalpy of combustion, the substance is normally ignited by an electric fuse in oxygen under pressure (to ensure complete combustion) in a stainless steel vessel called a bomb, so that the reaction takes place at constant volume. The whole apparatus is known as a bomb calorimeter (*Figure 6.2*) and the heat evolved is measured by observing the temperature rise and making due allowance for the thermal capacity of the calorimeter and contents. The heat evolved in this apparatus is at constant volume and is denoted by q_v; because w is zero under these conditions, it follows that:

$$\Delta U = q - w$$

becomes

$$\Delta U = q_v - 0$$

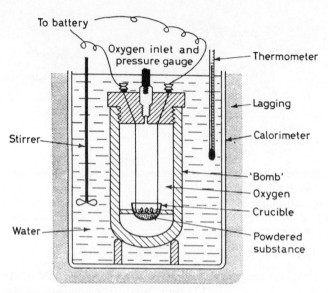

Figure 6.2 Bomb calorimeter

i.e.,

$$\Delta U = q_v$$

In a reaction such as:

$$C(s) + CO_2(g) \rightarrow 2CO(g)$$

carried out at constant volume, the heat absorbed, q_v, is 170 kJ. The equation shows that the volume change is caused by the formation of two moles of CO from one mole of CO_2. Using the ideal gas equation (p. 67), $p\Delta V = nRT$, we can write:

$$\Delta H = q_p = \Delta U + p\Delta V$$

$$= q_v + nRT$$

$$= +170\ \text{kJ mol}^{-1} + (2-1)\ \text{mol} \times 8.31 \times 10^{-3}\ \text{kJ K}^{-1}\ \text{mol}^{-1} \times 298\ \text{K}$$

i.e.,

$$\Delta H = +173\ \text{kJ mol}^{-1}$$

That is, q_p is greater than q_v (by 3 kJ), showing that this much thermal energy has been supplied from the surroundings to compensate for the work of expansion.

Hess's Law

Because the energy associated with a substance in a given state is always constant, there is always the same energy change involved in the conversion

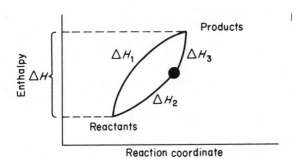

Figure 6.3 Hess's law

of one substance into another, no matter how the change is brought about (*Figure 6.3*); in terms of this Figure:

Total enthalpy change $\Delta H = \Delta H_1 = \Delta H_2 + \Delta H_3$

Formally, *Hess's Law of constant heat summation* states that the total enthalpy change for a reaction is independent of the path taken provided that the reactants and products start and finish in identical conditions every time. This law furnishes a method for calculating enthalpy changes of reactions which cannot be directly measured, for example, given the following thermochemical equations, where (l) denotes the liquid state:

$$C_2H_5OH(l) + 3O_2(g) \rightarrow 3H_2O(l) + 2CO_2(g) \qquad \Delta H = -1430 \text{ kJ} \qquad (1)$$

$$C(s) + O_2(g) \rightarrow CO_2(g) \qquad\qquad\qquad\qquad = -394 \text{ kJ} \qquad (2)$$

$$H_2(g) + \tfrac{1}{2}O_2(g) \rightarrow H_2O(l) \qquad\qquad\qquad\quad = -285 \text{ kJ} \qquad (3)$$

the enthalpy of formation of ethanol can be calculated, since the required equation is:

$$2C(s) + 3H_2(g) + \tfrac{1}{2}O_2(g) \rightarrow C_2H_5OH(l)$$

and this can be obtained from the combination of the given reactions:

$2 \times$ reaction $(2) + 3 \times$ reaction $(3) - 1 \times$ reaction (1)

Therefore the required enthalpy change is given by:

$$[2 \times (-394) + 3 \times (-285) - (-1430)] \text{ kJ}$$

i.e., the enthalpy of formation of ethanol is -213 kJ mol^{-1}. The enthalpy changes are shown schematically in *Figure 6.4*.

Since no absolute enthalpies are involved, but only changes in their values, the method of *intrinsic energies* can also be used. (ΔH for a reaction is equal to the sum of the intrinsic energies of the products minus the sum of the intrinsic energies of the reactants.) Elements in their standard states are arbitrarily assigned an intrinsic energy of zero. Where elements show polymorphism, only one form, the most stable, is given an intrinsic energy of zero. The energy associated with a compound relative to its elements is then referred to as its intrinsic energy which is identical with its enthalpy of formation. Thus using the above example:

$$C_2H_5OH(l) + 3O_2(g) \longrightarrow \quad 3H_2O(l) \quad + \quad 2CO_2 \qquad \Delta H = -1430 \text{ kJ}$$

Intrinsic
energies
$$x \qquad\qquad zero \qquad -(3 \times 285) \text{ kJ} \quad -(2 \times 394) \text{ kJ}$$

Therefore the intrinsic energy, or enthalpy of formation, of ethanol is given by

$$-1430 \text{ kJ} = -(3 \times 285) \text{ kJ} - (2 \times 394) \text{ kJ} - x \text{ kJ} - 0 \text{ kJ}$$

or

$$x = 1430 \text{ kJ} - 755 \text{ kJ} - 788 \text{ kJ}$$

$$= -213 \text{ kJ}$$

Therefore the enthalpy of formation of ethanol is -213 kJ mol^{-1}.

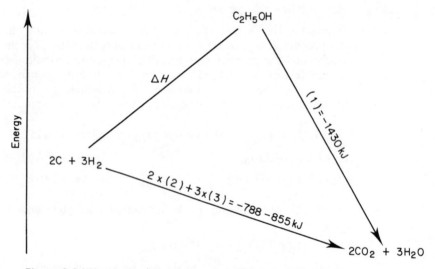

Figure 6.4 ΔH + reaction (1) = 2 × reaction (2) + 3 × reaction (3)

Bond energy

The bond energy is defined as the energy required to break 'one mole's worth' of the bond in question. That is to say, the bond energy of A—B is the enthalpy of the reaction:

$$A\text{—}B \rightarrow A + B$$

It is possible to calculate these values from a knowledge of the relevant enthalpies of dissociation and combustion. For example, in the case of the O—H bond the following figures can be used:

$$H_2(g) + \tfrac{1}{2}O_2(g) \rightarrow H_2O(l) \qquad \Delta H = -243 \text{ kJ}$$

$$H_2(g) \rightarrow 2H(g) \qquad \Delta H = +433 \text{ kJ}$$

$$O_2(g) \rightarrow 2O(g) \qquad \Delta H = +495 \text{ kJ}$$

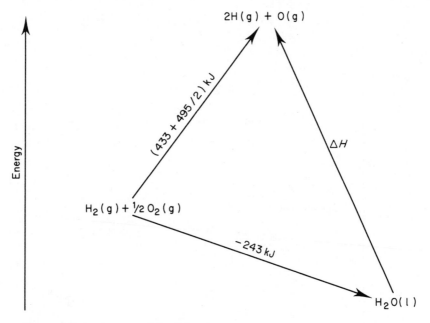

Figure 6.5 Bond energy of O—H bond

from which the enthalpy change for the reaction involving the dissociation of two O—H bonds:

$$H_2O(l) \rightarrow 2H(g) + O(g)$$

can be found by application of Hess's Law as in *Figure 6.5*,

i.e.,

$$-243 \text{ kJ} + \Delta H = 433 \text{ kJ} + 248 \text{ kJ}$$

Therefore

$$\Delta H = +924 \text{ kJ}$$

showing that the O—H bond energy is 924/2 kJ or 462 kJ mol^{-1}.

Resonance energy

Evidence for the non-localization of orbitals (p. 55, resonance) producing substances which are more stable than the simple structures suggest, is provided from thermochemical data. For example, the theoretical enthalpy of formation of carbon dioxide can be calculated by adding the appropriate enthalpies of dissociation and bond energies, i.e.,

Enthalpy of atomization of graphite $C(s) \rightarrow C(g)$ $\Delta H = +713 \text{ kJ}$

Enthalpy of dissociation of oxygen $O_2(g) \rightarrow 2O(g)$ $\Delta H = +502 \text{ kJ}$

Bond energy of $C = O$ $C(g) + O(g) \rightarrow C = O(g)$ $\Delta H = -741 \text{ kJ}$

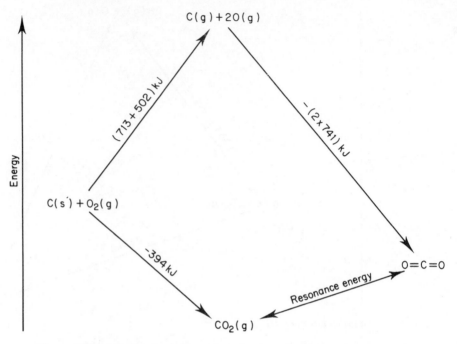

Figure 6.6 Energy diagram for formation of carbon dioxide

Therefore the energy required to form one mole of carbon dioxide from its elements, assuming that the molecule contains two simple C=O bonds, is given by (*Figure 6.6*):

$$C(s) + O_2(g) \rightarrow O=C=O(g) \qquad \Delta H = (+713 + 502 - 2 \times 741) \text{ kJ}$$

$$= -274 \text{ kJ}$$

However, the actual enthalpy of formation, as determined experimentally, is −394 kJ, so that each mole of carbon dioxide is stabilized by (394 − 274) kJ, i.e., the resonance energy is 120 kJ mol^{-1} and consequently carbon dioxide does not simply consist of multiple bonds.

Lattice energy

The formation of an electrovalent compound from its elements can be regarded as made up of several stages (p. 44):

(1) The dissociation (or sublimation) of the elements into single atoms.
(2) The loss or gain of electrons to form ions (ionization energies and electron affinities respectively).
(3) The arrangement of these ions to form one mole of solid. The lattice energy is the energy required to reverse the latter operation, that is to remove the ions in one mole of solid to an infinite distance apart in the gaseous state where mutual interactions are zero.

Provided that all the other quantities are known, the lattice energy can be calculated using Hess's Law. *Figure 6.7* shows the sequence for calcium oxide.

$$119 \text{ kJ} + 502/2 \text{ kJ} + 1730 \text{ kJ} + 640 \text{ kJ} = \Delta H - 635 \text{ kJ}$$

Therefore the lattice energy is $\Delta H = 3455 \text{ kJ mol}^{-1}$

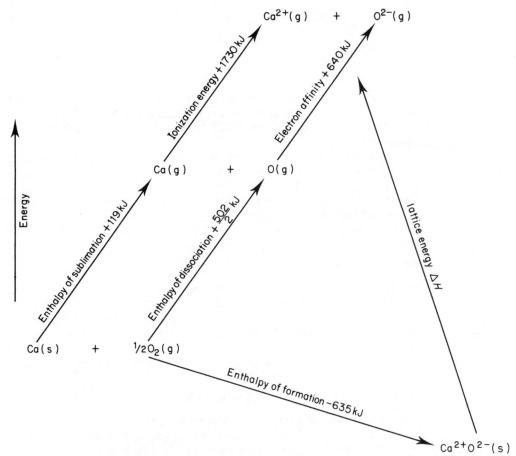

Figure 6.7 Calculation of the lattice energy of calcium oxide by a Born–Haber cycle. Positive values indicate that energy is absorbed by the system and negative values that energy is lost from it

Work, heat and entropy; the Second Law of Thermodynamics

Work and heat energy, although interconvertible, differ in one great respect; work is ordered energy and as such can be stored, e.g., in the form of a raised weight, whilst heat is disorganised energy and presents problems in conservation, readily becoming dissipated, and resulting in low-grade energy

Table 6.1 Standard entropies, $S/\text{J K}^{-1} \text{mol}^{-1}$, at 298 K and 100 kPa

Solids		Liquids		Gases	
C (diamond)	2.4	Br_2	152	N_2	192
C (graphite)	32			O_2	206
Cu	33			Cl_2	223
Zn	42			SO_2	249
				SO_3	256
				NO	211

which cannot be recovered easily for doing useful work. Therefore, although the First Law indicates conservation of energy, the real difficulty lies in retarding the natural tendency for disorder and the non-availability of the dissipated energy.

A measure of the degradation of energy is called *entropy*. Thus the larger the quantity of heat transferred in a process, the greater is the increase in entropy (disorder). At the same time more disorder results if the heat is transferred to a body at a low, rather than a high temperature, since kinetic energy will then be more obvious. If the process is reversible, as in the case, for example, of the vaporization of a liquid at its boiling point, then the entropy change, ΔS, is given by:

$$\Delta S = q_{rev}/T$$

where $q_{rev.}$ is the enthalpy change (latent heat in this case) and T is the constant absolute temperature (boiling point for this situation) at which transfer occurs. The notion that an increase in entropy represents an increase in disorder is revealed by considering the conversion of liquid water into steam, which is associated with an increase in randomness of structure; if the change takes place at the boiling point, the heat required to effect the change reversibly is 2260 J g^{-1} and therefore:

$$\Delta S = 2260 \times 18 \text{ J mol}^{-1}/373 \text{ K} = 109 \text{ J K}^{-1} \text{mol}^{-1}$$

The entropy of a perfect crystal at absolute zero, where there is no movement, is put at zero, and *Table 6.1* shows that gases, with their highly chaotic structure, have higher entropies than liquids, which in turn have higher entropies than solids.

The prospect of reaction

It is well known that systems tend to move from high to low energy, that is from low to high stability. For example, a ball runs of its own volition downhill, and not uphill. This is not to say that it could never run uphill, but it would only do so if given sufficient energy. We can say that systems tend to move spontaneously downhill, because this is the natural process leading to the dissipation of energy and an increase in entropy. Note that the *Second Law of Thermodynamics* states that the total entropy of any spontaneous process increases. Thus a living cell is striving to produce order so that it can survive and therefore the entropy of the system is decreasing, but at the same time, the cell is dissipating heat to the surroundings, so that there is an increase in entropy of the latter, and inevitably the *total* entropy change of the system and the surroundings together must increase. It is easy to overlook the entropy factor and conclude that all spontaneous processes must be exothermic but, if this were so, reversible reactions would be impossible, for if the change $A \rightarrow B$ were exothermic (i.e., ΔH negative) then the reverse reaction, $B \rightarrow A$ would be endothermic (ΔH positive).

The overriding principle which determines the feasibility of a reaction is the necessity for an increase in entropy (disorder) and this can be provided by a release of energy as well as the conversion of a solid into liquids or gases, or liquids into gases.

Free energy

To obtain the maximum work from a process, the system must operate reversibly, since by doing so the temperature can be maintained from heat supplied from the surroundings, i.e., the heat change must also be carried out reversibly. Therefore for the greatest efficiency the First Law becomes:

$$\Delta U = q_{rev.} - w_{max.}$$

However the useful work obtained will be less than the maximum, $w_{max.}$, because some work must be done against the constant pressure of the atmosphere if expansion occurs, i.e.,

$$w_{useful} = w_{max.} - p\Delta V$$

where p is the atmospheric pressure and ΔV is the expansion. Consequently

$$\Delta U = q_{rev.} - (w_{useful} + p\Delta V)$$

or

$$-w_{useful} = \Delta U + p\Delta V - q_{rev.}$$

But

$$\Delta H = U + p\Delta V$$

and

$$q_{rev.} = T\Delta S$$

Therefore

$$-w_{useful} = \Delta H - T\Delta S = \Delta G$$

where ΔG is the *free energy change*; if the system does work, then w_{useful} is positive and ΔG must be negative. The valuable conclusion is that a spontaneous reaction can do work and so a free energy decrease must then occur.

Consider the reaction:

$$H_2(g) + \tfrac{1}{2}O_2(g) \rightarrow H_2O(l)$$

which can be made to give useful work in the form of electrical energy if the process is carried out in a fuel cell (p. 79) giving $\Delta G = -237$ kJ mol^{-1}. However combustion of the gas mixture gives a value of $\Delta H = -286$ kJ. To some this may conjure up ideas of perpetual motion (*Figure 6.8*), but substitution in the equation:

$$\Delta G = \Delta H - T\Delta S,$$

should dispel such illusions, since at 298 K, $\Delta S = -0.16$ kJ K^{-1} mol^{-1}, the negative sign showing that ordering takes place during the reaction (gases →liquid), and therefore during the fuel cell reaction, heat is evolved.

For the electrolytic reaction:

$$H_2O(l) \rightarrow H_2(g) + \tfrac{1}{2}O_2(g)$$

$\Delta G \qquad\qquad = \Delta H \qquad\qquad -T \qquad \times \quad \Delta S$
$+237$ kJ mol^{-1} $= +286$ kJ mol^{-1} -298 K $\quad \times \quad 0.16$ kJ K^{-1} mol^{-1}
$+237$ kJ mol^{-1} $= +286$ kJ mol^{-1} -49 kJ mol^{-1}
$+237$ kJ mol^{-1} $= +237$ kJ mol^{-1}

ΔS is positive since gases are produced from the liquid.

It should be noted that the term 'reversible' in the thermodynamic sense applies to the fact that, depending on the direction and value of the potential difference across an electrolytic cell, the direction of the reaction can be reversed. The term *isothermal* means that the system is not isolated from the surroundings and that heat is able to flow into and out of the system at constant temperature.

The picture of spontaneous change representing an *overall* increase in stability and disorder is very useful to the chemist in appraising the feasibility of any reaction. Often these two factors supplement each other and there is little difficulty in predicting in which direction the reaction might go. In the case of the dissolution of zinc in dilute sulphuric acid:

$$Zn(s) + 2H^+(aq) \rightarrow Zn^{2+}(aq) + H_2(g) \qquad \Delta H = -152 \text{ kJ mol}^{-1}$$

[where (aq) denotes hydration] there is a considerable decrease in enthalpy and also an increase in entropy of the system, since the net change is from solid to gas. Therefore, as expected, the reaction is feasible, as the enthalpy and entropy changes both favour a decrease in free energy.

In the case of the reaction:

$$C(s) + H_2O(g) \rightarrow CO(g) + H_2(g) \qquad \Delta H = +132 \text{ kJ mol}^{-1}$$

heat is absorbed, but there is a considerable increase in entropy (a solid and a gas producing two gases) which, at white heat, more than compensates for the positive enthalpy change, making ΔG negative and allowing reaction to occur.

Standard molar entropies can be used, in conjunction with enthalpies, to calculate the free energy change of a reaction, and hence assess the prospect for a reaction occurring. For example, for the reaction:

$$\tfrac{1}{2}N_2(g) + \tfrac{1}{2}O_2(g) \rightarrow NO(g) \qquad \Delta H = +98 \text{ kJ mol}^{-1}$$

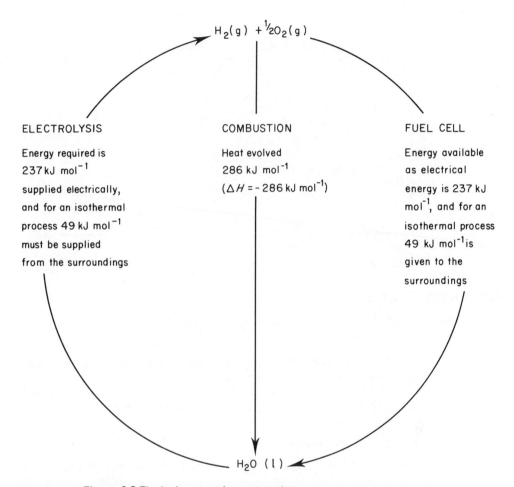

Figure 6.8 The hydrogen and oxygen system

the standard entropy values given in *Table 6.1* show that the change in entropy for the forward reaction is

$211 - \frac{1}{2}(206 + 192)$ or 12 J K^{-1} mol^{-1}

The free energy change:

$\Delta G = \Delta H - T \Delta S$

$= 98$ kJ mol$^{-1} - 298$ K $\times 12 \times 10^{-3}$ kJ K^{-1} mol^{-1}

$= 85.5$ kJ mol^{-1}

is positive and this reaction will not be spontaneous at 25 °C. To achieve such a reaction a high energy input is essential, as in the early manufacture of nitrogen oxide using electric arcs.

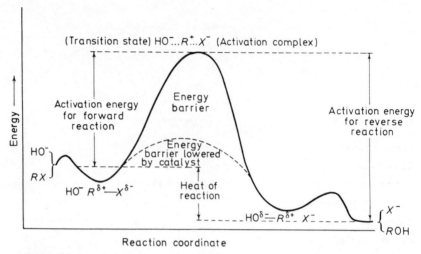

Figure 6.9 A typical reaction path for a reversible reaction

Chemical kinetics

The effect of temperature and catalysts on reaction velocity

We have so far considered the probability of spontaneous chemical reaction in terms of the balance between enthalpy and entropy, as represented by the free energy change. But not even if the free energy change is negative need the reaction be readily spontaneous. For example, the free energy change accompanying the reaction

$$H_2(g) + \tfrac{1}{2}O_2(g) \rightarrow H_2O(l)$$

is -237 kJ mol^{-1}, and yet the reaction, in the absence of a suitable third party, takes place at an incredibly slow rate at ordinary temperatures, because of the existence of a high energy barrier which has to be surmounted in the course of the reaction. *Figure 6.9* shows that when two reactive species approach one another, an initial decrease in potential energy occurs as induced dipoles are formed and the resulting orientation minimizes the energy. As the distance of approach becomes smaller the bonds are strained, resulting in an increase of potential energy at the expense of a decrease in kinetic energy of the system. Eventually a transition stage is reached where the combined *activation complex* can break down, either to give back the original reactants or to form new products. The energy required for the reactants to 'climb' the energy barrier is known as the *activation energy* and is given by the expression:

$$k = Ae^{-E/RT}$$

where k = the rate constant (p. 137)
$\quad\quad\quad A$ = a constant, known as the Arrhenius constant
$\quad\quad\quad T$ = the absolute temperature
$\quad\quad\quad E$ = the energy of activation.

It will be seen from this equation that the rate of reaction increases with temperature. This is in keeping with the kinetic theory which shows that the higher the temperature the greater is the energy associated with the molecules. The consequence of this is twofold; first, there will be more collisions per second, and secondly, and more importantly, more molecules will possess energy in excess of the activation energy. In fact, a rise in temperature of ten kelvins approximately doubles the rate of reaction, a matter of considerable interest to biologists, affecting as it does the response of cold-blooded animals to a change of temperature. Likewise states of fever can be equated to the faster rate of body reactions involved in responding to attack by bacteria or viruses. Conversely, many of the risks attending major surgery and its associated deep anaesthesia can be lessened by performing the operation under hypothermia. The patient is cooled to about 30 °C, thus slowing the metabolic, respiratory, circulation and excretory rates.

In discussing the synthesis of water earlier, reference was made to 'suitable third parties'. These are *catalysts*, which assist reaction by lowering the energy barrier (*Figure 6.9*). Although, with reference to *Figure 6.9*, ΔG for the conversion of A into B is favourable, reaction is prevented from being noticeably spontaneous by the energy barrier. For the catalysed reaction, the energy barrier is in some way lowered and there is accordingly a greater probability of some reacting molecules having sufficient activation energy to surmount the barrier. Catalysts contribute nothing permanently to the energy of a system and cannot therefore affect the position of equilibrium but it follows from the last equation that if E is lowered, then k is increased and consequently a catalyst will reduce the time required for equilibrium to be reached. Furthermore, they are often required in small quantities only and can be recovered at the end of a reaction, although they are generally changed physically, suggesting that they have become chemically involved at some stage and then reformed.

Catalysts can be classified as *homogeneous*, when both the catalyst and reactants are in the same phase, or *heterogeneous* when they constitute separate phases. As a consequence there appear to be at least two distinct types of mechanism in catalysed reactions. The *adsorption theory*, which applies to heterogeneous catalysis, suggests that the reacting species are adsorbed on to the surface of the catalyst, probably in the form of a monomolecular layer, producing an effective increase in the concentration of the reactants and thus enhancing the chances of reaction. Adsorption would also lead to a weakening of bonds inside the molecules, as secondary bonds are formed between the catalyst and the reacting species. The secondary bonds would ensure particular orientations of the molecules; both factors would affect the activation energy. In the case of metallic catalysts, their good thermal conductivity would permit any heat produced to be conducted away before it could bring about the dissociation of the product. Surface catalysts of this type are often highly specific, a fact that is attributed to the difference in the distances between the 'active centres' on the catalyst surface. For example, nickel can bring about the dehydrogenation of a primary alcohol to an aldehyde, in accordance with the adsorption of the hydroxyl bond on the metal and a resultant weakening of the oxygen–hydrogen link; on the other hand, alumina catalyses the dehydration of alcohol to alkene, suggesting the

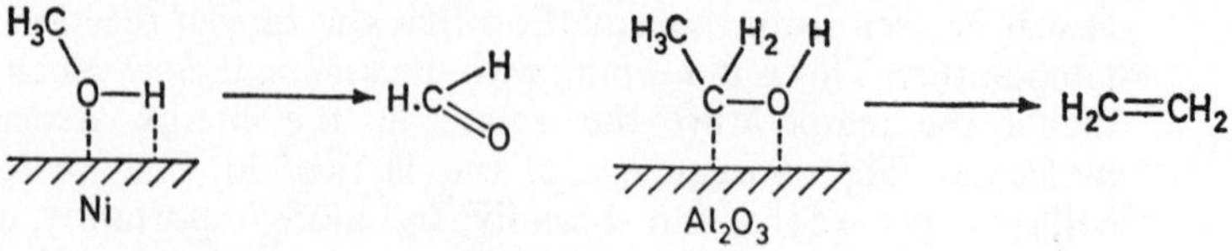

Figure 6.10 Catalytic specificity

weakening of the carbon–oxygen bond by adsorption of these two atoms because of the more favourable distances between active centres. (*Figure 6.10*). There are countless examples of surface catalysis, many of industrial importance and involving transition (*d*-block) metals. Examples include vanadium (as its pentoxide) in the Contact Process for sulphuric acid, iron in the Haber Process for ammonia, and nickel in the hydrogenation of unsaturated oils. It is believed that the class of complex proteins known as enzymes also function by surface adsorption, in keeping with their high specificity.

In homogeneous catalysis the general mechanism involves the formation and consequent decomposition of an intermediate compound. If C represents the catalyst and A and B are the reacting species, the reaction can be written as:

$$A + C \rightarrow AC$$

followed by:

$$AC + B \rightarrow AB + C$$

giving the overall reaction as:

$$A + B \rightarrow AB$$

For example, in the Friedel–Crafts reaction between chloroethane and benzene in the presence of anhydrous aluminium chloride, the reaction involves the steps:

$$C_2H_5Cl + AlCl_3 + C_2H_5^+(AlCl_4)^-$$
$$\text{(intermediate compound)}$$

$$C_6H_6 + C_2H_5^+(AlCl_4)^- \rightarrow C_6H_5C_2H_5 + HCl + AlCl_3$$

giving the overall reaction as:

$$C_6H_6 + C_2H_5Cl \rightarrow C_6H_5C_2H_5 + HCl$$

Many other examples of catalysis will be described in later Chapters.

Autocatalysis

Sometimes a catalyst is formed in the course of a reaction, so that the reaction becomes progressively faster. A common example of this phenomenon is the reaction between ethanedioate and manganate(VII) ions under acid conditions. Manganese(II) ions are formed as a result of the oxidation of the organic ion and the latter catalyse the reaction:

$$5C_2O_4^{2-} + 2MnO_4^- + 16H^+ \rightarrow 2Mn^{2+} + 8H_2O + 10CO_2(g)$$
$$\text{autocatalyst}$$

Poisoning of catalysts

Certain substances, often present as impurities, are preferentially adsorbed on to the surfaces of catalysts, preventing the latter from acting efficiently. The catalyst is then said to be poisoned, an apt description because in the case of living organisms, such adsorption, for example of arsenic on enzymes, can lead to death.

Inhibitors

Because of the general lowering of energy barriers, catalysts can allow side reactions to occur; it is thus possible for the presence of certain materials to lead to the reduction in the original reaction rate. Such substances are called *inhibitors*. Thus, whilst dibenzoyl peroxide provides the free radicals to initiate polymerization reactions, an inhibitor which destroys free radicals acts to terminate such chain reactions (p. 140).

Promoters

The addition of small quantities of a second substance may enhance the catalytic activity of the first. For example, aluminium oxide 'promotes' the activity of iron in the Haber process. Most promotors have been discovered empirically, a reminder that our theories of catalysis are far from complete.

The effect of concentration on reaction velocity

Chemical reaction is generally the result of the approach, in a favourably orientated manner, of suitably activated species and therefore is determined by the collision rate which in turn depends on the concentration and temperature. Clearly, the probability of reaction is raised if the concentration of the molecules is increased. For instance, in a simple reaction between molecules A and B, the rate is increased sixfold if the concentration of A is tripled and that of B doubled, i.e.,

Reaction rate $\propto [A][B]$

where $[A]$ and $[B]$ represent the 'active masses' of A and B respectively. For moderately dilute systems, 'active mass' is approximated to concentration, permitting us to state the *law of mass action* of Guldberg and Waage in the form: the rate of a chemical reaction is directly proportional to the product of the concentrations of the reacting species raised to suitable powers:

Reaction rate $\propto [A]^a[B]^b \dots$

Reaction rate $= k[A]^a[B]^b \dots$

where k is the rate constant (p. 134) and $a, b \dots$ are small numbers, usually 0, 1 and 2. The sum of $a, b, \dots$ is called the *order of the reaction*, although it is more important to recognize the individual orders. For example, if $a = 1$ the rate is said to be of the first order with respect to species A; if $b = 0$ the rate is of zero order with respect to B, and this means that the rate is independent of the concentration of B under the particular experimental conditions.

For a *first-order reaction*, the rate depends on the concentration of one substance, say A, and therefore we can write the reaction as:

$A \rightarrow$ products

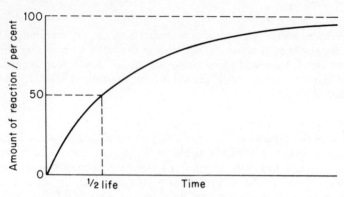

Figure 6.11 Rate of a first-order reaction

and the rate will eventually fall off as the reaction proceeds and the concentration of A decreases. Theoretically the complete reaction requires an infinitely long interval of time (*Figure 6.11*).

If, at any one instant, x mole of product have been formed from a mole of reactant A, the number of mole of reactant left will be $(a-x)$ and the rate of formation of x will be proportional to this, i.e.,

$$dx/dt = k[A]$$
$$= k(a-x)$$

where k is the first-order rate constant. Rearranging this equation gives:

$$dx/(a-x) = k.dt$$

This, on integration, gives:

$$-\ln(a-x) = kt + \text{constant}$$

When $t = 0$, $x = 0$ and therefore the constant $= -\ln a$, giving the final equation as:

$$\ln\frac{a}{-x} = kt$$

The time required to reduce the original concentration to one-half $(x = \frac{1}{2}a)$ is called the *half-life*, $t_{\frac{1}{2}}$. Substituting for $t_{\frac{1}{2}}$ in the last equation gives:

$$\ln\frac{a}{\frac{1}{2}a} = kt_{\frac{1}{2}}$$

and

$$2 = kt_{\frac{1}{2}}$$

or

$$t_{\frac{1}{2}} = 0.693/k$$

i.e., the half-life for a first order reaction is independent of the initial concentration. Consequently, it is a very significant quantity, for example, in radioactivity (*see* p. 28).

The overall reaction of most chemical changes is the sum total of several elementary steps and the rate of reaction is determined by the slowest of these, which acts as a 'bottleneck'. For example, kinetic studies have shown that the decomposition of dinitrogen pentoxide by heat is first order. This could be explained if the rate-determining (slowest) step involved one molecule only (i.e., if it were unimolecular):

$$N_2O_5 \xrightarrow{\text{slow}} NO_2 + NO + O_2$$

The nitrogen oxide formed would then react quickly with another molecule of dinitrogen pentoxide:

$$NO + N_2O_5 \xrightarrow{\text{fast}} 3NO_2$$

giving the overall (stoichiometric) equation:

$$2N_2O_5 \rightarrow 4NO_2 + O_2$$

Sometimes a reaction obeys first-order kinetics because one reactant is in such large excess that its diminution in concentration as reaction proceeds is negligible compared with that of the other constituents. An example of this is the hydrolysis ('inversion') of dilute sucrose solution in the presence of an acid catalyst:

$$C_{12}H_{22}O_{11} + H_2O \rightarrow 2C_6H_{12}O_6$$

The concentration of water in this dilute solution remains nearly constant and its small variation is not sufficient to affect the order of the reaction, which for such reactions is called 'pseudo-unimolecular'.

Consider now a *second-order reaction* of the type:

$$2A \rightarrow \text{products}$$

If after time t, x mole of product have been formed from an original concentration of a mole of each reactant, the rate law will give:

$$dx/dt = k[A]^2$$
$$= k(a-x)^2$$

or

$$\frac{dx}{(a-x)^2} = k \cdot dt$$

where k is the second-order rate constant.

Integration of this gives:

$$kt = \frac{x}{a(a-x)}$$

and for the half-life, $t_{\frac{1}{2}}$:

$$kt_{\frac{1}{2}} = 1/a$$

so that, unlike first-order reactions, the half-life of a second-order reaction is *not* independent of the initial concentration.

Many reactions are of second-order, for example, the thermal decomposition of hydrogen iodide under suitable conditions:

$$2HI \rightarrow H_2 + I_2$$

But it must be emphasized that a bimolecular reaction does not necessarily follow second-order kinetics, since the overall reaction may be made up of several steps, the slowest of which might be of first order. When the prerequisite for most chemical reactions is considered, i.e., collision of the correct species orientated in a favourable manner to one another, it will be realized that reactions of higher than second order are rare.

Determination of the order of reaction

Before the order of a reaction can be elucidated, it is imperative that the state of the system be known at various time intervals. The most obvious, but least satisfactory method, entails chemical sampling of the system at different times. This has the disadvantage that the system is disturbed by the sampling and that, even though the sample may be 'frozen', i.e., diluted considerably or cooled down as quickly as possible, a certain change after removal and before analysis is inevitable.

Far more satisfactory is the measurement of some physical property, such as light absorption, electrical conductivity, and pH, which is a function of the state of the system. A well known example is the measurement of the optical rotation of sucrose solution in the process of inversion.

If gradients are measured at different concentrations from a concentration–time curve, then a plot of the gradients (rates) against concentration will be linear for a first-order reaction. For a second-order reaction the square of the concentration will produce a linear curve.

If different concentrations of reactants are employed in different experiments and the time taken for reaction to become half complete is measured in each case, the order can be evaluated by comparison with the equations for half-life.

Alternatively, the change in concentration can be measured from time to time and appropriate plots made in accordance with the integrated rate equations to determine which produces a straight line relationship (*Figure 6.12*).

It must be stressed that in rate investigations, temperature must be maintained constant and it is much easier for mathematical analysis if only one of the reactants at a time is allowed to change, i.e., it should be arranged for the other reactant concentrations to remain constant by using them under 'swamping conditions'.

Chain reactions

The order of a reaction merely gives the kinetics of the slowest, rate-determining, step; it does not therefore provide an infallible indication of the

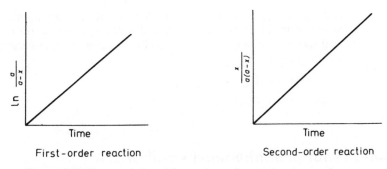

Figure 6.12 Characteristics of first-order and second-order reaction kinetics

other steps involved. This is particularly the case with chain reactions, where initiation is the measured stage and where propagation can continue until combination between free radicals eventually terminates the chain reaction. An example of a chain reaction is that between hydrogen and chlorine in the presence of light, the energy of which dissociates chlorine molecules into atoms, which attack hydrogen molecules to form hydrogen chloride and reactive hydrogen atoms, which can then attack more chlorine molecules and so perpetuate the reaction:

$$Cl_2 \xrightarrow[hv]{\text{light energy}} 2Cl^{\bullet} \qquad \qquad \text{Initiation}$$

$$Cl^{\bullet} + H_2 \rightarrow HCl + H^{\bullet}$$
$$\left.\vphantom{\begin{array}{c}a\\a\end{array}}\right\} \text{Propagation}$$
$$H^{\bullet} + Cl_2 \rightarrow HCl + Cl^{\bullet}$$

$$2Cl^{\bullet} \rightarrow Cl_2$$
$$\left.\vphantom{\begin{array}{c}a\\a\end{array}}\right\} \text{Termination}$$
$$2H^{\bullet} \rightarrow H_2$$

Parallel reactions (Figure 6.13)

A cause of much frustration in practical chemistry is the parallel reaction whereby a reactant may enter into different pathways, only one of which is perhaps desired. To obtain the maximum yield of the required product, it is important to recognize the situation and control the conditions, e.g.,

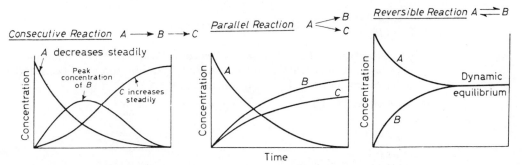

Figure 6.13 Types of reaction

concentration, temperature, catalyst, so that the required pathway is favoured to the partial exclusion of the others. This situation is encountered frequently in organic chemistry, as for example in the treatment of ethanol with concentrated sulphuric acid (p 482).

$$C_2H_5OH \rightarrow C_2H_4 + H_2O$$
$$C_2H_5OH \rightarrow C_2H_5OC_2H_5 + H_2O$$

Opposing reactions: chemical equilibrium

The products formed in a reaction often react together to give back the original reactants, i.e., the reaction is *reversible*:

$$A \rightleftharpoons X$$

The rate of the forward reaction decreases as the concentration of A decreases, and the rate of the backward reaction increases as the concentration of X increases so that at some point the two opposing rates become equal and the concentrations of A and X become constant: a *dynamic equilibrium* is established (*Figure 6.13*).

Because of the lack of truly elementary reactions, i.e., those whose mechanisms correspond to the overall equation, application of the law of mass action to reversible chemical changes reduces to a study of a hypothetical change, such as:

$$A + B \underset{k_2}{\overset{k_1}{\rightleftharpoons}} C$$

where the rate of the forward reaction $= k_1[A][B]$

and the rate of the backward reaction $= k_2[C]$

k_1 and k_2 are the rate constants. When equilibrium is reached, the concentrations of A, B and C remain constant and therefore the rates of the forward and backward reactions must be equal, i.e.,

$$k_1[A]_e[B]_e = k_2[C]_e$$

or

$$\frac{[C]_e}{[A]_e[B]_e} = \frac{k_1}{k_2} = K, \text{ the equilibrium constant}$$

The subscript e denotes the *equilibrium* concentrations. This *equilibrium law* is general for any reaction operating under equilibrium conditions and is independent of the actual mechanism, i.e., if

$$aA + bB \rightleftharpoons cC + dD$$

then

$$\frac{[C]_e^c[D]_e^d}{[A]_e^a[B]_e^b} = \text{a constant}$$

where $[A]_e$, etc. stand for the equilibrium concentrations. Note that the equilibrium constant is temperature-dependent and that the rate of a reaction cannot be deduced from the equilibrium constant as the latter generally refers to the sum of a number of elementary reactions, whilst the former refers to the slowest step.

For gaseous systems, partial pressures p can be conveniently substituted for molar concentrations. The equilibrium constant for molar concentrations is designated K_c, that for partial pressures, K_p.

In the case of the reaction:

$$N_2 + 3H_2 \rightleftharpoons 2NH_3$$

$$K_c = \frac{[NH_3]_e^2}{[N_2]_e[H_2]_e^3} \quad \text{and} \quad K_p = \frac{(p_{NH_3})^2}{p_{N_2} \times (p_{H_2})^3}$$

Let the total volume be v, and

a = initial number of moles of nitrogen

$3a$ = initial number of moles of hydrogen

$2x$ = number of moles of ammonia produced under a pressure p.

Then at equilibrium there are $(a-x)$ mole of nitrogen, $3(a-x)$ mole of hydrogen and $2x$ mole of ammonia, giving a total of $2(2a-x)$ mole of gas. Therefore:

$$K_c = \frac{(2x/v)^2}{(a-x)/v \times [3(a-x)/v]^3}$$

From Dalton's law of partial pressures, the pressure of nitrogen will be given by:

$$p_{N_2} = \frac{(a-x)}{2(2a-x)} \times p$$

where p is the total pressure of the system.
Similarly:

$$p_{H_2} = \frac{3(a-x)}{2(2a-x)} \times p \quad \text{and} \quad p_{NH_3} = \frac{2x}{2(2a-x)} \times p$$

Consequently

$$K_p = \frac{(p_{NH_3})^2}{p_{N_2} \times (p_{H_2})^3} = \frac{\left(\dfrac{2x}{2(2a-x)} \times p\right)^2}{\dfrac{(a-x)}{2(2a--x)} \times p\left(\dfrac{3(a-x)}{2(2a-x)} \times p\right)^3} = \frac{2^4 x^2 (2a-x)^2}{3^3 (a-x)^4 p^2}$$

In practice, x is much smaller than a, and so the equation can be simplified to:

$$K_p \simeq \frac{64x^2}{27p^2a^2} \quad \text{or} \quad x^2 \simeq p^2 \times \frac{27}{64} \times K_p a^2$$

so that the yield of ammonia is directly proportional to the pressure (approximately).

This conclusion is in accordance with *Le Chatelier's principle* which states that *if a constraint be applied to a system in equilibrium, then the system so shifts as to nullify as far as possible the effect of the constraint.* The formation of ammonia from its elements results in a contraction in volume (four volumes of gas producing two volumes) and therefore a diminution of pressure. Hence an increase of pressure on the system shifts the equilibrium in favour of the formation of ammonia. Also, since the reaction is exothermic, the principle predicts that lowering the temperature will also favour the forward reaction. Unfortunately under such conditions, the rate of attainment of equilibrium is very slow; therefore a catalyst and compromise temperature, giving a reasonable reaction velocity at the expense of the equilibrium position, are employed in the large-scale manufacture of ammonia.

From the above relationships:

$$\frac{K_p}{K_c} = \frac{\dfrac{2^4 x^4 (2a-x)^2}{3^3 (a-x)^4 p^2}}{\dfrac{(2x/v)^2}{(a-x)/v \times [3(a-x)/v]^3}} = \frac{2^2(2a-x)^2}{(pv)^2} = \frac{n^2}{(nRT)^2} = (RT)^{-2}$$

where n is the total number of moles present.

This is an example of the general relation:

$$K_p = K_c(RT)^{\Delta n}$$

where Δn is the change in the number of moles of gas.

Calculation of equilibrium constants

(1)   *From molar concentrations* 0.7 mole of ethanoic (acetic) acid and 0.8 mole of ethanol were kept at constant temperature until equilibrium was reached, when the mixture was found to contain 0.2 mole of the acid. Calculate the equilibrium constant.

From the equation

$$CH_3COOH + C_2H_5OH \rightleftharpoons CH_3COOC_2H_5 + H_2O$$

it is seen that for every mole of acid and alcohol used, one mole each of ester (ethyl ethanoate) and water are formed. In this case, the number of moles of acid used is $0.7 - 0.2 = 0.5$, which must equal the number of moles of alcohol used and the number of moles of ester and water produced, so that at equilibrium, where v is the volume of the system:

$$[\text{acid}] = 0.2/v \quad [\text{ethanol}] = (0.8 - 0.5)/v \quad [\text{ester}] = [\text{water}] = 0.5/v$$

Therefore

$$K_c = \frac{[\text{ester}][\text{water}]}{[\text{acid}][\text{alcohol}]} = \frac{0.5/v \times 0.5/v}{0.2/v \times 0.3/v} = 4.2$$

(2)   *From measurement of density|* The density of iodine vapour at 1473 K and 100 kPa pressure is 1.2 g dm^{-3}. Calculate the equilibrium constant, in terms of partial pressures, for the dissociation:

$I_2 \rightleftharpoons 2I$

i.e.

$$K_p = \frac{(p_I)^2}{p_{I_2}}$$

Let the degree of dissociation be α. Then the number of moles of undissociated iodine left after dissociation of one mole of iodine $= (1-\alpha)$ and the number of moles of dissociated iodine $= 2\alpha$. Hence the total number of moles after dissociation of one mole of iodine is

$$(1-\alpha)+2\alpha = 1+\alpha$$

Since at s.t.p. one mole (254 g) iodine occupies 22.4 dm^3, at 1473 K it will occupy $22.4 \times 1473/273$ dm^3 and if there is no dissociation the density will be

$$\frac{254 \times 273}{22.4 \times 1473} = 2.1 \text{ g dm}^{-3}$$

But

$$\frac{\text{Density assuming no dissociation}}{\text{Actual density}} = \frac{2.1}{1.2} = \frac{1+\alpha}{1}$$

(since by Avogadro's law, the volume is directly proportional to the mole present, and density is inversely proportional to the volume, the mass remaining constant for the closed system) giving

$$\alpha = (2.1-1.2)/1.2 = 0.75$$

For a total pressure, P

$$p_{I_2} = \frac{1-\alpha}{1+\alpha} \times P = \frac{1}{7} \times P \qquad \text{and} \qquad p_I = \frac{2\alpha}{1+\alpha} \times P = \frac{6}{7} \times P$$

Therefore

$$K_p = \frac{(p_I)^2}{p_{I_2}} = \frac{(6/7)^2 P^2}{(1/7)P} = 5.14P = 514 \text{ kPa}$$

Heterogeneous equilibria

So far the equilibrium law has been applied to homogeneous systems, but equally well it can be applied to the homogeneous phase of a heterogeneous system. For example, in the dissociation of solid calcium carbonate into solid calcium oxide and gaseous carbon dioxide:

$$CaCO_3 \rightleftharpoons CaO + CO_2$$

$$K_p = \frac{p_{CO_2} \times p_{CaO}}{p_{CaCO_3}}$$

The values p_{CaCO_3} and p_{CaO} are the vapour pressures of the solids, and provided that these are well established phases, they will have constant

Table 6.2 Dissociation of ammonium hydrogen sulphide

P_{NH_3}/kPa	P_{H_2S}/kPa	$(P_{NH_3} \times P_{H_2S})/kPa^2$
33.1	33.1	10 920
27.6	38.8	10 740
59.9	18.9	11 300

values at fixed temperatures (cf. saturated vapour pressures). The equation can therefore be modified

$$K_p = p_{CO_2} \times (\text{a constant})/(\text{a constant})$$

$p_{CO_2} = $ a constant, at a fixed temperature

This is found to be the case in practice, the latter constant being known as the *dissociation pressure*.

The same sort of reasoning for the system

$$NH_4HS(s) \rightleftharpoons NH_3(g) + H_2S(g)$$

leads to the conclusion that $(p_{NH_3} \times p_{H_2S})$ is constant at constant temperature. How closely theory accords with practice can be seen from *Table 6.2*.

Free energy and the position of equilibrium

The relation between the rate constant and the activation energy—the Arrhenius equation (p. 134) can be applied to both the forward and backward reactions of a reversible change.

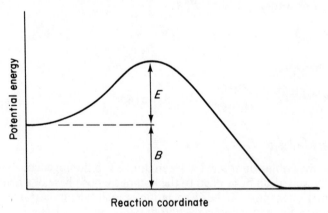

Figure 6.14 Potential energy changes during reaction

Figure 6.14 is a simplified version of *Figure 6.9* and, using the Arrhenius equation, we have:

For the forward reaction $k_1 = Ae^{-E/RT}$

For the backward reaction $k_2 = Ae^{-(E+B)/RT}$

Therefore

$$K = k_1/k_2 = Ae^{-E/RT}/Ae^{-(E+B)/RT}$$
$$= e^{B/RT}$$

But the free energy change for the forward reaction is given by:

$$\Delta G = -B$$

i.e.,

$$K = e^{-\Delta G/RT}$$

or

$$\Delta G = RT\ln K$$

so that if ΔG is large and negative, the reaction is virtually irreversible.

Summary

The prospect for a reaction occurring depends not only on the thermodynamic feasibility, which requires a negative value for the free energy change, but also on the kinetic feasibility as determined by the height of the energy barrier or activation energy. The latter can be reduced with the aid of catalysts which act either by a homogeneous or heterogeneous mechanism. For any reversible reaction:

$$aA + bB \rightleftharpoons cC + dD$$

the equilibrium law (a thermodynamic derivative) demands an equilibrium constant, K, which is temperature-dependent, and is given by:

$$K = \frac{[C]_e^c[D]_e^d}{[A]_e^a[B]_e^b}$$

where $[X]_e$ is the equilibrium concentration for species X. The law of mass action (a kinetic derivative) assumes that the rate of a reaction such as:

$$aA + bB \rightarrow \text{products}$$

is given by:

$$\text{Rate} = k[A]^x[B]^y$$

where k is the rate constant (temperature-dependent) and x and y are experimentally determined values, not necessarily related to a or b, and normally restricted to values of 0, 1, or 2.

Questions

(2)   Calculate the heat of formation of carbon disulphide from the following data:

$$\Delta H$$

$$CS_2(l) + 3O_2(g) \rightarrow CO_2(g) + 2SO_2(g) \qquad -1110 \text{ kJ}$$
$$C(s) \;\; + \;\; O_2(g) \rightarrow CO_2(g) \qquad\qquad\qquad -394 \text{ kJ}$$
$$S(s) \;\; + \;\; O_2(g) \rightarrow SO_2(g) \qquad\qquad\qquad -297 \text{ kJ}$$

(2)   Calculate the bond energy of the C—H link from

	ΔH
Heat of formation of carbon dioxide	-394 kJ
Heat of formation of water	-285 kJ
Heat of atomization of carbon	$+713$ kJ
Heat of dissociation of hydrogen	$+435$ kJ
Heat of combustion of methane, CH_4	-899 kJ

(3)   Suggest the roles of the catalyst in the following:

(a) Decomposition of potassium chlorate(V) in the presence of manganese(IV) oxide.

(b) Conversion of sulphur dioxide and oxygen to sulphur(VI) oxide in the presence of platinum.

(c) The esterification of ethanol by ethanoic (acetic) acid in the presence of mineral acid.

(d) The hydrolysis of ethyl ethanoate (acetate) by alkali.

Give any evidence you can to support your mechanisms.

(4)   Discuss the following statements:

(a) Magnesium reacts exothermically with oxygen only when heated.

(b) Magnesium oxide is reduced by carbon only at high temperatures.

(c) Sulphur dichloride oxide (thionyl chloride) reacts instantly and endothermically with acetic acid.

(d) Ammonium chloride dissolves in water with the absorption of heat.

(5)   A mixture of 21 g of ethanoic acid and 16 g of ethanol were kept at constant temperature until equilibrium was attained; analysis showed that 6 g of ethanoic acid remained. Calculate the equilibrium constant for the reaction.
What would be the composition of the equilibrium mixture if 5 g of ethanoic acid, 10 g of ethanol and 8 g each of water and ethyl ethanoate were kept at the same constant temperature until equilibrium was reached?

(6)   Nitrogen dioxide contains 20 per cent by volume of NO_2 molecules at a pressure of 100 kPa and a temperature of 30 °C, the remainder being dimeric molecules. Calculate the equilibrium constant, K_p, for the reaction

$$N_2O_4 \rightleftharpoons 2NO_2$$

at 30 °C and hence the percentage dissociation at the same temperature when the pressure is increased to 400 kPa.

(7)   The following values represent the change in optical rotation produced in the conversion of a solution of α-glucose containing 10 g dm^{-3} to β-glucose:

Time/min   5   10   20   30   40   50   60   1440 (∞ reading)
Rotation/deg.   97.6 96.8 95.2 94.2 93.3 92.5 91.9 87.4

   From these values show that the reaction is of first order, and hence evaluate the rate constant.

(8)   Suggest how you could investigate the effect of temperature and concentration on the reaction between thiosulphate(VI) ions and acid:

$$S_2O_3^{2-}(aq) + 2H^+(aq) \rightarrow SO_2(g) + S(s) + H_2O(l)$$

How could the order of the reaction be confirmed?

(9)   Derive the relationship $K_p = K_c(RT)^{\Delta n}$.

(10) Draw a graph of energy *versus* internuclear distance, to show the approach of molar quantities of sodium and chloride ions. Indicate the position of internuclear equilibrium.

(11) Use the following information [from Moelwyn-Hughes and Johnson, *Transactions of the Faraday Society*, **36**, 954 (1940)] to calculate the energy of activation of the first-order decomposition of benzenediazonium chloride:

Temp./°C	15.1	19.9	24.7	30.0
$10^5 \ k/s^{-1}$	0.930	2.01	4.35	9.92

(Plot balues of $\log k$ againat $1/T$ to get a straight line of slope $- E/2.30 \ R$: see the Arrhenius equation. Hence calculate E, the energy of activation.)

(12) Calculate the equilibrium constant for the reaction

$$C_6H_6 + 3H_2 \rightleftharpoons C_6H_{12}$$

given that the free energy change at $23\,°C$ is $-97.4 \text{ kJ mol}^{-1}$.

(13) From the bond energies given, determine which of the hydrogen halides has the greater ionicity. (Assume that the energy of a pure covalent bond is the average of the separate energies of the atoms forming the bond).

Bond	H—H	F—F	I—I	H—F	H—I
Energy/kJ mol^{-1}	435	159	152	433	300

(14) The following are some of the processes used in the petrochemical industry: catalytic cracking, catalytic reforming, hydrogenation, isomerization, alkylation, polymerization, dehydrogenation, and oxidation. Give examples of each of these processes, stating the general conditions required and discussing the physicochemical principles involved.

(15) An early value for the enthalpy change of the reaction:

$$Mg(s) + 2HCl(aq) \rightarrow MgCl_2(aq) + H_2(g)$$

was $-450\,\text{kJ mol}^{-1}$, whilst a recent value is $-461\,\text{kJ mol}^{-1}$. Suggest reasons for the discrepancy and outline the precautions you would take in redetermining the enthalpy change.

(16) An alternative mechanism for the thermal decomposition of dinitrogen pentoxide (p. 139) involves the steps:

$$N_2O_5 \underset{k_2}{\overset{k_1}{\rightleftharpoons}} NO_3 + NO_2$$

$$NO_3 + NO_2 \xrightarrow{k_3} NO_2 + NO + O_2$$

and

$$NO_3 + NO \xrightarrow{k_4} 2NO_2$$

where k_1, k_2, k_3 and k_4 are the rate constants.

(a) Show that these steps produce the stoichiometric (overall) equation:
$2N_2O_5 \rightarrow 4NO_2 + O_2$

(b) Write down expressions for the rates of (i) decomposition of N_2O_5, (ii) formation of NO_3, and (iii) formation of NO.

(c) In the steady state, the rates of formation of NO_3 and NO will be zero; explain why this assumption can be made.

(d) Combining the results of (b) and (c), show that the rate of decomposition of N_2O_5 is given by:

$$-\frac{d[N_2O_5]}{dt} = \frac{2k_2k_3[N_2O_5]}{k_2 + 2k_3}$$

which agrees with the experimental observation that the rate is of the first order with respect to the concentration of N_2O_5.

(17) Describe how you would measure the enthalpy change which accompanies any *one* reaction you may choose.

The enthalpy of combustion of fumaric acid ($C_4H_4O_4$) to $CO_2(g)$ and $H_2O(l)$ is exothermic by $1344\,\text{kJ mol}^{-1}$. The enthalpy of formation of the isomeric maleic acid is shown by the equation:

$4C + 2H_2 + 2O_2 \rightarrow C_4H_4O_4 \qquad \Delta H = -781\,\text{kJ mol}^{-1}$

Given that:

$H_2(g) + \tfrac{1}{2}O_2(g) \rightarrow H_2O(l) \qquad \Delta H = -286\,\text{kJ mol}^{-1}$

$C(s) + O_2(g) \rightarrow CO_2(g) \qquad \Delta H = -395\,\text{kJ mol}^{-1}$

calculate the enthalpy change for the isomerization:

maleic acid $\rightarrow$ fumaric acid

Is this reaction exothermic?

[O.]

(18)   (a) Define the term *order of a reaction*.
       (b) In an experimental study of a reaction in solution between two
            compounds A and B, the following information was obtained for the
            initial rate of the reaction:

10^4(*Initial rate*)/mol dm^{-3} s^{-1}	1.0	4.0	8.0
Initial concn. of A/mol dm^{-3}	0.10	0.20	0.20
Initial concn. of B/mol dm^{-3}	0.12	0.12	0.24

What is the order of the reaction with respect to (i) the reactant A, (ii) the
reactant B?
       (c) The decomposition of benzenediazonium chloride in aqueous solution is
            a reaction of first order which proceeds according to the equation:

$$C_6H_5N_2Cl \rightarrow C_6H_5Cl + N_2(g)$$

A certain solution of benzenediazonium chloride contains initially an
amount of this compound which gives 80 cm^3 of nitrogen on complete
decomposition. It is found that, at 30 °C, 40 cm^3 of nitrogen are evolved
in 40 minutes. How long after the start of the decomposition will 70 cm^3
of nitrogen have been evolved? (All volumes of nitrogen refer to the same
temperature and pressure.)

[Oxford 1977]

(19)   What do you understand by the term 'electron delocalization' (resonance)?
       How may enthalpies of hydrogenation be used to support the idea of electron
       delocalization in the benzene molecule? How does the electronic structure of
       a benzene molecule help to explain the following?

       (a) Concentrated sulphuric acid is added to concentrated nitric acid in the
            nitration of benzene.
       (b) Unlike iodomethane, iodobenzene does not react readily with aqueous
            sodium hydroxide.
       (c) Benzenecarboxylic (benzoic) acid is a stronger acid than ethanoic (acetic)
            acid in aqueous solution.

[J.M.B.]

(20)   The free energy data for the following reactions were obtained at one
       atmosphere pressure:

$Si(l) + O_2(g) \rightarrow SiO_2(s)$ $\qquad \Delta G = (-950 + 0.20T)$ kJ mol^{-1}

$2Mg(g) + O_2(g) \rightarrow 2MgO(s)$ $\qquad \Delta G = (-1520 + 0.44T)$ kJ mol^{-1}

$4/3Al(l) + O_2(g) \rightarrow \frac{2}{3}Al_2O_3(s)$ $\qquad \Delta G = (-1130 + 0.22T)$ kJ mol^{-1}

where T is temperature in kelvin.

       (a) Determine graphically, or by calculation, the minimum temperature at
            which magnesium metal can be obtained from magnesium oxide by
            reduction with silicon, at one atmosphere pressure.

(*b*) How could the minimum temperature at which the reaction in (*a*) proceeds, be lowered?

(*c*) Explain whether it would be possible to melt aluminium (melting point 660 °C) in a silica crucible without contamination resulting from a chemical reaction.

[J.M.B. 1978]

7 Electrochemistry

Electrolysis • Conductivity • Application of mass action: ionic equilibria • Aqueous systems: dissociation of water • Non-aqueous systems • Electrode potentials

Electrolysis

A flow of charged particles constitutes an electric current. If the particles are the mobile electrons of the metallic lattice, then the type of conduction is termed *electronic* and is not accompanied by any chemical side effects. If, on the other hand, the charged particles are *ions* that result from the transfer of electrons by electrovalency, then the conduction is *electrolytic*. Provided that the ions are rendered mobile by dissolving in a suitable solvent or by melting, the application of an electric potential causes, by attraction of opposite charges, migration of the negative ions (anions) to the positive electrode (the anode) and of positive ions (cations) to the negative electrode (the cathode). Electrons are transferred at the electrodes: *to* the anode from the anion and *from* the cathode to the cation. That is, chemical changes occur at the electrodes, and this is a characteristic feature of *electrolysis*.

For example, in the electrolysis of sodium chloride, where the crystal lattice is broken down by melting, sodium ions migrate to the cathode and are discharged, whilst chloride ions are discharged at the anode (*Figure 7.1*).

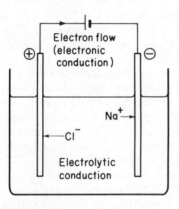

Figure 7.1 Electrolysis of fused sodium chloride

Cathode
$$Na^+ + e^- \rightarrow Na$$

Anode
$$Cl^- - e^- \rightarrow Cl$$
$$2Cl \rightarrow Cl_2\uparrow$$

It is clear that the precursor to electrolysis is the formation of mobile ions which, in the case of strongly electrovalent compounds, occurs either by melting or by solution in a suitable solvent. Evidence that ions are present when the substance is dissolved in a solvent like water is provided by the abnormal colligative properties of solutions of electrolytes (p. 117).

The course of electrolysis is sometimes difficult to predict when there is a mixture of ions present. It is normally true to say that the process which takes place is that which requires least electrical energy. Consequently, the least stable ions are discharged first; the electrochemical series (p. 179) can be used as a rough guide in this respect, but it should be remembered that an excessive potential (the *overpotential*) is required before a substance, especially a gas, is actually discharged: the amount varies from electrode to electrode and sometimes exceeds the electrode potential, preventing discharge. For example, in the electrolysis of sodium chloride solution using a mercury cathode, the following ions are present:

$Na^+ + Cl^-$ (present even in solid lattice)

$H^+ + OH^-$ (from the equilibrium $H_2O \rightleftharpoons H^+ + OH^-$)

It must be remembered that simple hydrogen ions do not exist in aqueous solution: the polarizing power of a proton is so great that it combines with water to form an oxonium ion, i.e.

$H^+ + H_2O \rightarrow H_3O^+$

and other more complex species. It is to be understood in this Chapter unless specified otherwise that H^+ implies such a hydrated proton, or $H^+(aq)$. Other ions are also hydrated or aquated in solution. It is the hydrated proton that has the typically 'acidic' properties of turning litmus red and dissolving magnesium to evolve hydrogen; pure ethanoic acid, for example, does not do these things, but only behaves 'acidically' when water is present.

At the anode, OH^- and Cl^- are attracted. It requires less energy to discharge OH^- than Cl^-, but water is a very weak electrolyte, and so there are far fewer hydroxide ions present than chloride ions. Therefore, unless alkali is added, there are virtually no hydroxide ions available at the electrode and chloride ions are accordingly discharged. Thus, concentration also affects the course of electrolysis; the greater the concentration, the greater the chance of discharge.

At the cathode, there is a surfeit of electrons, and electrolysis proceeds as these are donated to cations. The electrochemical series suggests that the less stable hydrogen ions will be discharged in preference to sodium ions but, if the cathode is mercury, the overpotential is so great for hydrogen that the sodium ions are preferentially discharged. The resulting atoms dissolve in the mercury to form sodium amalgam; the energy released in this latter reaction also favours this particular step.

This is an example of a metal being discharged and reacting with the electrode. More often, reaction takes place between the electrode and a newly-discharged gas; the initial product is a reactive atom, and unless the electrode is particularly inert (e.g., platinum), interaction is a distinct possibility. More often than not, this is a disadvantage, but in the case of the 'anodizing' of aluminium it can be turned to account. Dilute sulphuric acid is electrolysed using an aluminium anode; hydroxide ions are discharged at this electrode to give nascent oxygen, which attacks the aluminium to give a thick protective film of oxide, which can be dyed various colours.

Sometimes, instead of an ion being discharged at the anode to provide the electrons which are essential to a continuous flow of electricity, it is energetically favourable for the anode to dissolve as positive ions, leaving behind electrons. For example, in the electrolysis of copper(II) sulphate solution between copper electrodes, copper(II) ions dissolve from the anode, instead of anions being discharged: as copper is deposited at the cathode, the overall process represents the transfer of copper from anode to cathode and provides a means of purification:

Anode	*Cathode*
$Cu - 2e^- \rightarrow Cu^{2+}$	Cu^{2+} and H^+ attracted
$SO_4{}^{2-}$ and OH^- attracted but not discharged	Cu^{2+} selectively discharged (less stable and more concentrated than H^+)
	$Cu^{2+} + 2e^- \rightarrow Cu$

Electrolysis can be a very useful industrial process; apart from the above examples of anodizing and purification, it can be the basis for the extraction of elements (e.g., aluminium and chlorine) and for the manufacture of compounds [e.g., sodium hydroxide and potassium chlorate(V)] as well as a means of electroplating.

Faraday's Laws of Electrolysis

The actual quantities of substance liberated during electrolysis are summarized by Faraday's laws:

(1) The mass of substance liberated at an electrode is proportional to the quantity of electricity passed

i.e., to the number of coulombs (C) = amperes (A) × seconds

(2) The masses of different substances liberated by the same quantity of electricity are inversely proportional to the respective ionic charges.

These laws are the consequence of each ion carrying a definite charge and there being a constant number of ions (the Avogadro constant, L, equal to 6.023×10^{23}) in a mole of any ion.

If each ion has a valency of z, then a charge of $\pm z$ units will be associated with each ion. If m is the mass of an ion, then m/z is the mass of the substance which will be liberated for every unit of electricity supplied to or from the cell (First Law). For one ion the actual charge will be $\pm ze$, where e is the electronic charge (1.602×10^{-19} C) and for one mole of ion the total charge is $\pm zeL$. Now eL is a constant, known as the Faraday constant, F, and is found experimentally to equal 96·49 kC. Therefore the discharge of one mole of a z-valent ion requires zF coulombs and the number of moles liberated by F coulombs, is $1/z$ (Second Law). For example, one mole of copper ions of ionic charge two requires $2F$ coulombs for discharge and so F coulombs will liberate half a mole of copper. On the other hand, one mole of aluminium ions, of ionic charge three, require $3F$ coulombs and so F coulombs liberate one third of a mole of aluminium.

Conductivity

There is always a resistance to flow of current. Resistance is defined as the ratio of potential difference between two points to the current flowing:

$$\text{Resistance} = \frac{\text{Potential difference ('voltage')}}{\text{Current}}$$

and is measured in ohm (Ω). Thus

$$R/\Omega = \frac{P.D./V}{I/A} \qquad \text{or} \qquad \text{ohms} = \frac{\text{volts}}{\text{amperes}}$$

Many conductors such as metals and electrolyte solutions obey Ohm's law: *The current in a conductor is proportional to the potential difference between its ends.* The resistance of a length of material depends directly on the distance, l, along which the current flows and inversely on the cross-sectional area, A, i.e.,

Resistance, $R \propto l/a$ or $R = \rho l/A$

where ρ is the resistivity of the material and is constant at a fixed temperature.

In chemistry it is more usual to talk about the *conductance*, G, of a solution, so that we write

$G = 1/R$

where G has the unit of ohm^{-1}, Ω^{-1}, or siemens, S. The previous equation is

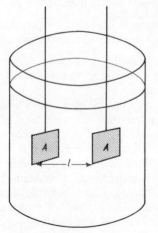

Figure 7.2 A conductivity cell

then written as

$$G = \kappa A/l$$

where κ is the *conductivity* (*electrolytic*) of the solution. This equation shows that the conductance of a solution between a pair of electrodes (*Figure 7.2*) is directly proportional to the number of conducting paths, i.e., the cross-sectional area of the electrodes, and inversely proportional to their separation.

Rearranging the last equation gives the units for the electrolytic conductivity

$$\kappa = \frac{G\ \Omega^{-1} \times l\ m}{A\ m^2} = \frac{Gl}{A}\ \Omega^{-1}\ m^{-1} \qquad \text{or} \qquad ohm^{-1}\ m^{-1}.$$

The resistance, and therefore the conductance, is measured using a Wheatstone bridge circuit and an alternating current of such frequency that electrolysis is negligible, usually 1 kHz. The balance point is measured using a galvanometer after rectification or a 'magic eye' indicator (*Figure 7.3*).

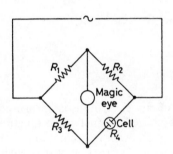

Figure 7.3 A conductivity bridge

When the bridge is balanced

$$R_1/R_2 = R_3/R_4$$

The conductance of the solution between the electrodes is then given by

$$G = 1/R_4 = R_1/R_3 R_2$$

If the dimensions of the electrodes are known, i.e., l/A has been measured, then κ can be computed. However l/A, called the *cell constant*, is generally obtained indirectly by first using a potassium chloride solution of known electrolytic conductivity and measuring its conductance.

As a solution is diluted, the electrolytic conductivity decreases, since the number of conducting particles (ions) between the electrodes diminishes, i.e.,

Electrolytic conductivity is proportional to the concentration

or

$$\kappa = \Lambda \times c$$

where c is the concentration and Λ is the constant of proportionality known as the *molar conductivity*.

In calculations involving conductivity, care must be exercised in the consistency of units, as shown in the following example.

Example The resistance of a 0.1M solution of an electrolyte in a conductivity cell of cell constant 1.5 cm^{-1} ($= 1.5 \times 10^2$ m^{-1}) is 50 Ω. Calculate the molar conductivity.

$$\text{Electrolytic conductivity } \kappa = \frac{\text{Cell constant}}{\text{Resistance}}$$

$$\kappa = \frac{1.5 \text{ cm}^{-1}}{50 \text{ } \Omega} = 0.03 \text{ } \Omega^{-1} \text{ cm}^{-1} \quad \text{or} \quad \frac{1.5 \times 10^2 \text{ m}^{-1}}{50 \text{ } \Omega} = 3.0 \text{ } \Omega^{-1} \text{ m}^{-1}$$

A 0.1M solution contains 0.1 mol dm^{-3} which is

0.1×10^{-3} mol cm^{-3} \quad or \quad 0.1×10^3 mol m^{-3}

Molar conductivity $\Lambda = \kappa/c$

Therefore

$$\Lambda = \frac{0.03 \text{ } \Omega^{-1} \text{ cm}^{-1}}{0.1 \times 10^{-3} \text{ mol cm}^{-3}} = 300 \text{ } \Omega^{-1} \text{ cm}^2 \text{ mol}^{-1}$$

or

$$\Lambda = \frac{3.0 \text{ } \Omega^{-1} \text{ m}^{-1}}{0.1 \times 10^3 \text{ mol m}^{-3}} = 0.03 \text{ } \Omega^{-1} \text{ m}^2 \text{ mol}^{-1}$$

If the molar conductivity of an electrolyte is measured at different concentrations and the values plotted against some function of the concentration, e.g., log or square root, in order to accommodate the large range in concentration required to obtain a curve, one of two types of curve will result (*Figure 7.4*).

Substances which follow curves of type *A* are characterized by a rapid rise in Λ to a limiting value, $\mathring{\Lambda}^{\infty}$, as the dilution is increased. They are called *strong electrolytes* and are ionic even in the solid state. The increase in conductivity on dilution is merely the effect of rendering the ions more mobile and freer from each other's influence. Substances which give curves of type *B*, on the other hand, do not give a clear limiting value for Λ^{∞}; they are *weak electrolytes*, e.g., organic acids, and are only slightly ionized in concentrated aqueous solution. Dilution in this case serves actually to increase the degree of ionization and, indeed, the degree of dissociation at concentration c is given by the fraction $\Lambda_c/\Lambda^{\infty}$ although Λ^{∞} can only be determined indirectly (p. 159). For strong electrolytes this fraction is more correctly called the apparent degree of dissociation, since it really represents a mobility ratio (p. 159).

The Law of Independent Migration of Ions (Kohlrausch)

Kohlrausch, in 1876, propounded his law that each ion of an electrolytic solution contributes a definite, constant amount to the molar conductivity:

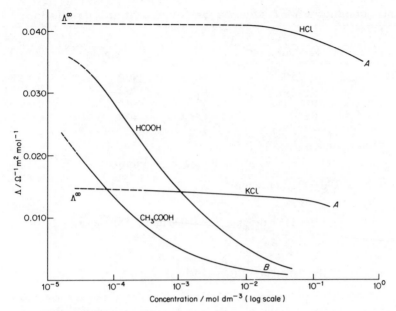

Figure 7.4 Conductivity curves

$$\Lambda^{\infty} = \Lambda_{+} + \Lambda_{-}$$

where Λ_{+} and Λ_{-} are the molar conductances of the ions at infinite dilution.

This law is based upon the following sort of evidence (all Λ given in Ω^{-1} cm^{2} mol^{-1}):

$\Lambda^{\infty}(NaCl) = 126.3$ $\Lambda^{\infty}(NaNO_3) = 121.5$
$\Lambda^{\infty}(KCl) \ = 149.8$ $\Lambda^{\infty}(KNO_3) \ = 145.0$
Difference   23.5   23.5

If we express these results in ionic form:

$$\Lambda^{\infty}(NaCl) = \Lambda(Na^{+}) + \Lambda(Cl^{-}) = 126.3 \qquad \Lambda^{\infty}(NaNO_3) = \Lambda(Na^{+}) + \Lambda(NO_3^{-}) = 121.5$$

$$\Lambda^{\infty}(KCl) \ = \Lambda(K^{+}) \ + \Lambda(Cl^{-}) = 149.8 \qquad \Lambda^{\infty}(KNO_3) \ = \Lambda(K^{+}) \ + \Lambda(NO_3^{-}) = 145.0$$

By subtraction

$$\Lambda(K^{+}) - \Lambda(Na^{+}) = 23.5 \qquad\qquad\qquad \Lambda(K^{+}) - \Lambda(Na^{+}) = 23.5$$

showing that the molar conductivity is the sum of the ionic conductivities.

Kohlrausch's law can be explained by assuming that different ions travel towards the electrodes at different speeds and hence make different contributions to the conductance. In fact, the law permits Λ^{∞} for a weak electrolyte to be calculated by adding together the relevant ionic conductivities; for example, at infinite dilution the molar conductivity of ethanoic acid is given by:

$$\Lambda^{\infty} = \Lambda(H^{+}) + \Lambda(CH_3CO_2^{-})$$

Table 7.1 Molar conductivities Λ of some ions at 25 °C at infinite dilution

Ion	$\dfrac{\Lambda_+}{\Omega^{-1}\,cm^2\,mol^{-1}}$	$\dfrac{\Lambda_+}{\Omega^{-1}\,m^2\,mol^{-1}}$	Ion	$\dfrac{\Lambda_-}{\Omega^{-1}\,cm^2\,mol^{-1}}$	$\dfrac{\Lambda_-}{\Omega^{-1}\,m^2\,mol^{-1}}$
H^+	350.0	0.035 00	OH^-	198.0	0.019 80
Na^+	50.0	0.005 00	Cl^-	76.3	0.007 63
K^+	73.5	0.007 35	$\frac{1}{2}SO_4^{2-}$	80.0	0.008 00
Ag^+	62.0	0.006 20	$CH_3CO_2^-$	40.0	0.004 00

From *Table 7.1*

$$\Lambda^\infty\ 350.0\ \Omega^{-1}\,cm^2\,mol^{-1}+40.0\ \Omega^{-1}\,cm^2\,mol^{-1}$$

$$=\ 390.0\ \Omega^{-1}\,cm^2\,mol^{-1}$$

or

$$\Lambda^\infty = 0.035\,00\ \Omega^{-1}\,m^2\,mol^{-1}+0.004\,00\ \Omega^{-1}\,m^2\,mol^{-1}$$

$$= 0.039\,00\ \Omega^{-1}\,m^2\,mol^{-1}$$

Solubility of sparingly soluble electrolytes

A further application of Kohlrausch's law is the determination of the solubility of sparingly soluble salts. A solution of such a salt will be so dilute as to give complete ionization. Λ^∞ may therefore be calculated from Kohlrausch's law. Determination of the electrolytic conductivity permits the concentration to be calculated and hence the solubility.

Example. The molar conductivity of silver chloride at infinite dilution is given from *Table 7.1* as:

$$\Lambda^\infty(AgCl) = \Lambda(Ag^+)+\Lambda(Cl^-)$$

$$= 62.0+76.3 = 138.3\ \Omega^{-1}\,cm^2\,mol^{-1}$$

$$= 0.006\,20+0.007\,63 = 0.013\,83\ \Omega^{-1}\,m^2\,mol^{-1}$$

From experiment, the electrolytic conductivity, κ, of saturated silver chloride solution at 25 °C is $1.85 \times 10^{-6}\ \Omega^{-1}\,cm^{-1}$ or $1.85 \times 10^{-4}\ \Omega^{-1}\,m^{-1}$. The concentration, c, of this solution is then given by

$$c = \frac{\kappa}{\Lambda} = \frac{1.85 \times 10^{-6}\ \Omega^{-1}\,cm^{-1}}{138.3\ \Omega^{-1}\,cm^2\,mol^{-1}} = 1.34 \times 10^{-8}\ mol\ cm^{-3}$$

or

$$c = \frac{\kappa}{\Lambda} = \frac{1.85 \times 10^{-4}\ \Omega^{-1}\,m^{-1}}{0.013\,83\ \Omega^{-1}\,m^2\,mol^{-1}} = 1.34 \times 10^{-2}\ mol\ m^{-3}$$

The relative molecular mass of silver chloride $AgCl = 108+35.5 = 143.5$ and therefore the solubility of silver chloride at 25 °C is

$$1.34 \times 10^{-8}\ mol\ cm^{-3} \times 143.5\ g\ mol^{-1} = 1.92 \times 10^{-6}\ g\ cm^{-3}$$

or

$$1.34 \times 10^{-2}\ mol\ m^{-3} \times 143.5\ g\ mol^{-1} = 1.92\ g\ m^{-3}$$

$$= 1.92 \times 10^{-3}\ g\ dm^{-3}$$

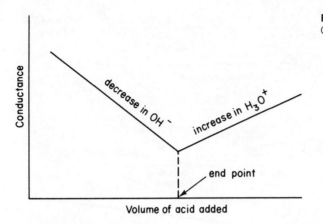

Figure 7.5
Conductometric titration

Conductometric titration

Reference to *Table 7.1* will reveal that hydrogen (or oxonium, H_3O^+) ions have by far the greatest conductivity. This fact can be utilized in conductometric titration, for instance between hydrochloric acid and sodium hydroxide (*Figure 7.5*). As hydrochloric acid is added to sodium hydroxide solution, hydroxide ions are removed:

$$H_3O^+ + OH^- \rightarrow 2H_2O$$

The conductance of the solution will therefore fall until the end point is reached. When excess of hydrochloric acid is then added, the conductance will rise again by virtue of the free oxonium ions; intersection of the two curves gives the end point of the titration.

Transport numbers

The greater the mobility of a particular ion, the greater will be the fraction of the total current carried by it. If t_+ is the fraction carried by the cation and t_- that carried by the anion, then $t_+ + t_- = 1$. These fractions are called the *transport numbers* of the ions, and they can be determined experimentally by measuring the changes in concentration around the electrodes during electrolysis. Consideration of a hypothetical case will make this clear. Imagine a simple binary electrolyte contained in a voltameter divided into three compartments, A, B, and C, such that there are equal numbers of ions in the anode and cathode compartments (*Figure 7.6a*). If an electric current is now passed through the apparatus, and if the cations move twice as rapidly as the anions, the fall in concentration in the anode compartment in a given time will be twice that in the cathode compartment (*Figure 7.6b*). In more general terms

$$\frac{\text{Current carried by cation}}{\text{Current carried by anion and cation}} = \frac{\text{Fall of concentration in anode compartment}}{\text{Total fall of concentration}}$$

$$= \text{Transport number of the cation, } t_+$$

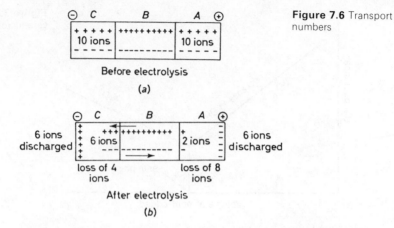

Figure 7.6 Transport numbers

The concentration of a particular ion, instead of decreasing, may actually increase, owing to the possibility of electrodes dissolving. By incorporating a coulometer—a cell used for measuring the quantity of electricity passed—in the circuit, the amount of electrode dissolved can be calculated and allowed for.

Example. There are 0.23 g of silver ions in the anode compartment of a voltameter (with silver anode) before electrolysis and 0.40 g afterwards. In a silver coulometer in series, 0.32 g of silver dissolves from the anode. Calculate the transport number of the silver ions.

If no migration from the anode compartment occurred, the amount of silver ions after electrolysis would be $(0.23 + 0.32) = 0.55$ g

$\therefore$ Fall due to migration $= (0.55 - 0.40) = 0.15$ g Ag

Total current $\equiv 0.32$ g Ag

$\therefore$ Transport number of $Ag^+ = 0.15/0.32 \doteq 0.47$

The total molar conductivity, Λ, is, by Kohlrausch's law, made up of the cationic component, Λ_+, and the anionic component, Λ_-, and it follows from the above that the differences in these values are due to the different mobilities of the ions; for instance, the hydrogen ion has a very high conductance because it is highly mobile and consequently carries a high proportion of the total current. Clearly, then, the transport numbers are related to the conductances in the following manner:

$$t_+ = \frac{\Lambda_+}{\Lambda} \qquad t_- = \frac{\Lambda_-}{\Lambda}$$

Indeed, one purpose of measuring transport numbers is to derive the molar conductances of ions.

Application of mass action: ionic equilibria

Ostwald, in 1888, applied the law of mass action to electrolytic dissociation. For a weak binary electrolyte, AB, in solution of dilution V and with a degree of ionization under these conditions of α, the following equilibrium exists between the undissociated molecule and its ions:

$$AB \rightleftharpoons A^+ + B^-$$

Concentrations $\dfrac{1-\alpha}{V}$ $\dfrac{\alpha}{V}$ $\dfrac{\alpha}{V}$

Applying the law of mass action

$$\frac{[A^+][B^-]}{[AB]} = \frac{(\alpha/V) \times (\alpha/V)}{(1-\alpha)/V}$$

$= K$, the dissociation constant, at constant temperature

that is,

$$K = \frac{\alpha^2}{(1-\alpha)V}$$

For a very weak electrolyte (i.e., when α is small) this can be simplified to

$$K = \frac{\alpha^2}{V}$$

Table 7.2 Application of Ostwald's dilution law to ethanoic acid

$\dfrac{V}{\text{dm}^3 \text{ mol}^{-1}}$	$\alpha = \dfrac{\Lambda_v}{\Lambda^\infty}$	$\dfrac{K = \alpha^2/[(1-\alpha)V]}{\text{mol dm}^{-3}}$
5.374	0.009 8	18.1×10^{-6}
10.573	0.013 8	18.0×10^{-6}
24.875	0.021 6	19.2×10^{-6}
63.26	0.033 6	18.5×10^{-6}

It has been pointed out before that the term 'degree of dissociation' has no real meaning for strong electrolytes, because these are always fully dissociated. In the case of weak electrolytes, however, there is considerable experimental support for the correctness of the above expression, which has come to be known as *Ostwald's dilution law*. Values for ethanoic acid are shown in *Table 7.2*.

Example. The molar conductivity of 0.1M ammonium hydroxide solution at 25 °C is 0.000 317 Ω^{-1} m^2 mol^{-1}. The molar conductivities for the ammonium and hydroxide ions at 25 °C are 0.007 35 and 0.019 85 Ω^{-1} m^2 mol^{-1} respectively. Calculate the value of the dissociation constant.

$$\Lambda^\infty = \Lambda_+ + \Lambda_- = 0.007\,35 + 0.019\,85 \ \Omega^{-1} \text{ m}^2 \text{ mol}^{-1}$$

$$= 0.027\,2 \ \Omega^{-1} \text{ m}^2 \text{ mol}^{-1}$$

Therefore

$$\alpha = \frac{\Lambda_v}{\Lambda^\infty} = \frac{0.000\,317}{0.027\,2} = 0.011\,7$$

$$\text{Dissociation constant} = \frac{(0.011\,7)^2}{(1-0.011\,7)\times 10} = 1.4\times 10^{-5} \text{ mol dm}^{-3}$$

Aqueous systems: dissociation of water

As water is progressively purified, its electrolytic conductance approaches a limiting value of $54 \text{ n}\Omega^{-1} \text{ cm}^{-1}$ or $5\cdot4\times 10^{-6}\,\Omega^{-1}\text{ m}^{-1}$ at $25\,°C$. Thus water itself is a weak electrolyte by virtue of the tendency towards ionization

$$H_2O \rightleftharpoons H^+ + OH^-$$

By Kohlrausch's law

$$\Lambda^\infty = \Lambda(H^+) + \Lambda(OH^-)$$

$$= 350 + 198.5 = 548.5\ \Omega^{-1}\text{ cm}^2\text{ mol}^{-1}$$

$$= 0.0350 + 0.019\,85 = 0.054\,85\ \Omega^{-1}\text{ m}^2\text{ mol}^{-1}$$

But

$$\Lambda^\infty = \kappa/c$$

Let V be the volume containing 1 mole of ions. Then

$$V = 1/c = \Lambda^\infty/\kappa$$

Therefore

$$V = \frac{548.5\ \Omega^{-1}\text{ cm}^2\text{ mol}^{-1}}{54\times 10^{-9}\ \Omega^{-1}\text{ cm}^{-1}} \approx 10^{10} \text{ cm}^3 \approx 10^7 \text{ dm}^3$$

or

$$V = \frac{0.054\,85\ \Omega^{-1}\text{ m}^2\text{ mol}^{-1}}{5.4\times 10^{-6}\ \Omega^{-1}\text{ m}^{-1}} \approx 10^4 \text{ m}^3 \approx 10^7 \text{ dm}^3$$

$$[H^+] = [OH^-] = \frac{1}{1\times 10^7} = 1\times 10^{-7} \text{ mol dm}^{-3}$$

Applying the mass action equation

$$K_{H_2O} = \frac{[H^+][OH^-]}{[H_2O]}$$

Because the ionization is so small $[H_2O]$ is approximately constant; therefore $[H^+][OH^-]$ is approximately constant and is called the *ionic product* of water, K_w.

Now

$$[H^+] = [OH^-] = 1\times 10^{-7} \text{ mol dm}^{-3}$$

thus

$$K_w = (1\times 10^{-7})^2 = 1\times 10^{-14} \text{ mol}^2 \text{ dm}^{-6} \text{ at } 25\,°C$$

One's immediate reaction may be to discount anything as small as this, but

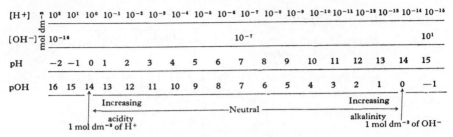

Figure 7.7 Relationships between $[H^+]$, $[OH^-]$, pH and pOH

to do so would be to ignore the essence of chemical equilibrium: in aqueous systems, the equilibrium will always adjust itself to maintain the ionic product, which can thus exert the most profound effect. For instance, if the hydrogen ion concentration is increased from 1×10^{-7} mol dm^{-3} (neutrality) to 1×10^{-1} mol dm^{-3} (decimolar acid), the concentration of hydroxide ions will be decreased from 1×10^{-7} to 1×10^{-13} to maintain the ionic product $(1 \times 10^{-1} \times 1 \times 10^{-13})$ at 1×10^{-14} mol^2 dm^{-6}.

Constant use of negative indices like these soon proves tiresome. To avoid this, the term pH has been introduced; it is defined as the negative logarithm of the hydrogen ion concentration. Similarly, pOH is the negative logarithm of the hydroxide ion concentration. In the above case, pH = 1 and pOH = 13. It follows from the definition that pH + pOH = pK_w (= 14 at 25 °C). Relationships among these quantities are displayed in *Figure 7.7*.

Examples
(1) Find the pH of 0.15M hydrochloric acid.
   A 0.15M solution contains 0.15 mol dm^{-3}. Since complete ionization of a strong acid may be assumed, 0.15 mole of hydrogen ions will be present in each dm^3, i.e., $[H^+] = 0.15$. Therefore

$$pH = -\log 0.15 = -(\bar{1}.18) = 1 - 0.18 = 0.82$$

(2) Find the pH of 0.1M solution of ethanoic acid, given that the dissociation constant at the temperature in question is 1.8×10^{-5} mol^2 dm^{-6}.

   For this weak electrolyte

$$K \sim \alpha^2/V \text{ and } V = 10$$

   Therefore

$$\alpha = \sqrt{(1.8 \times 10^{-5} \times 10)} = 1.25 \times 10^{-2}$$

   If the ethanoic acid were completely dissociated, the hydrogen ion concentration would be 0.1 mol dm^{-3}; therefore

$$[H^+] = 0.1 \times 1.35 \times 10^{-2} = 1.35 \times 10^{-3} \text{ mol dm}^{-3}$$

$$\log[H^+] = \bar{3}.1303 = -2.8697$$

$$pH = +2.87$$

Changes in pH during neutralization

Consider the progressive addition of 0.1M hydrochloric acid to 25 cm^3 of 0.1M sodium hydroxide.

(a) After addition of 24.0 cm^3 of acid, 1.0 cm^3 of 0.1M alkali (i.e., 1×10^{-4} mole) remains in 49 cm^3 of solution.

Therefore

$$[OH^-] = \frac{1000 \times 10^{-4}}{49} = 2 \times 10^{-3} \text{ mol dm}^{-3}$$

hence

$$pOH = -(3.301) = +2.699$$

and

$$pH = 14 - 2.699 = 11.3$$

(b) After addition of 24.5 cm^3 of 0.1M acid, 0.5 cm^3 of 0.1M alkali, containing 5×10^{-5} mole, remains in 49.5 cm^3 of solution.

Therefore

$$[OH^-] = \frac{1000 \times 5 \times 10^{-5}}{49.5} = 1 \times 10^{-3} \text{ mol dm}^{-3}$$

and

$$pOH = 3 \qquad pH = 11$$

(c) After addition of another 0.45 cm^3 of acid, 0.05 cm^3 of 0.1M alkali, containing 5×10^{-6} mole, remain in 49.95 cm^3 of solution; hence

$$[OH^-] = \frac{1000 \times 5 \times 10^{-6}}{49.95} = 1 \times 10^{-4} \text{ mol dm}^{-3}$$

$$pOH = 4 \qquad pH = 10$$

(d) After addition of a further 0.05 cm^3 of acid, the alkali is exactly neutralized, i.e.,

$$[H^+] = [OH^-] = 1 \times 10^{-7} \text{ mol dm}^{-3}$$

$$pH = 7$$

A similar situation prevails as excess of acid is gradually added. It can be seen, therefore, that, in the titration of strong acid with strong base, there is a considerable change in pH around the end point. In the case of weak electrolytes, the small degree of dissociation results in the initial pH or pOH being less; this, together with the fact that dissociation is encouraged as electrolyte is neutralized, results in the pH change being far less pronounced (*Figure 7.8*). Also, because of hydrolysis of the salt formed (p. 168) when either a weak base or a weak acid is used, the pH at the end point will not be 7.

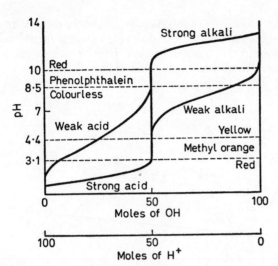

Figure 7.8 Neutralization curves

Indicators

The progress of a neutralization reaction can usually be followed by adding a small quantity of an indicator, often a weak organic acid or base where at least one ion has a different colour from the undissociated molecule. In the case of methyl orange, the undissociated molecule is red and the anion yellow:

$$(CH_3)_2\overset{+}{N}=\langle\underset{\text{red}}{\hspace{1.5cm}}\rangle=N-NH-\langle\hspace{0.5cm}\rangle-SO_3^-$$

$$\rightleftharpoons (CH_3)_2N-\langle\hspace{0.5cm}\rangle-N=N-\langle\hspace{0.5cm}\rangle-SO_3^- + H^+$$
$$\underset{\text{yellow}}{}$$

With phenolphthalein, the undissociated molecule is colourless and the anion red

colourless $\rightleftharpoons$ red $\quad +2H^+$

The amount of dissociation of the indicator, and hence the colour, depend upon the pH of the solution; the considerable changes in pH occurring around the end point in all titrations (except those between weak acid and weak base, for which there are no suitable indicators) bring about sharp changes in colour. The equilibrium constant for the reaction is given by

$$HX \underset{acid}{\overset{alkali}{\rightleftharpoons}} H^+ + X^-$$

$$K = \frac{[H^+][X^-]}{[HX]} \quad \text{i.e.,} \quad \frac{K}{[H^+]} = \frac{[X^-]}{[HX]}$$

Hence, when pH = pK*,

$$\frac{[X^-]}{[HX]} = 1$$

Accordingly, there are different amounts of dissociation at a certain pH, depending on the differing dissociation constants of the indicators. In the case of the titration of strong acid against strong base, the pH changes are such that a wide variety of indicators is available, for strong base against weak acid (when the end point is at pH > 7), phenolphthalein is often used whilst in titration of weak base against strong acid, methyl orange can be employed (*Figure 7.8*).

Hydrolysis

Reference has been made in the previous Section to hydrolysis occurring if one of the reactants is a weak electrolyte, and to the subsequent effect on pH. This requires further explanation; if reaction is between strong base and weak acid, there will be a tendency for the anion of the acid to react with water and so disturb the balance:

Strong base $BOH \rightarrow B^+ + OH^-$ (reaction virtually complete)

Weak acid $HA \rightleftharpoons H^+ + A^-$ (reaction reversible)

Therefore, for the salt AB formed, there will be the following interaction with water:

$$B^+ + A^- + H_2O \rightleftharpoons HA + B^+ + OH^-$$

that is, the delicate balance of the water dissociation is disturbed, with the result that the concentration of the hydroxide ion exceeds that of the hydrogen ion, and the aqueous solution, although containing equivalent quantities of acid and base, has an alkaline reaction.

Similarly, aqueous solutions of salts of weak base and strong acid have an acidic reaction.

This phenomenon is well illustrated by potassium cyanide and ammonium chloride solutions. The former, being a salt of the weak acid, hydrogen

* By analogy with pH, pK = −log K

cyanide, has a pH of 11·1 when in decimolar solution, whilst the latter, the salt of the weak base ammonia, in decimolar solution has a pH of 5.1.

Buffer solutions

The pH of a solution containing a weak acid or base and one of its salts can remain remarkably constant despite the addition of relatively large quantities of acid or alkali. In the case of ammonium hydroxide containing ammonium chloride, the large ammonium ion concentration resulting from the addition of the fully dissociated salt suppresses the ionization of the ammonium hydroxide (since K must remain constant):

$$NH_4OH \rightleftharpoons NH_4^+ + OH^-$$

$$K_b = \frac{[NH_4^+][OH^-]}{[NH_4OH]}$$

This phenomenon is known as the 'common ion' effect.

If acid is added to this system, the hydrogen ions will be removed by hydroxide ions to form water, and the ammonium hydroxide will dissociate further (and it has considerable potential for further dissociation, so much will its dissociation have been suppressed in the first place) to restore the equilibrium.

Hydroxide ions, on the other hand, are removed by combination with ammonium ions; at the same time, the ionization of ammonium hydroxide is suppressed. It can be seen, therefore, that systems like the above preserve the *status quo*; they are known as *buffer solutions*. Buffer solutions are very important in the metabolic processes of living organisms for the maintenance of a constant pH and for ensuring that the various chemical activities continue unimpaired. Body fluids have a pH value of 7.3 ~ 7.4 and a fall to 7.2 (acidosis) causes serious disturbances of cell function, whilst if the pH falls to 7.0 the ensuing catastrophic impairment produces coma and death. Alkalosis creates similar disturbances if the pH becomes in excess of 7.6. Buffer systems in the body comprise the plasma proteins, haemoglobin and proteins in tissue cells which all function because of their amphoteric nature, the sodium hydrogen carbonate–carbonic acid system in plasma and tissue fluids and the dihydrogen phosphate ($H_2PO_4^-$)–hydrogen phosphate (HPO_4^{2-}) system also present in blood and tissue fluids.

Since well-chosen buffer solutions provide a constant pH, they can be used, in conjunction with indicators, for comparison with unknown solutions.

The pH can be calculated as follows. Consider the equilibrium for a weak acid HA

$$HA \rightleftharpoons H^+ + A^-$$

$$K_a = \frac{[H^+][A^-]}{[HA]}$$

and

$$\log K_a = \log [H^+] + \log \frac{[A^-]}{[HA]}$$

Therefore, by rearranging and converting to pH

$$pH = pK_a + \log\frac{[A^-]}{[HA]}$$

If now a salt with the common ion A^- is added, it will provide virtually the entire quantity of this anion, so that $[A^-] \sim [\text{salt}]$. Furthermore, the ionization of the acid will be so suppressed that $[HA] \sim [\text{acid}]$.

The above equation can now be rewritten:

$$pH = pK_a + \log\frac{[\text{salt}]}{[\text{acid}]}$$

For example, if 50 cm³ of 0.15M sodium ethanoate solution is mixed with 20 cm³ of 0.1M ethanoic acid, for which $K_a = 1.7 \times 10^{-5}$ mol dm^{-3}

$$pH = -\log 1.7 \times 10^{-5} + \log\frac{50 \times 0.15}{20 \times 0.1}$$

$$= 4.76 + \log 3.75 = 5.33$$

Solubility products

It has been seen earlier that continuing dilution of an aqueous solution of a weak electrolyte encourages ionization. The saturated solution of a sparingly soluble salt (which will inevitably be of extremely high dilution) will consequently contain only ions and no undissociated molecules. In other words, the equilibrium that exists is between the ions in solution and the solid in contact with the solution:

$$AB \rightleftharpoons A^+ + B^-$$
solid   solution

The active mass of the solid is assumed to be constant, and use of the mass action equation gives

$$K = \frac{[A^+][B^-]}{[AB]_{\text{solid}}} \qquad \text{or} \qquad K' = S = [A^+][B^-]$$

where S is the *solubility product* (a term reminiscent of the ionic product of water).

The solubility of a sparingly soluble salt is therefore seen as an ionic rather than a molecular phenomenon; addition of either ion to a saturated solution of the salt will result in the solubility product being momentarily exceeded, with precipitation of the solid instantly following to restore the constant value for the solubility product.

Example. The solubility of silver chloride is 1·5 mg dm^{-3}. Calculate the solubility product and hence its solubility in M sodium chloride.

Rel. mol. mass of silver chloride = 143.5, therefore the solubility, in terms of mol dm^{-3}, will be $1.5 \times 10^{-3}/143.5 \sim 1 \times 10^{-5}$. One molecule of silver chloride gives one ion each of silver and chloride, so

$[Ag^+] = [Cl^-] = 1 \times 10^{-5}$ mol dm^{-3}

and solubility product $= [Ag^+][Cl^-] = (1 \times 10^{-5})^2 = 1 \times 10^{-10}$ mol^2 dm^{-6}

A solution of sodium chloride having a concentration of 1 mol dm^{-3}, which will be fully dissociated, will contain 1 mol dm^{-3} of both sodium and chloride. Hence the concentration of the common chloride ion will be 1 (neglecting the minute contribution from the silver chloride). As

$[Ag^+][Cl^-] = 1 \times 10^{-10}$ mol^2 dm^{-6}

$[Ag^+] = \dfrac{1 \times 10^{-10}}{1}$ mol dm^{-3}

that is, the amount of silver chloride now dissolved

$= 1 \times 10^{-10}$ mol dm^{-3}

$= 1 \times 10^{-10} \times 143.5$ g dm^{-3}

$= 1.435 \times 10^{-8}$ g dm^{-3}

Thus, the effect of sodium chloride solution of concentration 1 mol dm^{-3} is to reduce the solubility of silver chloride one hundred thousand-fold.

We have seen how the dissociation of water and the ionic product play a fundamental part in the equilibria of aqueous solutions. Let us now study the role of the solubility product.

Precipitation of sulphides

Sulphides are on the whole insoluble substances and are therefore readily precipitated by the passage of hydrogen sulphide through a solution of a salt of the metal. We can now say that precipitation will occur only when the solubility product has been exceeded, that is, when the concentrations of metallic ion and sulphide ion are high enough. If the latter can be reduced sufficiently, then there is the possibility that precipitation of the more soluble sulphide will be prevented and some separation of metals thereby effected.

This possibility is realized very simply by passing hydrogen sulphide into an acidified solution of the metal ions. The large concentration of hydrogen ions from the acid:

$$HCl \rightarrow H^+ + Cl^-$$

will, by the common ion effect, suppress the ionization of the hydrogen sulphide:

$$H_2S \rightleftharpoons H^+ + HS^-$$

$$HS^- \rightleftharpoons H^+ + S^{2-}$$

The presence of 1 mol dm^{-3} of H$^+$, in fact, reduces the value of $[S^{2-}]$ from about 10^{-10} to about 10^{-21} mol dm^{-3}. As the concentration of the metal ion is usually in the range $10^{-2} \sim 10^{-1}$ mol dm^{-3}, precipitation will only be effected if the solubility product for the sulphide is numerically less than about 10^{-22}. *Table 7.3* indicates which particular metals are most likely to be

Table 7.3 Some solubility products

Mercury(II) sulphide	Nickel(II) sulphide
(4×10^{-53}) $mol^2 dm^{-6}$	(1×10^{-19}) $mol^2\ dm^{-6}$
Copper(II) sulphide	Cobalt(II) sulphide
(8×10^{-45}) $mol^2\ dm^{-6}$	(2×10^{-20}) $mol^2\ dm^{-6}$
Cadmium sulphide	Iron(III) hydroxide
(4×10^{-29}) $mol^2\ dm^{-6}$	(1×10^{-36}) $mol^4\ dm^{-12}$
Lead(II) sulphide	Chromium(III) hydroxide
(4×10^{-28}) $mol^2\ dm^{-6}$	(3×10^{-29}) $mol^4\ dm^{-12}$
Zinc sulphide	Aluminium hydroxide
(1×10^{-20}) $mol^2\ dm^{-6}$	(8×10^{-23}) $mol^4\ dm^{-12}$
Manganese(II) sulphide	Manganese(II) hydroxide
(1×10^{-15}) $mol^2\ dm^{-6}$	(4×10^{-14}) $mol^3\ dm^{-9}$

precipitated in acid solution, i.e., in Group 2 in the conventional scheme of qualitative analysis.

More soluble sulphides will be precipitated more readily using alkaline rather than acid solution. Hydroxide ions from the alkali will remove hydrogen ions (to form largely undissociated water) and thus disturb the equilibrium in favour of the formation of more sulphide ions

$$H_2S \rightleftharpoons H^+ + HS^- \qquad\qquad HS^- \rightleftharpoons H^+ + S^{2-}$$
$$\downarrow OH^- \qquad\qquad\qquad\qquad\qquad \downarrow OH^-$$
$$H_2O \qquad\qquad\qquad\qquad\qquad\quad H_2O$$

Precipitation of hydroxides

In Group 3 of conventional qualitative analysis, ammonium chloride and ammonium hydroxide are added. Here, the large concentration of ammonium ions from the ammonium chloride prevents the weak base, ammonium hydroxide, from ionizing sufficiently for the solubility products of the hydroxides, other than those of iron(III), aluminium and chromium(III), to be exceeded (*Table 7.3*).

$$NH_4Cl \rightarrow NH_4^+ + Cl^-$$

$$NH_4OH \rightleftharpoons NH_4^+ + OH^-$$

$$\frac{[NH_4^+][OH^-]}{[NH_4OH]} = K_b$$

i.e., as $[NH_4^+]$ is increased by the addition of ammonium chloride, $[OH^-]$ is decreased.

Solubility of salts of weak acids

Many salts of weak acids which are virtually insoluble in water dissolve readily in acids. This, too, can be explained in terms of the law of mass action: hydrogen ions from the acid combine with the anions of the salt to form undissociated molecules of the weak acid, thus disturbing the equilibrium in favour of further dissolution of the salt; e.g., calcium phosphate

$$Ca_3(PO_4)_2 \rightleftharpoons 3Ca^{2+} + 2PO_4^{3-}$$

$$3H^+ + PO_4^{3-} \rightleftharpoons H_3PO_4$$

Non-aqueous systems

By now it should be clear that behind all reactions in aqueous systems lies the equilibrium

$$H_2O \rightleftharpoons H^+ + OH^-$$

The proton so formed has, by virtue of its small size, a considerable polarizing force and attacks a further molecule of water to form the oxonium ion, so that the complete self-ionization of water can be represented as

$$2H_2O \rightleftharpoons H_3O + OH^-$$

Most common chemical reactions revolve around this ionic equilibrium for water, and it is therefore not surprising to find that several concepts have stemmed from this source. Thus, an acid has been defined by Arrhenius as a substance that provides hydrogen ions in solution and an alkali as one providing hydroxide ions; neutralization is seen as combination of hydrogen and hydroxide ions to form a molecule of solvent, i.e., the reaction

$$HCl + NaOH \rightarrow NaCl + H_2O$$

is essentially

$$H_3O^+ + OH^- \rightarrow 2H_2O$$

By analogy with this, in liquid ammonia, which ionizes:

$$2NH_3 \rightleftharpoons NH_4^+ \quad + NH_2^-$$

ammonium   amide

ion   ion

ammonium compounds act as acids, and amides (or even imides or nitrides) as bases. They can be titrated together and the result is the formation of solvent, in this case ammonia:

$$NH_4Cl + NaNH_2 \rightarrow NaCl + 2NH_3 \qquad (NH_4^+ + NH_2^- \rightarrow 2NH_3)$$

$$2NH_4NO_3 + PbNH \rightarrow Pb(NO_3)_2 + 3NH_3 \qquad (2NH_4^+ + NH^{2-} \rightarrow 3NH_3)$$

$$3NH_4Br + Li_3N \rightarrow 3LiBr + 4NH_3 \qquad (3NH_4^+ + N^{3-} \rightarrow 4NH_3)$$

Nor does the similarity end there: amphoteric substances, e.g., zinc amide, exist

$$Zn(NH_2)_2 + 2KNH_2 \rightarrow K_2Zn(NH_2)_4 \ [cf. \ Zn(OH)_2 + 2KOH \rightarrow K_2Zn(OH)_4]$$

$$Zn(NH_2)_2 + 2NH_4Cl \rightarrow ZnCl_2 + 4NH_3$$
$$[cf. \ Zn(OH)_2 + 2HCl \rightarrow ZnCl_2 + 2H_2O]$$

and liquid ammonia can ammonolyse salts in a manner similar to that in which water can hydrolyse:

$$SbCl_3 + NH_3 \rightarrow SbN + 3HCl \ [cf. \ SbCl_3 + H_2O \rightarrow SbOCl + 2HCl]$$

Ammoniates also exist and can be compared with hydrates, e.g., $Co(NH_3)_6Cl_3$.

Similar situations are believed to exist in other non-aqueous media. For example, dinitrogen tetroxide self-ionizes

$$N_2O_4 \rightleftharpoons NO^+ + NO_3^-$$
$$\qquad\quad \text{nitrosyl} \quad \text{nitrate}$$

In this particular system, nitrosyl compounds will behave as acids and nitrates as bases, neutralization taking place between them.

As a great number of reactions have been studied in a wider variety of media, so the term 'acid' has been extended. Brønsted and Lowry defined an acid as a proton-donor and a base as a proton-acceptor; Lewis defined an acid as an electron-pair acceptor and a base as an electron-pair donor. It follows that every acid has its *conjugate base*, and every base its *conjugate acid*:

$$HNO_3 \qquad\qquad \rightleftharpoons \quad H^+ \quad + \quad NO_3^-$$

acid as proton donor   proton   conjugate base, for it can accept a proton to form the original acid

$$BF_3 \qquad\qquad\qquad + \quad :F^- \quad \rightleftharpoons \quad BF_4^-$$

acid as electron-pair acceptor   conjugate base

Protophilic solvents

Certain compounds containing oxygen and nitrogen can, by virtue of the lone pair of electrons on that atom, accept protons. They therefore function as Lewis bases and enhance the acidity of weak acids. Examples are ethers and amines:

An ether
$$\begin{array}{c} CH_3 \\ \diagdown \\ \quad O: + H-A \rightleftharpoons \\ \diagup \\ CH_3 \end{array} \left[\begin{array}{c} CH_3 \\ \diagdown \\ \quad O-H \\ \diagup \\ CH_3 \end{array}\right]^+ A^-$$

An amine
$$\begin{array}{c} CH_3 \\ | \\ CH_3-N: + H-A \rightleftharpoons \\ | \\ CH_3 \end{array} \left[\begin{array}{c} CH_3 \\ | \\ CH_3-N-H \\ | \\ CH_3 \end{array}\right]^+ A^-$$

These protophilic solvents can actually be used in the titration of a strong base against an acid so weak that it would be difficult to titrate in any other way. (Of course, the basic solvent will reduce the strength of the base, but if a very strong base is used the effect will be of no consequence.) For example, phenols can be titrated with methoxides in the presence of 1,2-diaminoethane as solvent

$$\text{C}_6\text{H}_5\text{-OH} + H_2NCH_2CH_2NH_2 \rightleftharpoons \text{C}_6\text{H}_5\text{-O}^- + H_3\overset{+}{N}CH_2CH_2NH_2$$

$$H_3\overset{+}{N}CH_2CH_2NH_2 + CH_3O^- \rightleftharpoons CH_3OH + H_2NCH_2CH_2NH_2$$

i.e., the overall reaction is

$$\text{(C}_6\text{H}_5)\text{—OH} + \text{CH}_3\text{O}^- \rightleftharpoons \text{(C}_6\text{H}_5)\text{—O}^- + \text{CH}_3\text{OH}$$

acid   base

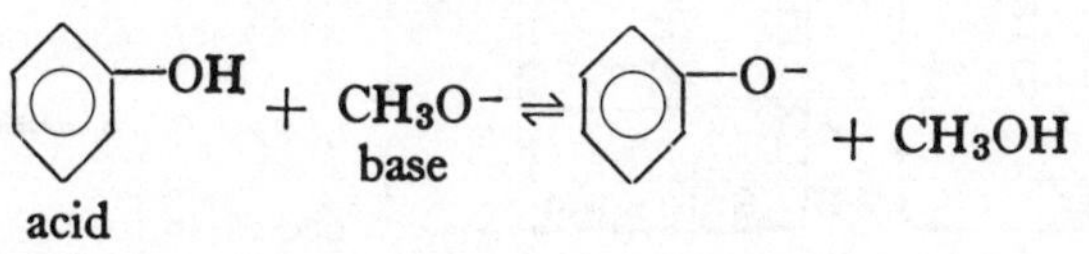

Protogenic solvents

These are solvents with the opposite quality to the protophilic; that is, they are proton donors; e.g., ethanoic acid, CH_3COOH

$$CH_3COOH \rightleftharpoons CH_3COO^- + H^+$$

The strength of an acid will decrease if it is transferred from a weaker to a stronger protogenic solvent. For example, nitric acid is still acidic, albeit more weakly, in ethanoic acid as well as in water because the solvent is still able to accept protons from it:

$$HNO_3 + CH_3COOH \rightleftharpoons CH_3COOH_2^+ + NO_3^-$$

but it is basic in sulphuric acid, because the latter is the stronger proton donor:

$$HNO_3 + H_2SO_4 \rightleftharpoons H_2NO_3^+ + HSO_4^-$$

The strengths of strong acids in a solvent such as water cannot be compared because dissociation is virtually complete in all cases, but by using a protogenic solvent, the strengths will be reduced (by different amounts), and a comparison becomes possible.

It follows that protogenic solvents, as well as suppressing the dissociation of acids, amplify the strengths of bases; weak bases can therefore be titrated against strong acids by using such a solvent; e.g., amines against chloric(VII) acid, with ethanoic acid as solvent

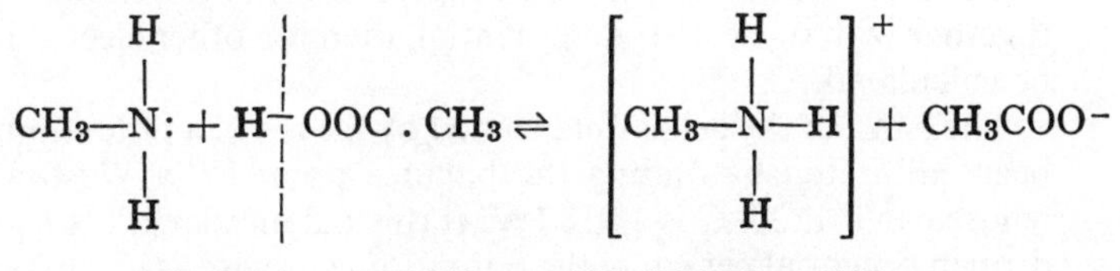

$$\begin{array}{c} \text{H} \\ | \\ \text{CH}_3\text{—N:} \\ | \\ \text{H} \end{array} + \text{H—OOC.CH}_3 \rightleftharpoons \left[\begin{array}{c} \text{H} \\ | \\ \text{CH}_3\text{—N—H} \\ | \\ \text{H} \end{array}\right]^+ + \text{CH}_3\text{COO}^-$$

$$CH_3COO^- + HClO_4 \rightleftharpoons CH_3COOH + ClO_4^-$$

The overall reaction is

$$\begin{array}{c} \text{H} \\ | \\ \text{CH}_3\text{—N:} \\ | \\ \text{H} \end{array} + \text{HClO}_4 \rightleftharpoons \left[\begin{array}{c} \text{H} \\ | \\ \text{CH}_3\text{—N—H} \\ | \\ \text{H} \end{array}\right]^+ + \text{ClO}_4^-$$

base   acid

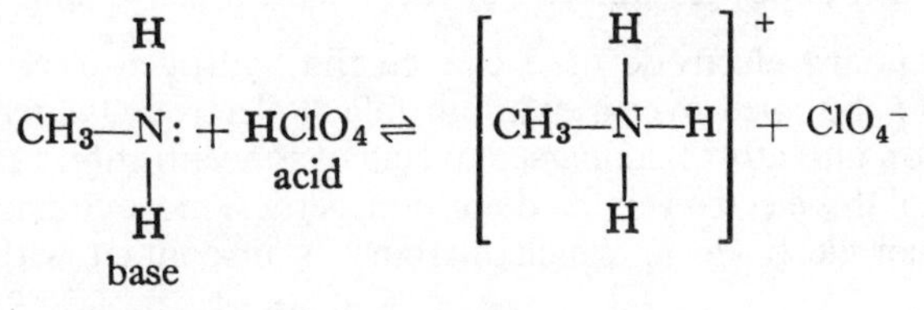

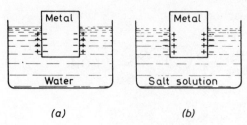

Figure 7.9 Electrical double layers. In (a) the metal is in contact with water and reaction

$$M + xH_2O - ne^- \rightleftharpoons M^{n+}(H_2O)_x$$

occurs. In (b) the metal is in contact with a solution of one of its salts and the reaction is

$$M(H_2O)_x^{n+} + ne^- \rightleftharpoons M^+ + xH_2O$$

Electrode potentials

When most metals are placed in water, the tendency which they have to form positive ions results in the metal becoming very slightly negatively charged with respect to the water. The negative charge residing on the metal attracts the positive metal ions towards it, giving rise to a 'double layer' (*Figure 7.9a*).

If the metal is placed in a concentrated solution of its ions, there is an opposing tendency for positive ions to leave the solution and to be deposited on the metal; if this 'condensing pressure' of the solution exceeds the 'solution pressure' of the metal, the result will be that the metal acquires a positive charge. This will attract negative ions from solution, and again a double layer will be formed (*Figure 7.9b*).

The actual potential between the metal and the solution will thus depend on two factors: the tendency of the metal to form ions (the less the electronegativity the greater the tendency to ionize) and the concentration of the metal ions in solution. The *standard electrode potential* is the potential of a metal in a solution of its ions, of concentration 1 mol dm^{-3}, compared with that of a hydrogen electrode.

Measurement of electrode potentials

The e.m.f. of a cell can be regarded as the potential difference between the two electrodes comprising the cell. Therefore, if the e.m.f. of the cell is known, together with one electrode potential, then the other electrode potential can be calculated.

The e.m.f. of the cell is determined by means of a potentiometer which has been calibrated by finding the balance point for a Weston standard cell, whose e.m.f. at 25 °C is 1.018 1 V. If this balance length is L_1 cm, then e, the drop in potential per cm of the potentiometer wire, is given by the expression

$$e = \frac{1.018\ 1}{L_1}\ V\ cm^{-1}$$

The e.m.f. of the cell will then be eL_2, where L_2 is the new balance point, using the cell under test instead of the standard cell (*Figure 7.10*).

The standard electrode used can be the hydrogen electrode, in which hydrogen is bubbled over a platinum black electrode (which catalyses the dissociation into atoms) immersed in acid of concentration 1 mol dm^{-3}. The potential of this electrode is, by definition, zero. A more common electrode is the calomel electrode in which mercury is in contact with dimercury(I)

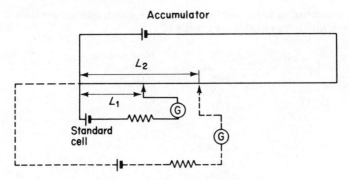

Figure 7.10 Circuit for measurement of electrode potentials

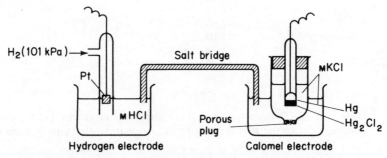

Figure 7.11 The calibration of a calomel electrode

chloride; the potential of this electrode can be calibrated against the hydrogen electrode and is found to be 0.242 V at 25 °C.

In the measurement of electrode potentials, two electrode half-cells are connected by a salt bridge which permits the flow of ions to complete the circuit (*Figure 7.11*).

If the calomel electrode is combined with a half-cell consisting of a metal electrode and an electrolyte containing the metal ions, then the potential difference will be given by

$$\text{e.m.f.} = E_M - E_{cal}$$

where E_M is the potential of the electrode in question.
The cell shown in *Figure 7.11* can be represented as:

$$\text{Pt,H}_2 \left|\begin{matrix}\text{HCl}\\ \text{(M)}\end{matrix}\right|\left|\begin{matrix}\text{KCl}\\ \text{(M)}\end{matrix}\right|\text{Hg}_2\text{Cl}_2, \text{Hg, Pt} \qquad E = +0.242 \text{ V}$$

Representation of cells and their sign convention

A typical cell can be written as

$$\text{Metal 1} \mid \text{Ion 1} \parallel \text{Ion 2} \mid \text{Metal 2} \qquad E = x \text{ V} \tag{1}$$

e.g.,

$$\text{Fe} \mid \text{Fe}^{2+} \parallel \text{Cu}^{2+} \mid \text{Cu} \qquad |E = +0.78 \text{ V} \tag{2}$$

Table 7.4 The electrochemical series; standard electrode potentials $E^{\ominus}$ (at 25 °C, referred to a standard hydrogen electrode)

Metals $[M^{n+}(aq) + ne^{-} \rightarrow M]$	$E^{\ominus}/V$	Non-metals $[X + ne^{-} \rightarrow X^{n-}(aq)]$	$E^{\ominus}/V$
Li^{+}/Li	-3.02	OH^{-}/O_2	$+0.40$
K^{+}/K	-2.92	$2I^{-}/I_2$	$+0.54$
Ba^{2+}/Ba	-2.90	$2Br^{-}/Br_2$	$+1.07$
Ca^{2+}/Ca	-2.87	$2Cl^{-}/Cl_2$	$+1.36$
Na^{+}/Na	-2.71	$2F^{-}/F_2$	$+2.87$
Mg^{2+}/Mg	-2.38		
Al^{3+}/Al	-1.67		
Zn^{2+}/Zn	-0.76		
Fe^{2+}/Fe	-0.44		
Sn^{2+}/Sn	-0.14		
Pb^{2+}/Pb	-0.13		
Fe^{3+}/Fe	-0.04		
$(2H^{+}/H_2$	0.00		
Cu^{2+}/Cu	$+0.34$		
Cu^{+}/Cu	$+0.52$		
Ag^{+}/Ag	$+0.80$		

It is a *convention* that the right-hand electrode is x volts positive with respect to the left. It is a further convention that this implies a cell reaction

$$\text{Metal } 1 + \text{Ion } 2 \rightarrow \text{Ion } 1 + \text{Metal } 2 \tag{3}$$

e.g.,

$$Fe + Cu^{2+} \rightarrow Fe^{2+} + Cu \qquad E = +0.72 \text{ V} \tag{4}$$

If we consider reaction (4), we can regard it as made up of

$$Fe = 2e^{-} + Fe^{2+} \qquad E = y \text{ V} \tag{5}$$

$$Cu^{2+} + 2e^{-} = Cu \qquad E = z \text{ V} \tag{6}$$

$$Fe + Cu^{2+} = Fe^{2+} + Cu \qquad E = (y+z) \text{ V}$$

It will be noticed that a positive cell e.m.f. is associated with the spontaneous direction of reaction. It would clearly be useful if we could have available a table of values such as y and z. This is done by *defining* the e.m.f. for the reaction

$$H^{+} + e^{-} \rightarrow \tfrac{1}{2}H_2 \text{ (or its reverse)} \tag{7}$$

as zero, or, as defining the standard electrode potential of the hydrogen electrode as zero, and measuring standard electrode potentials against it. The values in *Table 7.4* imply cells such as

$$\begin{array}{c|c||c|c} \text{Pt, H}_2 & H^{+} & M^{+} & M \\ 101 \text{ kPa} & \text{(M)} & \text{(M)} & \end{array} \quad E_M \text{ for metals} \tag{8}$$

$$\begin{array}{c|c||c|c} \text{Pt, H}_2 & H^{+} & Cl^{-} & \tfrac{1}{2}Cl_2, \text{ Pt} \\ 101 \text{ kPa} & \text{(M)} & \text{(M)} & \end{array} \quad E_{Cl} \text{ for a non-metal}$$

Looking at the diagram (8), we find that it means a reaction

$$\tfrac{1}{2}H_2 + M^+ = H^+ + M \qquad E \tag{9}$$

which can be broken down into two parts, with $E = E_M$:

$$\tfrac{1}{2}H_2 = H^+ + e^- \qquad E_H = \text{zero} \tag{10}$$

$$M^+ + e^- = M \qquad E_M \tag{11}$$

It is in this sense that the standard electrode potentials given in *Table 7.4* are related to the reactions specified in that Table. Note that if the standard electrode potential of reaction (11) is x V, then that for the reverse reaction is $-x$ V.

Electrochemical series

The chemistry of a metal is to a large extent determined by its electronegativity (the readiness with which it attracts electrons). Thus, the arrangement of the metals in order of the standard electrode potentials is a matter of great importance. It is this arrangement which comprises the *electrochemical series* (*Table 7.4*). (Strictly speaking, the electrode potentials are the products of 'wet' systems wherein the ions of the metal are probably hydrated; for 'dry' systems, the ionization energy may be a more relevant quantity.)

The high potential of metals at the top of the series is a measure of the readiness with which they lose electrons to form positive ions in aqueous solutions. It follows that these metals are very reactive and that their compounds are correspondingly stable, so much so that reduction to the metal can often prove a difficult operation. Consequently, many of these metals have been known in the elemental state only since the last century and the work of Davy on electrolysis.

From *Table 7.4*, reactions can be constructed by algebraic addition of the separate 'half-reactions', e.g.,

$$\begin{aligned}
Zn &\rightarrow Zn^{2+} + 2e^- & E^\ominus &= +0.76 \text{ V} \\
Pb^{2+} + 2e^- &\rightarrow Pb & E^\ominus &= -0.13 \text{ V} \\
\hline
Zn + Pb^{2+} &\rightarrow Pb + Zn^{2+} & E^\ominus &= +0.63 \text{ V}
\end{aligned}$$

The fact that the resultant potential is positive is indicative of the practicality of the reaction. Indeed, it is well known that if a metal higher in the Series is placed in contact with the ions of a metal lower in it, then the reaction is spontaneous, i.e., ΔG is negative. Now, since one volt is the potential difference when one joule of work is done in transferring one coulomb of electricity across it, E joules of work are performed per coulomb for a potential difference of E volts. For a valency change of z, the number of coulombs per mole $= zF$, where F is the faraday, and so the work done $= zFE$ joules per mole. That is,

Free energy change $= \Delta G = -zFE$

It follows that, if the algebraic addition of electrode potentials gives a positive result, the free energy change for the reaction will be negative and the reaction will tend to be spontaneous (p. 132). Caution is needed here. E.m.f. is an intensive, not an extensive, property, so when constructing the equations it must be remembered that the e.m.f. is not related to mass. Thus for the processes

$$K \rightarrow K^+ + e^- \qquad E^{\ominus} = +2.92 \text{ V}$$

$$2K \rightarrow 2K^+ + 2e^- \qquad E^{\ominus} = +2.92 \text{ V}, \text{ not } 5.84 \text{ V}$$

Metals high in the Series react vigorously with cold water to give the hydroxide, metals lower down tend to react when heated in steam, whilst metals at the bottom are virtually unaffected

Cold water $2Na + 2H_2O \rightarrow 2NaOH + H_2(g)$

Steam $3Fe + 4H_2O \rightleftharpoons Fe_3O_4 + 4H_2(g)$

Metals above hydrogen react in non-oxidizing acid (unless the over-potential exceeds the electrode potential, as is the case with lead) whilst those below are usually not reactive, e.g.,

	V			V
$Mg \rightarrow Mg^{2+} + 2e^-$	$E^{\ominus} = +2.38$		$Cu \rightarrow Cu^{2+} + 2e^-$	$E^{\ominus} = -0.34$
$2H^+ + 2e^- \rightarrow H_2$	$E^{\ominus} = 0.00$		$2H^+ + 2e^- \rightarrow H_2$	$E^{\ominus} = 0.00$
$Mg + 2H^+ \rightarrow Mg^{2+} + H_2$	$E^{\ominus} = +2.38$		$Cu + 2H^+ \rightarrow Cu^{2+} + H_2$	$E^{\ominus} = -0.34$
Reaction spontaneous			*Reaction not spontaneous*	

Corrosion and sacrificial protection

An electrical cell can be constructed by using two metals of different nobility placed in a salt solution. The potential of the cell will increase as the gap between the metals in the Series widens. Always the less noble metal dissolves, and such dissolution can be regarded as a form of corrosion, although perhaps it would not always be recognized as such. But if a metal is contaminated with another, more noble, metal, the phenomenon takes on the more familiar aspect of corrosion. For example, zinc can be contaminated by momentary immersion in copper(II) sulphate solution—it then consists of countless tiny cells and when placed in dilute sulphuric acid corrodes with great rapidity with evolution of hydrogen at the surface of the copper (*Figure 7.12*).

The fact that it is the less noble metal that dissolves can be utilized in the sacrificial protection of metals. In the case of galvanized iron, iron is covered by the less noble zinc: if the coating is not complete, then the formation of electrical cells is possible, and if the surface is covered with electrolyte such as rain water (containing carbonic acid), zinc passes into solution. This is important, because it means that the metal dissolves from the bulk of the surface and not from the spot of exposed iron. On the other hand, in the case of tinned iron, it is the iron that is the less noble metal so that, if a flaw develops and a small area of iron is exposed, contact with an electrolyte this time results in the iron, and not the tin, dissolving (*Figure 7.13*). As the dissolution takes place from a small area of metal, rapid 'pitting' occurs, with the eventual appearance of a hole. Tin is used in preference to zinc in food canning because tin, being less noble than zinc, is less liable to be attacked by the acids in food juices.

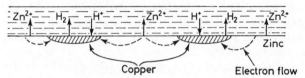

Figure 7.12 Electrolytic corrosion of zinc due to contamination with copper

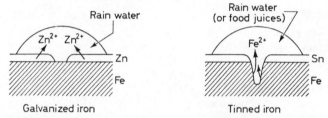

Figure 7.13 Comparison of electrolytic consequences of using zinc and tin in the coating of iron

Concentration cells

The electrode potential is a measure of the 'solution pressure' of the substance of which the electrode is made and the 'condensing pressure' of the ions of the substance present in the solution. The potential will therefore depend on the concentration of these ions, and so a potential difference will exist between two electrodes of the same material in solutions of different concentrations. This is the principle of the concentration cell. It can be shown that the e.m.f. of such a cell is given by

$$E = \frac{RT}{zF} \ln \frac{c_1}{c_2}$$

where z = charge on the ions, c_1, c_2 = concentrations ($c_1 > c_2$). Using the value $R = 8.314\,\mathrm{J\,K^{-1}\,mol^{-1}}$ and $F = 96.5\,\mathrm{kC\,mol^{-1}}$, and a temperature of 298 K, $RT/F = 0.025\,7$ V.

Measurements of pH can be carried out by using a concentration cell, for example with the standard hydrogen electrode combined with a hydrogen electrode dipping into a solution whose pH is being determined. As c_2, the concentration of the standard electrode solution, is 1, the above expression simplifies to

$$E = (RT/F) \ln c_{H^+}$$

$$pH = -\ln c_{H^+} = -FE/RT$$

so that, in the pH meter, pH can be calibrated in terms of the potential developed.

Reduction–oxidation potentials

An oxidizing agent removes electrons from a system whilst a reducing agent adds electrons. An inert electrode will therefore tend to acquire a positive charge when in contact with an oxidizing agent and a negative charge when in contact with a reducing agent. The resultant potential is known as the *redox potential. Table 7.5* shows the potential acquired by a platinum electrode when placed in an equimolecular solution of the two ions.

Table 7.5 Redox potentials for the system $M^{m+}(aq) + ne^- \rightarrow M^{(m-n)+}(aq)$

System	Potential/V	System	Potential/V
Co^{3+}/Co^{2+}	+1.82	Cu^{2+}/Cu^+	+0.17
Ce^{4+}/Ce^{3+}	+1.57	H^+/H	0.00
$MnO_4^-, H^+/Mn^{2+}$	+1.52	Ti^{4+}/Ti^{3+}	−0.06
$Cr_2O_7^{2-}, H^+/2Cr^{3+}$	+1.36	V^{3+}/V^{2+}	−0.2
Fe^{3+}/Fe^{2+}	+0.76	Cr^{3+}/Cr^{2+}	−0.4
$Fe(CN)_6^{3-}/Fe(CN)_6^{4-}$	+0.36		

Systems high in the series oxidize the electrode more strongly than those below; it therefore follows that if two systems are brought into contact, the one higher will oxidize the one below (i.e., the one below will reduce the one above). For example, *nascent* hydrogen will reduce iron(III) to iron(II)

$$Fe^{3+} + e^- \rightarrow Fe^{2+} \qquad\qquad E^\ominus = 0.76 \text{ V}$$

$$H - e^- \rightarrow H^+ \qquad\qquad E^\ominus = 0.00 \text{ V}$$

$$Fe^{3+} + H \rightarrow Fe^{2+} + H^+ \qquad\qquad E^\ominus = 0.76 \text{ V}$$

Acidified dichromate oxidizes iron(II) to iron(III)

$$Cr_2O_7^{2-} + 14H^+ + 6e^- \rightarrow 2Cr^{3+} + 7H_2O \qquad E^\ominus = 1.36 \text{ V}$$

$$6(Fe^{2+} \rightarrow Fe^{3+} + e^-) \qquad E^\ominus = -0.76 \text{ V}$$

$$Cr_2O_7^{2-} + 14H^+ + 6Fe^{2+} \rightarrow 2Cr^{3+} + 6Fe^{3+} + 7H_2O \qquad E^\ominus = 0.60 \text{ V}$$

Summary

The conductance of a solution of an electrolyte is the sum of two parts: the conductance of the anion plus that of the cation. These ion conductances are, in turn, dependent on the speed or mobility of the ions: the faster the ion moves, the greater its transport number or the fraction of the total current carried by it.

During electrolysis, the operation requiring least energy is normally carried out. Factors favouring discharge of an ion include high concentration, instability and exothermic reaction with the electrode after discharge. On the other hand, the existence of a high overpotential, high stability and low concentration reduce the possibility of discharge of that ion. One further possibility to consider is the dissolving of the anode, the electrons being released by this process replacing those that would have otherwise been lost by anions.

Electrolysis is dealt with quantitatively by Faraday's laws. Electrolytes can be divided into two classes: strong electrolytes, completely ionized even in the solid state, and weak electrolytes, the ionization of which is increased by dilution.

Application of the law of mass action to weak electrolytes gives rise to the notion of a dissociation constant, which is related to the degree of dissociation in the case of a binary electrolyte by the expression $K = \alpha^2/[(1-\alpha)V]$.

The law of mass action provides an explanation for phenomena such as the common ion effect, solubility product and hydrolysis. Although these terms are normally used in aqueous systems (where the ionic product is 10^{-14} mol^2 dm^{-6} and, therefore, pH + pOH = 14), there is no reason why they should not also be applied to non-aqueous systems; fruitful comparisons can be drawn between reactions in water and in solvents such as liquid ammonia.

The potential acquired by a metal when placed in a solution of its ions of concentration 1 mol dm^{-3}, is a measure of its readiness to form hydrated ions. The electrochemical series is an arrangement of elements in order of their electrode potentials and provides a means of comparing the elements with regard to their reactivities, including their tendency to displace each other from solution; the energy liberated in such reactions can be converted directly to electricity by making two metals the electrodes of a cell. A potential is also developed when an inert electrode is placed in a solution of an ion with reducing or oxidizing properties.

Questions

(1)   Cite the evidence for the existence of ions.

(2)   Using specific examples describe electrolytic methods for:
      (a) The isolation of a metal.
      (b) The isolation of a non-metal.
      (c) The formation of an oxidizing compound.
      (d) The formation of an organic compound.

(3)   Discuss the various possibilities during the electrolysis of sodium chloride solution: (a) with carbon electrodes far apart; (b) with a mercury cathode; (c) hot and concentrated and with carbon electrodes placed close together.

(4)   The following values for the resistance of potassium ethanedioate solution of

differing dilutions have been obtained, using a conductivity cell constant 140 m^{-1}

Dilution/dm^3 mol^{-1}	8	16	32	64	128	256	512
Resistance/Ω	215	375	710	1390	2700	5220	10430

The transport number of the potassium ion in potassium ethanedioate is 0.55. If the resistance of a solution f ethanedioic acid of concentration 1/32 mol dm^{-3} in the same conductivity cell is 595 Ω, calculate the degree of dissociation and hence the dissociation constant of the acid, given that the molar conductivity of the hydrogen ion is 0.0350 Ω^{-1} m^2 mol^{-1}.

(5) A saturated solution of silver chloride solution has an electrolytic conductivity of $1.5 \times 10^{-4} \, \Omega^{-1} \, m^{-1}$. If the molar conductivity of the silver ions is 0.0055 and of chloride 0.00655 Ω^{-1} m^2 mol^{-1} at this temperature, calculate the solubility product for silver chloride and hence the solubility in sodium chloride solution of concentration 3 mol dm^{-3}.

(6) Explain the following:
(a) Aluminium hydroxide is precipitated if ammonium sulphide is added to aluminium chloride solution.
(b) Magnesium hydroxide is not precipitated if ammonium hydroxide is added to a solution of magnesium sulphate and ammonium chloride.
(c) Sodium chloride is less soluble in concentrated hydrochloric acid than in water but lead chloride is more soluble.
(d) Zinc sulphide is precipitated from an alkaline but not from an acidic solution of hydrogen sulphide whilst tin(II) sulphide is precipitated from an acidic solution.
(e) Aluminium chloride solution is acidic and sodium carbonate solution is alkaline.

(7) Write an account of the development of the terms 'acid' and 'oxidation'.

(8) (a) If the e.m.f. of a cell with zinc and lead as the two electrodes is 0.63 V, calculate the free energy change associated with the replacement of one mole of lead by zinc.
(b) Show from the values in *Table 7.4* which of the following reactions is feasible:

$$2Br^- + I_2 \rightarrow Br_2 + 2I^-$$

$$2Br^- + Cl_2 \rightarrow Br_2 + 2Cl^-$$

(9) (a) Calculate the pH of a 0.01M solution of an acid whose dissociation constant is 4×10^{-6} mol dm^{-3}.
(b) If the osmotic pressure of a solution of a compound AB_2 is 2.2 times that expected in the absence of ionization, calculate the degree of ionization.

(10) If a solution containing c mol m^{-3} of a uni-univalent electrolyte is confined between electrodes of area 1 m^2 and separated by a distance of 1 m, show that for the application of a potential difference of 1 V, the current is given by

$cLe(u_+ + u_-)$, where u_+ and u_- are the velocities (mobilities) of the cations and anions respectively under a potential gradient of $1 \ V \ m^{-1}$, L is the Avogadro constant and e the elementary charge. Show also that under these conditions the current can be equated to the electrolytic conductivity. Hence deduce a relation for the molar conductivity and show that $\Lambda_+ = Fu_+$. What further assumptions have to be made to bring these equations into line with the experimental results shown in *Figure 7.4*?

(11) 300 V d.c. was applied across a U-tube containing an agar gel of sodium sulphate and phenolphthalein coloured with a few drops of alkali. Describe what would be observed and state what quantitative values could be obtained from such an experiment.

(12) Explain the following results:

(a) Zinc blocks attached to the side of a ship's steel hull prevent corrosion.
(b) The smaller the area at which corrosion occurs, the more intense the effect.
(c) Corrosion occurs where there is a lower concentration of oxygen.

Suggest experiments to investigate these effects quantitatively.

(13) (a) Show that for transfer between two half-cells of a mole of a z-valent ion under a potential difference of E volts, the free energy change is given by:

$$\Delta G = -zFE.$$

(b) The osmotic work done in diluting a solution is given by $\int \Pi dV$ (cf. *Figure 4.29*). Assuming that the solutions are dilute, substitute for V, using the general gas equation and integrate between pressures p_1 and p_2, corresponding to concentrations c_1 and c_2. Hence derive the e.m.f. for a concentration cell (p. 181).

(14) Write an account of some of the research being carried out into the development of new cells for powering an electric car.

(15) Calculate the transport number of the copper ion in copper(II) sulphate solution from the following data:

Current = 0.015 A; Time for which current flows = 17 580 s
Concentration of copper(II) sulphate solution before electrolysis = 0.1M
Volume of copper(II) sulphate solution in anode compartment = 40 cm^3
After electrolysis, titration of this solution against 0.1M sodium thiosulphate(VI) solution in the presence of excess iodide ion produced the following result:

25 cm^3 of anode solution required 29.0 cm^3 of the thiosulphate(VI) solution.

(16) Write an essay on corrosion.

(17)   The four types of ionic equilibria are: protolytic, complex ion, solubility and redox. Give one illustration of each type and describe briefly how the equilibria can be investigated.

(18)   Suggest structural formulae for $[H(H_2O)_4]^+$ and $[OH(H_2O)_3]^-$, the existence of which are to be found in strong aqueous solutions.

(19)   Show that the equilibrium constant for the reaction:

$$PbSO_4(s) + 2NaI(aq) \rightleftharpoons PbI_2(s) + Na_2SO_4(aq)$$

is equal to the ratio of the solubility products of lead sulphate and lead iodide.

(20)   (a) Sodium hydroxide solution was added to 25 cm^3 of an aqueous solution of ethanoic acid of concentration 0.1 mol dm^{-3}, and the pH was measured at intervals giving the following results:

cm³ of NaOH	0.0	4.0	8.0	12.0	16.0	18.0	20.0	22.0	22.5	23.0	23.5	28.0
pH	2.8	3.5	4.0	4.5	5.1	5.5	5.8	7.0	9.0	10.5	11.0	12.3

Plot these results and use this graph to determine the following:

(i) The pH at the end-point; account for this value.
(ii) The concentration in mol dm^{-3} of the sodium hydroxide solution.
(iii) The dissociation constant, K_a, of ethanoic acid.

(b) Name an indicator suitable for this titration, and give its colour in acid and alkaline solutions.

(c) Explain why ethanoic acid is called a *weak* acid.

[A.E.B. 1977]

(21)   (a) State Faraday's laws of electrolysis. Describe an experiment to illustrate the second law.
(b) Calculate the time in minutes necessary for a current of 10 A to deposit 1 g of copper from an aqueous solution of copper(II) sulphate.
(c) Show why the ratio of the masses of copper and sodium deposited under the appropriate conditions by the same quantity of electricity is 1.38.
(d) Calculate the charge on an electron given that the Avogadro constant is $6.02 \times 10^{23} \text{ mol}^{-1}$ and the Faraday constant is $96\,500 \text{ C/mol}^{-1}$.

8 Genesis, distribution and extraction of the elements

Genesis • Distribution • Extraction • Principles governing the extraction of the elements – chemical reduction, thermal decomposition, electrolytic reduction and oxidation

Genesis

Any theory about the evolution of the chemical elements must take account of the following facts:

(1) As far as is known, hydrogen and helium comprise about 76 and 23 per cent, respectively, of the total mass of the universe.
(2) Thermonuclear reactions only take place at very high temperatures.
(3) The 'iron group' of elements are about 10 000 times more plentiful than their neighbours and possess very stable nuclei.

Nonetheless there is no shortage of theories of the universe, and the one to be described is not without its competitors.

At some time, variously estimated to be between 12 and 20 thousand million years ago, the material of the universe came into being in a 'big bang', and consisted shortly thereafter of hydrogen and helium atoms in a 'ball' which has been expanding, according to Hubble, ever since. (Another theory suggests that hydrogen atoms are being created continuously.)

Gravitational attraction between hydrogen atoms produces a volume contraction to form a star and the conversion of potential energy into kinetic energy. When a temperature of about 5×10^6 K is reached, hydrogen atoms fuse together to form helium by some such sequence as

$$2 {}_1^1\text{H} \rightarrow {}_1^2\text{H} + {}_{+1}^0\text{e}^+$$

$${}_1^2\text{H} + 2 {}_1^1\text{H} \rightarrow {}_2^4\text{He} + {}_{+1}^0\text{e}^+$$

Cooling occurs as the hydrogen is used up, but this is followed by a further contraction and a resultant increase in the central temperature to about 100×10^6 K, accompanied by the formation of a much expanded envelope of gas. This situation is characteristic of the stars known as 'Red Giants', in which the following reactions probably occur

$$2 {}_2^4\text{He} \rightarrow {}_4^8\text{Be} \xrightarrow{{}_2^4\text{He}} {}_6^{12}\text{C} \xrightarrow{{}_2^4\text{He}} {}_8^{16}\text{O} \xrightarrow{{}_2^4\text{He}} {}_{10}^{20}\text{Ne}$$

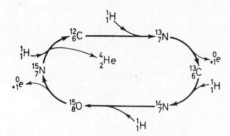

Figure 8.1 The carbon–nitrogen cycle in the fusion of hydrogen to helium

When most of the helium is used up, contraction occurs again and sufficient heat is generated to permit the products of the above reaction to interact, with the formation of elements up to and including iron and its neighbours, by which time a temperature of about 5×10^9 K has been reached (a 'White Dwarf'). It is the stability of the iron group of elements which is responsible for terminating this sequence of events, and with it the life of a 'first-generation' star.

Disintegration of such a star results in the mixing of the various elements with the debris of other, possibly younger, stars and with interstellar hydrogen ('Supernovae'). The operations described above can now be repeated with the 'second-generation' star that is formed from coalescence of this mixed material into a dense hot mass. This time, however, helium can be formed by the carbon–nitrogen cycle (*Figure 8.1*).

The ^{13}C isotope can also capture a proton and lose a neutron. Neutrons so released can then be captured by other elements, and it is believed that by neutron capture the stable iron group can be converted into the heavier elements. Such is one possible theory for the genesis of the elements in our own Sun. Elements of large relative atomic mass can also be produced by the dense neutron fluxes which occur during the stellar explosions called supernovae, e.g.,

$$^{186}_{74}W + {}^{1}_{0}n \rightarrow {}^{187}_{74}W \rightarrow {}^{187}_{75}Re + {}^{0}_{-1}e$$

$$^{187}_{75}Re + {}^{1}_{0}n \rightarrow {}^{188}_{75}Re \rightarrow {}^{189}_{76}Os + {}^{0}_{-1}e \quad \text{etc.}$$

The nucleosynthesis of most of the heavy elements has probably occurred in this way.

Distribution

For some reason the cosmic dust some 6×10^9 years ago condensed sufficiently to form a rotating sphere (our future Sun) accompanied by a rotating disc of matter which gave rise to the rest of the solar system. As the sphere contracted so the temperature rose until hydrogen fusion could occur within the Sun. Eventually cooling occurred and thermonuclear reactions in our own planet ceased. As cooling of the Earth continued, less volatile elements tended to liquefy and chemical bonds assumed some measure of permanence. At the same time, the small gravitational attraction due to the small size of the Earth allowed the lighter materials to escape into space. Although some hydrogen may have been absorbed into the molten metallic

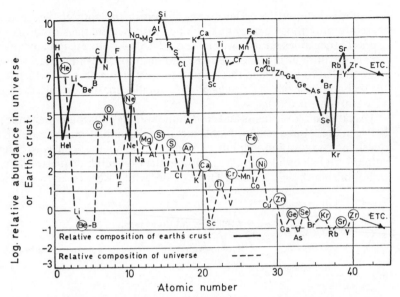

Figure 8.2 Comparison of the composition of the Universe and of the Earth's crust. Note that elements of even atomic number (ringed) tend to be more abundant in the Universe than those with odd atomic number, and that the Earth is very deficient in the noble gases

.core, much must have been lost in this manner. Oxygen, being a far denser gas, would be retained to a much larger extent, a very important factor in subsequent development. The escape velocity from the earth is 11 km s^{-1}, and hydrogen molecules have an average velocity $\frac{1}{4}$ of this value at 250°C. According to Jeans, a gas having $\frac{1}{4}$ escape velocity would take 5×10^4 years to escape, whilst if it has $\frac{1}{5}$ escape velocity the time required for complete escape increases to 2.5×10^9 years. Clearly the mean molecular velocity is a crucial factor in determining the course of events.

   In fact, the Earth can be regarded as an incompletely oxidized mixture of metals and silicon. Competition must have existed between the elements for the available oxygen (and sulphur, chlorine, etc.), and the greater the free energies of formation of the oxide (or sulphide, etc.) the greater the prospect of its formation (*Table 8.1*).

Table 8.1 Free energies of formation, ΔG_f, of oxides and sulphides

Oxides	ΔG_f/kJ mol^{-1}	Sulphides	ΔG_f/kJ mol^{-1}
CaO	−603	MnS	−189
MgO	−569	ZnS	−167
Al$_2$O$_3$	−527	MoS$_2$	−113
SiO$_2$	−398	PbS	−96
FeO	−251	FeS	−92
PbO '	−189	As$_2$S$_3$	−38

There is believed to be an excess of iron in the core of the Earth, and therefore only those oxides or sulphides with free energies of formation in excess of the values for iron would be expected to be formed in any quantity. The large amount of silicon present had an important influence on events; there tended to be formed three liquid phases, *molten iron*, on which floated the second phase of *iron*(II) *sulphide*, and finally, floating on this, the lightest phase of all, *silica (Figure 8.3)*.

Distribution of elements and compounds (*Table 8.2*) would then take place between these three phases. Noble metals (siderophils), unsuccessful in their quest for oxygen and sulphur, would dissolve in the iron core. Metallic sulphides (chalcophils) would dissolve in the iron sulphide phase and oxides (lithophils) in the silica phase. Those oxides which were basic enough would

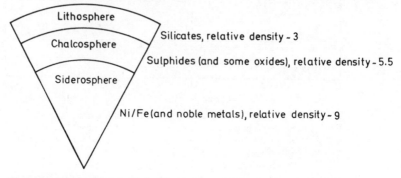

Figure 8.3 The structure of the Earth

Table 8.2 Geochemical distribution of the elements

Iron core (siderophil)	In sulphide layer (chalcophil)	In silicate layer (lithophil)	As gases (atmophil)	In organisms (biophil)
Transition metals, Ge, Sn	Ga, In Tl, Pb, As, Sb, Bi, S, Se, Te, Transition metals	Group I metals Group II metals B, Al, Lanthanides Si, Ti, Zr, Hf, Th, Group VII elements	H C N O Noble gases	H C N, P O, S K, I

do more than dissolve: they would react with the molten acidic silica to form silicates, which could subsequently solidify under vastly differing conditions. Rapid cooling at the Earth's surface produces fine-grained crystals, such as those present in basalt, whilst slow cooling in some intrusion beneath the surface gives the large crystals familiar to us in granite.

Eventually, when the surface of the Earth and its atmosphere had cooled sufficiently, water vapour condensed and fell as rain. Rivers appeared, to grow into oceans. Evaporation of water from the ocean surface maintained a

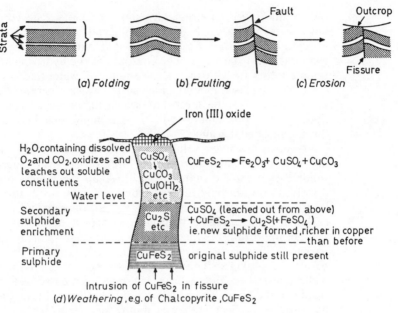

Figure 8.4 Changes in the Earth's crust

humidity in the atmosphere and perpetuated the 'water cycle', with more water being precipitated as rain to replenish the rivers and maintain the oceans. But the picture is not as symmetrical as it may at first appear. Soluble constituents in the Earth's crust were steadily dissolved, and sea water now contains 1.94 per cent chloride, 1.08 per cent sodium, 0.27 per cent sulphate, 0.13 per cent magnesium, 0.04 per cent each of potassium and calcium and smaller quantities of most other elements.

It was in this 'primeval soup', as it has been called, that life first appeared. There is some evidence for supposing that amino acids were formed by reaction between substances such as methane, water and ammonia, activated by electrical discharge. Amino acids were presumably then condensed into proteins, but it is difficult to see how the various substances could be integrated and coordinated into the activity of the living cell without the control and intervention of something akin to DNA (p. 467). The origin of life may always remain a mystery, but there is no doubting the marks that living things have made on the Earth's crust. Plants, by their photosynthetic processes, provide the atmosphere with oxygen; shells of aquatic animals are deposited on the ocean floor as carbonates which, under pressure and heat, can undergo metamorphosis into substances such as marble. Plant and animal remains can also be converted into coal and petroleum; coal itself can undergo metamorphosis, if exposed to sufficient heat, into coke and graphite. Humic acids, arising from plant decay, are even able to effect the dissolution of several metals whilst bacterial activity in tropical regions may result in the conversion of clays into bauxite.

Even now the story is far from complete. As the Earth diminished in size, the solid crust prevented the gradual and continuous readjustment necessary for the alleviation of stress. Growing strain was relieved dramatically by volcanic eruptions, earthquakes, folding and faulting; fresh material was thus brought to the surface. Seas were isolated and sedimentary rocks formed by evaporation of the water. The contours of the hydrosphere are dependent upon another variable factor: the amount of water fixed as ice at the polar caps. The increased melting of ice during warmer periods causes the oceans to rise, and the varying level of the sea may have played a part in the periodic flooding of tropical forests in the Carboniferous Period and so in the formation of coal seams.

As well as dissolving soluble minerals, water can effect physical 'weathering' of rock when it freezes; the force of expansion accompanying the solidification can be sufficient to break up the largest boulders. Oxygen dissolved in water oxidizes some sulphides as it passes through the surface layers of the Earth, whilst dissolved carbon dioxide converts insoluble carbonates into soluble hydrogencarbonates. Wind, as well as frost, is another important weathering agent and plays its part in reducing massive rock to the fine particles capable of supporting simple plants. Decay of these plants has provided the organic matter necessary for the sustenance of a wider range of organisms, from microbes on the one hand to sophisticated plants on the other. The proliferation of both terrestrial plants and animals has also had its effect on the distribution of certain elements, notably carbon, nitrogen and phosphorus. So it can be seen that the picture constantly changes (*Figure 8.5*).

Extraction

It has been estimated that the major constituents of the Earth as a whole are as shown in *Table 8.3*.

Fourteen elements (*Table 8.4*) make up 99.62 per cent of the Earth's crust (which is taken to comprise the atmosphere, hydrosphere and ten miles of lithosphere. The latter is believed to be made up of 95 per cent igneous rock, 4 per cent shale, 0.75 per cent sandstone and 0.25 per cent limestone).

It follows that many of the well known metals are in reality very scarce. They owe their familiarity to the conspicuous nature of their ores, which are concentrated in only a few places, and also to the fact that these ores are often readily reduced to the element. On the other hand, many elements, although not found naturally concentrated, are present as minor constituents of

Table 8.3 Constituents of the Earth as a whole

Element	mass per cent	Element	mass per cent
Iron	39.76	Aluminium	1.79
Oxygen	27.71	Sulphur	0.64
Silicon	14.53	Sodium	0.39
Magnesium	8.69	Potassium	0.14
Calcium	2.52	Phosphorus	0.11

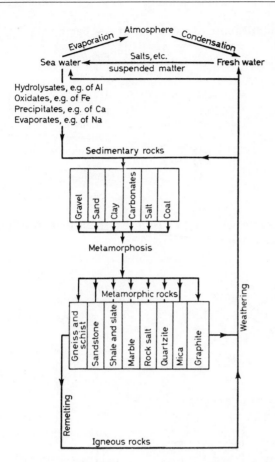

Figure 8.5 The geochemical cycle

common rocks and so are really quite abundant but too expensive to isolate because of their low concentration. (It should be noted that elements of even atomic number are more abundant than the adjacent elements of odd atomic number.)

Many of the ores used nowadays are of low grade, and the processes involved in extraction are sufficiently costly to warrant the concentration of the mineral prior to further treatment. This process is referred to as *mineral dressing* and consists of an initial crushing and grading of the ore, followed, if necessary, by making use of either density or surface differences to achieve greater concentration. In the former case, the crushed ore is washed in a suspension of suitable density to allow the separation of the mineral from the worthless rock (gangue) accompanying it. Surface differences are made use of in *froth flotation*, which depends on the difference in hydrophobic nature of the surfaces of the mineral and gangue particles. Even if only a small difference in 'wettability' exists, suitable surface-active substances (such as eucalyptus oil) can be added to enhance this effect; if air is then blown through the suspension of the finely crushed ore, the air bubbles become attached to the hydrophobic surfaces and carry the mineral particles to the

Table 8.4 Constituents of the crust of the Earth

Element	mass per cent	Element	mass per cent
Oxygen	48.60	Magnesium	2.00
Silicon	26.30	Hydrogen	0.76
Aluminium	7.73	Titanium	0.42
Iron	4.75	Chlorine	0.14
Calcium	3.45	Phosphorus	0.11
Sodium	2.74	Carbon	0.09
Potassium	2.47	Sulphur	0.06

top of the medium, where they pass into a stable foam which is continually removed (*Figure 8.6*).

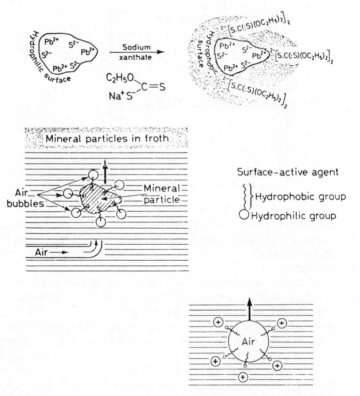

Figure 8.6 Froth flotation. Air bubbles adhere to hydrophobic surfaces and make them buoyant. Surface-active agents are widely used to make a hydrophilic surface hydrophobic and so attractive to air bubbles; selectivity can often be empirically achieved, e.g., by using sodium xanthate with galena (lead sulphide ore). Since hydrophilic groups have an affinity for ions, froth flotation can be employed for the separation of charged, as well as neutral, particles

More specialized physical methods of concentrating minerals depend on (*a*) magnetic separation, e.g., for iron(II) diiron(III) oxide, and (*b*) the readiness with which some substances acquire and retain electrostatic charges. This method is used in separating cassiterite, tin(IV) oxide, from gangue. Chemical methods involve separation by (*a*) dissolution, in which, for example, bauxite can be purified to aluminium oxide by dissolving in caustic alkali

$$(\text{impure})\ Al_2O_3 \xrightarrow{\ OH^-\ } Al(OH)_4{}^- \xrightarrow{\ CO_2\ } Al_2O_3\ (\text{pure})$$

and (*b*) precipitation, whereby, for example, magnesium can be removed from sea water by the addition of hydroxide ions

$$Mg^{2+} \xrightarrow{\ OH^-\ } Mg(OH)_2(s)$$

Principles governing the extraction of the elements

Since we live in an oxidizing atmosphere, the extraction of a metal from one of its ores must nearly always involve reduction

$$M^{n+} + ne^- \rightarrow M^0$$

This reduction is effected by pyrometallurgical operations using coke, by electrolysis, usually of fused mixtures, or, in a few cases, by thermal decomposition or reduction in a hydrogen atmosphere. The method employed depends on a number of factors, the two most important of which are the purity of the product required and the stability of the compound containing the element. Metals at the top of the electrochemical series are generally isolated electrolytically, whilst those lower down are prepared by chemical reduction.

It is becoming necessary to extract elements from ore-bearing rocks containing $<1\%$ of the required element as high grade ores are now more scarce. This has meant the increasing use of hydrometallurgical operations which typically involve the dissolution of the ore mineral in dilute acid, followed by extraction into a non-aqueous phase using a complexing agent (cf. p. 352), from which the required ions, now at an enhanced concentration, can be released by back extraction using mineral acid.

A few elements, notably the halogens, are normally present as anions and must be extracted by oxidative processes

$$X^{n-} \rightarrow X^0 + ne^-$$

although even of these, iodine is often obtained by reduction of sodium iodate(V).

Chemical reduction methods

The enthalpy of formation of a chemical bond depends on the electronegativity difference between the atoms involved in the linkage, and therefore the majority of metallic oxides have a large negative enthalpy of formation, e.g.

$$Mg + \tfrac{1}{2}O_2 \rightarrow MgO \qquad \Delta H = -622 \text{ kJ}$$

$$2Al + \tfrac{3}{2}O_2 \rightarrow Al_2O_3 \qquad \Delta H = -1677 \text{ kJ}$$

This implies that the energy required to reduce most of the metal oxides is large. However, from the equation $\Delta G = \Delta H - T\Delta S$ is can be seen that the free energy change, ΔG, associated with the formation of a solid oxide becomes *less* negative as the temperature is increased, since the system is going to a more ordered state, i.e. ΔS, the change in entropy or degree of disorder is negative, and therefore $-T\Delta S$ is positive.

On the other hand, for the reaction

$$2C + O_2 \rightarrow 2CO$$

an increase of one mole of gas is experienced and, accordingly, it is accompanied by a large increase in entropy (ΔS positive); thus, as the temperature is increased, ΔG will become *more* negative. At a certain temperature, therefore, the changes in free energy of the two systems will balance:

metallic oxide→metal+oxygen $\Delta G = +x$ kJ

carbon + oxygen→carbon monoxide $\Delta G = -x$ kJ

metallic oxide + carbon→metal + carbon monoxide $\Delta G = \quad 0$ kJ

At, and above, this temperature carbon will be capable of reducing the oxide under consideration. This is one reason for working at elevated temperatures in most metallurgical operations. These results are conveniently shown graphically, as in *Figure 8.7*, and the temperatures above which it is possible to reduce different oxides with carbon are then easily determined as in *Figure 8.8*.

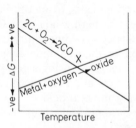

Temperature

Figure 8.7 Effect of entropy on the temperature coefficient of free energy and its application to the reduction of oxides by carbon. To the left of X the system is more stable (minimum ΔG at X) with metal in the form of oxide and carbon as element. To the right of X the system is more stable with metal in elemental form and carbon in the form of carbon monoxide, i.e. the metal can be reduced by carbon

Besides carbon, other reducing agents are used, but to a much smaller extent, e.g., aluminium in the Thermit process (see p. 356) and sodium, calcium or magnesium for reducing certain transition metal halides, e.g.,

$$TiCl_4 + 2Mg \rightarrow Ti + 2MgCl_2$$

Some sulphides are reduced *in situ* by reaction with part of the ore which has previously been oxidized (see later).

The chief difficulty which makes chemical reduction either impossible or uneconomic is the reaction of the metal produced with the reducing agent;

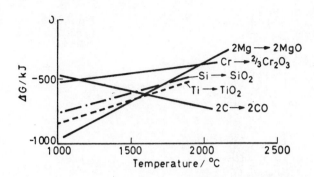

Figure 8.8 Ellingham diagram for reduction of oxides by carbon

e.g., if potassium is extracted with carbon, the metal reacts with the carbon monoxide formed to give the explosive compound (1)

$$
\begin{array}{c}
\text{OK} \qquad \text{OK} \\
\diagdown \qquad \diagup \\
\text{C}=\text{C} \\
\diagup \qquad \diagdown \\
\text{KO}-\text{C} \qquad\qquad \text{C}-\text{OK} \\
\diagdown\diagdown \qquad \diagup\diagup \\
\text{C}-\text{C} \\
\diagup \qquad \diagdown \\
\text{OK} \qquad \text{OK}
\end{array}
$$

Similarly, if alumina is reduced at high temperature by carbon, some reaction occurs between the aluminium and the carbon, forming aluminium carbide, Al_4C_3. As a result, this method has not been successful in replacing the electrolytic process.

Elements obtained by reduction with carbon (coke)
Foremost among these is iron (*Figure 8.9*). The chemistry of the blast furnace is very complex but the overall reactions can be simply represented as follows:

Excess of carbon reacts with hot oxygen from the air blast:

$$C + O_2 \rightarrow CO_2$$

But the heated carbon reduces carbon dioxide to carbon monoxide

$$CO_2 + C \rightarrow 2CO$$

Carbon monoxide reduces oxide to metal. Because of its exothermicity, this reaction takes place from left to right only in the cooler parts of the furnace:

$$Fe_2O_3 + 3CO \rightarrow 2Fe + 3CO_2(g) \qquad \Delta H = -23 \cdot 9 \text{ kJ}$$

Also in this part of the furnace, partial reduction occurs:

$$Fe_2O_3 + CO \rightarrow 2FeO + CO_2(g)$$

whilst in the hottest part of the furnace

$$FeO + C \rightarrow Fe(l) + CO(g)$$

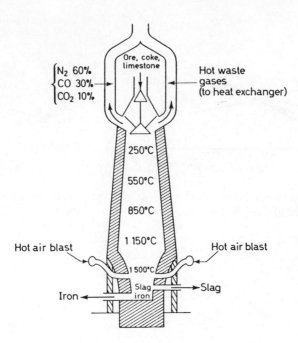

Figure 8.9 The blast furnace for the extraction of iron

A slag former, limestone, is also intimately mixed with the charge and the slag is removed from above the molten pig iron, which still, however, contains carbon, phosphorus and silicon. Purification by oxidation of iron oxides, for example, is used in producing wrought iron and steel (which contains other added elements such as manganese and chromium).

The oxides of manganese, chromium and zinc are all conveniently reduced by carbon. (When chromite, $FeCr_2O_4$, is so reduced, a series of chromium–iron alloys is obtained, which are collectively referred to as 'ferro-chromium' and are used for making stainless steel.)

Reference to *Figure 8.8* shows that magnesium oxide can be reduced at temperatures of about 2000 °C:

$$MgO + C \rightarrow Mg(l) + CO(g)$$

The products must be rapidly quenched to prevent the reverse reaction occurring, which at low temperatures is the thermodynamically favoured one.

Tin(IV) oxide is readily reduced by carbon, as are also the oxides of other elements at the lower end of the electrochemical series, such as those of lead, copper, mercury and the 'noble' elements. In the latter cases, the oxides first have to be obtained by roasting sulphide ores.

Two important non-metals extracted from their sources by carbon reduction are phosphorus and sulphur. The former is obtained by interaction between calcium phosphate, sand and coke in an electric furnace:

$$2Ca_3(PO_4)_2 + 2SiO_2 \rightarrow 2CaSiO_3 + P_4O_{10}$$

$$P_4O_{10} + 10C \rightarrow P_4(g) + 10CO(g)$$

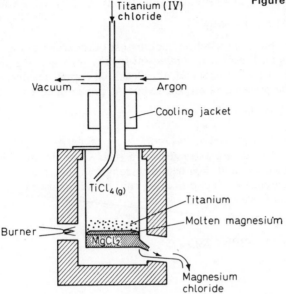

Figure 8.10 Extraction of titanium

whilst sulphur can be obtained from calcium sulphate via the intermediate calcium sulphide:

$$CaSO_4 + 4C \rightarrow CaS + 4CO(g)$$

$$CaS + 2H_2O \rightarrow Ca(OH)_2 + H_2S(g)$$

$$2H_2S + O_2 \rightarrow 2S(s) + 2H_2O$$

Elements obtained by reduction with metals

Manganese and chromium are conveniently prepared by reduction of manganese(II) dimanganese(III) oxide and chromium(III) oxide, respectively, with aluminium. The former is obtained from manganese(IV) oxide by the action of heat:

$$3MnO_2 \rightarrow Mn_3O_4 + O_2(g)$$

$$3Mn_3O_4 + 8Al \rightarrow 9Mn + 4Al_2O_3 \qquad \Delta H = -2510 \text{ kJ}$$

Titanium is obtained by reduction of its tetrachloride using magnesium or sodium (*Figure 8.10*):

$$TiCl_4(g) + 2Mg(l) \rightarrow Ti(s) + 2MgCl_2(s) \qquad \Delta G = -452 \text{ kJ}$$

The high negative value of ΔG shows that the free energy change is very favourable. The reaction rate is increased by bringing $TiCl_4$ vapour into contact with molten magnesium. The lower layer of molten magnesium chloride is run off as necessary, reaction being continued until there is 85 per cent conversion. Magnesium is removed from the titanium produced by vacuum distillation.

Silicon can also be obtained by reduction of its oxide with magnesium:

$$SiO_2 + 2Mg \rightarrow Si + 2MgO$$

Wet methods of reduction, making use of the order of the elements in the electrochemical series, are employed for the recovery of some metals from low-grade ores. For example, after leaching copper residues with sulphuric acid in the presence of air, the copper is precipitated by metallic iron

$$Fe + Cu^{2+} \rightarrow Cu(s) + Fe^{2+}$$

Elements obtained by other reductive processes

Lead and copper are extracted from their sulphide ores by first partially oxidizing the ore and then heating strongly in the absence of air to effect a mutual reduction. The processes can be summarized by the following sequences of reactions:

$$PbS + 2O_2 \rightarrow PbSO_4$$

$$PbS + PbSO_4 \rightarrow 2Pb + 2SO_2(g)$$

and

$$CuS \xrightarrow{\text{air}} CuO \rightarrow Cu_2O$$

$$\text{air} \searrow$$

$$Cu_2S + SO_2(g)$$

$$2Cu_2O + Cu_2S \rightarrow 6Cu + SO_2(g)$$

Mercury(II) oxide decomposes so readily on heating (ΔG is negative at temperatures above $500\,^{\circ}C$) that no mutual reduction between sulphide ore and oxidized product is necessary. Instead, the sulphide can be completely converted into the oxide and hence directly to the metal

$$HgS + O_2 \rightarrow Hg + SO_2(g)$$

An interesting reductive extraction of a non-metal is that of iodine from the sodium iodate(V) of Chile. There is partial reduction of iodate(V) to iodide by sodium hydrogensulphite; the iodide formed then reacts with the remaining iodate(V) to liberate iodine:

$$3OH^- + IO_3^- + 3HSO_3^- \rightarrow I^- + 3H_2O + 3SO_4^{2-}$$

$$5I^- + IO_3^- + 6H^+ \rightarrow 3I_2 + 3H_2O$$

Thermal decomposition

This is a method of preparing some metals in a high state of purity; for example, the decomposition of tetracarbonylnickel(0) as a means of purifying nickel, and the deposition of a metal from a halide—usually an iodide—by the method of van Arkel, in which the vapour of the halide is decomposed by a red-hot wire. This method is of special use for the purification of titanium and zirconium and is superior to the reduction with sodium or magnesium,

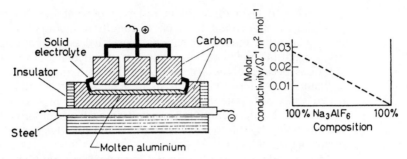

Figure 8.11 Electrolytic extraction of aluminium

as the metals are formed in a pure state and, although in the form of a powder, they can still be moulded and sintered by the application of pressure (the powder-metallurgical process).

Electrolytic reduction and oxidation

By using electrolytic methods, an unlimited amount of energy is available even for those reactions which are not thermodynamically favoured. Thus the reductive process

$$2Al_2O_3 + 3C \rightarrow 4Al + 3CO_2(g) \qquad \Delta G = +1465 \text{ kJ}$$

is not a feasible reaction, unless a considerable amount of energy is supplied to overcome the high positive value of the free energy change. The reaction can be accomplished electrolytically using carbon anodes and a current of about 4000 A at 5 V, i.e., a power of about 20 kW (*Figure 8.11*). Electrolytic methods can be used, provided that

(*a*) the ions in the crystal lattice of the mineral can be rendered mobile or, in the case of covalent molecules, the atoms can be made susceptible to electrolysis by dissolution in a suitable electrolyte. In the electrolytic reduction of Al_2O_3 the oxide melts at an impracticably high temperature, and $AlCl_3$ is not ionic. However, as the graph in *Figure 8.11* shows, solutions of Al_2O_3 in cryolite, Na_3AlF_6, have a reasonable ionic conductivity so electrolysis is possible.

(*b*) the required element can be discharged in a pure form. This means that either the mineral must be pure or that the discharge potential of the element must be lower than those of the impurities. Since the hydroxides of many metals are precipitated even in acid solutions (*Figure 8.12*), a low pH would

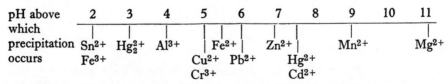

Figure 8.12 Chart showing effect of pH on precipitation of hydroxides in electrolysis of aqueous solutions

often have to be maintained in order to prevent the formation of such precipitates; unfortunately, such a condition favours the discharge of hydrogen rather than other cations. Nevertheless, copper and zinc can be discharged in preference to hydrogen, particularly from concentrated solutions, because of the large overpotentials for hydrogen on these elements.

Elements obtained by electrolysis are generally those with large positive or negative electrode potentials (p. 176). The electrolyte most commonly used is the molten halide, containing a small amount of impurity to lower the melting point. By this means both a metal and a halogen are obtained at the electrodes.

The cathode is generally iron but in some cases graphite, and the anode graphite; where possible the electrolytic vessel is made to act as one of the electrodes. The currents used are sufficiently great to keep the electrolyte molten once the external heating has lowered the resistance to such a value that a current can start to flow.

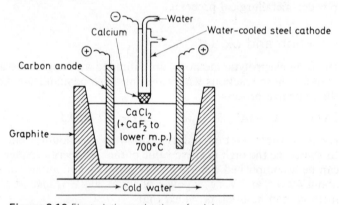

Figure 8.13 Electrolytic production of calcium

Elements extracted by this method are those of Groups I and II, aluminium, copper, zinc, the halogens and hydrogen. The apparatus used for the extraction of calcium, formed on a water-cooled steel cathode which is slowly withdrawn, is shown in *Figure 8.13*. At the anode, Cl^- and F^- are attracted, but Cl^- is selectively discharged in accordance with the electrochemical series:

$$2Cl^- - 2e^- = Cl_2(g)$$

The cathode reaction is

$$Ca^{2+} + 2e^- = Ca(s)$$

Summary

It is believed that the chemical elements were made by nuclear fusion of hydrogen atoms in stars at extremely high temperatures. Nuclear processes involving capture of neutrons under conditions likely to exist in a supernova are necessary for elements beyond iron. As temperature subsequently fell, so the likelihood of chemical reaction increased, since chemical bonds became more stable. In the case of the Earth, reaction took place selectively between metals and what oxygen there was. There was also reaction between metals and other non-metals, chiefly sulphur. Residual, unreacted metals formed the core of the Earth.

Compounds formed as above and located in the Earth's crust were, and still are, subject to various forms of weathering.

Extraction of the elements from their naturally occurring compounds involves reversing the processes by which they were formed: reduction in the case of metals and oxidation for non-metals.

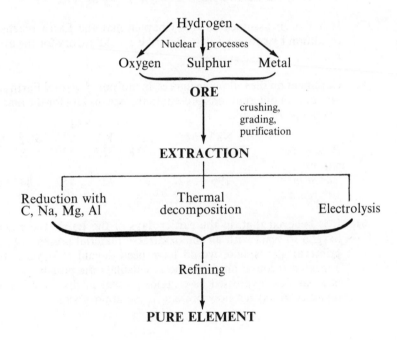

Questions

(1) How would you attempt to isolate potassium, aluminium, zinc and platinum from their chlorides?

(2) Give, with the aid of a diagram, the essential requirements for the isolation of sodium from sodium chloride electrolytically.

(3) (a) With reference to *Figure 8.8*, discuss the equilibrium between chromium(III) oxide, chromium, carbon and carbon monoxide.
(b) Describe the electrode reactions involved in the electrolysis of aluminium oxide dissolved in molten sodium hexafluoroaluminate between carbon electrodes.

(4)   'Most of the heat lost by radiation from the Earth is accounted for by the disintegration of natural radioactive sources'. Discuss this statement in relation to cosmological theories.

(5)   What further evidence would you seek to test the theories of evolution and distribution described in this Chapter?

(6)   What can you deduce from the following information?

Mass of Earth	$\approx 1 \times 10^{25}$ kg
Radius of Earth	$\approx 6 \times 10^6$ m
Density of the Earth's crust	$\approx 3 \times 10^3$ kg m^{-3}

(7)   It has been assumed in this Chapter that the Earth reached its present condition by cooling down. Describe a 'cold' theory for the evolution of the Earth.

(8)   Comment on the following figures, in mol cm^{-3} of total Earth surface, for the amounts of different ions present in the oceans and for the amount added by rivers per 10^8 years:

	Na$^+$	Mg^{2+}	Ca^{2+}	K$^+$	Cl$^-$	SO$_4^{2-}$	CO$_3^{2-}$	NO$_3^-$
Present in oceans	129	15	2.8	2.7	130	8	0.3	0.01
Added in 10^8 years	196	122	268	42	157	84	342	11

(9)   It is believed that, in the earlier days of the Earth, there was insufficient oxygen to react with all the oxidizable material present. On this view, the primeval atmosphere would have been devoid of oxygen. If it has since appeared through photosynthesis, calculate the possible amount of carbonaceous fuel (expressed as carbon) present in the Earth's crust, from the quantity of oxygen now present in the atmosphere.

9 Hydrogen

Occurrence, preparation and properties • Hydrides

	Protium	Deuterium	Tritium
Symbol	H or ${}^{1}_{1}H$	D or ${}^{1}_{2}H$	T or ${}^{1}_{3}H$
Relative atomic mass (${}^{12}C = 1$)	1.008	2.015	3.006
Ionization energy/kJ mol^{-1}	1315		
Electron affinity/kJ mol^{-1}	75		
Atomic radius/pm	30		
Melting point/K	14	18	
Boiling point/K	20	23	
Specific latent heat of fusion/J mol^{-1}	118	197	

Occurrence, preparation and properties

The hydrogen atom is the simplest of all atoms. The most abundant isotope (protium) consists simply of a proton and an extranuclear electron. Two further isotopes are known: *deuterium*, occurring to the extent of about one part in 5000, has a neutron as well as a proton in the nucleus, and *tritium*, present only in minute quantities, contains two neutrons. As is always the case with isotopes, these three forms of hydrogen are virtually identical in chemical properties, but there are significant differences in physical properties owing to the large percentage changes in mass.

The element occurs widely combined with other elements in the Earth's crust, but not to any significant extent in elemental form because the Earth's gravitational field is too small to retain such a light molecule for any length of time. It accounts for about 1 per cent of the lithosphere by mass (e.g., as petroleum and other organic remains), but on the basis of the number of atoms it is one of the most abundant of elements. As a constituent atom of water it forms about 10 per cent by mass of the hydrosphere.

Preparation

(i) From acids

Hydrogen can be very conveniently prepared, albeit in a somewhat impure state, by the action of a non-oxidizing acid on any metal above hydrogen in the electrochemical series, provided that the electrode potential is greater than the overpotential. In the case of zinc, for example, the overpotential is almost as great as the electrode potential, and accordingly pure zinc is attacked only very slowly. However, if the metal surface is contaminated by a substance of low overpotential (for example, with copper by treatment with copper(II) sulphate solution), reaction is brisk, the hydrogen being evolved at the impurity

$$Cu^{2+} + Zn \rightarrow Cu(s) + Zn^{2+}$$
$$Zn + 2H^+ \rightarrow Zn^{2+} + H_2(g)$$

From water

Very pure hydrogen is obtained when barium hydroxide is electrolysed, using nickel electrodes. Hydrogen ions from the water present are selectively discharged at the cathode

$$H_2O \rightleftharpoons H^+ + OH^-$$

$$H^+ + e^- \rightarrow H\cdot$$

$$2H\cdot \rightarrow H_2$$

Any carbon dioxide that may be present is absorbed by the barium hydroxide, whilst any traces of oxygen are removed by passing the gas over heated platinum gauze which catalyses the reduction of the oxygen by some of the hydrogen to water. The remaining gas can be subsequently dried over phosphorus(V) oxide.

The more electropositive metals will react with water or steam with the evolution of hydrogen, for example, calcium with cold water and iron with steam

$$Ca + 2H_2O(l) \rightarrow Ca(OH)_2 + H_2(g)$$

$$3Fe + 4H_2O(g) \rightleftharpoons Fe_3O_4 + 4H_2(g)$$

From alkalis

Amphoteric metals such as zinc and aluminium react with strong alkali to produce hydrogen:

$$Zn + 2OH^- + 2H_2O \rightarrow Zn(OH)_4^{2-} + H_2(g)$$
$$\text{zincate}$$

$$2Al + 2OH^- + 6H_2O \rightarrow 2Al(OH)_4^- + 3H_2(g)$$
$$\text{aluminate}$$

Industrial methods

In industry, hydrogen is obtained by the electrolysis of sodium hydroxide or brine (p. 334), or from water gas by the Bosch process. The *water gas* is produced by passing steam over white-hot carbon; the reaction is endothermic and the temperature rapidly falls below the optimum

$$C + H_2O \rightarrow CO(g) + H_2(g) \qquad \Delta H = +131 \text{ kJ}$$

If, however, the hydrogen is required with nitrogen for the manufacture of ammonia, the current of steam is replaced by air at this juncture and an exothermic reaction takes place, with the formation of *producer gas*:

$$C + \tfrac{1}{2}O_2 + 2N_2 \rightarrow CO(g) + 2N_2(g) \qquad \Delta H = -110 \text{ kJ}$$

When the temperature is once again high enough, steam replaces air, and the procedure is repeated. If now the mixture of gases, containing hydrogen, nitrogen and carbon monoxide, is passed, together with more steam, over iron(III) oxide at about 500°C, the carbon monoxide reduces the steam to hydrogen and is itself oxidized to carbon dioxide (the Shift reaction)

$$CO + H_2O + N_2 + H_2 \rightarrow CO_2(g) + N_2(g) + 2H_2(g)$$

The carbon dioxide produced can be removed by passage through water under pressure and any remaining carbon monoxide absorbed in ammoniacal copper(I) methanoate (formate).

Hydrogen is also formed in large quantities in the petroleum industry as a by-product in the catalytic dehydrogenation of alicyclic compounds to aromatic structures of higher octane number (p. 420), e.g.,

$$C_6H_{12} \rightarrow C_6H_6 + 3H_2$$
Cyclohexane   Benzene

A further by-product of the petroleum industry is methane, and in recent years methods have been developed for the use of this as a feedstock for water gas processes, e.g.,

$$CH_4 + H_2O \xrightarrow[900\,°C]{\text{Ni}} CO + 3H_2$$

Similar processes are used on a large scale for the production of hydrogen from heavy hydrocarbon oil-refinery residues and steam.

Uses

The hydrogen produced commercially is used chiefly in the Haber process for the synthesis of ammonia (p. 278), for the catalytic reduction of carbon monoxide to methanol (p. 484), for the manufacture of hydrochloric acid and for the hydrogenation of fats and coal. Long-term interest centres on the possibility of the controlled fusion of hydrogen atoms at very high temperatures, with the consequent production of much energy.

Properties

Hydrogen is colourless, odourless and tasteless and is the lightest gas known, its low mass being responsible for its rapid diffusion.

It is possible for the two protons in the diatomic molecule to spin either in the same direction (*ortho*-hydrogen) or in opposite directions (*para*-hydrogen). At room temperature, the ratio of *ortho*- to *para*-hydrogen is about three to one but, because the internal energy of the latter is the lower, as the temperature falls, the amount of *para*-hydrogen increases until, at around absolute zero, the amount of *ortho*-hydrogen is negligible.

The chemical versatility of hydrogen is indicated by its forming more compounds than any other element.

If the hydrogen atom loses an electron (i.e., is oxidized), a proton is left. The ionization energy is fairly high, so that the change is not brought about very easily; however, the large polarizing power of the proton, consequent upon its small size, results in its reacting with easily polarizable substances like water, and this subsequent reaction assists the ionization of hydrogen:

$$H(g) \rightarrow H^+(g) + e^- \qquad \Delta H = +1315 \text{ kJ}$$

$$H^+ + H_2O \rightarrow H^+(aq) \qquad \Delta H = -1070 \text{ kJ}$$

This means that hydrogen is a reducing agent, particularly when in the atomic form. (Several reactions can be brought about by reagents which produce hydrogen *in situ*, but not by hydrogen which has been prepared separately. It is possible that this reactive 'nascent' hydrogen consists of atomic hydrogen of momentary existence.)

The hydrogen atom can also achieve stability by gaining an electron to give the helium configuration, either by electrovalency or by covalency:

Electrovalency
It is clear that hydrogen could only gain an electron by chemical means from a metal more electropositive than itself. Thus, if it is passed over heated calcium, an electrovalent hydride is formed

$$Ca + H_2 \rightarrow Ca^{2+}(H^-)_2$$

Covalency
This is undoubtedly the chief mode of reaction of the hydrogen atom, and the covalent link with the carbon atom forms the basis of organic compounds. It is also the type of link present in the hydrogen molecule, the bond being formed by the interaction of the s orbitals:

$$\text{Ⓗ} + \text{Ⓗ} \rightarrow \text{Ⓗ Ⓗ}$$

or

$$H\cdot + \cdot H \rightarrow H:H$$

An indication of the strength of this bond is provided by the enthalpy of dissociation

$$H_2(g) \rightarrow 2H\cdot \qquad \Delta H = +435 \text{ kJ}$$

An electric arc between tungsten electrodes produces sufficient energy to dissociate hydrogen molecules into individual atoms. In the presence of a third body (to absorb the energy given out) these atoms recombine, with the evolution of much heat. This reaction has been utilized in welding, where the material to be welded constitutes the third body; there is the further advantage that the area to be welded is surrounded by an inert atmosphere of hydrogen which prevents oxidation.

Hydrogen reacts directly with many non-metals on the right of the Periodic Table. With fluorine, reaction takes place spontaneously and explosively in the cold, but the vigour declines with the other Group 7 elements. Thus, light is required to activate reaction between hydrogen and chlorine and a catalyst is needed for iodine to bring about even partial reaction.

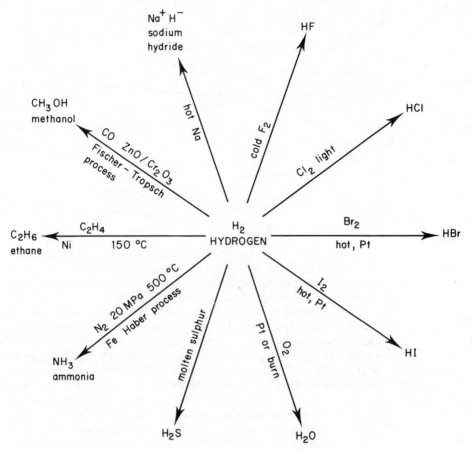

Figure 9.1 Reactions of hydrogen

In the case of Group 6 elements, reaction occurs with oxygen on heating, provided moisture is present, and hydrogen sulphide is formed if hydrogen is bubbled through molten sulphur, although the latter is a reversible reaction. The strong affinity between hydrogen and oxygen is the basis for many reducing actions of hydrogen; for example, the reduction of hot copper(II) oxide to copper

$$CuO + H_2 \rightarrow Cu + H_2O$$

Direct reaction with other non-metals tends to be slight. For instance, nitrogen reacts with hydrogen only under carefully controlled, specialized conditions (p. 278), cf. *Figure 9.1*.

Hydrides

Hydrides are binary compounds containing hydrogen and one other element; if the latter is less electronegative than hydrogen, then the name of

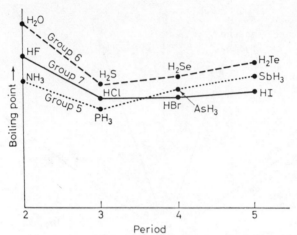

Figure 9.2 Comparison of
boiling points of covalent
hydrides

the compound ends in 'hydride'; for example, NaH is called sodium hydride. On the other hand, if the second element is more electronegative than hydrogen, 'hydrogen' usually forms the first part of the name; thus HCl is hydrogen chloride rather than chlorine hydride.

If the second element is a metal of Group 1 or 2, its valency electrons are transferred to the hydrogen, and salt-like hydrides containing the negative hydride ion result. As the second element becomes more non-metallic (i.e., increases in electronegativity), the hydrogen atom carries a relatively greater positive charge until, in the case of the very electronegative fluorine, oxygen and nitrogen, there is sufficient separation of charge to give rise to hydrogen bonding (e.g. ... H—F ... H—F ...) and an approach towards a macromolecular state, so that the hydrides NH_3, H_2O and HF are *less* volatile than those of their group members of higher relative atomic mass (*Figure 9.2*).

It is convenient to classify hydrides as ionic (with elements of Groups 1 and 2) and covalent (with elements of Groups 3 to 7), although there are also interstitial hydrides, of vague composition, formed by many of the transitional metals, in which the small hydrogen atoms occupy interstices in the crystal lattice of the metal. Hydrogen is therefore occluded, often in very large amounts, on coming into contact with the metal and is expelled on heating.

Table 9.1 illustrates the typical hydrides formed by members of the various groups of the Periodic Table.

Ionic hydrides

These are normally prepared by passing a stream of dry hydrogen over the heated metal. That the saline product contains the hydride ion, H^-, is shown by the fact that hydrogen is evolved at the *anode* during the electrolysis of the fused compound

$$Ca + H_2 \rightarrow CaH_2$$

At anode

$$H^- - e^- \rightarrow H; \qquad 2H \rightarrow H_2$$

Table 9.1 Hydrides

1	2		3	4	5	6	7	
Li^+H^-	$Be^{2+}(H^-)_2(?)$		$\overset{\delta-}{B_2H_6}$	$\overset{\delta+}{CH_4}$	$\overset{\delta+}{NH_3}$	$\overset{\delta+}{H_2O}$	$\overset{\delta+}{HF}$	↑
Na^+H^-	$Mg^{2+}(H^-)_2(?)$		$\overset{\delta-}{(AlH_3)_2}$	$\overset{\delta-}{SiH_4}$	$\overset{\delta-}{PH_3}$	$\overset{\delta+}{H_2S}$	$\overset{\delta+}{HCl}$	*Positive charge on hydrogen increasing*
K^+H^-	$Ca^{2+}(H^-)_2$		$\overset{\delta-}{Ga_2H_6}$	$\overset{\delta-}{GeH_4}$	$\overset{\delta-}{AsH_3}$	$\overset{\delta+}{H_2Se}$	$\overset{\delta+}{HBr}$	
	ionic; solid			$\overset{\delta-}{SnH_4}$	$\overset{\delta-}{SbH_3}$	$\overset{\delta+}{H_2Te}$	$\overset{\delta+}{HI}$	
		Transitional metals; non-stoichiometric, interstitial hydrides		$\overset{\delta-}{PbH_4}$	$\overset{\delta-}{BiH_3}$			

covalent; gases (except $(AlH_3)_n$)

Positive charge on hydrogen increasing

The hydride ion instantly reacts with water to liberate hydrogen

$$H^- + H_2O \rightarrow OH^- + H_2(g)$$

The readiness with which the hydride ion relinquishes its negative charge is related to its reducing power; it will, for instance, reduce carbon dioide to methanoates (formates):

$$H^- + O{=}C{=}O \rightarrow {}^-O{-}C{=}O$$
$$\underset{H}{|}$$

The complex hydrides, sodium tetrahydridoborate, $NaBH_4$, and lithium tetrahydridoaluminate, $LiAlH_4$, are particularly important reducing agents in organic chemistry on this account.

Covalent hydrides

A general method of preparation of these compounds is the action of water or acid on binary compounds of the elements with metals:

$$Mg_2Si + 2H_2O \rightarrow 2MgO + SiH_4(g) \text{ (and other silicon hydrides)}$$

$$AlN + 3H_2O \rightarrow Al(OH)_3 + NH_3(g)$$

$$FeS + 2HCl \rightarrow FeCl_2 + H_2S(g)$$

$$PI_3 + 3H_2O \rightarrow H_3PO_3 + 3HI(g)$$

Group 3
Boron forms a very interesting series of hydrides when magnesium boride is treated with dilute hydrochloric acid. All of the products are electron-deficient, i.e., it is impossible to formulate conventional structures in which

the atoms of boron have an octet of electrons in their outer quantum shells. The parent member of the series is diborane, B_2H_6, which is a gas rapidly hydrolysed to boric acid and hydrogen by the action of water:

$$B_2H_6 + 6H_2O \rightarrow 2H_3BO_3 + 6H_2(g)$$

The shape of the molecule is known to be as shown, but the method of bonding in the bridge between the boron atoms is uncertain.

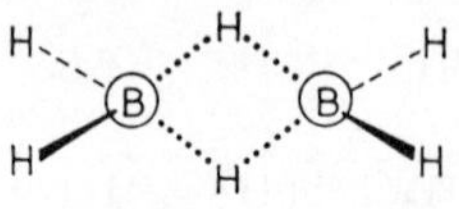

Group 4
Carbon exhibits to a unique degree the ability to form chains of atoms, which can be 'straight', branched or cyclic:

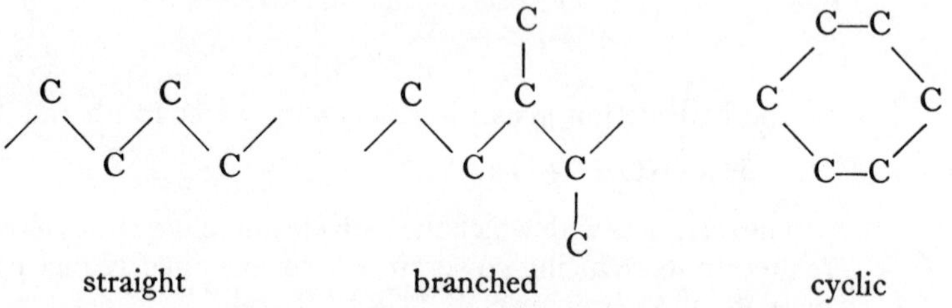

straight   branched   cyclic

Furthermore, carbon can form multiple bonds with ease. These factors give rise to a very large number of hydrides of carbon (discussed in Chapter 19).

Silicon chains are also possible but are not so stable; as a result, the number of hydrides of silicon is very limited, the more so as multiple bonds between silicon atoms are not stable.

Most of the hydrides of silicon—the silanes—are more readily flammable than hydrocarbons in air; they are also rapidly hydrolysed by water in the presence of alkali, whilst the hydrocarbons are inert to water. This can be attributed to the presence of unoccupied $3d$ orbitals in the silicon atom, with which the water molecules can coordinate to initiate reaction:

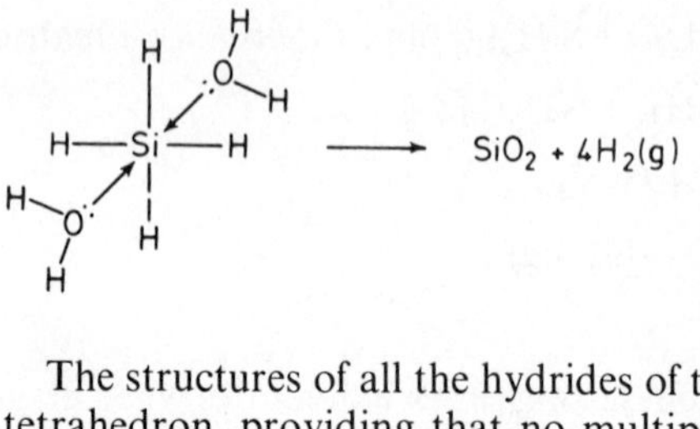

The structures of all the hydrides of the Group 4 elements are based on a tetrahedron, providing that no multiple bonds are present, e.g., methane, CH_4:

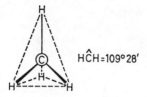

$H\hat{C}H = 109°28'$

Group 5

The representative hydrides of nitrogen and phosphorus are ammonia, NH_3 and phosphine, PH_3, respectively. Both can be prepared by the decomposition of appropriate salts:

$$NH_4^+ + OH^- \rightarrow NH_4OH \rightarrow NH_3(g) + H_2O$$

$$PH_4^+ + H_2O \rightarrow H_3O^+ + PH_3(g)$$

The structures of ammonia and phosphine are both pyramidal, with the lone pair directed to the corner of a tetrahedron:

$H\hat{N}H = 106°45'$

The nitrogen atom of the ammonia molecule donates its lone pair far more readily than the phosphorus atom of phosphine, so that ammonia is far more basic—the ammonium ion formed by coordination of ammonia with a proton closely resembles the ions of the alkali metals:

$$H^+ + :NH_3 \rightarrow [NH_4]^+$$

Hydrogen bonding in ammonia results in its being less volatile than phosphine. Ammonia is also far more soluble than phosphine in water because of the ability of the nitrogen atoms to form hydrogen bonds with water molecules, giving rise to molecules of NH_4OH. Ammonia does not burn in air, whereas phosphine readily burns to phosphorus(V) oxide

$$4PH_3 + 8O_2 \rightarrow P_4O_{10} + 6H_2O$$

Group 6

Both water, H_2O, and hydrogen sulphide, H_2S, can be synthesized by direct combination of the elements. Their structures are angular, with bond angles rather less than tetrahedral because of the repulsive force exerted on the bond pairs by the lone pairs

$104°31'$ $92°20'$

As a result of association caused by hydrogen bonding, the melting and boiling points of water are much larger than would normally be expected from a molecule of such a size. That oxygen can form these bonds with hydrogen more readily than nitrogen can be seen from the differences

Table 9.2 Hydrogen bonding in water

Hydride	Relative molecular mass	Melting point/°C	Boiling point/°C
Ammonia	17	-78	-33
Phosphine	34	-132	-87
Water	18	0	100
Hydrogen sulphide	34	-85	-60

between the physical constants of water and of ammonia. Comparable information for phosphine and for hydrogen sulphide shows the considerable part played by the hydrogen bonding in the former two compounds (*Table 9.2*).

The appreciable separation of charge in the water molecule, together with the asymmetric, angular structure, results in considerable polar character and a very high dielectric constant. Water is also amphoteric, able both to accept and to donate protons:

$$H_2O + H^+ \rightleftharpoons H_3O^+$$

$$H_2O \rightleftharpoons H^+ + OH^-$$

Consequently, water has remarkable solvent powers. In contrast, hydrogen sulphide is able only to donate protons and is only mildly polar

$$H_2S + H_2O \rightleftharpoons H_3O^+ + HS^- \qquad pK_a = 7$$

Group 7
All the halogen elements form hydrides of reasonable stability. Hydrogen fluoride and chloride are prepared by the action of hot, concentrated sulphuric acid on fluorides and chlorides of metals, e.g.,

$$F^- + H_2SO_4 \rightarrow HSO_4^- + HF$$

Hydrogen bromide and iodide are less stable and are oxidized by concentrated sulphuric acid; they are therefore seldom made by methods analogous to the above but by the action of water on the bromide or iodide of phosphorus, prepared *in situ* from the elements:

$$2P + 3Br_2 \rightarrow 2PBr_3$$

$$PBr_3 + 3H_2O \rightarrow H_3PO_3 + 3HBr$$

All four hydrides ionize exothermically in the presence of water to form acidic solutions:

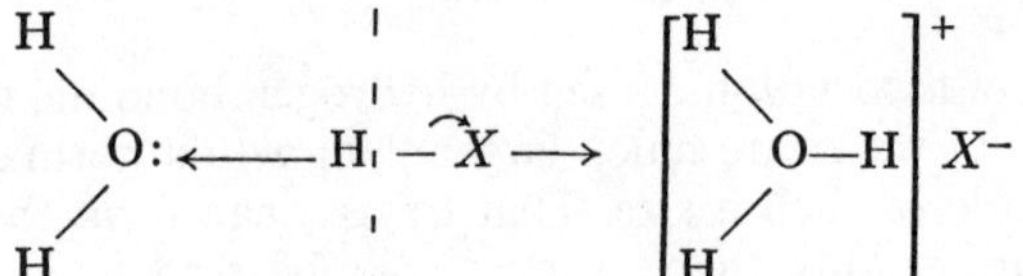

The strength of the acid solutions is in the order $HF < HCl < HBr < HI$, in accordance with the decreasing bond strength on going from HF to HI.

(Hydrofluoric acid is a particularly weak acid, because of the occurrence of hydrogen bonding which tends to prevent hydrogen escaping as a solvated proton.)

The hydrides of Group 7 elements are all, to a greater or lesser extent, reducing agents. It also follows from the respective bond energies that hydrogen iodide is the strongest reducing agent (as least energy is required to break the bond and provide hydrogen) and hydrogen fluoride the weakest. Hydrogen iodide is also the least stable thermally; it is about 20 per cent dissociated at 350 °C and becomes progressively more so as the temperature is increased:

$$2HI(g) \rightleftharpoons H_2(g) + I_2(g) \qquad \Delta H = 2 \times 5.9 \text{ kJ}$$

Summary

For a summary of this Chapter, see Figure 9.1 (p. 209).

Questions

(1)   The bond angles between the constituent atoms in water, hydrogen sulphide, ammonia and phosphine are 104°31′, 92°20′, 106°45′, 93°50′, respectively. Comment on the differences in these angles.

(2)   What conclusions can be drawn from the fact that hydrogen chloride is evolved when sodium chloride is heated with concentrated sulphuric acid, but sulphuric acid is not evolved when sodium sulphate is heated with concentrated hydrochloric acid?

(3)   Write down reactions which illustrate the reducing power of hydrogen, hydrogen iodide and lithium tetrahydridoaluminate.

(4)   In what reactions does hydrogen function as an oxidizing agent?

(5)   Write an essay on the isotopy of hydrogen.

(6)   Write an account of the role of water as a solvent.

(7)   Where do you think hydrogen should be placed in the Periodic Table; in Group 1, Group 7, or neither?

10 Group 7: the halogens

Atomic properties • Occurrence and extraction of the halogens – fluorine, chlorine, bromine, iodine and astatine • Reactions of the halogens • The hydrogen halides • Halides • Oxygen compounds of Group 7 elements • Cationic halogen compounds • Interhalogens • Analysis of halogens and their compounds

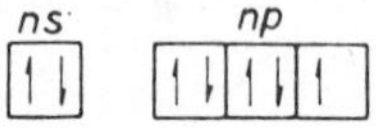

Group 6	ns np		Group 0

Group 6			Group 0
O	Fluorine, F	2.7	Ne
S	Chlorine, Cl	2.8.7	Ar
Se	Bromine	2.8.18.7	Kr
Te	Iodine, I	2.8.18.18.7	Xe
Po	Astatine, At	2.8.18.32.18.7	Rn

Atomic properties

Electronic configuration and oxidation states

The feature common to all elements of this group is an outer electronic structure of s^2p^5, i.e., one electron is needed to completely fill the p orbitals and thus to attain the noble-gas type configuration either through electrovalent or covalent bond formation. Thus the elements all exhibit an oxidation state of -1. Positive oxidation states are also known for all the halogens except fluorine, and in accordance with the decrease in ionization energies (*see Table 10.1*) this feature is most pronounced with iodine, which can even completely lose an electron in certain chemical combinations.

Atomic and ionic radii

The atomic and ionic radii increase in the manner expected as the Atomic Number increases as shown in *Table 10.2*. The ionic radii are much greater than the corresponding atomic radii because of the extra electron accommodated and are larger than those of the alkali metal (Group 1) ions from the corresponding Periods, since the latter have lost the single electron from their outer shells (p. 80).

Table 10.1 Ionization energies and oxidation states

Halogen	Ionization energy/kJ mol^{-1}	Oxidation states
F	1685	−1, 0
Cl	1255	−1, 0, 1, 3, 4, 5, 6, 7
Br	1142	−1, 0, 1, 5, 6
I	1010	−1, 0, 1, 3, 5, 7

Table 10.2 Properties of halogen atoms and ions

	Fluorine	Chlorine	Bromine	Iodine
Atomic radius/pm	72	99	114	133
Ionic radius of X^-/pm	136	181	195	216
First ionization energy/kJ mol^{-1}	1685	1255	1142	1010
Electron affinity/kJ mol^{-1}	−335	−365	−331	−305
Electronegativity	3.9	3.0	2.8	2.5

Electron affinities and electronegativities (*Table 10.2*)

As expected from the electronic configuration, the halogens have high electron affinities, the maximum being at chlorine. Similarly, the electronegativities are large, fluorine, with its small atomic volume, being the most electronegative of all the elements. As the sizes of the atoms increase, so the influence of the positively charged nucleus on the periphery of the atoms decreases, and electrons are then less readily attracted. That this is so is indicated by the decrease in electronegativity as one passes down the Group, and by the accompanying appearance of some metallic properties with iodine.

It is clear that the halide ions, X^-, having the configuration s^2p^6, are readily formed and that the elements normally exist as diatomic molecules in which the s^2p^6 configuration is attained by electron sharing

or in terms of orbitals

The stability of this configuration is indicated by the large enthalpies of dissociation (bond energies) of the molecules, although that of fluorine is

Table 10.3 Some thermal properties of the halogens

	F_2	Cl_2	Br_2	I_2
Enthalpy of dissociation/kJ mol^{-1}	155	239	188	147
Melting point/°C	−223	−102	−7	114
Boiling point/°C	−187	−35	59	183
Colour in gaseous state	Pale yellow	Green	Red	Violet

relatively low because its small size cannot prevent some repulsion between pairs of non-bonding electrons (*Table 10.3*).

The melting and boiling points (*Table 10.3*) of the halogens are low, as expected of structures consisting of covalent diatomic molecules held together by weak van der Waals forces. The rise in the melting points is brought about mainly by the increase in the volume of the molecules (and, hence, in van der Waals bonding) and partly through the greater availability of orbitals as the atomic number increases, tending to give stability to the diatomic state as a result of possible electron delocalization. This latter phenomenon is also indicated by the increase in intensity of the colours of the halogen vapours on passing from fluorine to iodine, showing the increasing ease with which electrons are excited.

Electrode potentials

The reactivities of the elements are related to their redox potentials (*Table 10.4*),

$$X^2 + 2e^- \rightleftharpoons 2X^- \text{ (aq)}$$

Fluorine is readily reduced to the fluoride ion and is consequently a very powerful oxidizing agent. The redox potentials of the remaining elements decrease rapidly; thus the oxidizing power of chlorine, bromine and iodine gets progressively less.

Table 10.4 Redox potentials of the halogens

	Fluorine	*Chlorine*	*Bromine*	*Iodine*
Redox potential/V	2.87	1.36	1.06	0.54

Occurrence and extraction of the halogens

Fluorine

This element occurs principally as the insoluble calcium salt, $Ca^{2+}(F^-)_2$, *fluorspar*, from which hydrogen fluoride is obtained by treatment with concentrated sulphuric acid:

$$2F^- + H_2SO_4 \rightarrow SO_4^{2-} + 2HF(g)$$

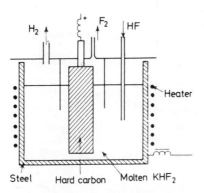

Figure 10.1 Electrolytic preparation of fluorine

As fluorine is the most electronegative of all elements known, it is impossible to oxidize hydrogen fluoride to the free element by any other element or compound. Sufficient energy can be supplied electrically, however, and the element may therefore be obtained by an electrolytic process in which the non-conducting hydrogen fluoride, after reaction with potassium fluoride to produce potassium hydrogen difluoride, $K^+HF_2^-$, acts as an electrolyte in the molten state (*Figure 10.1*).

The reaction at the carbon anode is

$$F^- \rightarrow \tfrac{1}{2}F_2(g) + e^-$$

and that at the steel cathode

$$H^+ + e^- \rightarrow \tfrac{1}{2}H_2(g)$$

(Periodic additions of anhydrous hydrogen fluoride must be made to the electrolyte to replace that decomposed.)

Fluorine has not found many uses until recent years, but it is rapidly assuming a considerable importance in the preparation of *fluorocarbons*, in which the fluorine atoms, which are much larger than those of hydrogen, protect the carbon chains from attack, making such compounds particularly valuable as lubricants and refrigerants (p. 448). Fluorine also finds application in uranium technology, partly because uranium(VI) fluoride is a convenient compound for the separation of the isotopes of the metal by diffusion processes (p. 67), and partly because the metal itself is obtained by the reduction of the tetrafluoride.

Chlorine

Chlorine, bromine and iodine exist as salts of the Group 1 metals and magnesium, the extreme solubility of which results in their main source being either the sea or salt deposits on the sites of dried-up seas. The main sources of chlorine are *rock salt*, Na^+Cl^-, and *carnallite*, $K^+Mg^{2+}(Cl^-)_3 . 6H_2O$, from which it is obtained by electrolysis. A concentrated solution of brine can be electrolysed in a Solvay cell (*Figure 10.2*), the position of the anodes being adjustable so that the anode–cathode distance can be maintained at about 2 mm. The reaction at the mercury cathode is

$$Na^+ + e^- \rightarrow Na \text{ (which dissolves to form an amalgam)}$$

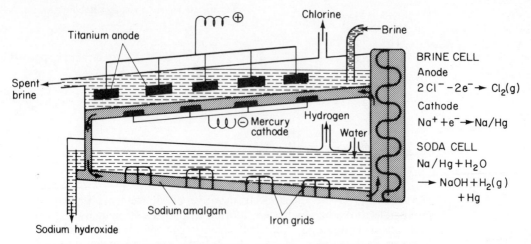

Figure 10.2 Manufacture of chlorine in the Solvay cell

and at the anode

$$2Cl^- \rightarrow Cl_2(g) + 2e^-$$

In the U.K., 75 per cent of the annual chlorine output is at present obtained by this method. On the other hand, in the U.S.A. about 80 per cent is produced in diaphragm or Downs cells (pp. 334, 340), and only 20 per cent by the Solvay process.

Chlorine is extensively used in metallurgical operations, e.g., for the recovery of tin and aluminium from scrap. Many organic solvents and plastics (p. 446) are chloro-derivatives of hydrocarbons, and chloroethane is an intermediate in the preparation of additives for petrol such as tetraethyl-lead(IV) for 'anti-knock' (p. 444). Chlorine is also used for sterilizing water and in the chemical industry for the manufacture of hydrochloric acid, bromine and bleaching powder:

$$\underset{\text{cold, solid}}{Ca(OH)_2} \xrightarrow[\substack{\text{current} \\ \text{of } Cl_2}]{\text{counter}} \underset{\text{bleaching power}}{Ca(OCl)_2 + \text{basic chloride}}$$

Bromine

Bromine is chiefly obtained from sea water, in which it occurs as bromides to the extent of about 0·015 per cent. It is liberated by passing chlorine into the water, which is maintained at a pH of 3·5 by the addition of a suitable quantity of sulphuric acid

$$2Br^- + Cl_2 \rightarrow 2Cl^- + Br_2(g)$$

(acid is necessary to minimize the loss of bromine by reaction with the water, as indicated by the equation $Br_2 + H_2O \rightarrow HBr + HOBr$). The bromine is absorbed in a suitable solvent such as benzene, from which it is subsequently extracted by fractional distillation.

Bromine is used for making 1,2-dibromoethane, an additive for petrol containing tetraethyl-lead(IV); it ensures that after combustion the lead is removed in the exhaust as volatile lead(IV) bromide instead of forming a deposit on the cylinder walls of the engine. The element is also extensively used in photography, in the form of its silver salt, photochemical reduction of silver bromide to metallic silver being particularly easily accomplished. The manufacture of potassium bromide for medicinal purposes is another outlet for the element, and it is also used in the dyestuffs industry.

Figure 10.3 Aerial view of a bromine processing plant in Anglesey. (By courtesy of the Associated Octel Co. Ltd)

Iodine

The main sources of iodine are certain seaweeds in which it is present to the extent of about 3 per cent by mass, and *caliche*, which is impure sodium nitrate containing about 0.2 per cent of sodium trioxoiodate(V) [iodate(V)].

Extraction from seaweed is effected by burning and leaching the ash, which contains alkali metal iodides, with water. The solution so obtained is evaporated to remove the chlorides and sulphates of sodium and potassium by crystallization, after which sulphuric acid is added and any sulphur precipitated (from sulphides) removed. Finally, iodine is liberated by distillation of the liquor in the presence of manganese(IV) oxide

$$2I^- + Mn^{4+}(O^{2-})_2 + 4H^+ \rightarrow Mn^{2+} + 2H_2O + I_2(g)$$

Most iodine is obtained from caliche. The ore is leached with water and pure sodium nitrate is crystallized out. The mother liquors are then concentrated and treated with the amount of sodium hydrogensulphite necessary to reduce five-sixths of the iodate(V) to hydriodic acid. This then reacts with unchanged iodate(V) to liberate iodine. The process can be summarized by the equations

$$IO_3^- + 3HSO_3^- \rightarrow HI + 3SO_4^{2-} + 2H^+$$

$$5HI + IO_3^- + H^+ \rightarrow 3I_2(g) + 3H_2O$$

The iodine so obtained is purified by sublimation over potassium iodide, which reacts with any iodine chloride present

$$ICl + I^- \rightarrow Cl^- + I_2(g)$$

Iodine and many of its compounds find use in medicine, both as antiseptics and for the prevention of goitre, an abnormal growth of the thyroid gland caused by iodine deficiency. Iodides and iodates are also used in photography and in analytical chemistry.

Astatine

This element does not occur naturally to any appreciable extent but has been synthesized by bombardment of bismuth with alpha-particles:

$$^{209}_{83}Bi + {}^4_2He \rightarrow {}^{111}_{85}At + 2{}^1_0n \qquad (t_{\frac{1}{2}} = 2.7 \times 10^4 \text{ s})$$

Laboratory preparation of the halogens

Chlorine, bromine and iodine can all be prepared by oxidation of the corresponding anion with a vigorous oxidizing agent, such as manganese(IV) oxide, in the presence of concentrated sulphuric acid:

$$MnO_2 + 4H^+ + 2e^- \rightarrow Mn^{2+} + 2H_2O$$

$$2X^- \rightarrow X_2 + 2e^-$$

In the case of chlorine, it is often more convenient to drop cold, concentrated hydrochloric acid on to crystals of potassium manganate(VII) (permanganate):

$$MnO_4^- + 8H^+ + 5e^- \rightarrow Mn^{2+} + 4H_2O$$

$$2Cl^- \rightarrow Cl_2 + 2e^-$$

Reactions of the halogens

As the atomic number increases, so the chemical reactivity decreases; this is particularly so with regard to the -1 oxidation state and is in agreement with the calculated decrease in electronegativity.

Fluorine will react readily with all the elements except nitrogen, oxygen and the noble gases, most non-metals igniting spontaneously in the gas. Chlorine also reacts with most elements except carbon, nitrogen and oxygen, although not as readily as fluorine; with both fluorine and chlorine, the highest valency state of the combining metal is usually involved. Bromine often appears to be more reactive than chlorine because of its concentration in the liquid state, although it is really less reactive and combines spontaneously only with a few elements. Iodine is the least reactive of the halogens but it will combine directly with many metals on warming.

Reactions with hydrogen

The reactions represented by the equation

$$H_2 + X_2 \rightarrow 2HX$$

can be summarized as in *Table 10.5*.

Table 10.5 Reactions of the halogens with hydrogen

Fluorine	Chlorine	Bromine	Iodine
explodes even at 20 K	explodes in direct sunlight	requires a catalyst (Pt) at 500 K	reversible, requires a catalyst and temperatures > 500 K

The heat of formation of hydrogen iodide is positive for solid iodine but slightly negative for gaseous iodine; in the latter case, therefore, the reaction is exothermic. By Le Chatelier's principle the degree of thermal dissociation should increase with rise in temperature (*Figure 10.4*). The ready reversibility of this reaction makes hydrogen iodide, alone of the hydrogen halides, a useful reducing agent.

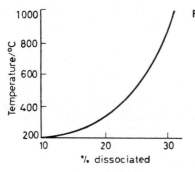

Figure 10.4 Dissociation of hydrogen iodide

Reaction with water

The reactions of the halogens with water can be considered in terms of the redox potentials involved as shown in *Table 10.6*. The signs show that the first three halogens can liberate oxygen from water, and in fact fluorine cannot be kept in aqueous solution, as immediate reaction occurs. When chlorine and

Table 10.6 Redox potentials and the reactions of halogens with water

(1) $2H_2O \rightarrow 4H^+ + O_2(g) + 4e^-$	$E^\ominus = -0.81V$			
	F	Cl	Br	I
(2) $E^\ominus/V$ for $2X_2 + 4e^- \rightarrow 4X^-$	2.87	1.36	1.06	0.54
(3) $E^\ominus/V$ for $2H_2O + 2X_2 \rightarrow H^+ + O_2(g)$ by addition of processes (1) and (2)	2.06	0.55	0.25	-0.27

bromine liberate oxygen—with much less vigour than fluorine—a different equilibrium is normally established, particularly in diffuse daylight

$$X_2 + 2H_2O \rightarrow H_3O^+ + X^- + HOX \quad \text{[hydrogen oxohalate(I) or hypohalous acid]}$$

Both reactions will be encouraged by the presence of hydroxide ions, which can then remove the oxonium ions (H_3O^+) as neutral water molecules. Thus, fluorine reacts with dilute alkali to give oxygen difluoride

$$2F_2 + 2OH^- \rightarrow F_2O + H_2O + 2F^-$$

which decomposes in strong alkali to give oxygen

$$F_2O + 2OH^- \rightarrow 2F^- + H_2O + O_2(g)$$

Chlorine with cold alkali gives the chloride and oxochlorate(I) ('hypochlorite') ions, which latter disproportionate in hot alkali to form chloride and trioxochlorate(V) (chlorate) ions

$$Cl_2 + 2OH^- \rightarrow Cl^- + OCl^- + H_2O$$

$$3OCl^- \rightarrow 2Cl^- + ClO_3$$

Bromine likewise gives oxobromate(I) and trioxobromates(V).

Because of the unfavourable electrode potentials, iodine will produce any significant amount of oxoiodate(V) ions only in the presence of dilute alkali, and even then they are so unstable that disproportionation readily occurs to give iodide and trioxoiodate(V) ions.

Halogens as oxidizing agents

Besides the above oxidizing reactions, the halogens find several applications as oxidizing agents, as the following examples indicate:

$$2Fe^{2+} + Cl_2 \rightarrow 2Fe^{3+} + 2Cl^-$$

$$Sn^{2+} + I_2 \rightarrow Sn^{4+} + 2I^-$$

$$4Br_2 + S_2O_3{}^{2-} + 5H_2O \rightarrow 2SO_4{}^{2-} + 8Br^- + 10H^+$$

$$I_2 + \quad\quad 2S_2O_3{}^{2-} \quad\quad \rightarrow 2I^- + \quad\quad\quad S_4O_6{}^{2-}$$
$$\text{Trioxothiosulphate(VI)} \quad\quad \mu\text{-Dithio-bistrioxosulphate(VI)}$$
$$\text{('thiosulphate')} \quad\quad\quad\quad\quad \text{('tetrathionate')}$$

$$AsO_3{}^{3-} \;\; +I_2 + H_2O \rightleftharpoons AsO_4{}^{3-} \; +2HI$$
$$\text{arsenate(III)} \quad\quad\quad \text{arsenate(V)}$$
$$\text{('arsenite')} \quad\quad\quad\quad \text{('arsenate')}$$

The reactions involving iodine are particularly useful for quantitative estimations because of the colour changes involved and of the ready detection of iodine by a blue-black complex it forms with starch.

The hydrogen halides

The more important reactions of these compounds have been considered already in the comparative study of hydrides (p. 209 et seq.); it remains to describe their preparation.

Hydrogen fluoride

This is prepared by distilling finely ground fluorspar, calcium fluoride, with concentrated sulphuric acid in a lead retort

$$2F^- + H_2SO_4 \rightarrow 2HF(g) + SO_4{}^{2-}$$

Hydrogen fluoride is evolved and absorbed in water until about an 80 per cent solution of hydrofluoric acid is obtained. Alternatively, anhydrous hydrogen fluoride may be obtained by condensation of the vapour below 19 °C to a volatile liquid. The aqueous solution attacks glass and is therefore usually kept in gutta-percha or poly(ethene) bottles:

$$SiO_2 + 4HF \rightarrow SiF_4(g) + 2H_2O$$
(from glass)

$$SiF_4 + 2HF \rightarrow H_2SiF_6$$
$$\text{Hexafluorosilicic acid}$$

Hydrogen fluoride should be handled with the utmost care as it causes unpleasant wounds on the skin.

Hydrogen chloride

This is prepared by the action of hot, concentrated sulphuric acid on sodium chloride, dried by passage through concentrated sulphuric acid and then collected by the upward displacement of air. In the laboratory it is usual to utilize only half of the available acid:

$$Cl^- + H_2SO_4 \rightarrow HSO_4{}^- + HCl(g)$$

but on the industrial scale this is not sufficiently economical and, by heating the mixture further, the whole of the available acid is utilized:

$$Cl^- + HSO_4^- \rightarrow SO_4^{2-} + HCl(g)$$

Because of the very great solubility of hydrogen chloride in water (500 volumes of gas in one volume of water at s.t.p.), care must be taken when the gas is dissolved to prevent water from sucking back on to the concentrated sulphuric acid as the pressure is reduced.

Hydrogen chloride is a by-product of the chlorination of organic compounds and is synthesized industrially by direct combination of the two elements in the presence of activated charcoal:

$$H_2 + Cl_2 \rightarrow 2HCl$$

Hydrogen bromide

This is conveniently prepared by dropping bromine on to red phosphorus in a little water. Bromides of phosphorus are immediately formed and hydrolysed, e.g.,

$$2P + 3Br_2 \rightarrow 2PBr_3$$

$$PBr_3 + 3H_2O \rightarrow H_3PO_3 + 3HBr$$

Traces of bromine carried over with the hydrogen bromide are removed by passing the mixture over glass beads coated with a paste of red phosphorus and water.

Hydrogen iodide

This can be obtained in a similar manner, except that in this case water is dropped on to a mixture of red phosphorus and iodine. The most convenient way to prepare the aqueous solution, hydriodic acid, is to pass hydrogen sulphide into a suspension of iodine in water:

$$I_2 + H_2S + aq \rightarrow S(s) + 2HI(aq)$$
$$\text{Hydriodic acid}$$

All of the hydrogen halides dissolve readily in water and produce azeotropes of maximum boiling point (p. 107); with the exception of hydrofluoric acid, the solutions are strongly acidic:

Halides

The metals of Groups 1 and 2 (excluding beryllium) form ionic halides. Most other elements, particularly in their highest valency states, give halides which are predominantly covalent. As expected, the degree of covalence increases with the size of the halogen atom, fluorides having considerably more ionic character than the other halides.

Preparation

The most versatile method of preparation is by direct halogenation, when, because of the high electronegativity of fluorine and chlorine and, to a lesser extent, bromine, the maximum valency state of a metal of variable valency is exerted, e.g.,

$$2Fe + 3Cl_2 \rightarrow Fe_2Cl_6$$

$$2Al + 3Cl_2 \rightarrow Al_2Cl_6$$

$$Sn + 2Br_2 \rightarrow SnBr_4$$

$$C + 2F_2 \rightarrow CF_4$$

Halides which are mainly covalent can also be prepared from the oxide by mixing with carbon as reducing agent and heating in the presence of the halogen, e.g.,

$$Al_2O_3 + 3C + 3Cl_2 \rightarrow Al_2Cl_6 + 3CO(g)$$

Dry methods such as these are essential for the preparation of easily hydrolysed halides, particularly when they are required in the anhydrous form.

Wet methods can be used when the product is not susceptible to hydrolysis (either because it is highly ionic or insoluble). The metal, its oxide, hydroxide or carbonate can be dissolved in aqueous halogen acid and the resultant solution evaporated, e.g.,

$$Fe + 2HCl \rightarrow FeCl_2 + H_2(g)$$

$$MgO + 2HBr \rightarrow MgBr_2 + H_2O$$

$$Na_2CO_3 + 2HI \rightarrow 2NaI + H_2O + CO_2(g)$$

Insoluble halides, which include the fluorides of calcium, strontium and barium together with the chlorides, bromides and iodides of silver, copper(I) and lead, the chloride of dimercury(I) and the iodides of mercury, can all be made by precipitation reactions of the type

$$M^{n+} + nX^- \rightarrow MX_n(s)$$

Hydrolysis

Most of the covalent halides are hydrolysed to a greater or lesser extent by water or alkali. There is no single explanation for this, but the likely reaction in individual cases can often by deduced. Some different possibilities are discussed below.

(1) Beryllium chloride, $BeCl_2$, has considerable covalent character, since the small size and double charge of the beryllium ion confer a large polarizing power, which gives rise to the reaction between the ion and water molecules, producing an equilibrium of the type

$$Be^{2+} + 4H_2O \rightarrow [Be(H_2O)_4]^{2+} \rightleftharpoons [Be(OH)(H_2O)_3]^+ + H^+$$

Similar reasoning applies to the hydrolysis of silicon tetrachloride, except that here the oxygen atoms of the water molecules can donate electrons to the favourably situated $3d$ orbitals of the silicon atom, as is the case with silane (p. 212). In contrast to this, tetrachloromethane is not susceptible to hydrolysis under normal conditions because the outer (second) quantum level of the carbon atom is already full, and accordingly there are no empty orbitals of suitably low energy available for coordination of water molecules.

The more the covalent character exhibited by the original compound, the greater is the extent of the hydrolysis, provided that suitable empty orbitals are available to accept the electrons of the oxygen in potentially coordinated water molecules. An excellent example of a readily established equilibrium is that afforded by bismuth(III) chloride

$$BiCl_3 + H_2O \rightleftharpoons BiOCl(s) + 2HCl$$

(2) The halides (excluding fluorides) of nitrogen and oxygen are readily hydrolysed by a different mechanism. The lone pair(s) of electrons associated with these atoms can be donated to one of the hydrogen atoms of the water molecule. This results in an electronic rearrangement to yield the products of hydrolysis, e.g.,

Oxygen compounds of the Group 7 elements

There are several binary compounds of oxygen and the halogens, although very few of them are of any great importance. Compounds of oxygen with fluorine are correctly known as fluorides, since fluorine is more electronegative than oxygen, whilst the remainder (other than I_2O_4 and I_4O_9, see p. 231) are classed as oxides. *Table 10.7* lists all the known oxygen compounds of the halogens.

Oxygen difluoride is prepared by the action of fluorine on dilute sodium hydroxide solution (p. 224). Other oxygen fluorides can be formed by passing electric discharges through mixtures of oxygen and fluorine, but they readily decompose even below $0\,°C$.

The oxides of chlorine are all endothermic, unstable and explosive. Dichlorine oxide, Cl_2O, chlorine dioxide, ClO_2, and dichlorine heptoxide, Cl_2O_7, are all gases at room temperature whilst dichlorine hexoxide, Cl_2O_6, is a liquid.

Chlorine dioxide is of some importance as an industrial oxidizing agent

Table 10.7 Binary compounds of oxygen and the halogens (important compounds in bold type)

OH_2			
O_2F_2			
O_3F_2	**Cl_2O**	Br_2O	
	ClO_2	BrO_2	
		BrO_3	
O_4F_2		(or **Br_3O_8**)	I_2O_4
			I_2O_5
	Cl_2O_6		I_4O_9
	Cl_2O_7		

and is also of theoretical interest, as the molecule contains an odd electron and is accordingly paramagnetic

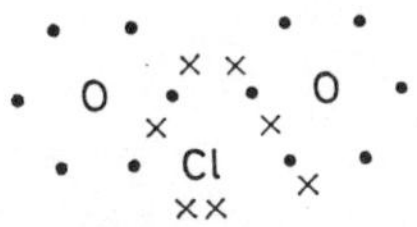

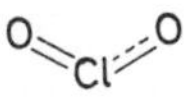

All the oxides of chlorine are acid anhydrides and react in the following way with water:

$$Cl_2O + H_2O \rightarrow 2HOCl$$

$$2Cl^{IV}O_2 + H_2O \rightarrow HCl^{III}O_2 + HCl^{V}O_3$$

$$Cl_2^{VI}O_6 + H_2O \rightarrow HCl^{V}O_3 + HCl^{VII}O_4$$

$$Cl_2O_7 + H_2O \rightarrow 2HClO_4$$

The structures of the acids and of the corresponding anions are shown in *Table 10.8*. Like the oxides, the acids and their salts have powerful oxidizing

Table 10.8 The oxoacids of chlorine

Acid	Structure	Stability	Structure of anion
Oxochloric(I) (hypochlorous)	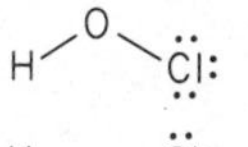		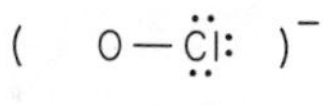
Dioxochloric(III) (chlorous)	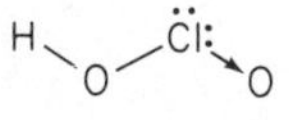	unstable	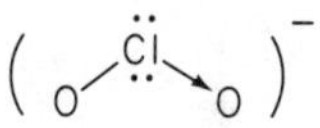
Trioxochloric(V) (chloric)	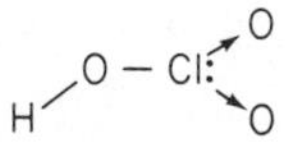		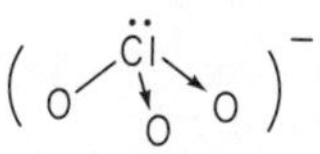
Tetraoxo-chloric(VII) (perchloric)	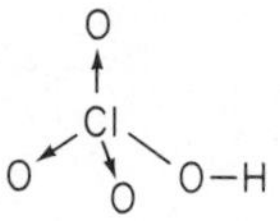	isolable, but explosive on heating	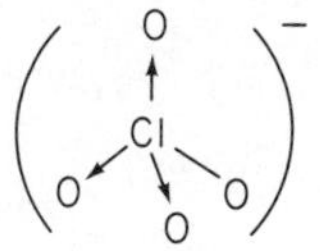

properties, the sodium and calcium salts of oxochloric(I) acid being particularly well known as bleaching agents ('hypochlorites'). A notable exception to this oxidizing character is afforded by aqueous solutions of tetraoxochloric(VII) acid; while the pure covalent acid is an exceptionally powerful oxidizing agent, solutions of the acid are quite stable, possibly because of the perfect symmetry of the ion and the completed orbitals of its constituent atoms.

Only two oxoacids of bromine have been well characterized: oxobromic(I) acid, HOBr, and tribromic(V) acid, $HOBrO_2$, both similar to their chlorine analogues.

The only simple oxide of iodine is the one of empirical formula I_2O_5. This is a stable white solid and apparently has a polymeric structure. It is easily obtained by direct oxidation of iodine with concentrated nitric acid, the initial product, trioxoiodic(V) acid, being easily dehydrated by the application of heat:

$$I_2 \xrightarrow{HNO_3} HIO_3 \xrightarrow{-H_2O} I_2O_5$$

or structurally:

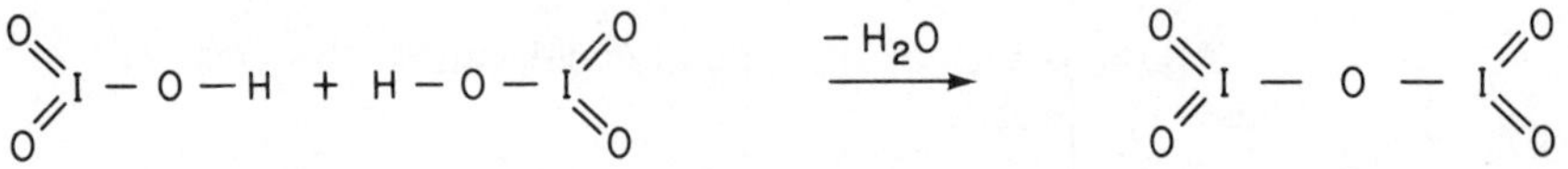

It is a fairly good oxidizing agent and is used as such for the quantitative estimation of even small amounts of carbon monoxide (and, hence, of oxygen in organic compounds)

$$I_2O_5 + 5CO \rightarrow I_2 + 5CO_2$$

The iodine liberated is estimated in the usual way with standard sodium thiosulphate solution.

Characteristically, because of its large size, iodine also forms 'periodates', the parent acid of which may be regarded as a derivative of $I(OH)_7$. Strong oxidation of iodate(V) by oxochlorate(I) yields, on acidification, an acid of formula H_5IO_6 [$I(OH)_7 - H_2O$], hexaoxoiodic(VII) acid, but commonly known as *ortho*periodic acid and having the structure

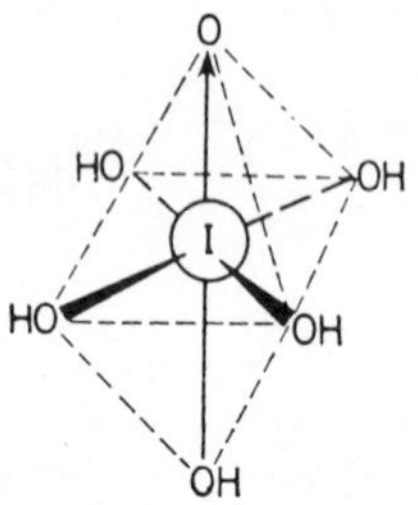

Dehydration of this compound can be carried out in stages to give, first, $H_4I_2O_9$ (i.e. $2H_5IO_6 - 3H_2O$), and then HIO_4 ($H_4I_2O_9 - H_2O$), tetraoxoiodic(VII) or periodic acid. It should be noted that in all these compounds the iodine is present in an oxidation state of $+7$.

Cationic halogen compounds

Several compounds of iodine (and a few of bromine and chlorine) are known in which the halogen is in the form of a positively charged ion, typifying the increase in metallic character on descending a group. The two 'oxides', I_2O_4 and I_4O_9 are thought to contain such ions, the former being formulated as $IO^+IO_3^-$, ioxyl(III) iodate(V), and the latter as $I^{3+}(IO_3^-)_3$, iodine(III) iodate(V). Iodine(III) phosphate(V), ethanoate (acetate) and a few other salts are also known.

Iodine also occurs as a unipositive ion. Thus the unstable oxoiodic(I) acid may better be regarded as iodine hydroxide, since the following equilibria have been determined

$$HOI \rightleftharpoons H^+ + OI^- \qquad K \sim 10^{-13} \text{ mol dm}^{-3}$$

$$HOI \rightleftharpoons HO^- + I^+ \qquad K \sim 10^{-10} \text{ mol dm}^{-3}$$

However, it appears that the ion I^+ is stable only when complexed, e.g., with pyridine (py) (p. 464). In this state several salts are known, such as

Pyridyliodine(I) nitrate $[Ipy]^+NO_3^-$

Pyridyliodine(I) tetraoxochlorate(VII) $[Ipy]^+ClO_4^-$

(The coordination of a pyridine molecule to the iodine cation stabilizes it by completing an octet of electrons in the outer electronic shell.)

The interhalogens

Compound formation between halogen atoms themselves results in formation of the interhalogens (*Table 10.9* and *Figure 10.5*).

Table 10.9 The interhalogens

ClF	BrF		BrCl	ICl	IBr	AtI
ClF_3	BrF_3	IF_3		ICl_3		
ClF_5	BrF_5	IF_5				
		IF_7				

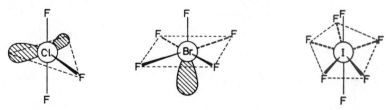

Figure 10.5 Structures of representative interhalogens: (a) ClF_3, (b) BrF_5, (c) IF_7

All these compounds can be prepared by direct combination of the elements in appropriate proportions under suitable conditions. Perhaps the most important at present is bromine trifluoride. This exists to a large extent in the form of associated molecules linked by fluorine bridges

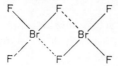

but it also ionizes to some extent in accordance with the equation

$$2BrF_3 \rightleftharpoons BrF_2^+ + BrF_4^-$$

Bromine trifluoride provides a very useful medium for fluorinating reactions, and many metals, oxides and even other halides are readily converted to fluorides by this reagent.

Analysis of halogens and their compounds

Halogens are detected in their elementary state by their characteristic colours and odours. With the exception of iodine they will also bleach damp litmus paper.

The halide ions are recognized by the formation of the fuming, pungent hydrogen halides on treatment with concentrated sulphuric acid. Because of the increasing reducing power of these latter compounds as the atomic number of the halogen increases, those of bromine and iodine produce other products also in the presence of sulphuric acid as shown in *Table 10.10*.

Table 10.10 Products of reaction of halide ions with concentrated sulphuric acid

Halide	F^-	Cl^-	Br^-	I^-
Products	HF*	HCl	HBr	HI
			Br_2	I_2
			SO_2	$SO_2 + H_2S^\dagger$

* Attacks damp glass, etching it
† Produces sulphur: $SO_2 + 2H_2S \rightarrow 2H_2O + 3S$

Alternatively, the halide ions (except fluoride and chloride) can be directly oxidized using, for example, sodium chlorate(I). The free halogen produced can be extracted by tetrachloromethane and identified by the colour imparted—yellow indicating bromine and violet indicating iodine.

Halide ions other than fluoride can also be confirmed by treatment with nitric acid, followed by addition of silver nitrate solution (*see Table 10.11*)

$$Ag^+ + X^- \rightarrow AgX(s)$$

The difference in the ability of the precipitates to dissolve in ammonia in accordance with the equation

$$Ag^+ + 2NH_3 \rightleftharpoons [Ag(NH_3)_2]^+$$

Table 10.11 Properties of silver halide precipitates

Halide	AgCl	AgBr	AgI
Colour of precipitate	White	Cream	Yellow
Reaction with ammonia	Soluble	Soluble with difficulty	Insoluble

is caused by the decreasing solubility products of the halides, in the order chloride > bromide > iodide.

Halogens in higher oxidation states may be identified by reducing to the halide, for example with sulphur dioxide, followed by application of one of the above techniques.

Quantitatively, fluorides may be estimated gravimetrically as the insoluble calcium fluoride, whilst the other halides are determined by titration with silver nitrate solution, using sodium chromate (or an adsorption indicator) in neutral conditions

$$Ag^+ + X^- \rightarrow AgX(s)$$

Then, at the end point,

$$2Ag^+ + CrO_4{}^{2-} \rightarrow Ag_2CrO_4(s)$$
$$\text{yellow} \qquad \text{red}$$

Summary

The general chemistry of the halogens is summarized in *Figure 10.6* (p. 234), chlorine being taken as representative of elements of Group 7.

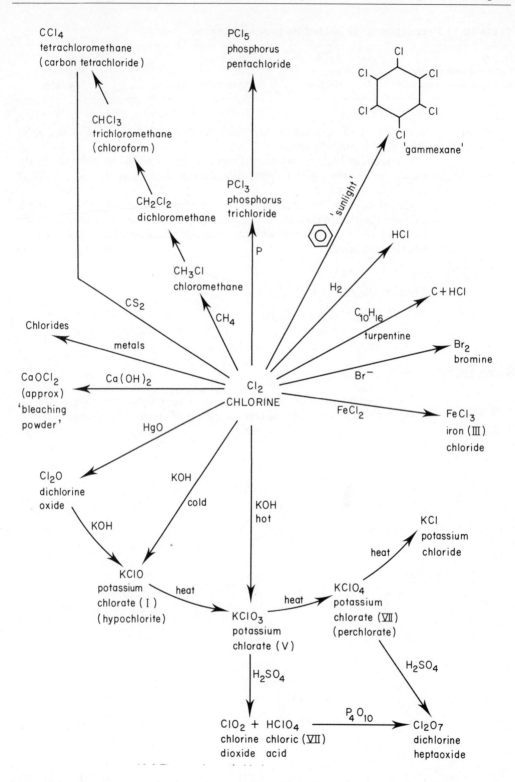

Questions

(1) In which respects is fluorine not a typical member of Group 7?

(2) Boron trifluoride is acidic in anhydrous hydrofluoric acid, whilst ethanol is basic. Suggest reasons for these observations.

(3) What sort of compound would you expect to contain halogen in a cationic form? How would you show the existence of halogen in this condition?

(4) In preparing dichlorine oxide, chlorine is passed at 0 °C over dry mercury(II) oxide, and for preparing aqueous oxochloric(I) acid, chlorine is shaken with a suspension of mercury(II) oxide. Discuss the use of the metal oxide in these reactions.

(5) Comment on the structure and reactions of bleaching powder.

(6) Suggest reasons for the existence of SF_6, whereas the highest chloride of sulphur is SCl_4.

(7) What would you expect to be the chemical properties of astatine?

(8) Under what conditions could an arsenate(III) be estimated volumetrically?

(9) Discuss the merits and limitations of various methods of classification of the halides.

(10) Write an account of the functions of halogen compounds in the human body.

(11) The average concentration of bromide in sea-water is 0.0008 mol kg^{-1}. Calculate the volume of sea-water of relative density 1.03 required to produce 1000 kg of bromine, if the bromine is completely extracted. How much chlorine would be needed to displace this amount of bromine?

(12) Outline the chemistry of an industrial process in which chlorine is one of the important products. (N.B. Detailed diagrams of industrial plant are not required.)
      Describe the action of the reagents listed below on (i) sodium chloride, (ii) sodium bromide, and (iii) sodium iodide, giving equations and explanations.
      (a) Concentrated sulphuric acid is added separately to each of the solids (i), (ii) and (iii) and the mixture warmed.
      (b) to aqueous solutions of (i), (ii) and (iii) aqueous silver nitrate is added, followed by aqueous ammonia.
      (c) To aqueous solutions of (i), (ii) and (iii) chlorine water is added followed by tetrachloromethane (carbon tetrachloride); the mixture is then shaken and allowed to settle.
                                                                                   [A.E.B.]

(13) Some reactions of iodine and chlorine compounds are described below. Using these data and stating clearly your reasoning, identify, write structural formulae for, and discuss the oxidation states of the halogens in the compounds A–G. Where possible write balanced equations for the reactions involved.

The reaction between iodine and an excess of chlorine at $-78\,°C$ leaves a yellow compound A when the excess of chlorine is removed. A has the composition $I = 54.4$ per cent, $Cl = 45.6$ per cent by mass and a relative molecular mass of 467. When A is heated, a liquid B and a gas C are produced. The composition of B is $I = 78.2$ per cent, $Cl = 21.8$ per cent by mass. Dilute hydrochloric acid dissolves A to give a yellow solution from which a yellow compound D of stoichiometry $(butyl)_4NICl_4$ may be isolated by the addition of tetrabutylammonium chloride. D is an ionic compound containing one anion and one cation per stoichiometric unit. The reaction of A with an excess of aqueous potassium iodide gives a brown solution E, from which F, CsI_3, may be obtained by the addition of solid caesium iodide. Solution E can be decolourized by aqueous sodium thiosulphate(VI). The gas C reacts with mercury(II) oxide to give another gas G which has a relative molecular mass of 87. On heating G decomposes into C and molecular oxygen. When C is passed into aqueous potassium iodide a solution with the same properties as E results.

[Cambridge 1978]

11 Group 6 elements

The elements • Reactions • Structures and allotropy • Occurrence and extraction – oxygen, sulphur, selenium, tellurium, polonium • Oxides • Peroxides • Hydroxides and oxoacids • Oxo salts • Sulphides Sulphur dioxide and sulphites • Sulphur(VI) oxide • Sulphuric acid

	ns	np	
Group 5	↑↓	↑↓ ↑ ↑	Group 7

N	**Oxygen, O**	2.6	F
P	**Sulphur, S**	2.8.6	Cl
As	**Selenium, Se**	2.8.18.6	Br
Sb	**Tellurium, Te**	2.8.18.18.6	I
Bi	**Polonium, Po**	2.8.18.32.18.6	At

The elements

The first member of this group is *oxygen*, by far the most abundant element in the earth's crust, and the last *polonium*, occurring only to the extent of one part in 2.5×10^{10} parts of pitchblende and with a half-life of 1.19×10^7 s

$$^{210}_{84}\text{Po} \rightarrow\, ^{206}_{82}\text{Pb} + {}^{4}_{2}\text{He}\ (\alpha\text{-particle})$$

Not surprisingly, then, polonium is the most recent member of this group to have been discovered. Oxygen has been known since the eighteenth century, although it was first regarded as dephlogisticated air. In view of its abundance, it might be argued that the discovery of oxygen was very late; doubtless, the fact that it is an invisible gas and is not the major constituent of the atmosphere is largely responsible. In contrast, sulphur, existing as a solid in a fairly pure state, has been known since well before the time of Christ.

As the elements of this group are just two electrons short of the nearest noble gas and the stable electronic configuration which that implies, much of their chemistry is associated with the acquisition of two electrons, either by transfer or sharing. In addition, compounds are formed in which the coordination is in excess of two: oxygen can have a maximum coordination number of four only, whereas sulphur can have one of six and tellurium even of eight. This difference between the elements is a function of the outer shell; in the case of oxygen, this is complete with eight electrons, i.e., there are no *d* orbitals of suitable energy available. But the outer shell of tellurium can be

Table 11.1 Atomic size and ionization energy

	Atomic radius/pm	Ionic radius, X^{2-}/pm	First ionization energy/kJ mol^{-1}
O	74	140	1318
S	104	180	1001
Se	117	200	940
Te	137	220	871

expanded to include five *d* orbitals. The absence of *d* orbitals from the second shell is responsible for oxygen standing apart from the other elements of the Group, and sulphur is usually preferred as the typical element.

The increase in the number of electron shells on passing from oxygen to polonium results in a progressive increase in atomic volume. There is consequently less attraction between the nucleus and the outer electrons with the latter elements, so that there is a fall in both the ionization energy and electronegativity on descending the Group (*Table 11.1*). Tellurium and polonium even exhibit some metallic behaviour, both possessing high densities and forming, for example, basic nitrates and sulphates.

Reactions

The elements of this group react with most metals. Only the most electropositive metals give ionic compounds and these are hydrolysed in aqueous solution

$$O^{2-} + H_2O \rightarrow 2OH^-$$

$$S^{2-} + 2H_2O \rightarrow H_2S + 2OH^-$$

Transitional metals give compounds of appreciable covalent character. Frequently, non-stoichiometric (Berthollide) compounds exist: iron(II) oxide can vary in composition between $Fe_{0.91}O$ and $Fe_{0.95}O$, with iron positions in the lattice being occasionally empty and with some iron atoms being in oxidation state $+3$ to maintain electrical neutrality. Perhaps the most striking example of non-stoichiometry is provided by cobalt telluride which can exist between the composition CoTe and $CoTe_2$.

In reactions with halogens, oxygen is clearly atypical. Although hemi-oxides exist (e.g. Cl_2O), the more stable compounds contain more oxygen than halogen (e.g., I_2O_5). On the other hand, tellurium and polonium form tetrahalides, even with iodine. Sulphur, selenium and tellurium even give hexafluorides, and it is significant that, whilst SF_6 and SeF_6 (maximum coordination number 6) are very stable, TeF_6 (maximum coordination number 8) is rapidly hydrolysed by water. In several of these and other compounds, the electronegativity of the later elements is less than that of the second constituent, so that a positive oxidation number results. For example, in the hexafluorides, sulphur, selenium and tellurium all exhibit an oxidation state of $+6$.

The elements of this group react together to some extent. They all react with oxygen to form di- or tri-oxides. Furthermore, sulphur(VI) oxide can dissolve selenium to give $SeSO_3$ and tellurium to give $TeSO_3$.

Structures and allotropy

Polymorphism (or allotropy) is very conspicuous in this Group.

Oxygen exists both as the dimer, O_2, and trimer, O_3 (and at low temperatures, as O_4). The small mass of the molecules results in these polymorphs being volatile. In the case of the remaining elements, however, the tendency towards chain formation or metallic lattices means that they are all solid at room temperature (*Table 11.2*).

Table 11.2 Structure, volatility and density

Element	Structure	Volatility; m.p./°C	Density/g cm^{-3}
Oxygen	diatomic	gas; -220	1.27 (of solid)
Sulphur	puckered 8-ring	solid; 114 (α)	2.06 (α)
Selenium	zig-zag chains	solid; 217 (grey)	4.80 (grey)
Tellurium	zig-zag chains	solid; 450 (metal)	6.24 (metal)
Polonium	cubic lattice	solid; 254 (β)	9.51 (β)

Oxygen and trioxygen

The diatomic molecule, O_2, is paramagnetic; the orbitals fill with pairs of electrons until the last two are reached when, because of the lower energy content, they remain single and unpaired.

The trimer, *trioxygen* or *ozone*, on the other hand, has no unpaired electron and is therefore diamagnetic. Its structure can be regarded as a hybrid of

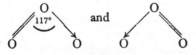

It is prepared endothermically from oxygen by photochemical or electrical stimulus

$$3O_2 \rightarrow 2O_3 \qquad \Delta H = 2 \times 142 \text{ kJ}$$

Trioxygen is present in the upper atmosphere, where ultra-violet light is probably the activating agent. Commercially, it is produced by the action of a silent electrical discharge on air or dioxygen (*Figure 11.1*). In Brodie's 'ozonizer' the acid on either side of the gas stream is kept at a potential difference of several thousand volts; the glass prevents sparking. The yield is never higher than 80 per cent and often much lower. Pure trioxygen is obtained as a dark-blue liquid when the mixture is liquefied and fractionally distilled.

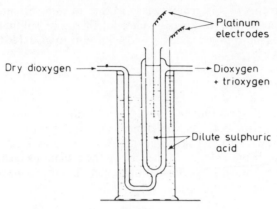

Figure 11.1 Preparation of trioxygen: Brodie's apparatus

Trioxygen is a powerful oxidizing agent, especially in acid solution:

Neutral $O_3 + H_2O + 2e^- \rightarrow O_2(g) + 2OH^-$ $E^{\ominus} = +1.2$ V

Acid $O_3 + 2H^+ + 2e^- \rightarrow O_2(g) + H_2O$ $E^{\ominus} = +2.01$ V

As the redox potential for iodine/iodide is only 0·54 V

$$2I^- \rightarrow I_2 + 2e^- \qquad\qquad E^{\ominus} = -0.54 \text{ V}$$

it follows that potassium iodide is oxidized to iodine by trioxygen, even in neutral solution:

$$2I^- + O_3 + H_2O \rightarrow I_2 + O_2(g) + 2OH^- \qquad E^{\ominus} = 1.2 - 0.54 = 0.66 \text{ V}$$

Dioxygen is usually evolved when trioxygen reacts, for example

$$H_2O_2 + O_3 \rightarrow H_2O + 2O_2(g)$$

$$PbS + 4O_3 \rightarrow PbSO_4 + 4O_2(g)$$

The oxidizing power of trioxygen is utilized in bleaching and sterilization. On a more academic level, it is important because of its ability to attack unsaturated carbon–carbon bonds to form 'ozonides' (p. 425). Turpentine contains such unsaturated links and, because it absorbs trioxygen, is used in the determination of its molecular formula (*Figure 11.2*).

A fixed volume of air is 'ozonized' and the diminution in volume on cooling to the original temperature is observed. The trioxygen present is then removed by breaking the phial of turpentine. The subsequent contraction as the trioxygen is absorbed by the turpentine is found to be twice the original contraction accompanying 'ozonization'.

Let the formula of the allotrope be O_n. Then, on 'ozonization'

$$nO_2 \rightarrow 2O_n$$

or n vols. of dioxygen give 2 vols. of product (by Avogadro's principle). But there is a contraction of 1 volume during 'ozonization', i.e.,

$n - 2$ vols. = 1 vol.   or $n = 1 + 2 = 3$

Hence the formula is O_3.

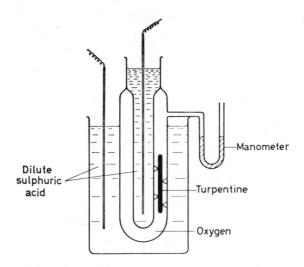

Figure 11.2 Confirmation
of the trioxygen formula

Dilute
sulphuric
acid

Manometer

Turpentine

Oxygen

Sulphur

Sulphur illustrates its power of catenation by forming the S_8 molecule:

Below 96 °C, the most stable arrangement of these molecules is in the form
of a rhombic crystal (α-S); above 96 °C, in the form of a monoclinic crystal (β-
S). Consequently, the rhombic variety is usually obtained when sulphur
crystallizes from a cold solution (for example, in carbon disulphide) and
monoclinic crystals result from the solidification of molten sulphur. This type
of allotropy, where a definite transition temperature exists between the two
allotropes, is called *enantiotropy*. (The equilibrium between the two allotropes and liquid sulphur is represented diagrammatically on page 91).

Rhombic sulphur melts at 114 °C, monoclinic sulphur at 120 °C, to give a
honeycoloured liquid consisting of mobile S_8 molecules (S_λ). At about
190 °C, however, the liquid becomes dark and viscous as the molecules open
and form long, tangled chains (S_μ). Sudden cooling of this variety gives plastic
sulphur (γ-S). Eventually, at 444 °C, the liquid boils, giving a vapour
consisting initially of S_8 molecules; these dissociate progressively at higher
temperatures until, at 2000 °C, they are monatomic. (Diatomic sulphur,
formed in the course of this heating, resembles oxygen in being paramagnetic.) These changes can be summarized as in *Figure 11.3*.

Selenium, tellurium, polonium

There are at least five polymorphs of *selenium*, two of which, like sulphur, are
composed of octatomic molecules and exist as rhombic and monoclinic

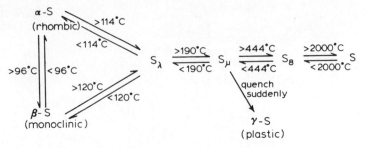

Figure 11.3 Phases of sulphur

crystals. The most stable form, however, is the grey 'metallic', which appears to be made up of infinite chains arranged spirally.

At least one crystalline form of *tellurium* is known, a 'metallic' type corresponding to grey selenium.

Two forms of *polonium* have so far been discovered, cubic (α) and rhombohedral (β).

Occurrence and extraction

Oxygen

There is not sufficient oxygen available to give complete oxidation of all the elements present on earth. It is reasonable to suppose that most of it was initially 'fixed' by reaction with more electropositive elements. Subsequent photochemical decomposition of water vapour and photosynthetic processes of plants have released large quantities of oxygen into the atmosphere, from which it is removed by respiration and other combustion processes, as well as by the weathering of rocks (particularly when the oxygen is dissolved in water). There are, then, two opposing influences tending to maintain a rough balance of atmospheric oxygen. At present, elemental oxygen comprises about 20 per cent of the atmosphere; in combined form, it accounts for about 50 per cent of the lithosphere, chiefly as silicates, whilst, in the form of water, it makes up about 90 per cent of the hydrosphere.

Oxygen is obtained commercially by the fractional distillation of liquid air or by the electrolysis of sodium hydroxide solution, using iron electrodes and a high current density. In the case of the latter method, hydroxide ions are attracted to the anode where they relinquish their valency electrons and decompose into water and oxygen

$$OH^- - e^- \rightarrow OH$$

$$4OH \rightarrow 2H_2O + O_2(g)$$

It can be prepared in small quantities in the laboratory either by the electrolysis of alkali solutions or aqueous solutions of electrolytes containing anions more resistant to discharge than the hydroxide ion, for example dilute sulphuric acid

$$H_2SO_4 \rightleftharpoons 2H^+ + SO_4^{2-}$$

$$H_2O \rightleftharpoons H^+ + OH^-$$

Sulphate and hydroxide ions are attracted to the anode and the hydroxide ions are selectively discharged. An inert anode must be used, for otherwise it may react with the nascent oxygen or even dissolve itself to provide the circuit with the electrons that would otherwise be provided by the hydroxide ions.

Oxygen is also liberated by the action of heat on various oxocompounds.

Chlorates, e.g. potassium chlorate(V) [preferably with manganese(IV) oxide as catalyst]:

$$2KClO_3 \rightarrow 2KCl + 3O_2(g)$$

Chlorine and chlorine dioxide are also formed in small quantities.

Nitrates, e.g., potassium nitrate or copper(II) nitrate:

$$2KNO_3 \rightarrow 2KNO_2 + O_2(g)$$

$$2Cu(NO_3)_2 \rightarrow 2CuO + 4NO_2(g) + O_2(g)$$

Oxides of noble metals, e.g., mercury(II) oxide:

$$2HgO \rightarrow 2Hg + O_2(g)$$

Peroxides, e.g., barium peroxide:

$$2BaO_2 \rightarrow 2BaO + O_2(g)$$

Peroxides also readily release oxygen on treatment with acidified potassium manganate(VII)

$$2MnO_4^- + 5O_2^{2-} + 16H^+ \rightarrow 2Mn^{2+} + 8H_2O + 5O_2(g)$$

Oxygen is used in large quantities for oxyhydrogen and oxyacetylene welding, and to assist respiration in illness, mountaineering and space and underwater exploration.

Sulphur

Sulphur is found abundantly as sulphides, the high density of which has caused them to sink to low levels in the silicate phase. Where they have come into contact with water containing dissolved oxygen, oxidation to sulphates has taken place and this has often been followed by leaching out into the oceans of the world. Subsequent disturbance of the earth's crust has in some cases resulted in seas being isolated and sulphate deposited. There is the further possibility of sulphates being reduced by bacterial activity or by carbonaceous matter to elementary sulphur, which is also deposited by volcanic activity. Being a constituent of some protein, sulphur is found in organic residues such as petroleum and coal.

Large deposits of sulphur occur in Texas and Sicily. In Texas, the sulphur is melted by superheated steam and forced to the surface by compressed air (*Figure 11.4*).

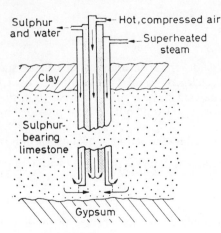

Sulphur — — — Hot, compressed air
and water — — — Superheated steam

Clay

Sulphur-bearing limestone

Gypsum

Figure 11.4 Production of sulphur in Texas: the Frasch pump

In Sicily, the sulphur-bearing rock is ignited. The heat generated causes the remaining sulphur to melt and run into moulds; it is then purified by sublimation. Sulphur can also be extracted from petroleum and from the 'spent oxide' of the gas works (p. 304) although this latter source has diminished in significance in the U.K. since the discovery and utilization of 'North Sea Gas'.

Sulphur finds many uses, for a variety of reasons. Its toxic nature, particularly towards lower organisms, renders it invaluable as a fungicide, either in elementary form or combined in organic compounds. In the vulcanization of rubber, it increases the strength by establishing cross-linkages between adjacent rubber fibres. Its tendency towards exothermic oxidation results in its use as the element in gunpowder and as phosphorus sulphide in matches. Large quantities are also consumed in the manufacture of carbon disulphide and sulphuric acid.

Selenium

Selenium (ionic radius, 200 pm) can replace sulphur (ionic radius, 180 pm) in many sulphide lattices and is often found associated with sulphides. In a concentrated state it is contained in the 'anode mud' obtained during the extraction of copper from copper pyrites. If the anode mud is roasted, selenium dioxide is formed; this volatilizes and the vapour is dissolved in water to give selenic(IV) acid, which is then reduced to the element by adding hydrochloric acid and passing sulphur dioxide:

$$\text{Se (impure)} \longrightarrow \text{SeO}_2 \xrightarrow{\text{H}_2\text{O}} \text{SeO}_3{}^{2-} \xrightarrow{\text{SO}_2} \text{Se(s) (pure)}$$

The last reaction is essentially between the selenate(IV) and sulphite ions. As the reducing tendency declines with increase in atomic number, it is the sulphite which is oxidized and the selenate(IV) which is reduced.

The electrical conductivity of grey selenium is increased in a remarkable manner by exposure to light, and one of its chief uses is in photoelectric cells.

Tellurium

Not much is known of the geochemistry of tellurium, but it is the only element found so far in chemical combination with gold. It is also associated with lead and bismuth ores, but the main commercial source is the anode mud mentioned above. Fusion of this with sodium cyanide converts the sulphur and selenium present into sodium hexacyanosulphate(IV) and hexacyanoselenate(IV), respectively, whereas tellurium is converted into sodium telluride from which tellurium is precipitated by the passage of air through the aqueous solution:

$$\text{Anode mud} \xrightarrow{\text{NaCN}} \text{Na}_2\text{Te} \xrightarrow{\text{air}} \text{Te(s)}$$

Tellurium is required for the colouring of glasses and for alloying with lead to improve its tensile strength and resistance to corrosion.

Polonium

Polonium occurs in pitchblende as a product of radioactive decay and can be prepared synthetically by the neutron irradiation of bismuth:

$$^{209}_{83}\text{Bi} + ^{1}_{0}\text{n} \rightarrow ^{210}_{83}\text{Bi} \qquad ^{210}_{83}\text{Bi} \xrightarrow{-\beta} ^{210}_{84}\text{Po}$$

Hydrides

All the elements of this Group form hydrides of formula H_2X which, with the exception of H_2O, are poisonous, pungent and volatile. The relative lack of volatility of water is the result of association of molecules through hydrogen bonding. Some properties are shown in *Table 11.3*. The thermal stability decreases from H_2O to H_2Te.

Oxygen also forms a peroxide, H_2O_2, whilst sulphur reveals its ability to form chains by continuing the sequence as far as H_2S_6

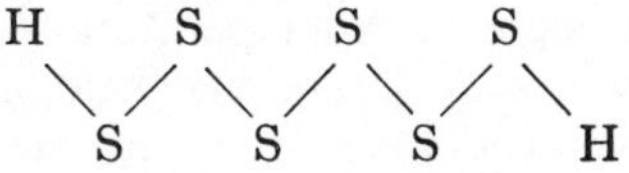

```
H    S    S    S
 \  / \  / \  / \
  S    S    S    H
```

Table 11.3 Hydrides of Group 6 elements

Hydride	K_a/mol dm^{-3} $= \dfrac{[H^+][HX^-]}{[H_2X]}$	M.p./°C	B.p./°C
H_2O	$\sim 10^{-15}$	0	100
H_2S	$\sim 10^{-7}$	-85	-60
H_2Se	$\sim 10^{-4}$	-66	-41
H_2Te	$\sim 10^{-3}$	-51	$+2$

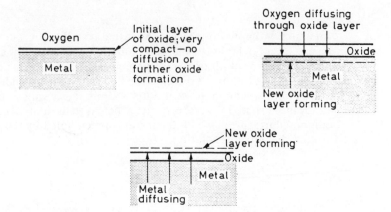

Figure 11.5 Oxide formation

Oxides

Oxygen combines directly with most elements to give oxides of low energy, so that the reactions are exothermic; e.g.

$$Al_2O_3, \quad \Delta H_f = -1590 \text{ kJ mol}^{-1} \qquad SO_2, \quad \Delta H_f = -297 \text{ kJ mol}^{-1}$$

Higher valency states are usually formed, and several elements even give peroxy compounds, e.g. sodium:

$$2Na + O_2 \rightarrow Na_2O_2$$

The ease of combustion depends on the state of subdivision, the partial pressure of oxygen and the energy of activation. Thus some finely divided metals are pyrophoric, i.e., burst into flame on exposure to air, but most elements require heating to effect complete oxidation, and even then actual burning is usually necessary. Otherwise, surface oxidation or tarnishing of the solid is the rule, although the nature of the surface film is important. If the film is very porous, then further oxidation can take place by diffusion of either oxygen or solid through the film, the rate of diffusion (and hence of reaction) being increased with increase in temperature (*Figure 11.5*).

Oxides can also often be prepared by the action of heat on oxycompounds, for example peroxides, carbonates, hydroxides, nitrates and sulphates. Generally, the less electropositive the metal, the greater the ease of decomposition of these compounds:

$$2BaO_2 \rightarrow 2BaO + O_2(g)$$

$$CaCO_3 \rightarrow CaO + CO_2(g)$$

$$Cu(OH)_2 \rightarrow CuO + H_2O$$

$$2Pb(NO_3)_2 \rightarrow 2PbO + 4NO_2(g) + O_2(g)$$

$$2FeSO_4 \rightarrow Fe_2O_3 + SO_2(g) + SO_3(g)$$

Dehydration of weak acids also often gives an oxide. Boron oxide, for instance, is produced by heating boric acid:

$$2H_3BO_3 \rightarrow B_2O_3 + 3H_2O$$

(Many of these so-called acids are, however, possibly no more than hydrated oxides in the first place, so this may not always be a genuine method of preparation.)

Some oxides are precipitated by adding acid or alkali to an appropriate solution. The acidification of a silicate produces silica, whilst alkali converts a solution of a silver salt into a precipitate of silver oxide:

$$SiO_3{}^{2-} + 2H^+ \rightarrow SiO_2(s) + H_2O$$

$$2Ag^+ + 2OH^- \rightarrow Ag_2O(s) + H_2O$$

Iron(III) oxide is precipitated when an aqueous solution of an iron(III) salt is boiled. This illustrates the preparation of an oxide by hydrolysis:

$$2Fe^{3+} + 3H_2O \rightarrow Fe_2O_3(s) + 6H^+$$

Concentrated nitric acid is a powerful oxidizing agent and often produces an oxide directly from the elements, as with tin and carbon:

$$3Sn + 4HNO_3 \rightarrow 3SnO_2 + 2H_2O + 4NO$$

$$3C + 4HNO_3 \rightarrow 3CO_2 + 2H_2O + 4NO$$

Structure of oxides

Oxides of the more electropositive metals are chiefly ionic and belong to the more common crystal systems, the actual system depending on the ratio of the radius of the cation to that of the anion, and upon the empirical formula. For example, oxides of type M_2O_3 have the corundum structure whilst type MO crystallizes either in the rock salt (6:6 coordination) or zinc blende (4:4 coordination) system (*see* p. 81).

Covalent oxides can exist in the form of
(a) Discrete molecules, e.g., carbon dioxide
(b) Single chains, e.g., selenium dioxide
(c) Double chains, e.g., antimony(III) oxide
(d) 3-Dimensional lattices, e.g., β-cristobalite (*Figure 11.6*).

Oxides are usually classified as

(1) Acidic, neutral, basic and amphoteric
(2) Simple, higher and sub-oxides

These distinctions are often of degree rather than kind and are not always adequate. It is more rewarding to search for reasons behind differences in behaviour.

Ionic oxides are basic because the oxide ion can react with water to give hydroxide, or with acid to produce water:

$$O^{2-} + H_2O \rightarrow 2OH^-$$

$$O^{2-} + 2H^+ \rightarrow H_2O$$

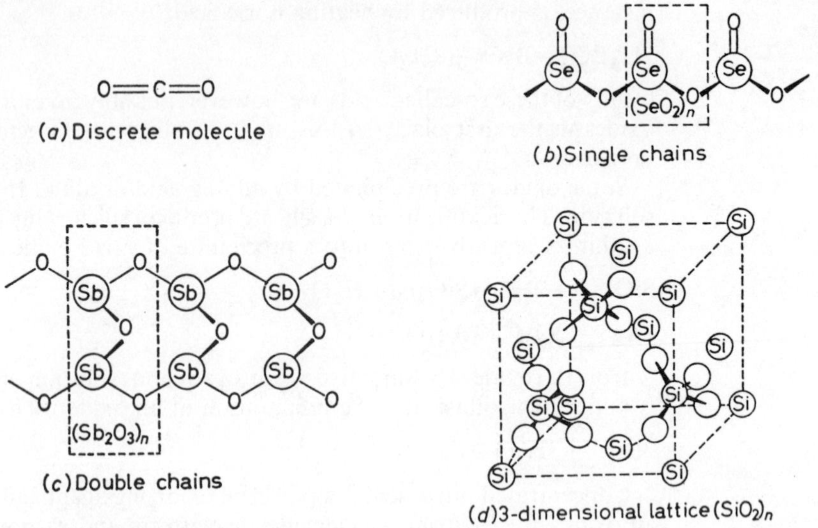

$$O=C=O$$

(a) Discrete molecule

(b) Single chains

(c) Double chains

(d) 3-dimensional lattice $(SiO_2)_n$

Figure 11.6 Covalent oxides

Some metals, with a tendency to complex-ion formation, give reactions with both acids and alkalis and are said to be amphoteric, e.g., zinc oxide

$$ZnO + 2H^+ \rightarrow Zn^{2+} + H_2O$$

$$ZnO + 2OH^- + H_2O \rightarrow Zn(OH)_4^{2-}$$

Oxides of non-metals tend to be acidic by virtue of the large electronegativity of oxygen; the inductive effect results in a shift of electrons towards the oxygen, leaving the other element deficient in electrons and able to coordinate with oxide or hydroxide ions:

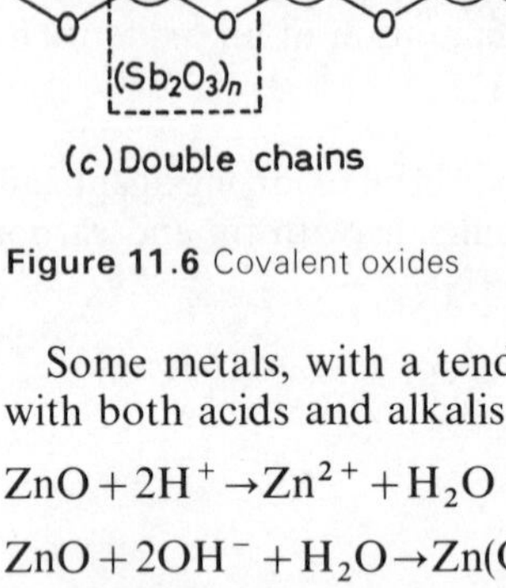

$$\overset{\delta+}{X} \rightarrowtail \overset{\delta-}{O} \qquad \longrightarrow \qquad \begin{bmatrix} X-O \\ | \\ O \end{bmatrix}^{2-}$$
$$\uparrow$$
$$\ddot{O}^{2-}$$

Consequently, the more oxygen in the oxide, the more acidic it is, e.g.,

CrO	Cr_2O_3	CrO_3
basic	amphoteric	acidic

There is a tendency for oxides of metals and non metals to react together to form salts, the extent depending on the readiness with which the metallic oxide provides the oxide ion required by the non-metal, e.g.,

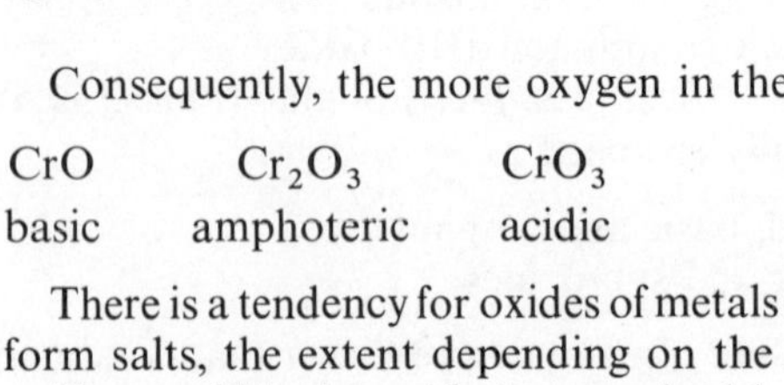

$$\overset{\delta-}{O}=\overset{\delta+}{C}=\overset{\delta-}{O}$$
$$\uparrow \qquad \longrightarrow Na^+_2 \begin{bmatrix} O & O \\ & \diagdown\diagup \\ & C \\ & \| \\ & O \end{bmatrix}^{2-}$$
$$Na^+_2\ddot{O}^{2-}$$

Table 11.4 Free energy changes in reactions between oxides

Reaction	ΔG/kJ mol^{-1}	Notes
$K_2O + H_2O$	-193	Increase in acidity of acidic oxides increases
$K_2O + CO_2$	-352	prospect of reaction
$K_2O + SO_3$	-659	↓
$BaO + CO_2$	-218	Decrease in basicity of basic oxides reduces
$CaO + CO_2$	-130	prospect of reaction
$MgO + CO_2$	-67	↓

This point is illustrated by the free energy changes accompanying the reaction between various oxides as shown in *Table 11.4.*

Peroxides

All true peroxides contain the linkage —O—O— and consequently give hydrogen peroxide when treated with cold acids

$$[-O-O-]^{2-} + 2H^+ \rightarrow H_2O_2$$

Hydrogen peroxide, H_2O_2

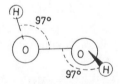

Hydrogen peroxide is required commercially as a propellant (either alone or with substances such as hydrazine) and as a bleaching agent. It is manufactured by two methods:

(1) By electrolysis of a cool solution of ammonium hydrogensulphate, using a platinum anode and a carbon cathode and high current density. Discharge of the hydrogensulphate ion gives rise to peroxodisulphuric(VI) acid

$$HSO_4^- - e^- \rightarrow HSO_4$$

$$2HSO_4 \rightarrow H_1S_2O_8$$

This solution is then distilled and hydrogen peroxide formed by hydrolysis

$$S_2O_8^{2-} + 2H_2O \rightarrow 2HSO_4^- + H_2O_2$$

(2) By passage of hydrogen into 2-ethylanthraquinone dissolved in a mixture of benzene and cyclohexanol, in the presence of, for example, nickel as catalyst. Reduction to the quinol takes place. If now a current of

oxygen replaces the hydrogen, the quinol is oxidized back to the quinone as the hydrogen is removed from it in the form of hydrogen peroxide:

The overall reaction can therefore be summarized as $H_2 + O_2 \rightarrow H_2O_2$.

Dilute solutions of hydrogen peroxide in water can be concentrated by distillation under reduced pressure. Final traces of water are removed by freezing, leaving a pale blue liquid of fairly high density (relative density $= 1.47$) and high relative permittivity.

Hydrogen peroxide is thermodynamically unstable and is therefore very susceptible to catalytic decomposition, e.g., by manganese(IV) oxide.

$$H_2O_2(l) \rightarrow H_2O(l) + \tfrac{1}{2}O_2(g) \qquad \Delta G = -122 \text{ kJ}$$

Hydrogen peroxide is thus a powerful oxidizing agent, especially in acid solution. Its standard redox potential exceeds that for iodine/iodide and for hexacyanoferrate(III)/hexacyanoferrate(II) so that both iodides and hexacyanoferrates(II) are readily oxidized in acid solution:

$$2I^- + H_2O_2 \rightarrow 2OH^- + I_2$$

$$2[Fe(CN)_6]^{4-} + H_2O_2 + 2H^+ \rightarrow 2[Fe(CN)_6]^{3-} + 2H_2O$$

In alkaline solution, though, its redox potential is less than that of the latter system and so, under these conditions, hexacyanoferrate(III) can oxidize hydrogen peroxide:

$$2[Fe(CN)_6]^{3-} + H_2O_2 + 2OH^- \rightarrow 2[Fe(CN)_6]^{4-} + 2H_2O + O_2(g)$$

Even in acid solution, the redox potential for the oxidation of hydrogen peroxide is less than that of manganate(VII) (permanganate)/manganese(II), and so acidified potassium manganate(VII) oxidizes it:

$$5H_2O_2 - 10e^- \rightarrow 10H^+ + 5O_2(g)$$

$$2MnO_4^- + 16H^+ + 10e^- \rightarrow 2Mn^{2+} + 8H_2O$$

This last reaction is used to estimate the concentrations of hydrogen peroxide solutions

$$2MnO_4^- + 5H_2O_2 + 6H^+ \rightarrow 2Mn^{2+} + 8H_2O + 5O_2(g)$$

that is, one mole (34 g) of hydrogen peroxide requires 2 dm^3 of 0.2M KMnO$_4$.

Hydrogen peroxide is marketed as 'x volumes' in terms of the reaction

$$2H_2O_2 \rightarrow 2H_2O + O_2(g)$$

2×34 g   22.4 dm^3 at s.t.p.

so a hydrogen peroxide solution containing 68 g dm^{-3} is a '22.4 volumes' solution.

Hydroxides and oxoacids

Hydroxides of the very electropositive metals are formed simply by dissolving their oxides in water. This is essentially a reaction of the oxide ion:

$$O^{2-} + H_2O \rightarrow 2OH^-$$

Those hydroxides that are insoluble in water can be conveniently prepared by precipitation, by treating a solution of a salt of a metal with an alkaline solution, e.g.,

$$M^{2+} + 2OH^- \rightarrow M(OH)_2(s)$$

This reaction is of importance in qualitative analysis and is made selective by using ammonium hydroxide as the alkali and suppressing its ionization with ammonium chloride (the common ion effect, p. 172), so that only the very insoluble hydroxides of iron(III), chromium(III) and aluminium are precipitated.

Salts of strong acids and weak bases always hydrolyse to some extent unless prevented by lowering the pH value; in fact, many hydroxides can be regarded as formed from the hydrated cations: e.g.,

$$[Al(H_2O)_6]^{3+} \underset{+H^+}{\overset{-H^+}{\rightleftharpoons}} [Al(H_2O)_5(OH)]^{2+} \underset{+H^+}{\overset{-H^+}{\rightleftharpoons}} [Al(H_2O)_4(OH)_2]^+ \rightleftharpoons \text{etc.}$$

Hydroxides of the more electropositive metals are ionic and give hydroxide ioion solution (that is, they are alkalis). As the electronegativity of the element increases, so also does the covalent character; consequently, as the Periodic Table is crossed from left to right, there is a progressive decrease in alkalinity and an increase in acidity:

Period 3

NaOH	Mg(OH)$_2$	Al(OH)$_3$	(HO)$_2$SiO	(HO)$_3$PO	(HO)$_2$SO$_2$	HOClO$_3$
alkaline	basic	amphoteric	weak acid		strong acids	$\longrightarrow$

Consideration of Fajans' rules leads to the same general conclusions: if a cation of a non-metal were present, it would have such a large polarizing power on account of its small size that the structure would instantly revert to the covalent form. If there *is* any ionization, it is the result of the weakening of the O—H link by the polarizing power of the central atom so that hydrogen ions are released:

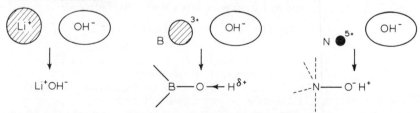

It follows that the acidity of a substance will be increased by an increase in the number of oxygen atoms in the molecule; the effect is particularly marked if some of the oxygen atoms, instead of being attached to hydrogen as well as the element in question, are linked solely to the latter (*Table 11.5*).

Table 11.5 First acid dissociation constants $K_1 = [H^+][X^-]/[HX]$ for some hydroxides and oxoacids

Substance	K_1/mol dm^{-3}	Substance	K_1/mol dm^{-3}
$B(OH)_3$	$\sim 4 \times 10^{-10}$	$O_2I(OH)$	$\sim 2 \times 10^{-1}$
$Cl(OH)$	$\sim 3 \times 10^{-8}$	$O_2N(OH)$	$\sim 1 \times 10^{-1}$
$OC(OH)_2$	$\sim 4 \times 10^{-7}$	$O_2S(OH)_2$	$\sim 1 \times 10^{1}$
$ON(OH)$	$\sim 4 \times 10^{-4}$	$O_3Cl(OH)$	$\sim 1 \times 10^{8}$
$OS(OH)_2$	$\sim 2 \times 10^{-2}$		

The more hydroxy groups there are in an acid, the closer are they together and the greater the prospect of elimination:

Sometimes, one molecule of water is eliminated from two molecules of acid, and an oxygen 'bridge' results:

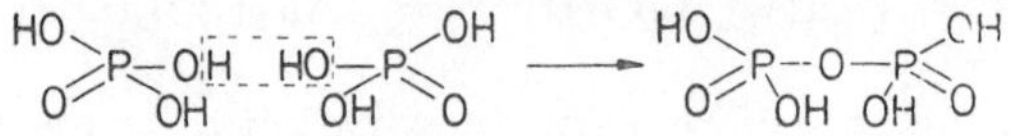

Oxo salts

The stability of oxo salts is a function of both the polarizability of the oxyanion and the polarizing power of the cation. The cations of Groups 1, for instance, with their low polarizing power, exist in solid hydrogencarbonates, sulphates(IV) (sulphites) and nitrates(III) (nitrites)*.

Basic salts are very common and, in the case of salts of strong acids, arise through the presence of polarizing cations

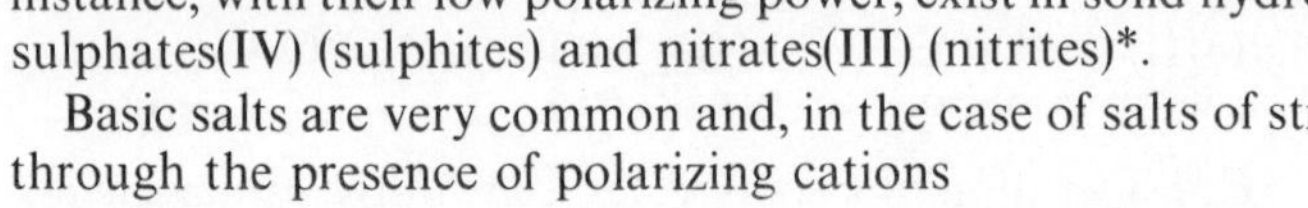

In the case of salts of weak oxyanions, hydrolysis leads to a rise in pH and subsequent reaction between liberated hydroxide ions and the cations

$$MX + 2H_2O \rightleftharpoons M^{2+} + H_2X + 2OH^-$$

* In recent years much confusion has been caused by conflicting and contradictory rulings about the naming of some oxoions. Strictly, according to IUPAC rulings along the lines of the Extended Coordination Principle, the ion SO_4^{2-} should be tetraoxosulphate(VI) and NO_3^- should be trioxonitrate(V). Such names are cumbersome for ions and their related acids in very common use, and for some time the forms sulphate(VI), nitrate(V), nitric(III) acid, etc. were encouraged. There now appears to be some moving back to a less rigorous position in regard to some of these substances and the forms shown in the right-hand column of *Table 11.6* are also used in this book.

Table 11.6 Synonyms for some common oxoacids, oxides, and salts

Formula	Synonym 1	Synonym 2
SO_2	sulphur(IV) oxide	sulphur dioxide
H_2SO_3	sulphuric(IV) acid	sulphurous acid
SO_3^{2-}	sulphate(IV) ion	sulphite ion
SO_3	sulphur(VI) oxide	sulphur trioxide
H_2SO_4	sulphuric(VI) acid	sulphuric acid
SO_4^{2-}	sulphuric(VI) ion	sulphate ion
HNO_2	nitric(III) acid	nitrous acid
NO_2^-	nitrate(III) ion	nitrite ion
HNO_3	nitric(V) acid	nitric acid
NO_3^-	nitrate(V) ion	nitrate ion

If the acid produced by hydrolysis is volatile and is lost from the system. the equilibrium shifts in favour of further hydrolysis and basic salt formation. (The situation is often made more complicated by water of crystallization which may shield the anion from polarization.)

It can be seen, then, that there is a wide range of stability of oxo salts towards water. There is similar variation in their stability to heat. The more electropositive the metal and the lower its polarizing power, the more ionic the oxo salts and the greater the resistance to heat although, of course, provided the temperature is high enough, all compounds will decompose.

Nitrates are, on the whole, vulnerable to heat; those of the heavy metals decompose to oxide; e.g.,

$$2Zn(NO_3)_2 \rightarrow 2ZnO + 4NO_2(g) + O_2(g)$$

and, in the case of metals of very low electropositivity, even the oxide decomposes

$$Hg_2(NO_3)_2 \rightarrow 2Hg + 2NO_2(g) + O_2(g)$$

Nitrates of the alkali metals decompose initially to the nitrite

$$2NaNO_3 \rightarrow 2NaNO_2 + O_2(g)$$

Carbonates are also generally decomposed by moderate heat, although it is customary to regard the carbonates of the alkali metals as thermally stable. The decomposition temperatures given in *Table 11.7* illustrate the correlation of thermal instability with the extent of covalent character.

Sulphates decompose to the oxide only at high temperatures. Sulphur dioxide and sulphur trioxide are usually evolved

$$2FeSO_4 \rightarrow Fe_2O_3 + SO_2(g) + SO_3(g)$$

Sulphides

Because sulphur is considerably less electronegative than oxygen and the sulphide ion has a very large ionic radius and is easily polarized, most

Table 11.7 Decomposition temperatures of carbonates

Carbonate	Decomposition temperature/° C
Potassium carbonate	810
Magnesium carbonate	350
Zinc carbonate	300
Mercury(II) carbonate	130

Significantly, pure aluminium carbonate has not yet even been isolated.

sulphides are covalent. Only those of the alkali and alkaline earth metals are ionized to any extent and, because the sulphide ion is attacked by water, they are readily hydrolysed in aqueous solution:

$$2H_2O + S^{2-} \rightarrow H_2S(g) + 2OH^-$$

(cf. the hydrolysis of ionic oxides).

Most of the other sulphides of metals are insoluble in water and have a metallic lustre (iron pyrites is known as 'fool's gold') and electrical conductivities comparable with those of metals. They are usually prepared by passing hydrogen sulphide through aqueous solutions of their salts. Selective precipitation is achieved to some extent in qualitative analysis by controlling the pH of the solution and therefore the extent of the ionization of the hydrogen sulphide (p. 172).

The very high electronegativity of oxygen produces compounds in which the highest valency states are stabilized. This is not the case with sulphur; for example, whilst Bi_2O_5 and PbO_2 exist, Bi_2S_5 and PbS_2 are unknown.

Hydrogen sulphide

$$b.p. - 61°C$$

This colourless gas (b.p. $-61\,°C$), of offensive odour, is usually prepared by the action of hydrochloric acid on iron(II) sulphide, but a purer product results from antimony(III) sulphide

$$Sb_2S_3 + 6H^+ \rightarrow 2Sb^{3+} + 3H_2S(g)$$

The hydrogen sulphide given off can be washed with water and dried over anhydrous magnesium sulphate. It is a reducing agent and, in the course of reduction, is usually oxidized to sulphur, $H_2S \rightarrow 2H^+ + S(s) + 2e^-$:

$$H_2S + 2Fe^{3+} \rightarrow 2Fe^{2+} + 2H^+ + S(s)$$

$$H_2S + Cl_2 \rightarrow 2HCl + S(s)$$

$$H_2S + H_2SO_4 \rightarrow 2H_2O + SO_2(g) + S(s)$$

The sulphur dioxide in the last example can be further reduced

$$2H_2S + SO_2 \rightarrow 2H_2O + 3S(s)$$

Hydrogen sulphide burns in air to form sulphur dioxide and water, provided there is sufficient oxygen to give complete combustion; usually some sulphur is formed as well

$$H_2S \xrightarrow{O_2} H_2O + S(s) + SO_2(g)$$

Hydrogen sulphide is acidic in aqueous solution

$$H_2S \rightleftharpoons H^+(aq) + HS^- \qquad K_1 = 9 \times 10^{-8} \text{ mol dm}^{-3}$$

$$HS^- \rightleftharpoons H^+(aq) + S^{2-} \qquad K_2 = 1 \times 10^{-15} \text{ mol dm}^{-3}$$

The extent of ionization is influenced by the acidity of the solution. In concentrated acid, the ionization is suppressed by the common ion effect (p. 172) and the sulphide ion concentration is then only sufficient to precipitate the very insoluble sulphides. In alkaline solution, on the other hand, further ionization is encouraged because of the removal of hydrogen ions (to form water), and more soluble sulphides can then be precipitated.

Sulphides of most non-metals exist. As with oxygen, sulphides of the non-metals of Period 2 are present as discrete molecules, e.g. carbon disulphide and sulphur nitride

S=C=S
Carbon disulphide

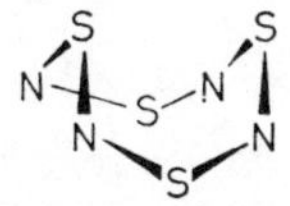

Sulphur nitride

Non-metals of the remaining periods usually form sulphides of macromolecular structure

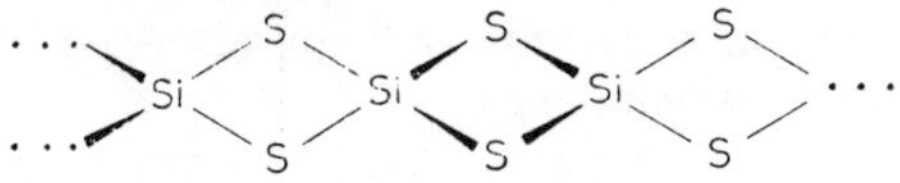

Polysulphides

If sulphur is heated with solutions of sulphides of the alkaline earth and alkali metals, a mixture of polysulphides is formed. Addition of the cold product to hydrochloric acid at 10 °C or less results in the formation of a yellow oil which on distillation yields hydrides up to H_2S_6. In all of these compounds, sulphur reveals its capacity for catenation:

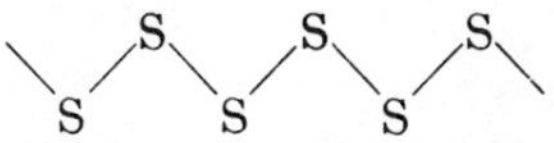

Sulphur dioxide and sulphites [sulphates(IV)]

Sulphur dioxide, SO_2

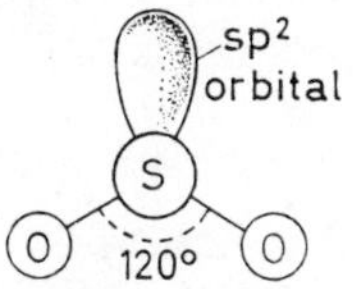

sp² orbital

S

O 120° O

This is obtained as a colourless, pungent gas (b.p. $-10\,^{\circ}C$) when sulphur-containing materials are burnt in air. In the case of coal, sulphur dioxide is formed incidentally, and it is the escape of this gas into the atmosphere which plays such a great part in its pollution. Commercially, sulphur dioxide is produced when pyrites and 'spent oxide' are burnt, e.g.,

$$4FeS_2 + 11O_2 \rightarrow 2Fe_2O_3 + 8SO_2(g)$$

Sulphur dioxide is also a by-product of the extraction of metals from sulphide ores. Large quantities result, for example, from the roasting of zinc blende

$$2ZnS + 3O_2 \rightarrow 2ZnO + 2SO_2(g)$$

The chief uses of sulphur dioxide are in the manufacture of sulphuric acid, as a bleaching agent, disinfectant and preservative and, together with calcium hydroxide [that is, as $Ca(HSO_3)_2$], for removing lignin from wood and leaving the cellulose in a condition suitable for paper manufacture.

In the laboratory, sulphur dioxide is conveniently prepared by the action of either hot, concentrated sulphuric acid on copper (or of cold dilute acid on sulphites), and collected by downward delivery

$$Cu + 2H_2SO_4 \rightarrow CuSO_4 + 2H_2O + SO_2(g)$$

Sulphur dioxide dissolves readily in water to give a solution commonly called 'sulphurous acid' but which is now believed to be made up largely of a clathrate (p. 392), of composition $SO_2.7H_2O$, with sulphur dioxide impri-soned in a water lattice. There is, however, a certain amount of the dibasic acid, H_2SO_3, present as well and the corresponding acid and normal salts (hydrogensulphites and sulphites) are well characterized.

Sulphites contain the pyramidal ion:

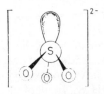

and, because sulphurous acid is not very strong, their aqueous solutions are extensively hydrolysed

$$SO_3^{2-} + 2H_2O \rightleftharpoons H_2SO_3 + 2OH^-$$

Sulphites also react with acids to liberate sulphur dioxide

$$SO_3^{2-} + 2H^+ \rightarrow H_2SO_3 \rightarrow H_2O + SO_2(g)$$

In solution, sulphur dioxide can be both a reducing and an oxidizing agent: when it reduces, it is often oxidized itself to sulphate, and when it is reduced itself by stronger reducing agents, the end product is often sulphur.

Reducing reactions

$$H_2SO_3 + H_2O - 2e^- \rightarrow SO_4^{2-} + 4H^+ \qquad E^{\ominus} = -0.20 \text{ V}$$

The electrons provided by this system can be accepted by an oxidizing system such as

$$MnO_4^- + 8H^+ + 5e^- \rightarrow Mn^{2+} + 4H_2O \qquad E^{\ominus} = 1.52 \text{ V}$$

Combination of the two systems gives

$$2MnO_4^- + 5H_2SO_3 \rightarrow 2Mn^{2+} + 4H^+ + 5SO_4^{2-} + 3H_2O \qquad E^{\ominus} = 1.32 \text{ V}$$

Similarly, sulphur dioxide reduces an acidified solution of potassium dichromate, the colour changing from orange to green

$$Cr_2O_7^{2-} + 3H_2SO_3 + 2H^+ \rightarrow 2Cr^{3+} + 3SO_4^{2-} + 4H_2O$$

orange   green

The colour change accompanying this reaction is often used as a test for sulphur dioxide.

Oxidizing reactions

$$H_2SO_3 + 4H^+ + 4e^- \rightarrow S(s) + 3H_2O \qquad E^{\ominus} = 0.45 \text{ V}$$

The system is capable of oxidizing iron to iron(II)

$$Fe - 2e^- \rightarrow Fe^{2+} \qquad E^{\ominus} = 0.44 \text{ V}$$

that is

$$2Fe + H_2SO_3 + 4H^+ \rightarrow 2Fe^{2+} + S(s) + 3H_2O$$

Zinc is also capable of being oxidized by a solution of sulphur dioxide; this time, the sulphite is reduced to dithionate:

$$Zn + 2SO_3^{2-} + 4H^+ \rightarrow Zn^{2+} + S_2O_4^{2-} + 2H_2O \qquad E^{\ominus} = 0.89 \text{ V}$$

As the critical temperature of sulphur dioxide is 158 °C, it can be liquefied merely by the application of pressure. The liquid formed is of high relative permittivity, and it has been suggested that it may self-ionize slightly:

$$2SO_2 \rightleftharpoons SO^{2+} + SO_3^{2-}$$

(compare with water and liquid ammonia).

Sodium thiosulphate [sodium trioxothiosulphate], Na$_2$S$_2$O$_3$.5H$_2$O

If sodium sulphite solution is boiled with sulphur, sodium thiosulphate(VI) is formed:

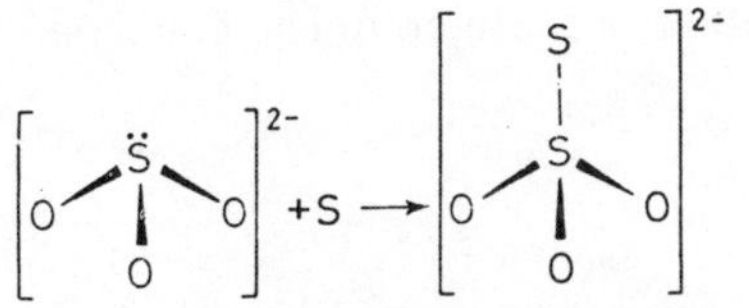

The reaction corresponds to the oxidation of sulphite to sulphate

$$\left[\ddot{S}(O)(O)(O) \right]^{2-} + O \longrightarrow \left[S(O)(O)(O)(O) \right]^{2-}$$

Sodium thiosulphate is very important in the laboratory as a means of reducing iodine in volumetric analysis:

$$2S_2O_3^{2-} + I_2 \rightarrow \quad S_4O_6^{2-} \quad + 2I^-$$
$$\text{Tetrathionate}$$

Commercially it is of value (as 'hypo') for the 'fixing' of photographic film: it reacts with the silver ions which have not been exposed to light (and which have therefore not been reduced during development) to form complex ions which can be washed out of the gelatine emulsion on the film, e.g.,

$$2S_2O_3^{2-} + Ag^+ \rightarrow [Ag(S_2O_3)_2]^{3-}$$

Thiosulphates react with acids to give sulphur dioxide and a deposit of sulphur; they are therefore readily recognized:

$$S_2O_3^{2-} + 2H^+ \rightarrow H_2O + SO_2(g) + S(s)$$

Sulphur(VI) oxide

Sulphur(VI) oxide can be made on the small scale by heating sulphates, for example iron(III) sulphate

$$Fe_2(SO_4)_3 \rightarrow Fe_2O_3 + 3SO_3(g)$$

On the large scale it is manufactured by the *contact process* (*Figure 11.7*). Sulphur dioxide and purified air are passed over a catalytic melt of vanadium(V) oxide in an alkali metal disulphate at a temperature of about 450 °C:

$$2SO_2 + O_2 \rightleftharpoons 2SO_3 \qquad \Delta H = -(2 \times 96) \text{ kJ}$$

Application of Le Chatelier's principle leads to the conclusion that the forward reaction will be favoured by high pressure and low temperature. In practice, the slight increase in yield resulting from the application of excessive pressure is not thought to warrant the increased capital cost, whilst low temperature means that the equilibrium is reached only slowly. A compromise temperature of 450 °C together with a catalyst ensures a fairly high reaction velocity and a not-too-unfavourable equilibrium. Excess of air is also used, as this brings about a more complete conversion of the sulphur dioxide:

$$K = \frac{[SO_3]^2}{[SO_2]^2[O_2]}$$

Sulphur(VI) oxide is a very hygroscopic solid. It is polymeric and the more common form melts at 18 °C and resembles asbestos in appearance.

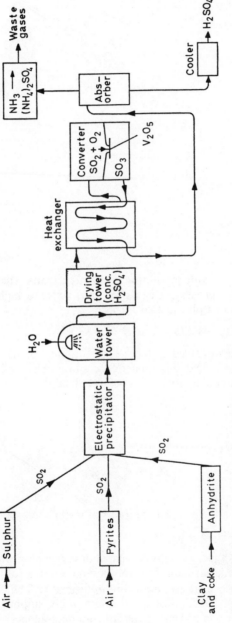

Figure 11.7 Manufacture of sulphuric acid by the Contact Process

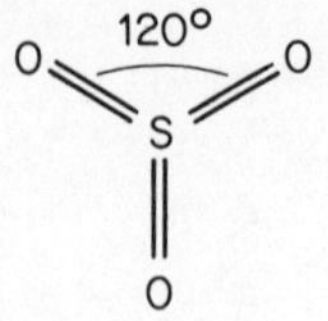

Vapour

(planar monomer)

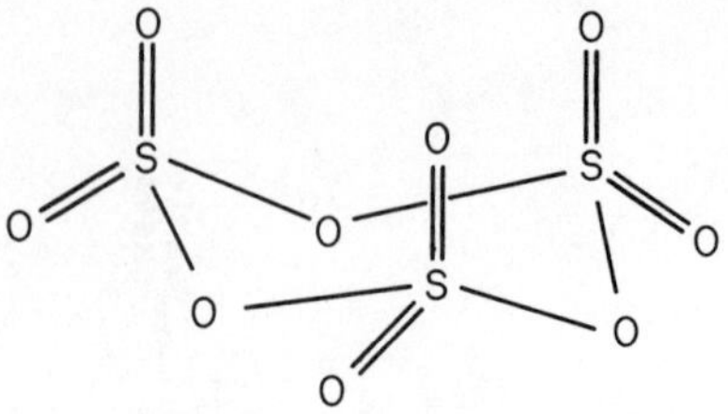

γSO_3 (solid)

(cyclic trimer)

Sulphuric acid

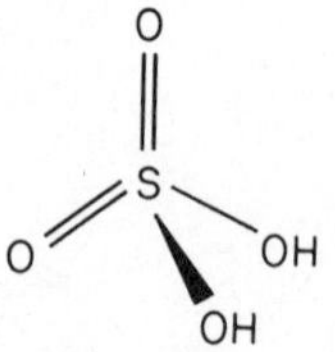

Sulphur(VI) oxide is 'sulphuric anhydride' and when, after manufacture by the contact process, it is passed into water, a highly exothermic reaction results and sulphuric acid forms:

$$H_2O + SO_3 \rightarrow H_2SO_4 \qquad \Delta H = -88 \text{ kJ}$$

The heat evolved reduces the solubility of the sulphur(VI) oxide and produces an acid mist. Therefore sulphur(VI) oxide is passed into cold, concentrated sulphuric acid to give disulphuric (pyrosulphuric) acid:

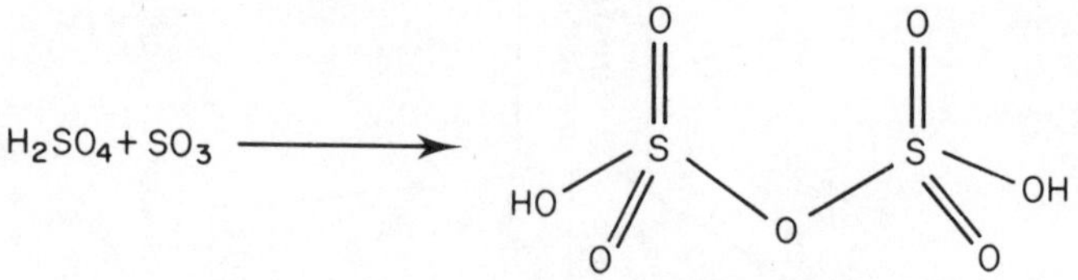

$$H_2SO_4 + SO_3 \longrightarrow$$

Addition of the requisite amount of water then yields more sulphuric acid:

$$H_2S_2O_7 + H_2O \rightarrow 2H_2SO_4$$

The contact process allows an acid of very high purity and concentration to be obtained, particularly if the raw material is sulphur.

Sulphuric acid is important industrially. The larger uses include the manufacture of fertilizers (e.g., ammonium sulphate and superphosphate), paints, fibres (e.g., viscose rayon), and detergents (which are often based on sulphonic acids). Smaller quantities are used in the manufacture of plastics, dyestuffs and other acids, and in the removal of mill-scale from iron (iron pickling).

Whilst natural sulphur is easily the major source of sulphur dioxide for the manufacture of sulphuric acid, sulphides such as pyrites and zinc blende are also utilized, sulphur dioxide being produced by roasting the mineral in air:

$$2ZnS + 3O_2 \rightarrow 2ZnO + 2SO_2$$

In Britain, the *anhydrite process* is important. Anhydrite, $CaSO_4$, clay and coke are finely ground together and heated strongly; the calcium sulphate is reduced by the coke to calcium oxide which combines with silica and alumina to give cement clinker as a by-product. The sulphur dioxide produced simultaneously is then converted into sulphuric acid by the contact process

$$CaSO_4 + C \rightarrow CaO + CO_2(g) + SO_2(g)$$
$$\downarrow$$
$$CaSiO_3, \text{ etc.}$$

Sulphuric acid, when pure, is a colourless, oily liquid of high density and relative permittivity. It undergoes some self-ionization

$$2H_2SO_4 \rightleftharpoons H_3SO_4^+ + HSO_4^-$$

The addition of water results in the evolution of considerable heat as oxonium ions are formed

$$H_2SO_4 + H_2O \rightarrow H_3O^+ + HSO_4^-$$

This reaction takes place so readily that the concentrated acid is very hygroscopic and is therefore useful as a drying agent. It can also, by removing the elements of water from compounds, act as a dehydrating agent. For example, glucose can be dehydrated to carbon, and methanoic (formic) acid to carbon monoxide

$$C_6H_{12}O_6 - 6H_2O \rightarrow 6C$$

$$HCOOH - H_2O \rightarrow CO(g)$$

When hot and concentrated, sulphuric acid is a strong oxidizing agent. For instance, carbon is oxidized to carbon dioxide and copper to copper(II) ions, the acid in each case being reduced to sulphur dioxide

$$C + 2H_2SO_4 \rightarrow CO_2(g) + 2SO_2(g) + 2H_2O$$

$$Cu + 2H_2SO_4 \rightarrow CuSO_4 + SO_2(g) + 2H_2O$$

Being dibasic, sulphuric acid gives rise to two series of salts, the normal sulphates and the acid hydrogensulphates.

Most sulphates and hydrogensulphates are soluble in water and are therefore prepared usually by the neutralization of sulphuric acid by base, followed by crystallization.

Sulphates are, on the whole, fairly stable to heat, especially in the case of the more electropositive metals. Strong heating of sulphates of the transitional metals results in the oxide being formed, e.g., nickel(II) sulphate:

$$2NiSO_4 \rightarrow Ni_2O_3 + SO_2(g) + SO_3(g)$$

The sulphate ion is tetrahedral in shape and often associated with a

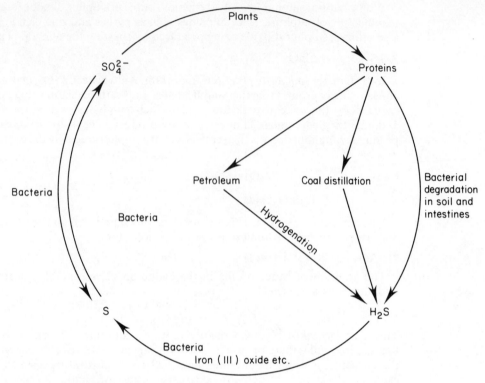

Figure 11.8 The sulphur cycle in nature

molecule of water. As the cation is also frequently hydrated, sulphates may contain appreciable water of crystallization, e.g., $FeSO_4.7H_2O$.

Isomorphous groups of double sulphates exist. The 'alums' have the general formula $M^IN^{III}(SO_4)_2.12H_2O$ where M^I is the ammonium, thallium(I), or alkali metal ion and N^{III} an ion such as aluminium, chromium(III), iron(III) or gallium(III). There is also a series $M^I_2N^{II}(SO_4)_2.6H_2O$, typified by ammonium iron(II) sulphate.

One of the few insoluble sulphates is barium sulphate. This is therefore precipitated in the identification and estimation of soluble sulphates:

$$Ba^{2+} + SO_4^{2-} \rightarrow BaSO_4(s)$$

Thionic acids, of general formula $H_2S_xO_6$, exist up to $x = 6$ and illustrate further the ability of sulphur to form chains. The sodium salt of tetrathionic acid, $HO.SO_2.S.S.SO_2.OH$, is obtained as a by-product of the reaction between thiosulphate and iodine.

Figure 11.9 (*opposite*) The general chemistry of sulphur; although sulphur is taken as the typical element of Group 6, the reactions shown within the broken circle, where the element is in a negative oxidation state, can be compared with those of oxygen

Summary

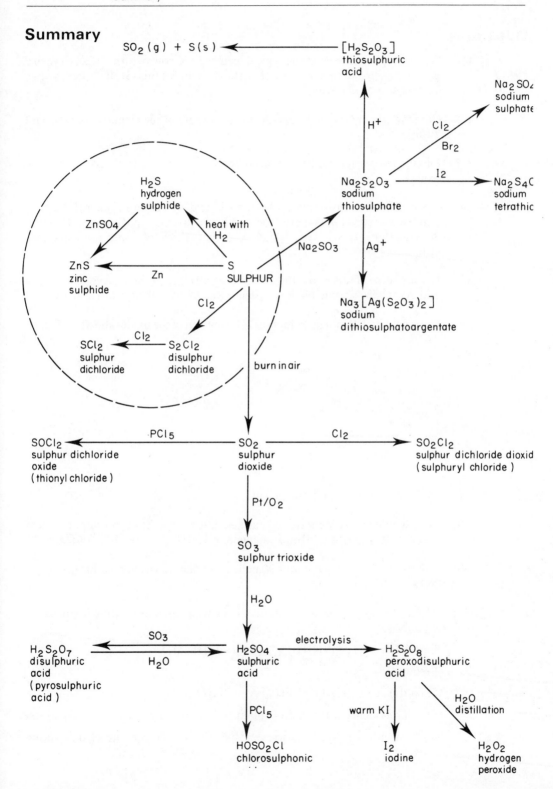

Questions

(1) Why is sulphur chosen as the typical member of Group 6 rather than oxygen? By comparing the salient features of the elements of this Group, discuss how far this choice is justified.

(2) Compare the hydrides of this Group with those of the elements of adjacent groups.

(3) Discuss the merits and limitations of the various methods of classification of oxides.

(4) In the past, the consumption of sulphuric acid by a country was held to be a reliable yardstick for assessing the standard of living of that country. How far do you think this is true today? Can you think of some more reliable criterion?

(5) In the course of this book, the terms Avogadro's hypothesis, Avogadro's law and Avogadro's principle have been used. Discuss their relative merits.

(6) Comment on the molar enthalpies of formation of the oxides shown in *Figure 11.10*.

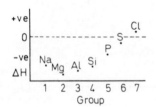

Fig. 11.10

(7) Write an essay on polymorphism.

(8) How would you account for the marked difference in acidity of sulphuric acid $(K_{a(1)} = 10^{-1}\,\text{mol}\,\text{dm}^{-3})$ and ethanedioic acid $(K_{a(1)} = 6 \times 10^{-3}\,\text{mol}\,\text{dm}^{-3})$?

(9) Compare the chemistry and structures of the iso-electronic ions, ClO_4^{-}, SO_4^{2-}, PO_4^{3-} and SiO_4^{4-}.

(10) The mechanism of the dissolution of iron in an aqueous solution involves the equilibria:

$$Fe + OH^{-} - e^{-} \rightleftharpoons [Fe(OH)]_{adsorbed} \rightleftharpoons [Fe(OH)(aq)]^{+} + e^{-}$$

followed by

$$[Fe(OH)(aq)]^{+} + H_3O^{+} \rightarrow Fe(aq)^{2+} + 2H_2O$$

On this evidence attempt to explain the following results:

(*a*) Concentrated hydrochloric acid produces a slower rate of dissolution than the moderately concentrated acid.

(b) Dilute and concentrated sulphuric acid produce different results.

(11) X is a colourless gas. Its aqueous solution has the following properties:

(a) It reduces iron(III) to iron(II) ions.
(b) It decolorizes a solution of iodine.
(c) It is reduced to Y in the presence of zinc and sulphuric acid.
(d) It reacts with Y to form a pale yellow precipitate.

A given volume of 9 diffused through a porous plug in 13.75 s, while the same volume of the gas Y (relative molecular mass = 34) required only 10 s.

Identify X and Y and write equations for the reactions (a) to (d). How would you expect gas X to react with a solution of potassium permanganate and gas Y with a solution of iron(III) chloride?

[O.]

(12) X is an element. If 2.56 g of X is dissolved in 50 g of carbon disulphide the boiling point is raised by 0.46 kelvins. X forms oxides P and Q containing 50 per cent and 60 per cent by mass of oxygen respectively. Both oxides are soluble in water to give acidic solutions. Iron will react with X to form a compound R of composition Fe_mX_n. Action of acid on R gives a gas S with a density $1.52\,g\,dm^{-3}$ and containing 5.88 per cent by mass of hydrogen. Make what deductions you can about X and its compounds, and write equations for the reactions described.

(1 mol of solute in 1 kg of CS_2 raises the boiling point by 2.3 kelvins. The molar volume of a gas is $22.4\,dm^3\,mol^{-1}$ at 273 K and 1 atm.)

[O]

12 Group 5 elements

The elements • Structures • General chemistry • Occurrence and
extraction – nitrogen, phosphorus, arsenic, antimony and
bismuth • Hydrides of nitrogen and phosphorus • Oxides of
nitrogen • Oxoacids of nitrogen • Phosphoric(V) acids and
phosphates(V)

Group 4	ns np		Group 6
C	Nitrogen, N	2.5	O
Si	Phosphorus, P	2.8.5	S
Ge	Arsenic, As	2.8.18.5	Se
Sn	Antimony, Sb	2.8.18.18.5	Te
Pb	Bismuth, Bi	2.8.18.32.18.5	Po

The elements

A common characteristic of the elements of Group 5 is the existence of five
electrons in the outer shell. In the ground state, these consist of two of
opposite spin in the s orbital and one in each of the p orbitals. The energy
levels of the outermost electrons of the nitrogen and phosphorus atoms are
indicated by the relevant ionization energies shown in *Figure 12.1*. It can be
seen that the ease with which an electron can be removed becomes much less
with each successive one and that there is a very large increase in the energy
required when an electron is removed from the penultimate shell.

An increase in relative atomic mass and molecular complexity on
descending the Group results in nitrogen, with a simple diatomic molecule,
being its only gaseous member (cf. Group 6). Another consequence of
increasing atomic number is an increase in atomic volume: there is thus less
attraction between the nucleus and the outermost electrons and therefore a
general decrease in electronegativity in passing from nitrogen to bismuth,
particularly between nitrogen and phosphorus. Nitrogen is one of the most
electronegative of all elements and can gain three electrons to form the
nitride ion, N^{3-}, whereas bismuth is sufficiently electropositive to show
metallic properties, sometimes losing three electrons to give Bi^{3+} (*Table
12.1*).

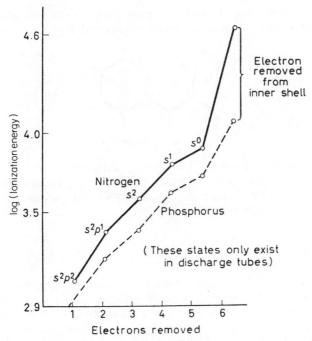

Figure 12.1 Successive ionization energies of nitrogen and phosphorus

Table 12.1 Some properties of the elements of Group 5

Element	Electronic configuration	Atomic radius/pm	Ionic radius (M^{3+})/pm	M.p./°C	B.p./°C
N	$2s^2 2p^3$	74		−210	−196
P	$3s^2 3p^3$	110		44 (white)	280 (white)
As	$4s^2 4p^3$	120	70	610 (sublimes)	
Sb	$5s^2 5p^3$	140	90	630	1380
Bi	$6s^2 6p^3$	150	120	271	1451

Structures

Only atoms of small size are capable of forming multiple bonds, and therefore nitrogen is unlike the rest of the Group in that it forms a diatomic molecule, $N \equiv N$, of considerable stability ($\Delta H_f = -921$ kJ mol^{-1}). Doubtless, this fact is chiefly responsible for the general non-reactivity of the element and the comparative instability of many of its compounds, e.g., the oxides.

Phosphorus, arsenic and antimony are all polymorphic, the less dense form in each case being translucent, tetrahedral in molecular shape and soluble in organic solvents, and the denser form opaque, metallic in form, insoluble in organic liquids and an electrical conductor.

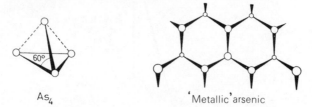

As₄   'Metallic' arsenic

The tetrahedral molecules, having very small inter-bond angles of 60 degrees, are strained and unstable; thus white phosphorus readily ignites in air to form the stable compound, phosphorus(V) oxide, in which the smallest inter-bond angles are 101 degrees (p. 272).

General chemistry

Members of this Group have oxidation states ranging from -3 to $+5$, nitrogen showing the greatest versatility in this respect as shown in *Table 12.2*. Nitrogen, although with little of the facility of its neighbour carbon towards catenation, does form multiple bonds such as those shown in *Table 12.3*. It should be noted that imines and diazo compounds are more readily reduced than nitriles and nitrogen, respectively.

Table 12.2 Oxidation states of nitrogen in various compounds

Compound	Oxidation number	Compound	Oxidation number
NH_3	-3	N_2O	$+1$
N_2H_4	-2	NO	$+2$
NH_2OH	-1	N_2O_3	$+3$
N_2	0	N_2O_4	$+4$
		N_2O_5	$+5$

Molecules containing double-bonded nitrogen are angular, and there are possibilities of geometric isomerism (p. 402). For example, azobenzene exists in two forms

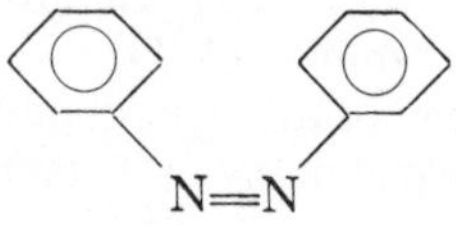

N=N

cis

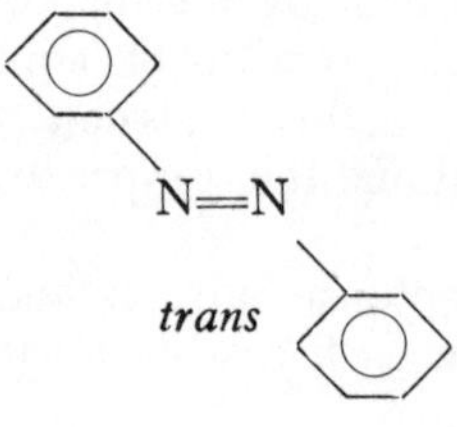

N=N

trans

Table 12.3 Comparison of bond enthalpies of C—N and N—N bonds

Bond	$\Delta H/\text{kJ mol}^{-1}$	Bond	$\Delta H/\text{kJ mol}^{-1}$
C—N	-305	N—N	-163
C=N (imine)	-616	N=N (diazo)	-418
C≡N (nitrile)	-891	N≡N	-921

The maximum covalency of nitrogen is four (because of the absence of d orbitals in the second quantum shell) and the resultant 4-covalent molecules, like those for the rest of the Group, are based on a tetrahedral structure. In the other elements, 6-covalent compounds are common and have an octahedral structure, e.g., $Sb(OH)_6^-$ and PCl_6^-.

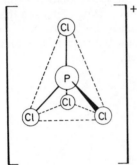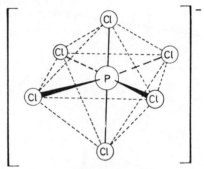

tetrachlorophosphorus(III)
4-covalent phosphorus
(tetrahedral)

hexachlorophosphate(V)
6-covalent phosphorus
(octahedral)

The decrease in electronegativity on descending the Group results in the appearance of some cationic chemistry with antimony and bismuth, even though it normally means the antimonyl ion, SbO^+, in the case of the former. Bismuth, on the other hand, gives rise to the bismuth ion, Bi^{3+}, and to a fairly stable series of salts with oxoacids.

The elements of this group, like those of Groups 4 and 6, feature prominently in organic chemistry. Nitrogen is an essential component of protein, and phosphorus, in the form of phosphoric(V) acid, is instrumental in the manufacture of protein in the living cell and in respiration and photosynthesis (p. 519). Nitrogen, phosphorus, arsenic and antimony form compounds of the type MR_3 (where R = alkyl or aryl radical), whereas bismuth participates chiefly in compounds of the type OMR_3.

Hydrides

All the elements form hydrides of the type MH_3, the stability decreasing markedly from nitrogen to bismuth. These are volatile compounds (*Table 12.4*), although ammonia, NH_3, is sufficiently polar to give rise to hydrogen bonding (p. 57), with a consequently lower volatility than would otherwise

Table 12.4 Boiling points and free energies of dissociation of Group 5 hydrides

	NH_3	PH_3	AsH_3	SbH_3	BiH_3
B.p./°C	-34	-87	-55	-18	$+20$
$\Delta G/kJ\ mol^{-1}$ for $2MH_3 \rightarrow M_2 + 3H_2$	$+33$	-26	-136	-292	-400

be expected. The effect of the lone pair on the central atom is to give a pyramidal molecule (p. 52), the angles between the bonds decreasing as the electronegativity of the element decreases. Thus, the HÑH angle in ammonia is almost tetrahedral, whilst the HP̂H angle in phosphine is only 93 degrees. As nitrogen is much more electronegative than phosphorus, the electron pairs of the covalent bonds are much nearer nitrogen than is the case with phosphorus. There are consequently much stronger mutually repulsive forces between the bonds in ammonia than in phosphine, and so the influence of the lone pair is less pronounced.

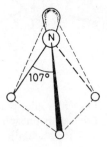 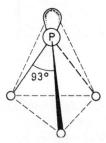

Nitrogen also forms N_2H_4, hydrazine, and N_3H, hydrogen azide, whilst phosphorus forms P_2H_4, diphosphane.

Halides

Nitrogen forms only trihalides, NX_3. Of these, only NF_3 is stable. NCl_3 is explosive

$$2NCl_3 \rightarrow N_2(g) + 3Cl_2(g) \qquad \Delta H = -250\ kJ\ mol^{-1}$$

The existence of NBr_3 is doubtful and the tri-iodide is always associated to some extent with ammonia. Trihalides of the remaining elements have been characterized; like the hydrides they are all pyramidal. Although there is an increase in ionic character on descending the Group, they are all hydrolysed to a greater or lesser extent by water

$$PCl_3 + 3H_2O \rightarrow 3HCl + P(OH)_3 \text{ (i.e., } H_3PO_3)$$

$$4AsCl_3 + 6H_2O \rightarrow 12HCl + As_4O_6$$

$$SbCl_3 + H_2O \rightleftharpoons 2HCl + SbOCl$$

$$BiCl_3 + H_2O \rightleftharpoons 2HCl + BiOCl$$

Phosphorus and antimony form penta- as well as tri-halides, the pentachlorides having trigonal bipyramidal structures in the vapour state

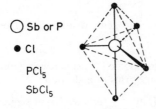

○ Sb or P

● Cl

PCl$_5$

SbCl$_5$

Solid phosphorus pentachloride consists of tetrahedral PCl$_4{}^+$ and octahedral PCl$_6{}^-$ ions (*see* diagram p. 269) whilst solid phosphorus pentabromide exists as PBr$_4{}^+$Br$^-$. These halides are also hydrolysed by water, e.g.,

PCl$_5$ + H$_2$O → 2HCl + POCl$_3$

POCl$_3$ + 3H$_2$O → 3HCl + H$_3$PO$_4$ [i.e., PO(OH)$_3$]

Oxides and oxoacids

The characteristic oxides have the empirical formula M_2O_3. With the exception of bismuth(III) oxide they can be regarded as acid anhydrides (*Table 12.5*).

N$_2$O$_3$ ionizes when in the liquid state

N$_2$O$_3$ ⇌ NO$^+$ + NO$_2{}^-$

The dimeric molecules of the phosphorus, arsenic and antimony compounds are tetrahedral in shape

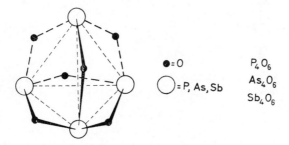

● = O

○ = P, As, Sb

P$_4$O$_6$

As$_4$O$_6$

Sb$_4$O$_6$

There is a general increase in basic character on descending the Group, As$_4$O$_6$ reacting with concentrated sulphuric acid to give a basic sulphate, whilst Sb$_4$O$_6$ yields a normal sulphate under the same conditions and bismuth(III) oxide even reacts with phosphoric(V) acid.

All give oxides of empirical formula M_2O_5, although in the case of bismuth, the compound has apparently never been prepared in the pure state. With the exception of the latter, these oxides too can be regarded as acid anhydrides, e.g.,

P$_4$O$_{10}$ + 6$_2$O → 4H$_3$PO$_4$ [phosphoric(V) acid]

Table 12.5 Group 5 trioxides and related oxoacids

Typical reactions	*Synonyms*
$N_2O_3 + H_2O \rightarrow 2HNO_2$	Nitrous acid, hydrogen dioxonitrate(III), or nitric(III) acid
$P_4O_6 \rightarrow H_3PO_3$ $H_4P_2O_5$ HPO_2 $\Big\}$	Forms of phosphonic [phosphorous or phosphoric(III)] acid in various forms of hydration
$As_4O_6 \rightarrow X_3AsO_3$	Arsenates(III), trioxoarsenates(III) (arsenites)
$Sb_4O_6 \rightarrow XSbO_2$	Antimonates(III), dioxoantimonates(III) (antimonites)
$Bi_2O_3 \rightarrow Bi(OH)_3$	Bismuth(III) hydroxide

although the antimony 'acids' are probably no more than hydrated forms of the oxide.

Like the trioxides, the pentoxides of phosphorus, arsenic and antimony are dimeric molecules

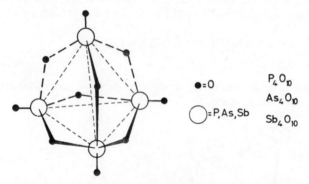

Both the trioxides and the pentoxides can be obtained by direct combination of the relevant elements under suitable conditions.

Sulphides

The sulphides are formed in an analogous manner to the oxides. As would be expected from simple valency considerations, arsenic, antimony and bismuth give trisulphides, the first two reacting with alkali and sulphides to form trithioarsenate(III) and trithioantimonate(III), respectively, e.g.,

$$As_2S_3 + 3Na_2S \rightarrow 2Na_3AsS_3$$

(cf. $As_2O_3 + 3Na_2O \rightarrow 2Na_3AsO_3$).

In addition, arsenic, like nitrogen, forms a tetrasulphide. Both have the structure

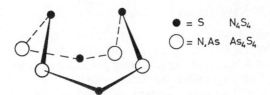

Four sulphides of phosphorus are so far known, all of which contain four phosphorus atoms in the molecule and are based on the P_4 tetrahedron. The most stable of these is P_4S_3

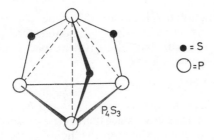

Nitrides and phosphides

Binary compounds of nitrogen and phosphorus with elements of lower electronegativity are known as nitrides and phosphides, respectively, whilst similar compounds with the elements of Groups 6 and 7 are normally classified as chalkogenides or halides.

Nitrides (cf. carbides)
There are three possibilities:
Covalent nitrides are prepared by direct combination, by the action of nitrogen on a mixture of an oxide and carbon or of ammonia on a halide, e.g.,

$$3SiCl_4 + 4NH_3 \rightarrow Si_3N_4 + 12HCl(g)$$

Of particular interest is boron nitride, a solid of remarkable stability, existing in two forms having the structures of the isoelectronic graphite and diamond (p. 302). It is manufactured by heating boron oxide and carbon in an atmosphere of nitrogen at 1400 °C.

$$B_2O_3 + 3C + N_2 \rightarrow 2BN + 3CO(g)$$

Electrovalent nitrides contain the nitride ion, N^{3-}. Lithium, sodium, the metals of Group 2, and possibly also potassium, combine directly with nitrogen on heating, to form compounds of high melting point which are readily hydrolysed by water with the evolution of ammonia, e.g.,

$$N^{3-} + 3H_2O \rightarrow 3OH^- + NH_3(g)$$

The metals of Group 1, excluding lithium, are also capable of replacing part of the hydrogen of ammonia, on heating in the gas, to give so-called amides, e.g.,

$$2K + 2NH_3 \rightarrow 2KNH_2 + H_2(g)$$

Interstitial nitrides also exist, in which the small nitrogen atom occupies holes in the metallic lattice: the metallic structure persists in most cases, so that the product is hard, with high melting point and electrical conductivity. For example, vanadium nitride has a m.p. of 2570 °C and a hardness of 9–10 on the Moh scale (a hardness scale ranging from 1 for talc to 10 for diamond). Not surprisingly, these compounds are rarely stoichiometric but usually nitrogen-deficient; they are essentially derivatives of the transitional metals, formed by heating the metal in ammonia to above 1000 °C.

Phosphides

Phosphides can be roughly divided into the same classes as nitrides, although there are fewer volatile covalent compounds, complex polymers usually being formed instead. The electrovalent phosphides, although hydrolysed to give phosphine, e.g.,

$$P^{3-} + 3H_2O \rightarrow 3OH^- + PH_3(g)$$

are undoubtedly less ionic than the corresponding nitrides. The interstitial hydrides are better regarded as alloys (since there is less prospect of the larger phosphorus atom being accommodated in the 'holes' of the lattice).

Ammonolysis of phosphorus pentachloride gives an interesting series of fairly inert polymers, the *phosphonitriles*, of empirical formula $PNCl_2$, such as $(PNCl_2)_3$:

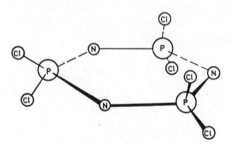

Occurrence and extraction

Nitrogen

The extreme stability of the nitrogen molecule, mentioned already in this Chapter, leads one to the expectation that nitrogen compounds would be readily converted into the element. Indeed, molecular nitrogen comprises about 78 per cent (4×10^{15} tonnes) of the atmosphere and represents the balance of several biological processes. Bacteria play a vital part in the 'nitrogen cycle' (*Figure 12.2*). In the course of their various metabolic processes, some remove nitrogen from the atmosphere whilst others reverse the procedure. The presence of nitrate in the soil is essential for plant growth, as plants require nitrogen in this form for the synthesis of proteins. Man, having realized this, now manufactures enormous quantities of nitrogenous

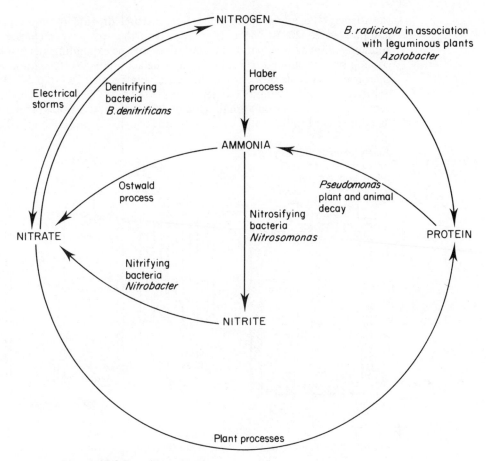

Figure 12.2 The nitrogen cycle

fertilizers for addition to the soil. Chief among these is ammonium sulphate, but ammonium nitrate, sodium nitrate and calcium nitrate (partly in the form of 'nitro-chalk') are also used. Ammonia itself is used in glasshouses, and the organic substances urea and calcium cyanamide are popular.

Little nitrogen is found in either the hydrosphere or the lithosphere. In the latter, the most important commercial source is Chile saltpetre, 'caliche', $NaNO_3$. The arid climate of Chile is no doubt responsible for the fact that this very soluble material has not been leached out, but there are many theories as to its origin. One of the most interesting is that it is the consequence of the local existence of a catalyst for the reaction

$$\tfrac{1}{2}N_2 + \tfrac{1}{2}H_2O + \tfrac{5}{4}O_2 \rightarrow HNO_3$$

This leads to the intriguing thought that if this, so far undiscovered, catalyst were of general occurrence, the atmosphere would be devoid of oxygen and the sea a dilute solution of nitric acid!

Nitrogen, required chiefly for the manufacture of ammonia (p. 278), can be extracted from the atmosphere by both physical and chemical methods. In the former, the air is dried, freed from carbon dioxide by cooling, and compressed to 20–30 MPa. It is cooled further by heat exchange and suddenly allowed to expand: the work done by the expanding gas results in cooling intense enough to bring about liquefaction (p. 88). The liquid air is then fractionally distilled, when the more volatile nitrogen (b.p. $-195\,°C$) distils off, leaving behind oxygen (b.p. $-182\,°C$). Chemically, oxygen is removed from the air on a small scale by passing it over a heated metal such as copper, and industrially by passing it through red-hot coke (p. 207).

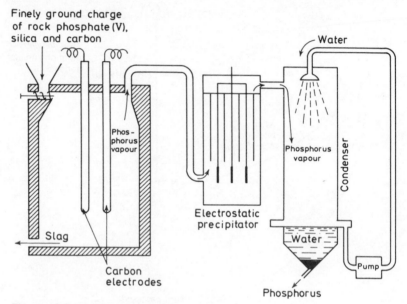

Figure 12.3 Electrothermal extraction of phosphorus

Phosphorus

Phosphorus is found as phosphate(V) in the lithosphere, the most common form being apatite, $CaF_2.3Ca_3(PO_4)_2$. It plays a vital part in many processes of the living cell (p. 558) and is a fundamental constituent, in the form of calcium phosphate(V), of the bones of vertebrates. It can, in fact, be extracted from bones, but it is more usual, on the large scale, to use rock phosphate(V), which is well mixed with sand and carbon, and heated to about $1500\,°C$ in an electric furnace: white phosphorus distils over and is solidified under water (*Figure 12.3*).

The reaction can be represented by the equations

$$2Ca_3(PO_4)_2 + 6SiO_2 \rightarrow 6CaSiO_3 + P_4O_{10}$$

$$P_4O_{10} + 10C \rightarrow P_4(g) + 10CO(g)$$

The silicon dioxide (sand), although only weakly acidic, can displace phosphorus(V) oxide because of the non-volatility of the former compared with the latter at the high temperature of the furnace.

Phosphorus, in the elemental red form, is used in safety matches and in rat poison. As the sulphide, it comprises part of the head of the 'non-safety' match, and in the form of phosphate(V) and 'superphosphate' it is used as a fertilizer.

Arsenic

Arsenic is closely associated with sulphide ores; consequently it is found in the flue-dust deposits from the extraction of metals from these ores, particularly of nickel, copper and tin. The arsenic(III) oxide present can be reduced to arsenic by heating with carbon

$$As_4O_6 + 6C \rightarrow As_4 + 6CO(g)$$

Arsenic is added in very small quantities to molten lead, especially in the manufacture of lead shot, to increase the tensile strength. Use is made of its toxic character in the preparation of drugs and insecticides.

Antimony

Antimony is found chiefly as the sulphide, stibnite, Sb_2S_3, but it sometimes partly replaces sulphur in galena. Roasting of stibnite gives a sublimate of oxide which is then reduced to the element by heating with carbon

$$Sb_4S_6 + 9O_2 \rightarrow Sb_4O_6 + 6SO_2(g)$$

$$Sb_4O_6 + 6CO \rightarrow 4Sb + 6CO(g)$$

Antimony confers greater hardness on Group 4 metals. It is alloyed with lead in the manufacture of accumulators and piping, and with lead and tin in pewter and type metal. [In the latter, an important aspect is the extension of the range of temperature throughout which it is plastic (*Figure 12.4*)]. Antimony has been used for adornment since prehistoric times, and today lead antimonate(V) is a constituent of some paints.

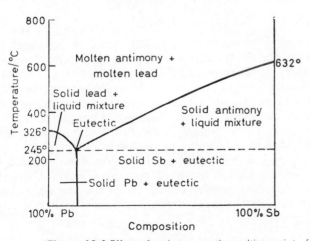

Figure 12.4 Effect of antimony on the melting point of lead

Bismuth

In the upper lithosphere this element is found chiefly as sulphide or as the trioxide formed by weathering. Galena acts as host for it and the flue dusts from the smelting of galena can be 'worked up' for bismuth: the oxide present is reduced by heating with iron or carbon. Bismuth is particularly useful for making readily fusible alloys; e.g., 'Wood's fusible metal', consisting of 4 parts bismuth, 2 parts lead and 1 part each of tin and cadmium, melts at 71 °C. Bismuth telluride shows the rare combination of low thermal but high electrical conductivity and is therefore used in thermoelectric materials.

Hydrides of nitrogen and phosphorus

Ammonia, NH_3, b.p. $= -33.4\,°C$

This colourless gas of characteristic odour is prepared on the small scale by hydrolysing nitrides with acid (sometimes, water) or ammonium compounds with alkali

$$N^{3-} + 4H^+ \rightarrow NH_4^+ \xrightarrow{\;OH^-\;} NH_3(g)$$

$$NH_4^+ + OH^- \rightarrow H_2O + NH_3(g)$$

On the large scale, it is made by the commercially more attractive method of direct synthesis

$$N_2 + 3H_2 \rightleftharpoons 2NH_3 \qquad \Delta H = -46 \text{ kJ mol}^{-1}$$

It is mainly used in the manufacture of fertilizers and nitric acid. Smaller quantities are used in the manufacture of nylon, wood pulp, explosives, and organic chemistry. Application of Le Chatelier's principle (p. 144) leads to the conclusion that the yield of ammonia will be increased by the use of low temperature and high pressure. Unfortunately, reaction velocities are low at low temperature, and therefore equilibrium would be approached only very slowly; in practice, a compromise of about 500 °C is used. A very high pressure of 20 MPa is also employed, together with a catalyst of iron, the action of which may be promoted by the addition of molybdenum. Under these conditions the yield is about 12 per cent (*Figures 12.5* and *12.6*).

Ammonia is extremely soluble in water, giving a solution often known as ammonium hydroxide but which appears to consist mainly of physically dissolved ammonia

$$NH_3 + H_2O \rightleftharpoons NH_3.H_2O \rightleftharpoons NH_4OH \rightleftharpoons NH_4^+ + OH^- \qquad pK = 4.7$$

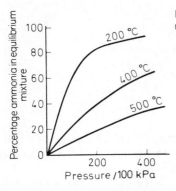

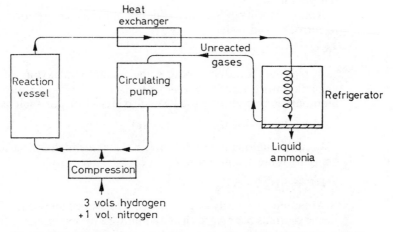

Figure 12.5 Effect of pressure and temperature in the manufacture of ammonia

Figure 12.6 Flow sheet for manufacture of ammonia by the Haber process

The nitrogen of the ammonia molecule is a proton acceptor

$$H^+ + :NH_3 \rightleftharpoons NH_4{}^+$$

and thus ammonia is basic and gives rise to a series of ammonium salts, of which the sulphate and nitrate are used in large quantities as fertilizers. The overwhelming use of ammonia is in the manufacture of such products, including urea (p. 557). It is also important as a means of manufacturing diamines prior to the synthesis of nylon (p. 560) and as a feedstock in nitric acid manufacture. In the laboratory, ammonia solution is very useful as a source of hydroxide ions in qualitative analysis, being preferred to, e.g., sodium hydroxide because strong heating removes all the ammonium ion and prevents any possible interference at later stages of the analysis. The common ion effect can also be used to control the pH of the solution.

The pyramidal molecule of ammonia is capable of inversion, i.e., the nitrogen atom can pass through the base of the pyramid to form an apex at the other side. If ammonia is exposed to heterogeneous radio waves, radio energy of a certain fixed frequency is absorbed as the nitrogen atoms oscillate

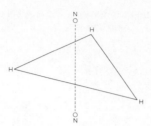

Figure 12.7 Inversion of ammonia molecule; in this diagram, lines do not represent bonds but simply indicate relative spatial positions

to and fro; this constant frequency can be used as a means of stabilizing the frequency of a microwave oscillator and thus of increasing the accuracy of what is known as the 'atomic clock'.

The fact that ammonia is readily liquefied and has a high enthalpy of vaporization (1.4 kJ g^{-1} at the boiling point) results in its being used as a refrigerant.

Ammonia will not burn in the air but pure oxygen supports its combustion to nitrogen and water

$$4NH_3 + 3O_2 \rightarrow 2N_2(g) + 6H_2O$$

A platinum catalyst modifies the reaction in favour of the far more valuable nitrogen oxide (p. 286). Oxidation to nitrogen also takes place if ammonia is passed over the heated oxides of metals such as copper

$$3CuO + 2NH_3 \rightarrow 3Cu + 3H_2O + N_2(g)$$

Liquid ammonia is a useful solvent, having a high dielectric constant and being to some extent ionized (p. 173).

$$2NH_3 \rightleftharpoons NH_4^+ + NH_2^-$$

Ammonia can easily be recognized by its smell and by the fact that it will turn red litmus blue; since all ammonium compounds, on heating with alkali, liberate ammonia, these too can be similarly identified. Many nitrogenous compounds, such as proteins, can be estimated by the methods of Kjeldahl and Dumas. In Kjeldahl's method, the nitrogen is converted into ammonium sulphate by reaction with concentrated sulphuric acid

$$\text{nitrogenous compound} \xrightarrow{H_2SO_4} (NH_4)_2SO_4$$

The salt is then treated with excess of alkali

$$(NH_4)_2SO_4 + 2OH^- \rightarrow 2NH_3(g) + SO_4^{2-} + 2H_2O$$

and the ammonia generated is absorbed in standard acid. The excess of acid is determined volumetrically. In Dumas' method, the ammonia (produced as before) is passed over heated copper oxide and the volume of nitrogen liberated is measured

$$2NH_3 \rightarrow N_2$$

34 g   22·4 dm^3 at s.t.p.

The composition of ammonia can be established by reacting excess of ammonia with a known volume of chlorine in the apparatus shown (*Figure*

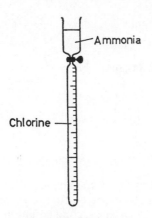

Figure 12.8 Volumetric composition of ammonia

Ammonia

Chlorine

12.8). Nitrogen and hydrogen chloride are formed and the latter is absorbed in water. The remaining volume of nitrogen is found to have one-third that of the original chlorine when measured under the same conditions of temperature and pressure, i.e.,

3 vols. chlorine→1 vol. nitrogen

But

3 vols. chlorine combine with 3 vols. hydrogen

Therefore

1 vol. nitrogen is associated with 3 vols. hydrogen

By Avogadro's law, then

1 molecule (2 atoms) nitrogen is associated with
3 molecules (6 atoms) hydrogen

Thus the empirical formula of ammonia is N_1H_3 and the molecular formula is $(NH_3)_x$. The vapour density is 8·5 and the relative molecular mass therefore 17. Thus, $x = 1$, and the empirical formula, NH_3, is also the molecular formula.

Hydrazine, $\begin{array}{c} H \quad\quad H \\ \diagdown\text{N}-\text{N}\diagup \\ \diagup \quad\quad \diagdown \\ H \quad\quad\quad H \end{array}$ |m.p. 2 °C, b.p. 113 °C

Hydrazine is nowadays in much demand, and great interest centres on the possible methods of manufacture.

Direct synthesis

$N_2 + 2H_2 \rightarrow N_2H_4$ $\Delta G^{\ominus} = +159$ kJ

The large positive free energy change in this reaction means that it is most unlikely to be a feasible method of preparation, especially as the free energy change for the formation of ammonia is negative

$$\tfrac{1}{2}N_2 + \tfrac{3}{2}H_2 \rightarrow NH_3 \qquad \Delta G^{\ominus} = -17 \text{ kJ}$$

This latter reaction is accordingly far more likely to occur.

Oxidation of ammonia

$$2NH_3 + \tfrac{1}{2}O_2 \rightarrow N_2H_4 + H_2O(g) \qquad \Delta G^{\ominus} = -38 \text{ kJ}$$

But

$$2NH_3 + \tfrac{3}{2}O_2 \rightarrow N_2 + 3H_2O(g) \qquad \Delta G^{\ominus} = -653 \text{ kJ}$$

An increase in pressure would improve the chance of the first reaction occurring but nevertheless it must be expected that most of the ammonia would still be oxidized to nitrogen, as in the second equation, because of the considerable decrease in free energy.

Reduction of dinitrogen oxide

$$N_2O + 3H_2 \rightarrow N_2H_4 + H_2O(g) \qquad \Delta G^{\ominus} = -172 \text{ kJ}$$

$$N_2O + H_2 \rightarrow N_2 + H_2O(g) \qquad \Delta G^{\ominus} = -331 \text{ kJ}$$

$$N_2O + 4H_2 \rightarrow 2NH_3 + H_2O(g) \qquad \Delta G^{\ominus} = -364 \text{ kJ}$$

This time, the energy picture is more encouraging; furthermore, increase in pressure would particularly favour the first reaction. Indeed, high yields have been claimed for this method, using an iron catalyst.

Despite the research undertaken on the various possibilities described, the usual method of manufacture remains the oxidation of ammonia with oxochlorate(I) in the presence of a little gelatin or edta (p. 354), which probably catalyses the relevant reaction and inactivates, by chelation (p. 353), various metallic ions which would otherwise catalyse decomposition reactions:

$$OCl^- + NH_3 \rightarrow OH^- + NH_2Cl$$

$$NH_2Cl + OH^- + NH_3 \rightarrow N_2H_4 + Cl^- + H_2O$$

This reaction produces a weakly alkaline solution of hydrazine, mainly in the form of its monohydrate (cf. NH_3), from which the anhydrous compound can be obtained by distillation. It is a diacid base but occurs largely in the form of the univalent hydrazinium cation when in acid solution:

$$N_2H_4 + H^+ \rightleftharpoons N_2H_5^+ \qquad (\text{cf. } NH_3 + H^+ \rightleftharpoons NH_4^+)$$

Hydrazine is a powerful reducing agent, especially in alkaline solution:

$$N_2H_4(aq) + 4OH^- \rightarrow N_2(g) + 4H_2O + 4e^- \qquad E^{\ominus} = +1.16 \text{ V}$$

Under alkaline conditions it reacts vigorously with hydrogen peroxide:

$$N_2H_4 + 2H_2O_2 \rightarrow N_2(g) + 4H_2O$$

This reaction, as well as that with oxygen, has been utilized in rocket propulsion and in 'fuel cells'. Such cells convert the chemical energy of a

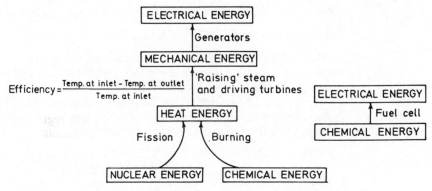

Figure 12.9 Energy transformations in power stations and fuel cells. Note the efficiency of the conversion of heat to mechanical energy is given by

$$\frac{(\text{Temp. at inlet}) - (\text{Temp. at outlet})}{\text{Temp. at inlet}}$$

conventional fuel directly into electrical energy rather than into the intermediate forms of heat and mechanical energy, when limitations of efficiency, expressed in the Second Law of Thermodynamics, are inevitably present (*Figure 12.9*). In one type of fuel cell, hydrazine hydrate is present in alkaline solution. At the fuel electrode the hydrazine decomposes:

$$N_2H_4 \rightarrow N_2(g) + 4H^+ + 4e^-$$

The electrons travel externally to the oxygen electrode and form hydroxide ions which react with hydrogen ions from the fuel electrode (*Figure 12.10*)

$$O_2 + 2H_2O + 4e^- \rightarrow 4OH^-$$

$$4H^+ + 4OH^- \rightarrow 4H_2O$$

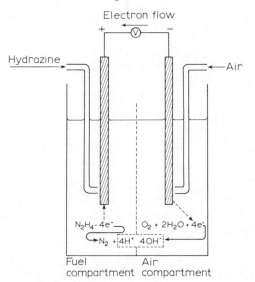

Figure 12.10 A fuel cell

The reactions taking place can therefore be summarized as

$$N_2H_4 + O_2 \rightarrow N_2(g) + 2H_2O$$

The affinity of hydrazine for oxygen also results in its being used for removing the latter from boiler water and for stabilizing various oxygen-susceptible liquids such as phenylamine.

Hydrazine, like its derivatives such as 2,4-dinitrophenylhydrazine, is also of great importance in organic chemistry (p. 504). It condenses with carbonyl compounds to give hydrazones and then azines, e.g.,

$$R.CHO + H_2N.NH_2 \rightarrow R.CH{:}N.NH_2 \rightarrow R.CH{:}N.N{:}CH.R$$

and it also reacts with acid chlorides:

$$R.COCl + H_2N.NH_2 \rightarrow R.CO.HN.NH_2 + HCl$$

The resultant hydrazides have various applications. For instance, *cis*-butenedioic hydrazide is used as a plant-growth regulator and benzenesulphonyl hydrazide as a blowing agent in the manufacture of foam plastics because of the nitrogen gas released on heating.

Hydrogen azide, HN$_3$, $\overset{H}{\underset{N \rightleftharpoons N \rightleftharpoons N}{\big\downarrow^{111^\circ}}}$ b.p. 35.7 °C

Hydrogen azide is a colourless liquid of unpleasant smell, prepared by heating sodium in dry ammonia gas and treating the sodamide so formed with dinitrogen oxide:

$$2Na + 2NH_3 \rightarrow 2NaNH_2 + H_2(g)$$

$$NaNH_2 + N_2O \rightarrow NaN_3 + H_2O$$

Distillation of the product with sulphuric acid gives hydrogen azide

$$N_3{}^- + H^+ \rightarrow HN_3(g)$$

Derivatives of this acid, the *azides*, are either ionic and fairly stable, e.g., $Na^+N_3{}^-$, or covalent and explosive, e.g. lead azide, $Pb(N_3)_2$, which is used as a detonator.

Hydroxylamine, NH$_2$OH, H$---$(N)$-$O$\overset{H}{\diagup}$ m.p. 33 °C

Hydroxylamine can be regarded as being derived from ammonia by replacement of one hydrogen atom by a hydroxyl group. It is a weaker base than ammonia, presumably owing to the lone pair of electrons on the nitrogen atom not being so available because of its attraction towards the more electronegative oxygen atom. However, hydroxylamine can react with a proton to give hydroxyammonium compounds, e.g., $[NH_3OH]^+Cl^-$, which, being more stable than the parent base, are the usual sources of the latter. They are normally prepared by cathodic reduction of nitrates or nitrites, e.g.,

$$CH_3ONO \rightarrow NH_2OH + CO(g)$$

$$HONO_2 + 6H^+ + 6e^- \rightarrow NH_2OH + 2H_2O$$
Nitric acid

but increasing amounts come from the petrochemicals industry, nitration of hydrocarbons being followed by reduction and distillation with sulphuric acid, e.g.,

$$RCH_3 \rightarrow RCH_2NO_2 \rightarrow RCH_2NHOH \rightarrow NH_2OH$$

Hydroxylamine and its compounds are reducing agents, e.g.,

$$4Fe^{3+} + 2NH_2OH \rightarrow 4Fe^{2+} + N_2O(g) + 4H^+ + H_2O$$

Under certain circumstances, however, they can act as oxidizing agents; for instance, the above reaction takes place in acid solution, but in alkaline solution the base can oxidize iron(II) to iron(III):

$$2Fe^{2+} + NH_2OH + H_2O \rightarrow 2Fe^{3+} + NH_3(g) + 2OH^-$$

In organic chemistry, hydroxylamine is of interest because of its ability to condense with carbonyl compounds to form oximes (p. 505), e.g.,

$$R.CHO + NH_2OH \rightarrow R.CH:NOH + H_2O$$

Phosphine, PH_3,

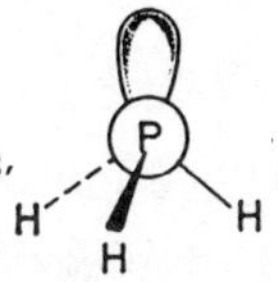

b.p. $= -87.4\,°C$

This gas, which has a smell of rotting fish, is normally obtained by the action of strong alkali on white phosphorus

$$P_4 + 3OH^- + 3H_2O \rightarrow PH_3(g) + 3H_2PO_2^-$$

The gas is usually contaminated with hydrogen and with diphosphane, P_2H_4, formed in side-reactions. A purer product can be made by the action of dilute acid on calcium phosphide:

$$P^{3-} + 3H^+ \rightarrow PH_3(g)$$

The phosphorus atom, being considerably less electronegative than that of nitrogen, cannot partake in hydrogen bonding. As a result of this, phosphine (in contrast to ammonia) has a very low solubility in water. For the same reason it is also much less basic than ammonia: phosphonium salts have been prepared, but they are rapidly hydrolysed by water.

The P—H bond is very weak, and phosphine is easily oxidized either to phosphorus or one of its oxides or oxyacids.

Oxides of nitrogen

The most common oxides are dinitrogen oxide (nitrous oxide, N_2O), nitrogen oxide (nitric oxide, NO) and nitrogen dioxide, NO_2 (which is usually in equilibrium with its dimer, N_2O_4). All these can be obtained by the reduction of nitric acid under differing conditions. Dinitrogen trioxide and dinitrogen pentoxide are also known but are of little importance.

The great stability of the nitrogen molecule means that these oxides are formed endothermically from their elements and thus dissociate fairly readily on heating. (With dinitrogen oxide, the dissociation into nitrogen and oxygen is so marked that it can easily be mistaken for oxygen.) Consequently, all three oxides will support combustion to some extent, although nitrogen oxide, the most stable of them in this respect, does so only if the substance is burning at a temperature of at least 1000 °C. There is thus the apparent paradox of burning sulphur being extinguished by this gas whilst strongly-burning phosphorus burns even more brightly than in air

$$P_4 + 10NO \rightarrow P_4O_{10} + 5N_2(g)$$

Advantage is taken of their thermal instability in establishing the molecular formulae. A measured volume of the gas confined by mercury is decomposed into its elements by means of an electrically heated iron wire. The oxygen formed is then removed from the gaseous phase by combination with the iron. When the reaction is complete, the volume of the remaining nitrogen (at the original temperature and pressure) is measured. A knowledge of the vapour density is then sufficient to establish the molecular formula (see p. 69).

Dinitrogen oxide, N_2O, b.p. $-88.5°C$

Dinitrogen oxide is obtained by the reduction of nitric acid with zinc:

$$4Zn + 10HNO_3 \rightarrow 4Zn^{2+} + 8NO_3^- + 5H_2O + N_2O(g)$$

or by the action of heat on ammonium nitrate:

$$NH_4^+NO_3^- \rightarrow N_2O(g) + 2H_2O \qquad \Delta H = -21\ kJ$$

The molecule is linear, $N\equiv N=O$, and is isoelectronic with that of carbon dioxide.

Dinitrogen oxide is used as an anaesthetic in minor surgery.

Nitrogen oxide, NO, b.p. $-151.7°C$

This colourless, insoluble gas can be conveniently made in small quantities by the action of 50 per cent nitric acid on copper, and collected over water:

$$3Cu + 8HNO_3 \rightarrow 3Cu^{2+} + 6NO_3^- + 4H_2O + 2NO(g)$$

On the large scale, it is made as an intermediate in the manufacture of nitric acid, by the catalytic oxidation of ammonia (p. 280). Small quantities are formed in the atmosphere by direct synthesis in the vicinity of lightning flashes.

Nitrogen oxide is paramagnetic, and therefore an unpaired electron is present in the molecule. The structure appears to be intermediate between the two formulations

$$:\dot{N}\!=\!\ddot{O}. \quad \text{and} \quad :\overset{+}{N}\!=\!\ddot{O}:$$

Dimerization, involving the unpaired electron, takes place increasingly at very low temperatures, especially in the solid.

Nitrogen oxide is a reactive compound and shows considerable versatility:

(1) *Loss of one electron*, forming the nitrosyl cation, NO^+. This is isoelectronic with the molecule of carbon monoxide and can react similarly with transitional metals to form nitrosyls, corresponding to carbonyls. An example of this type of reaction is afforded by the formation of the brown complex $[Fe^I(NO^+)(H_2O)_5]^{2+}$ when nitrogen oxide is passed into a solution of iron(II) ions, as in the 'brown ring' test for a nitrate. Nitrosyl hydrogensulphate, $NO^+HSO_4^-$, was formed during the manufacture of sulphuric acid by the lead chamber process.

(2) *Gain of one electron*, to give NO^-. When nitrogen oxide is passed into a solution of sodium in liquid ammonia, the compound Na^+NO^- is formed.

(3) *Sharing of electrons*, as for example when nitrogen oxide and chlorine are passed over charcoal to form nitrosyl chloride:

$$2NO + Cl_2 \rightarrow 2Cl\!-\!N\!=\!O$$

Reaction takes place instantly when nitrogen oxide comes into contact with air or oxygen at room temperature, brown fumes of nitrogen dioxide being formed

$$2NO + O_2 \rightarrow 2NO_2 \rightleftharpoons N_2O_4$$

Dinitrogen tetraoxide, N_2O_4, b.p. 21.2 °C

Commonly called nitrogen dioxide, because there is endothermic dissociation into the paramagnetic monomer, NO_2, at room temperatures and above (*Figure 12.11*), it is prepared in the laboratory by the action of heat on lead nitrate

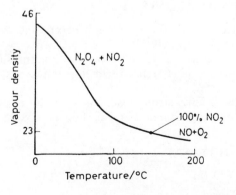

Figure 12.11 Equilibrium between NO_2 and N_2O_4

$$2Pb(NO_3)_2 \rightarrow 2PbO + 4NO_2(g) + O_2(g)$$

Cooling of the gas evolved results in the condensation of a diamagnetic liquid, which is green because of traces of blue N_2O_3 with the pale yellow N_2O_4. Present evidence suggests that the structure of the N_2O_4 is chiefly

This liquid boils near 21 °C, giving a pale yellow vapour made up almost entirely of N_2O_4 molecules. As the temperature is increased so is dissociation, accompanied by a steady darkening of colour until, at 150 °C, dissociation is just about complete and the vapour is almost black. The structure of the monomer which, like nitrogen oxide, contains an unpaired electron, is perhaps best represented as

Decomposition into nitrogen oxide and oxygen begins to take place if the temperature is increased further

$$N_2O_4 \underset{\text{cool}}{\overset{\text{heat}}{\rightleftharpoons}} 2NO_2 \underset{\text{cool}}{\overset{\text{heat}}{\rightleftharpoons}} 2NO + O_2$$

The electrical conductivity of liquid dinitrogen tetraoxide is very low, suggesting that any self-ionization is very slight. Dissolution in liquids of high dielectric constant, however, encourages ionization into nitrosyl and nitrate ions, especially if subsequent reaction takes place

$$N_2O_4 \rightleftharpoons NO^+ + NO_3^-$$

[cf. the isoelectronic ethanedioate (oxalate) ion: $C_2O_4^{2-} \rightarrow CO + CO_3^{2-}$].

Liquid dinitrogen tetraoxide will dissolve various metals and salts, the course of reaction indicating self-ionization of the above type. In the case of zinc, the zinc nitrate initially formed gives rise to an interesting complex

$$Zn + 2N_2O_4 \rightarrow Zn(NO_3)_2 + 2NO(g)$$

$$Zn(NO_3)_2 + 2N_2O_4 \rightarrow (NO^+)_2[Zn(NO_3)_4]^{2-}$$

This should be compared with the reaction between zinc and sodium, hydroxide which yields $(Na^+)_2[Zn(OH)_4]^{2-}$.

Several previously unknown anhydrous nitrates of the transition metals have also been isolated from systems involving anhydrous dinitrogen tetraoxide as solvent.

The compound is also a fairly strong oxidizing agent; for example, it will oxidize hydrogen sulphide to sulphur and carbon monoxide to carbon dioxide:

$$NO_2 + H_2S \rightarrow NO(g) + H_2O + S(s)$$

$$NO_2 + CO \rightarrow NO(g) + CO_2(g)$$

Oxoacids of nitrogen

Nitrous acid [nitric(III) acid]

$$\begin{array}{c} H \\ \diagdown \\ \quad O{-}N \\ \qquad \diagdown\!\!\diagdown \\ \qquad\quad O \end{array}$$

Nitrous acid is known only in dilute aqueous solution, in which state it behaves as a weak acid ($K_a \sim 10^{-5}$ mol dm^{-3}). Attempts to concentrate the solution result in its decomposition

$$3HNO_2 \rightleftharpoons HNO_3 + H_2O + 2NO(g)$$

The back reaction affords a method of preparing a dilute solution. An equivalent method is the addition of water to a mixture of nitrogen oxide and dinitrogen tetraoxide.

A 'pure' dilute solution of nitrous acid can be obtained by carefully treating barium nitrite with sulphuric acid

$$Ba(NO_2)_2 + H_2SO_4 \rightarrow BaSO_4(s) + 2HNO_2$$

For most practical purposes, however, the presence of extraneous stable ions is no limitation, so that by 'nitrous acid' is often meant the product of the mixing of cold, dilute solutions of sodium nitrite and hydrochloric acid

$$NaNO_2 + HCl \rightarrow HNO_2 + NaCl$$

The nitrites are made by heating the corresponding nitrates, either alone or with a reducing agent such as lead; addition of the latter allows a lower temperature for reaction and hence prevents decomposition of heavy metal nitrites to oxides, e.g.,

$$Ba(NO_3)_2 + 2Pb \rightarrow Ba(NO_2)_2 + 2PbO$$

Nitrites of the electropositive metals are ionic and fairly stable. Several covalent nitrites also exist, including organic derivatives.

Nitrous acid can be oxidized to nitric acid and reduced to various oxides and hydrides of nitrogen so that, under different conditions, it will act as a reducing or as an oxidizing agent. For example, in acid solution, iodides are oxidized to iodine:

$$2I^- + 4H^+ + 2NO_2^- \rightarrow I_2 + 2H_2O + 2NO(g)$$

whereas, in neutral solution, iodine is reduced to iodide:

$$I_2 + H_2O + NO_2^- \rightarrow 2H^+ + 2I^- + NO_3^-$$

A very important reaction involving a nitrite is that of 'diazotization', by which dyestuffs can be prepared (p. 456).

It has been said already that nitrous acid is unstable. Evolution of brown fumes of nitrogen dioxide on the addition of dilute mineral acid is indicative of the presence of a nitrite, as is the formation of the characteristic brown complex, $[Fe(NO^+)(H_2O)_5]^{2+}$, upon the addition of iron(II) sulphate solution. These reactions, as well as that of diazotization, can be used in identifying nitrites.

Quantitative estimation can be effected either volumetrically, by oxidation with acidified potassium manganate(VII),

$$2MnO_4^- + 5NO_2^- + 6H^+ \rightarrow 2Mn^{2+} + 5NO_3^- + 3H_2O$$

or colorimetrically by formation of the red azo dye with 4-aminobenzenesulphonic acid (sulphanilic acid) and 1-aminonaphthalene.

Nitric acid and nitrates [Nitric(V) acid and nitrates(V)]

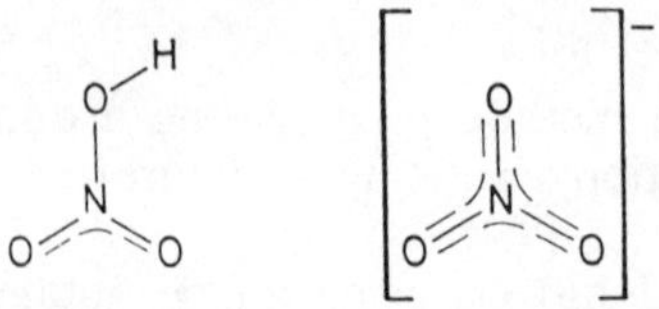

Nitric acid is prepared in the laboratory by the action of hot, concentrated sulphuric acid on sodium nitrate, followed by condensation of the vapour evolved

$$NO_3^- + H_2SO_4 \rightarrow HSO_4^- + HNO_3$$

It is manufactured by the catalytic oxidation of ammonia. Dry ammonia, together with excess of air, is passed through a platinum gauze catalyst at about 500 °C: this causes nitrogen oxide to be formed:

$$4NH_3 + 5O_2 \rightarrow 4NO(g) + 6H_2O$$

The reaction mixture is then cooled to favour formation of nitrogen dioxide (see Figure 12.11), which is dissolved in water in the presence of air to give nitric acid (Figure 12.12):

$$2NO + O_2 \rightarrow 2NO_2$$

$$4NO_2 + 2H_2O + O_2 \rightarrow 4HNO_3$$

Nitric acid is required in considerable quantities for the manufacture of fertilizers, e.g., ammonium nitrate, and explosives, e.g., TNT. The formation of these compounds illustrates various features of the acid: the pure substance, which is a colourless, volatile liquid of boiling point 84 °C, self-ionizes to some extent:

$$2HNO_3 \rightleftharpoons H_2NO_3^+ + NO_3^-$$

One molecule of the acid is here acting as a proton donor and the other as a proton acceptor. Other substances can affect this position: for example, water can assume the role of proton acceptor to give the oxonium ion:

$$HNO_3 + H_2O \rightleftharpoons H_3O^+ + NO_3^-$$

The acidity of nitric acid is thus enhanced and becomes the most obvious feature of dilute aqueous solutions, so that, e.g., neutralization of ammonia

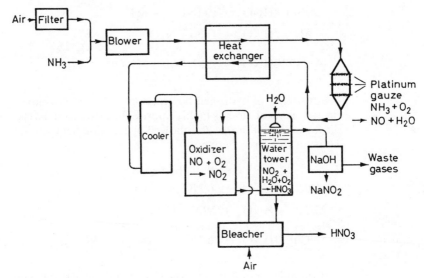

Figure 12.12 Manufacture of nitric acid

solution by dilute nitric acid gives ammonium nitrate. Concentrated sulphuric acid, on the other hand, encourages dehydration of the $H_2NO_3{}^+$ ion:

$$H_2NO_3{}^+ \rightarrow H_2O + NO_2{}^+$$

Aromatic hydrocarbons and their derivatives are particularly prone to 'nitration' by this system:

$$R.H + NO_2{}^+ \rightarrow R.NO_2 + H^+$$

Thus methylbenzene gives rise to methyl-2,4,6-trinitrobenzene (T.N.T.) (*see also* p. 429).

Nitric acid is also a powerful oxidizing agent, especially in concentrated solution. On heating alone, it decomposes into water, nitrogen dioxide and oxygen, and with most non-metals produces oxoacids. For example, phosphorus is converted into phosphoric(V) acid

$$P_4 + 20HNO_3 \rightarrow 4H_3PO_4 + 20NO_2(g) + 4H_2O$$

and sulphur to sulphuric acid

$$S + 2HNO_3 \rightarrow H_2SO_4 + 2NO(g)$$

This oxidizing capacity is responsible for the almost complete absence of hydrogen from the evolved gases when metals are dissolved in nitric acid, although magnesium gives a reasonable yield of hydrogen with cold, very dilute acid. In the case of metals above hydrogen in the electrochemical series, nascent hydrogen is possibly first formed and then instantly oxidized by the remaining nitric acid, the extent of reduction of the acid depending to a large extent on its concentration. In the case of zinc, dinitrogen oxide is evolved and might be the result of the sequence

$$4Zn + 8HNO_3 \rightarrow 4Zn(NO_3)_2 + 8H$$

$$8H + 2HNO_3 \rightarrow 5H_2O + N_2O$$

$$4Zn + 10HNO_3 \rightarrow 4Zn(NO_3)_2 + 5H_2O + N_2O(g)$$

Formation of hydrogen by reaction with metals below hydrogen in the electrochemical series is highly improbable. Here, initial reaction is probably the formation of the oxide, followed by reaction between this and more acid. Thus, in the case of copper

$$Cu + HNO_3 \rightarrow CuO + HNO_2$$

$$CuO + 2HNO_3 \rightarrow Cu(NO_3)_2 + H_2O$$

$$HNO_2 + HNO_3 \rightarrow 2NO_2 + H_2O$$

$$Cu + 4HNO_3 \rightarrow Cu(NO_3)_2 + 2H_2O + 2NO_2(g)$$

It should be pointed out, however, that there is much conflicting evidence on this subject and the reactions given are certainly oversimplified.

Metal nitrates are very soluble salts, prepared in the usual way by the action of the acid on metal, base or carbonate. They are all unstable to heat, particularly those of the more noble metals, and the course of the reaction depends upon the electronegativity of the metal. Thus, nitrates of the alkali metals decompose initially to give nitrites, whereas the heavy metal nitrates form the oxides: if the oxide is itself thermally unstable, then the metal is formed as well:

$$2NaNO_3 \rightarrow 2NaNO_2 + O_2(g)$$

$$2Pb(NO_3)_2 \rightarrow 2PbO + 4NO_2(g) + O_2(g)$$

$$Hg_2(NO_3)_2 \rightarrow 2Hg + 2NO_2(g) + O_2(g)$$

Ammonium nitrate is unique in that violent shock or heating results in explosion, whilst gentle heating causes it to melt and decompose into dinitrogen oxide and water:

$$NH_4NO_3 \rightarrow N_2O(g) + 2H_2O(g)$$

Covalent nitrates are less stable than the ionic; like the covalent azides they tend to be explosive, such as propane-1,2,3-triyl trinitrate ('nitroglycerine') and 'fluorine nitrate', NO_3F. The halogen nitrates can, however, be stabilized by pyridine, dative covalence from the nitrogen of the latter enhancing the ionic character of the nitrate:

$$C_5H_5N: + Cl.NO_3 \longrightarrow C_5H_5N \rightarrow Cl^+NO_3^-$$

Inorganic nitrates can be recognized by means of the 'brown ring test': a solution of the nitrate is treated with iron(II) sulphate solution in dilute sulphuric acid; concentrated sulphuric acid is then carefully introduced to give two liquid phases. A brown ring develops at the junction of the two liquids (cf. test for nitrites, p. 289):

$$NO_3^- + 3Fe^{2+} + 4H^+ \rightarrow 3Fe^{3+} + 2H_2O + NO$$

$$Fe^{2+} + NO \rightarrow [Fe^I(NO^+)]^{2+}$$

This test is invalid in the presence of bromides, since concentrated sulphuric acid oxidizes bromides to bromine, which also gives a brown ring at the interface. In this case, the nitrate is identified by reduction to ammonia by Devarda's alloy (45 per cent Al, 5 per cent Zn, 50 per cent Cu) in the presence of alkali:

$$8Al + 5OH^- + 18H_2O + 3NO_3^- \rightarrow 8Al(OH)_4^- + 3NH_3(g)$$

The ammonia evolved can be identified by the usual qualitative methods or absorbed in standard acid if it is desired to estimate the nitrate quantitatively. (If ammonium ions are originally present, it is necessary to boil with excess of sodium hydroxide solution until no more ammonia is evolved, before introducing Devarda's alloy.)

Phosphoric(V) acids and phosphates(V)

Tetraoxophosphoric(V) or orthophosphoric acid, H_3PO_4, can be manufactured by digesting rock phosphate(V) with sulphuric acid for several hours

$$Ca_3(PO_4)_2 + 3H_2SO_4 \rightarrow 3CaSO_4 + H_3PO_4$$

When made by this process, it is contaminated with soluble calcium dihydrogen phosphate(V). (This 'superphosphate', of great importance as a source of soluble phosphorus in fertilizers, can be manufactured by a modification of this procedure.) A purer product is obtained by oxidizing phosphorus to phosphorus(V) oxide and dissolving this in hot water. In the laboratory, it can be made by oxidizing phosphorus with concentrated nitric acid (p. 291). Evaporation of the resultant solution gives a syrupy liquid, the high viscosity being caused by association through hydrogen bonds (cf. propane-1,2,3-triol or glycerol):

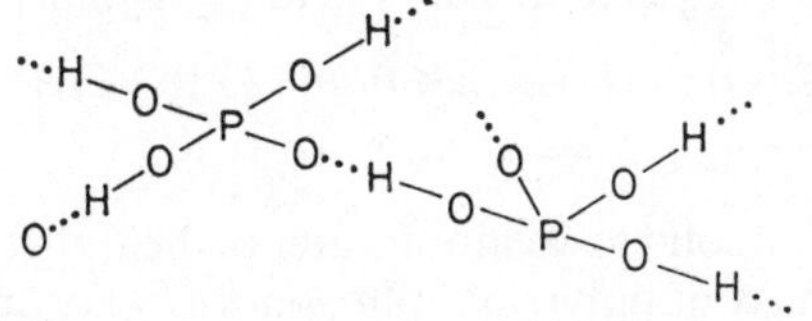

Deliquescent crystals of the compound (m.p. 42 °C) can be obtained by vacuum desiccation. It is a tribasic acid, and progressive neutralization with sodium hydroxide gives in turn the three sodium salts:

$$H_3PO_4 \xrightarrow[(-H_2O)]{NaOH} NaH_2PO_4 \xrightarrow[(-H_2O)]{NaOH} Na_2HPO_4 \xrightarrow[(-H_2O)]{NaOH} Na_3PO_4$$

Disodium hydrogenphosphate is commonly called sodium phosphate as it is the one giving the most nearly neutral solution (*Figure 12.13*). Orthophosphoric is a weak acid with the three dissociation constants:

$$K_1 = [H^+][H_2PO_4^-]/[H_3PO_4] = 7 \times 10^{-3} \text{ mol dm}^{-3}$$

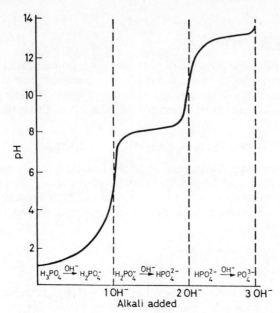

Figure 12.13 Neutralization of phosphoric(V) acid with strong alkali

$$K_2 = [\mathrm{H}^+][\mathrm{HPO_4}^{2-}]/[\mathrm{H_2PO_4}^-] = 6 \times 10^{-9} \text{ mol dm}^{-3}$$

$$K_3 = [\mathrm{H}^+][\mathrm{PO_4}^{3-}]/[\mathrm{HPO_4}^{2-}] = 5 \times 10^{-13} \text{ mol dm}^{-3}$$

so that the salts in aqueous solution are extensively hydrolysed.

Phosphoric(V) acid also forms esters with alcohols, e.g.,

$$\mathrm{(HO)_3PO + 3C_2H_5OH} \longrightarrow \mathrm{(C_2H_5O)_3PO} + 3\mathrm{H_2O}$$
$$\text{ethanol} \qquad \text{triethyl phosphate(V)}$$

On heating to about 220 °C, tetraoxophosphoric(V) acid steadily loses water and forms heptaoxodiphosphoric(V) acid (pyrophosphoric acid):

$$\mathrm{(HO)_2PO.\overline{OH} + \overline{H}OPO(OH)_2} \longrightarrow \mathrm{(HO)_2PO.O.PO(OH)_2 + H_2O}$$
$$2\mathrm{H_3PO_4} \qquad \longrightarrow \qquad \mathrm{H_4P_2O_7} + \mathrm{H_2O}$$

This colourless, crystalline solid loses more water on heating to about 320 °C, to form a translucent mass of polytrioxophosphoric(V) or metaphosphoric acid

$$n\mathrm{H_4P_2O_7} - n\mathrm{H_2O} \rightarrow \mathrm{(HPO_3)_{2n}}$$

All phosphates(V) and phosphoric(V) acids have structures based on the $\mathrm{PO_4}$ tetrahedron, although only the tetra- and hepta-oxophosphates(V) give discrete ions as shown in *Figure 12.14*.

Tetraoxophosphoric(V) acid is an important constituent of living matter, playing a vital and varied part in many metabolic processes (p. 558).

Phosphates(V) also play a diverse part in industrial life. It is apparent from the occurrence of phosphorus in plant and animal life that it is necessary as a plant fertilizer; it is usually added to the soil in the form of the rather insoluble

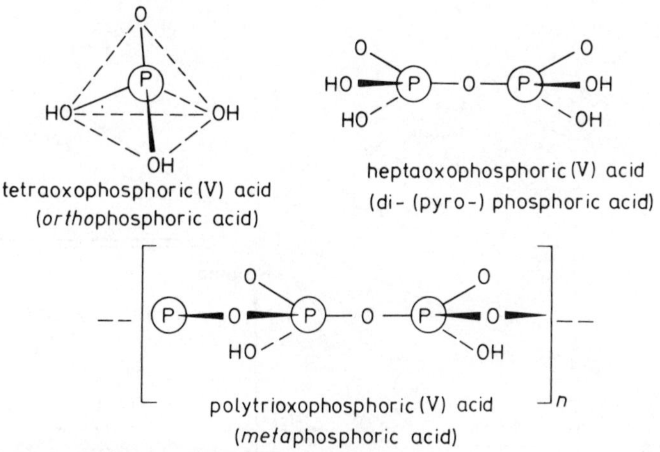

tetraoxophosphoric (V) acid
(*ortho*phosphoric acid)

heptaoxophosphoric (V) acid
(di- (pyro-) phosphoric acid)

polytrioxophosphoric (V) acid
(*meta*phosphoric acid)

Figure 12.14 The PO_4 tetrahedron in the structures of the phosphoric(V) acids: (a) tetraoxophosphoric(V) acid (orthophosphoric acid), (b) heptaoxophosphoric(V) acid (diphosphoric or pyrophosphoric acid), and polytrioxophosphoric(V) acid (metaphosphoric acid)

calcium phosphate(V) (as bone meal) or as the more soluble 'superphosphate'.

Polytrioxophosphates(V) of the alkali metals are used in water softening. Calcium and magnesium ions are 'sequestered', i.e., rendered inactive, by reaction with the colloidal polytrioxophosphate(V) polymer, giving a structure:

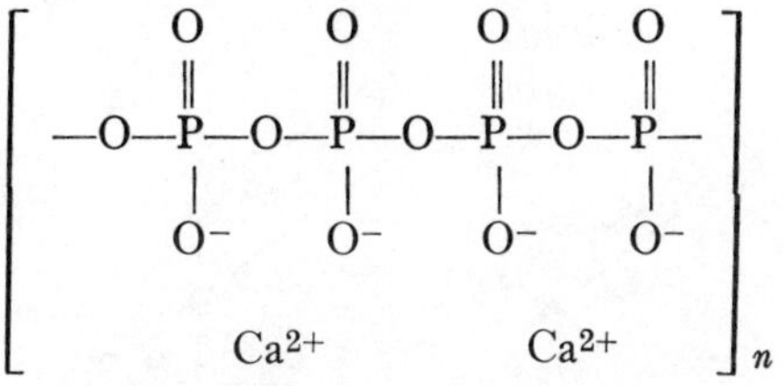

Esters of alcohols and phosphoric(V) acid, such as tributyl phosphate(V), are often used as plasticizers in the production of rubber and plastics articles, and also for the solvent extraction of metals.

Phosphates(V) can be identified by the slow formation of a yellow precipitate of the complex ammonium phosphomolybdate on mixing a solution of the salt with dilute nitric acid and ammonium molybdate(VI) solution. [Arsenates(V) form a similar precipitate but only on warming the mixture.]

Magnesium ammonium phosphate(V) is quantitatively precipitated when an aqueous solution of a magnesium salt is added to an ammoniacal solution of a phosphate(V); this precipitate can be filtered off, washed and ignited to constant weight:

$$2MgNH_4PO_4 \rightarrow Mg_2P_2O_7 + 2NH_3(g) + H_2O$$

In this way, phosphates(V) are quantitatively estimated.

Summary

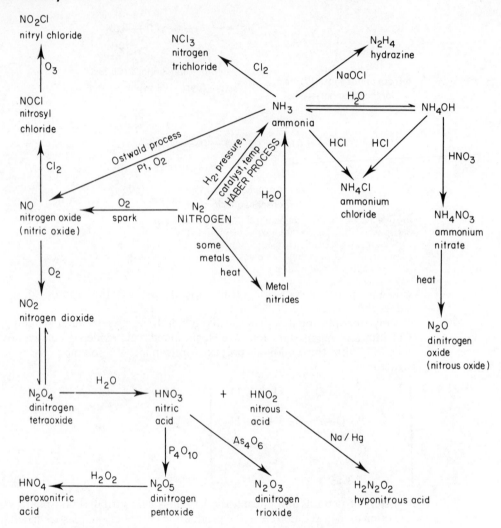

Figure 12.15 Reactions of nitrogen

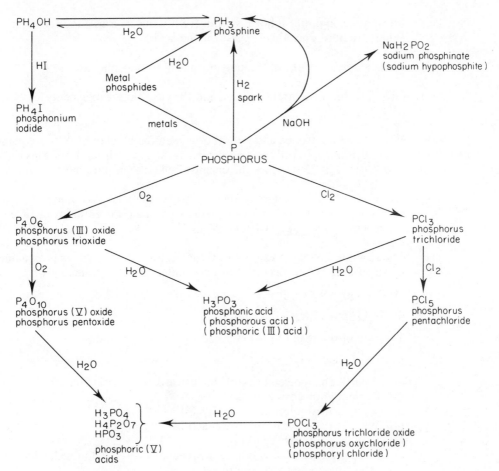

Figure 12.16 Reactions of phosphorus

Questions

(1) Suggest why the compound H_3NO_3 is unknown, whereas H_3PO_3 is well known.

(2) Why do you think phosphine is much less basic than ammonia?

(3) What reactions would you expect to occur when bromine is passed into nitrous acid? How could you test your conclusions?

(4) Discuss, and as far as possible explain, the changes in properties of the compounds $X(OH)_3$, where X is N, P, As, Sb, Bi respectively.

(5) How might traces of arsenic in a mixture be detected?

(6)   Phosphinic (hypophosphorous) acid, H_3PO_2, behaves as a monobasic acid. What structure does this suggest for the acid?

(7)   Write a brief account of the polymorphism of phosphorus.

(8)   Give possible structural formulae and the electron arrangements for N_4S_4, $(PNCl_2)_3$, P_4S_3.

(9)   Why are both nitrogen and phosphorus required by plants? Why are plants generally unable to utilize these elements when in their natural forms as N_2 and $Ca_3(PO_4)_2$? How does the chemist make them more freely available?

(10) Using the bond energies given on page 269 and assuming that the electronegativities of N and Cl are equal, calculate the enthalpy change for the reaction: $2NCl_3(g) \rightarrow Cl_2(g) + N_2(g)$

(11) The reaction $2NO(g) + O_2(g) \rightarrow 2NO_2(g)$ at normal pressures was explained by suggesting that it occurred in two fundamental steps as follows:

(i) $2NO(g) \rightleftharpoons N_2O_2(g)$

(ii) $N_2O_2(g) + O_2(g) \rightarrow 2NO_2(g)$

The rate of the reaction is described by the relationship:

Rate $= k[NO]^2[O_2]$

(a) What is the molecularity of the forward reaction in step (i)?
(b) Explain your answer to (a).
(c) What is the molecularity of step (ii)?
(d) What is the overall order of reaction?
(e) Explain your answer to (d).
(f) If the pressure of only one of the reactants is to be doubled, which will cause the greater increase in rate?
(g) Suggest how the rate of the reaction may be increased without changing the pressure or the temperature.
(h) Explain briefly how the suggestion in (g) causes an increase in rate.
                                                                    [J.M.B.]

(12) On addition of a solution of 0.280 g of sodium nitrite in water to a solution of 0.2475 g of hydroxylammonium chloride (hydroxylamine hydrochloride) $NH_3OH^+Cl^-$, a colourless, odourless gas was evolved.
      After the reaction had ceased the solution was mixed with 10 cm^3 of 0.05 mol dm^{-3} tetraoxomanganate(VII) (permanganate) solution. After heating and acidifying the mixture, whereupon the permanganate oxidized the excess of nitrite, the excess of permanganate was estimated by titration with an iron(II) solution. It was found that 5.97 cm^3 of the permanganate had not been consumed by the nitrite.
      Establish an equation for the initial reaction and identify the gas evolved.
                                                                    [C.]

(13) (a) By writing equations for suitable reactions, illustrate the behaviour of ammonia as (i) a base, (ii) a reducing agent (reductant), (iii) a ligand.

(b) For the elements, nitrogen, phosphorus and arsenic, (i) give formulae for their characteristic hydrides, (ii) state how the thermal stability of the characteristic hydride varies from nitrogen to phosphorus to arsenic.

(c) (i) Give the oxidation number (state) of nitrogen in the following compounds: nitric acid; nitrous acid; ammonia.
(ii) Explain why nitric acid can be reduced but cannot be oxidized; give an example of its reduction.
(iii) Explain, giving one example in each case, why nitrous acid can act both as an oxidizing agent and as a reducing agent.

[A.E.B. 1979]

13 Group 4 elements

The elements • Structures • Occurrence and extraction – carbon, silicon, germanium, tin and lead • Reactions of Group 4 elements • Hydrides • Halides • Oxoacids and oxosalts • Organic compounds

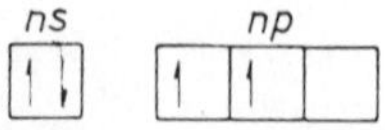

Group 3	ns	np		Group 5

Group 3			Group 5
B	Carbon, C	2.4	N
Al	Silicon, Si	2.8.4	P
Ga	Germanium, Ge	2.8.18.4	As
In	Tin, Sn	2.8.18.18.4	Sb
Tl	Lead, Pb	2.8.18.32.18.4	Bi

The elements

There is *superficially* a closer relationship between the elements of this Group than has been the case hitherto: all the elements are solid at room temperature and all are capable, in one form or another, of conducting electricity. Carbon electrodes are used in dry cells, silicon and germanium in transistors and lead in accumulators. Carbon, in the form of diamond, could be mistaken at a cursory glance for silicon dioxide and various silicates, whilst in the form of graphite it is used in 'lead' pencils.

There is, however, a profound difference between the chemistry of carbon and that of the rest of the group. Carbon is unique in the extent to which it can combine with itself, either with single or multiple bonds. Its chemical versatility is exploited by the living cell, and over a million organic compounds are already known: in this profusion it is surpassed only by hydrogen (which is able to do this only because it is attached to carbon in organic compounds).

The outer shell of all these elements contains four—two s and two p—electrons. The earlier elements are able to hybridize their valency electrons after promotion of an s electron to a p orbital to give a tetrahedral configuration with a valency of four:

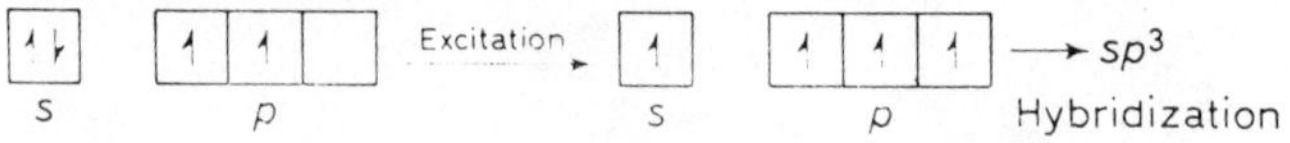

Excitation ——► sp^3

s p s p Hybridization

This ability declines, however, as the series is descended: the main valency state of lead is two. This is attributed to an 'inert pair' effect, the two p electrons being lost, leaving behind the s electrons of lower energy

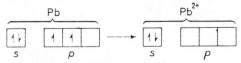

Although tin also gives an ion, Sn^{2+}, it is easily hydrolysed and is a powerful reducing agent, e.g.,

$$Hg^{2+} + Sn^{2+} \rightarrow Hg(l) + Sn^{4+}$$

The carbon atom readily forms π-bonds with other carbon atoms (p. 61). The fact that its outer shell can hold only eight electrons means that the compounds it forms are quite stable. Silicon not only shows great reluctance to π-bonding, but its outer shell contains available d orbitals, so that its compounds are readily hydrolysed (p. 212). Because of the possibility of expansion of the outer shells of silicon, germanium, tin and lead, complex ions are numerous, e.g., $SiF_6{}^{2-}$, $SnO_3{}^{2-}$, $Pb(OH)_6{}^{2-}$.

The progressive increase in atomic volume on passing from carbon to lead, together with an increase in shielding power from the inner shells, produces a fall in electronegativity and a decrease in ionization energy (*Table 13.1*).

Table 13.1 Atomic size and ionization energies

Atom	Outer electronic configuration	Atomic radius/pm	Ionic radius/pm M^{2+}	Ionic radius/pm M^{4+}	Ionization energies/kJ mol^{-1} 1st	Ionization energies/kJ mol^{-1} 2nd
C	$2s^2 2p^2$	77			1090	2360
Si	$3s^2 3p^2$	117		39	788	1580
Ge	$4s^2 4p^2$	122		53	762	1540
Sn	$5s^2 5p^2$	141		71	706	1418
Pb	$6s^2 6p^2$	154	132	84	715	1458

The bond energies for single bonds between atoms of the same elements from carbon down to lead fall sharply, and permit some prediction to be made about the likely reactivity of the elements. The bond energy for the carbon–oxygen bond is only slightly larger than that for carbon–carbon, whereas the silicon–oxygen bond has twice the energy of silicon–silicon. It is to be expected then that hydrocarbons will be far more resistant to oxidation than silanes (especially as the carbon–hydrogen bond is much stronger than the silicon–hydrogen bond), and this is found to be the case in practice. The decline in bond energies from carbon to lead for the $M–M$ bonds also suggests that the ability to catenate will become less pronounced (*Table 13.2*).

There is a marked increase in density on descending the Group, in accordance with an increase in metallic character. On the other hand, the volatility increases, owing to a weakening of the $M–M$ bond as the Group is descended (*Table 13.3*); both density and volatility are to some extent dependent on the structures assumed by the elements.

Table 13.2

Bond energies/kJ mol^{-1}

C—C 344	Si—Si 209
C–O 356	Si—O 418
C—H 415	Si—H 315

Table 13.3 Density and melting and boiling points

	Density g cm^{-3}	Melting point/°C	Boiling point/°C
Carbon (graphite; diamond)	2.2; 3.5	3575	4200
Silicon	2.3	1414	2300
Germanium	5.4	958	2700
Tin (grey; white)	5.7; 7.3	232	2360
Lead	11.3	327	1755

Structures

Carbon

Carbon exists in the crystalline condition as both diamond and graphite (*Figure 13.1*). In diamond, the four covalent bonds are directed tetrahedrally, giving a three-dimensional framework of high melting point and hardness.

Graphite has a layer structure with sheets of carbon atoms held together by van der Waals forces at a distance of 340 pm. It is therefore less dense than diamond and also very soft: application of stress causes the layers to slide over each other, which helps explain its usefulness as a lubricating agent. The C—C distances in the hexagonal sheets are uniform, and it is believed that the remaining p orbitals participate in π-bonding; the increased electron availability results in graphite being a good conductor of heat and electricity. The distance between the carbon layers is large enough to permit the entry of various atoms; for example, alkali metals can be directly absorbed to give substances of increased conductivity, whilst formation of 'graphite oxide' removes the π-bonds and the product is non-conducting.

Graphite is more stable than diamond at ordinary temperatures and pressures; this type of allotropy (polymorphism) is *monotropy* (p. 90). Conversion of diamond into graphite is a very slow process though. 'Amorphous' carbon, such as lampblack and charcoal, contains minute crystals of graphite.

Silicon and germanium

The only known forms of silicon and germanium correspond to the diamond structure, with sp^3 hybridized orbitals directed tetrahedrally.

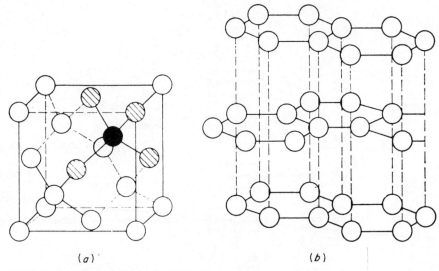

Figure 13.1 Crystal structures of allotropes of carbon: (a) diamond, showing tetrahedral distribution of bonds around each carbon atom; (b) the layer structure of graphite

Tin

Tin, like carbon, is polymorphic. Grey (α-)tin has the diamond structure. At 13 °C it is in equilibrium with white (β-)tin (which has a metallic structure), i.e., the allotropy is *enantiotropy* (p. 90).

Lead

As might be expected from the increase in metallic character on descending the Group, lead exists only in the metallic form, with cubic close packing.

Occurrence and extraction

Carbon

Carbon occurs in the atmosphere to the extent of about 0.03 per cent as carbon dioxide. This is formed as a result of respiration and other combustion processes and removed by photosynthesis; recent evidence from investigations of the upper atmosphere suggests that the proportion of carbon dioxide is increasing.

In the lithosphere, carbon is found both combined as carbonates and elementally as diamond and graphite. It is a vital constituent of living tissue and is consequently present in their products of decay, for instance as coal and petroleum.

Carbon is extracted in the form of coke by the carbonization of coal. Coal is heated in the absence of air and the volatile material is removed. Coke remains behind in the retorts, whilst condensation and washing of the gases

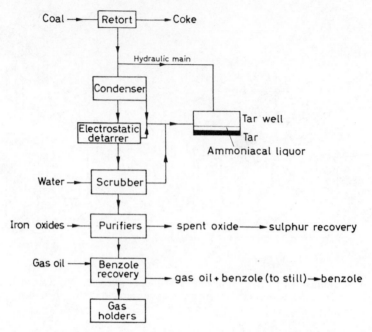

Figure 13.2 Flow sheet for the distillation of coal in the manufacture of coal gas and coke

evolved removes coal tar (p. 416), ammonia, naphthalene, etc. (*Figure 13.2*). Hydrogen sulphide is removed from the gas by passage over iron(III) oxide, which itself is converted into sulphide and from which sulphur dioxide can later be obtained by roasting when the oxide has become 'spent'. Any benzole that is present can be dissolved in gas oil which is subsequently distilled. The remaining gas (coal gas) has the approximate composition: hydrogen 50 per cent, methane 32 per cent, carbon monoxide 8 per cent, nitrogen 6 per cent, ethene 4 per cent, and is used as a gaseous fuel although this traditional route has more recently been supplemented by the reforming of hydrocarbons from the petroleum industry (where 'rich' gas is converted to larger quantities of 'lean' gas) and by the increasing supply of natural gas (*see Figure 13.3*). Some of the many products obtained from coal are shown in *Figure 13.4*.

Low-temperature carbonization gives a less pure product than coke but one which burns more readily (smokeless fuel).

It is predicted that coal will be used to diminish the misuse of oil and natural gas and the hierarchy for the use of coal in the future should be (*a*) blast furnaces, (*b*) domestic boilers, (*c*) power stations and large industrial users, (*d*) conversion to methanol/synthetic natural gas, and (*e*) liquefaction.

Silicon

Silicon comprises about a quarter of the Earth's crust in the form of silica and silicates. It is extracted from silicon(IV) oxide (silica) by reduction with

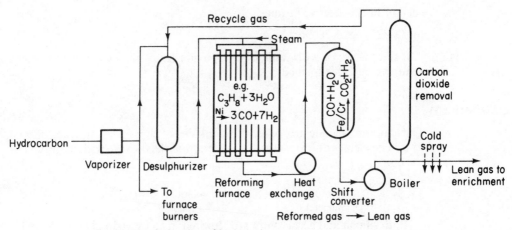

Figure 13.3 Reforming of hydrocarbons

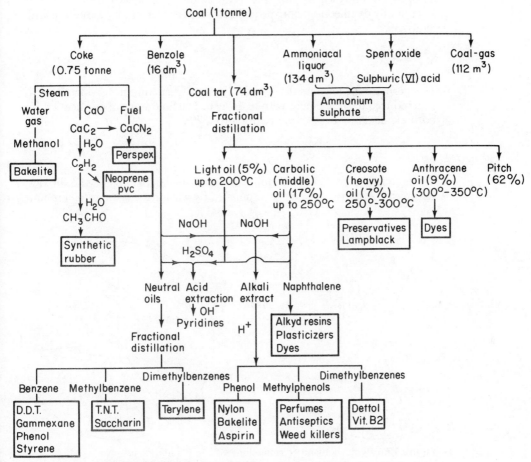

Figure 13.4 Distillation of coal

carbon in the electric furnace

$$SiO_2 + 2C \rightarrow Si + 2CO(g)$$

If silica is mixed with some iron(III) oxide, ferro-silicon, a very important reducing agent, is formed.

Germanium

Germanium is associated with the sulphides of copper, silver, lead, tin and zinc. The element is extracted by reduction of the dioxide with hydrogen or carbon

$$GeS_2 \xrightarrow[H_2SO_4]{HNO_3} \underset{(impure)}{GeO_2} \xrightarrow{HCl} GeCl_4 \xrightarrow{H_2O} \underset{(pure)}{GeO_2} \xrightarrow{H_2} Ge$$

Both silicon and germanium are required in an exceedingly pure form for use as semiconductors in transistors. In each case, the element is heated with halogen to convert it into the tetrahalide. The silicon tetrahalide is fractionally distilled and converted back into the element by reduction with hydrogen

$$Si \xrightarrow{X_2} SiX_4 \xrightarrow{H_2} Si$$

The germanium tetrahalide is hydrolysed to germanium dioxide which is also reduced to the element with hydrogen. The final stage of purification for both elements involves zone refining (p. 570).

Tin (*Figure 13.5*)

Tin is found native in small quantities, but its chief source is the dioxide, cassiterite. After the ore has been purified by washing, roasting (to eliminate

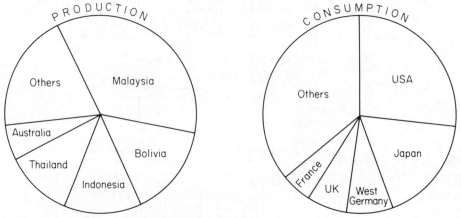

Figure 13.5 Production and consumption of tin in the non-communist world; note the position of countries from the first and third worlds

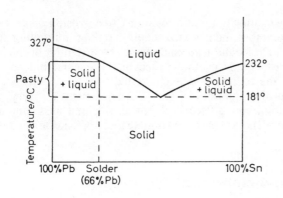

Figure 13.6 Phase diagram for solders

arsenic and sulphur as volatile oxides) and by removing any iron magnetically, the metal is obtained by smelting with carbon (together with some calcium oxide, which removes silica as slag) in a reverberatory furnace

$$SnO_2 + 2C \rightarrow Sn(l) + 2CO(g)$$

As tin is not attacked to any extent by organic acids, it is used for coating iron in making tin plate for food preservation, even though this material is very susceptible to corrosion. Its low melting point is utilized and enhanced by alloying with lead to make solder. The phase diagram (*Figure 13.6*) shows that not only does tin depress the melting point of lead but gives a range of temperature through which the solder has a desirable 'pasty' quality, with solid and liquid present together.

Lead

Although lead occurs as sulphate, carbonate and chromate(VI), its main ore is galena, PbS. This is roasted in air to convert it partially to oxide and sulphate:

$$2PbS + 3O_2 \rightarrow 2PbO + 2SO_2$$

$$PbS + 2O_2 \rightarrow PbSO_4$$

The oxidizing atmosphere is then replaced by one of reduction by cutting off the air supply; more lead sulphide or carbon (and calcium oxide if the ore is contaminated with silica) is added and the temperature is raised. The following reactions take place:

$$2PbO + PbS \rightarrow 3Pb(l) + SO_2(g)$$

$$PbSO_4 + PbS \rightarrow 2Pb(l) + 2SO_2(g)$$

$$PbO + C \rightarrow Pb(l) + CO(g)$$

If the quantity of silver in the lead so obtained warrants it, it is then 'desilvered' (p. 105). The product may be finally purified by electrolysis, with the impure lead as anode of the cell and lead(II) hexafluorosilicate as electrolyte.

Its softness, low melting point and resistance to corrosion make lead very popular with the plumber and it is used widely in piping and roofing. It is also used for lining tea chests and lead chambers. The invention of the internal combustion engine has had a profound effect on the demand for lead; it forms the electrodes of the battery and is converted into tetraethyllead(IV) for improving the burning qualities of petrol. The element is a very efficient absorber of radiation and is used in X-ray shields and as a container for radioactive isotopes. Its oxides and carbonate find application in paints, e.g., as red and white lead.

Reactions of the Group 4 elements

Tin and lead reveal their metallic character by tarnishing to some extent in air at ordinary temperatures. If all the elements of Group 4 are heated sufficiently strongly in air or oxygen, they burn to give various oxides.

Lead also reacts with water at room temperature, especially if the water is soft and oxygenated: under these conditions lead(II) hydroxide is formed, and as this is fairly soluble, the reaction continues. However, in hard waters, the presence of carbonate or sulphate ions produces protective coatings of carbonate or sulphate. Carbon and silicon will react on heating strongly in steam

$$C + H_2O \rightarrow CO(g) + H_2(g)$$

$$Si + 2H_2O \rightarrow SiO_2 + 2H_2(g)$$

Concentrated nitric acid attacks all the elements except silicon; in the cases of carbon, germanium and tin, the dioxide is formed, e.g.,

$$Sn + 4HNO_3 \rightarrow SnO_2 + 2H_2O + 4NO_2(g)$$

Lead reveals its more metallic nature by forming the nitrate

$$3Pb + 8HNO_3 \rightarrow 3Pb(NO_3)_2 + 4H_2O + 2NO(g)$$

Neither tin nor lead are as reactive to acids as their electrode potentials might indicate, probably owing to overpotential and surface protection; the only acid to which silicon is vulnerable is hydrofluoric:

$$Si + 6HF \rightarrow \underset{\substack{\text{hexafluorosilicic} \\ \text{acid}}}{H_2SiF_6} + 2H_2(g)$$

Fused sodium hydroxide attacks all except carbon to give oxoanions, e.g.,

$$Si + 2NaOH + H_2O \rightarrow Na_2SiO_3 + 2H_2(g)$$

Hydrides

Carbon is exceptional in the way in which it can form compounds with hydrogen; it gives straight-chain and cyclic compounds in which there may be single bonds and localized or non-localized π orbitals (p. 55). The profusion of hydrides of the remaining elements falls off as the Group is descended, and the only type so far prepared with certainty are straight-chain compounds corresponding to the alkanes (p. 394). In the case of silicon and germanium, compounds up to M_6H_{14} have been obtained by treating magnesium silicide or germanide with dilute hydrochloric acid, e.g.,

$$Mg_2Si + 4H^+ \rightarrow 2Mg^{2+} + SiH_4(g), \text{ etc.}$$

The lower members of the *silane* and *germane* series are gases at room temperature; as with hydrocarbons, their volatility decreases with increase in relative molecular mass. Where they differ sharply from alkanes, though, is in their ease of hydrolysis by water or alkali, particularly with silanes: this reactivity can be ascribed to the availability of *d* orbitals able to accept electrons from the oxygen atom (p. 213).

The silanes react with hydrogen halides to give halogen derivatives, e.g.,

$$SiH_4 + HCl \rightarrow SiH_3Cl(g) + H_2(g)$$

and they are also powerful reducing agents, converting hydroxide ions to hydrogen:

$$Si_2H_6 + 8OH^- \rightarrow 2SiO_4^{4-} + 7H_2(g)$$

Only one hydride of tin, stannane, SnH_4, has so far been identified, as the result of treating tin(IV) chloride with lithium tetrahydridoaluminate in ether at low temperature. This hydride is unstable at room temperature, although it is not readily hydrolysed.

Plumbane, PbH_4, is believed to be formed transiently by the action of dilute acid on magnesium–lead alloys.

Halides

All the possible tetrahalides are known except $PbBr_4$ and PbI_4, the bond energies of these presumably being too small to allow of their existence.

As with the hydrides, the halides of carbon are on the whole comparatively inert and, in marked contrast with the rest, resist hydrolysis. (The tetrachloride is prepared commercially by chlorinating carbon disulphide in the presence of an iron catalyst:

$$CS_2 + 3Cl_2 \rightarrow CCl_4(g) + S_2Cl_2(g)$$

Catenation of carbon atoms is also much in evidence. Poly(tetrafluoroethene) (PTFE), for example, is made up of macromolecules with the repeating unit

$$\begin{bmatrix} & F & F & \\ & | & | & \\ -C & - & C & - \\ & | & | & \\ & F & F & \end{bmatrix}$$

All the tetrahalides of the remaining elements are very susceptible to hydrolysis and are therefore made by reactions under anhydrous conditions, usually by direct combination of the two elements.

Both carbon and silicon form oxohalides; characteristically, carbon gives discrete molecules, e.g., $COCl_2$, carbonyl chloride, obtained by direct union of carbon monoxide and chlorine, whilst silicon gives more complex molecules with union through Si—O bonds; for instance, treatment of silicon halides with moist ether gives compounds containing the structure

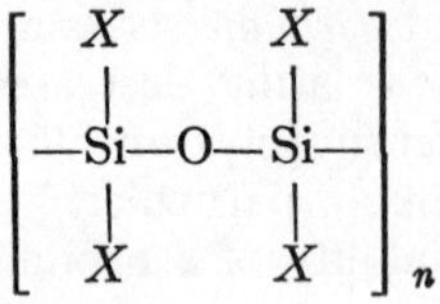

$$\left[\begin{array}{cc} X & X \\ | & | \\ -Si-O-Si- \\ | & | \\ X & X \end{array}\right]_n$$

Germanium, tin and lead form dihalides with some ionic character, as befits the increase in metallic nature, although methods of preparation differ greatly: germanium dihalides can be made by disproportionation reactions [i.e., by heating germanium(IV) halides with germanium] whilst, because they are fairly insoluble, lead dihalides are obtained by double decomposition

$$GeCl_4 + Ge \rightleftharpoons 2GeCl_2$$

$$Pb^{2+} + 2Cl^- \rightarrow PbCl_2(s)$$

Oxides

All the elements give monoxides and dioxides. There is great variation in the stability of these compounds, and only carbon oxides are present as discrete molecules corresponding to the empirical formulae.

Carbon monoxide

Carbon monoxide has a structure intermediate between the two forms

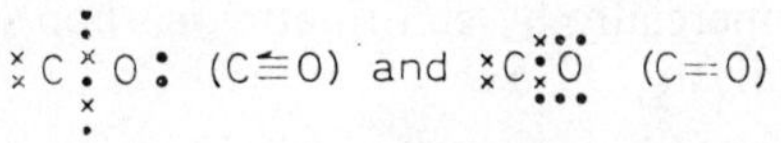

$$\overset{\times}{\underset{\times}{C}} \overset{..}{\underset{..}{.}} O \overset{..}{\underset{..}{.}} \quad (C\equiv O) \quad \text{and} \quad \overset{\times}{\underset{..}{C}} \overset{\times\times\times}{\underset{...}{O}} \quad (C=O)$$

Although it is obtained by dehydration of methanoic (formic) acid, it cannot be regarded as the true anhydride of this acid, as the reverse reaction does not occur, presumably because the electronic structures of the two compounds are not compatible

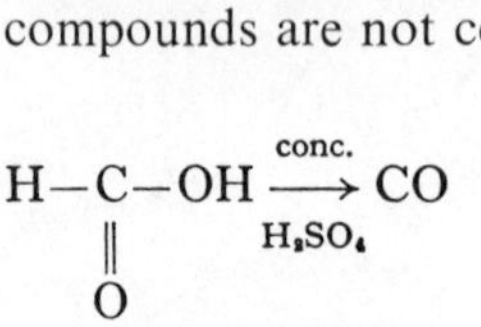

$$H-\underset{\underset{O}{\|}}{C}-OH \xrightarrow[H_2SO_4]{\text{conc.}} CO$$

Carbon monoxide is formed when carbon dioxide is reduced by carbon at high temperature, and so it tends to predominate when carbon burns in a limited amount of oxygen

$$C + O_2 \rightarrow CO_2(g)$$

$$CO_2 + C \rightarrow 2CO(g)$$

It is also obtained on the large scale by passing steam over white-hot coke: the mixture of hydrogen and carbon monoxide so formed is a useful fuel and is often used unchanged as 'water gas'

$$C + H_2O \rightarrow CO(g) + H_2(g) \qquad \Delta H = +126 \text{ kJ}$$

Carbon monoxide is a colourless, odourless gas which burns in air with a characteristic blue flame. It is extremely poisonous, as it combines with haemoglobin in the blood to form a compound more stable than oxyhaemoglobin, so that the normal process of respiration is impaired.

It is a powerful reducing agent and much of its large-scale use can be attributed to this, e.g., the reduction of iron ores

$$Fe_2O_3 + 3CO \rightarrow 2Fe + 3CO_2(g)$$

It combines with many non-metals; for example, with sulphur it gives carbonyl sulphide, COS, and with chlorine, in the presence of light, carbonyl chloride (phosgene) is formed. With hydrogen, in the presence of zinc oxide and chromium(III) oxide, the multiple bond is saturated and methanol formed

$$CO + 2H_2 \rightarrow CH_3OH$$

Under different conditions, the carbon monoxide can be reduced completely to methane

$$CO + 3H_2 \xrightarrow{\text{Ni}} CH_4(g) + H_2O$$

With certain of the transitional metals, carbon monoxide serves as a *ligand* and forms carbonyls which are covalent and volatile. For example, when it is passed over heated nickel, tetracarbonylnickel(0), $Ni(CO)_4$, is formed by coordinating to the metallic atom:

$$
\begin{array}{c}
O \\
\downarrow\| \\
C \\
{\scriptstyle\cdot\cdot} \\
\downarrow \\
O \equiv C: \rightarrow \ Ni \ \leftarrow \ :C \equiv O \\
\uparrow \\
{\scriptstyle\cdot\cdot} \\
C \\
\|\uparrow \\
O
\end{array}
$$

This can be removed as vapour (leaving impurities behind) and decomposed to the pure metal at a higher temperature

$$Ni + 4CO \xrightarrow{90\,°C} Ni(CO)_4(g) \xrightarrow{180\,°C} Ni(s) + 4CO(g)$$

When carbon monoxide is passed under pressure into fused sodium hydroxide, it reacts to form sodium methanoate (formate)

$$Na^+OH^- + CO \rightarrow H.COO^-Na^+$$

The gas is insoluble in water but it dissolves in ammoniacal copper(I) chloride to form complexes with the copper, e.g.,

$$Cu(NH_3)_2Cl \xrightarrow{CO} Cu(CO)Cl$$

Carbon dioxide, O=C=O

Carbon dioxide is the true anhydride of carbonic acid

$$O=C=O \xrightarrow{H_2O} O=C\begin{smallmatrix}OH\\\\OH\end{smallmatrix}$$

Because the equilibrium lies well over to the left, carbon dioxide is evolved when carbonates are treated with acids stronger than carbonic acid:

$$CO_3^{2-} + 2H^+ \rightarrow H_2CO_3 \rightarrow CO_2(g) + H_2O$$

It also results from heating hydrogencarbonates and most carbonates:

$$HCO_3^- \rightarrow CO_2(g) + OH^-$$

$$CO_3^{2-} \rightarrow CO_2(g) + O^{2-}$$

The colourless, heavy gas can be dried by passage through concentrated sulphuric acid and collected by upward displacement of air.

Industrially, it is a by-product of fermentation processes and of the manufacture of calcium oxide

$$C_6H_{12}O_6 \xrightarrow{enzymes} 2C_2H_5OH + 2CO_2(g)$$

$$CaCO_3 \rightarrow CaO + CO_2(g)$$

It is used in the Solvay process for making sodium hydrogencarbonate (p. 336) and, when solid, as a mobile refrigerant on account of its large enthalpy of sublimation

$$CO_2(s) \rightarrow CO_2(g) \qquad \Delta H = +25 \text{ kJ at } -56\,°C$$

The evolution of carbon dioxide when acid reacts with carbonate is utilized in baking powders (the acid used is solid and so only reacts to produce carbon dioxide when there is moisture present to dissolve it), in sherbets and in fire extinguishers. Carbon dioxide is also of great physiological importance in the regulation of respiration, blood circulation and acid-base balance. It reacts with ammonia and adenosine triphosphate (ATP) to form carbanoyl phosphate, which itself is a precursor of the nucleoprotein constituents arginine and pyrimidines.

Oxides of silicon, germanium and tin

Silicon monoxide is claimed to be formed by the reduction of silica by silicon at high temperature. *Silica* itself, silicon dioxide, occurs widely and is formed when silicon burns in oxygen. It exists in three crystal forms, each with a high- and low-temperature modification:

quartz $\rightleftharpoons$ tridymite $\rightleftharpoons$ cristobalite

In all these forms, silicon is surrounded tetrahedrally by four oxygen atoms in a macromolecular lattice.

Although silica is not hydrated to form silicic acid, it does react on fusion with alkalis to give silicates, and in this sense is an acid anhydride.

Germanium monoxide is unstable and, like all the bivalent compounds of germanium, tends to disproportionate on heating:

$$2Ge^{II} \rightarrow Ge^{IV} + Ge$$

Tin(II) oxide is prepared by heating the ethanedioate or by carefully dehydrating the compound formed when tin(II) chloride solution is treated with alkali. It is amphoteric, dissolving in caustic alkalis to form stannates(II), e.g.,

$$SnO + H_2O + OH^- \rightarrow Sn(OH)_3^-$$

The stable oxide of tin is *tin(IV) oxide*, which is commonly made by treating tin with concentrated nitric acid and strongly heating the white residue obtained (see above), but is better prepared by the hydrolysis of tin(IV) compounds, e.g.,

$$SnCl_4 + 2H_2O \rightarrow SnO_2 + 4HCl$$

This, too, is amphoteric, although one of the few acids in which it will dissolve is concentrated sulphuric acid. With caustic alkalis is dissolves, giving the stannate(IV) ion:

$$SnO_2 + 2H_2O + 2OH^- \rightarrow Sn(OH)_6^{2-}$$

Oxides of lead

Lead gives three oxides. *Lead(II) oxide* is obtained by the general methods for metallic oxides, such as the action of heat on the metal (in air), the carbonate, hydroxide or nitrate.

This oxide is also amphoteric and dissolves both in acids [giving lead(II) salts] and in caustic alkalis [giving plumbates(II)]:

$$PbO + 2H^+ \rightarrow Pb^{2+} + H_2O$$

$$PbO + 2OH^- \rightarrow PbO_2^{2-} + H_2O$$

Dilead(II) lead(IV) oxide, red lead, Pb_3O_4, is usually obtained by heating lead(II) oxide in air at about 400 °C:

$$6PbO + O_2 \rightarrow 2Pb_3O_4$$

If this is treated with nitric acid, which produces a soluble salt, *lead(IV) oxide* is left behind

$$Pb_3O_4 + 4H^+ \rightarrow 2Pb^{2+} + 2H_2O + PbO_2(s)$$

Lead(IV) oxide is also precipitated by the action of oxidizing agents on solutions of lead(II) compounds. It is a strong oxidizing agent, converting, e.g., hydrochloric acid into chlorine, and it forms the positive plate of the lead accumulator; as an accumulator discharges, the lead(IV) oxide gains electrons and the lead electrode loses them:

$$Pb^{4+} + Pb \rightarrow 2Pb^{2+}$$

If lead(IV) oxide is treated with, for example, cold, concentrated hydrochloric acid or glacial ethanoic (acetic) acid, it dissolves to form essentially covalent compounds:

$$PbO_2 + 4HCl \rightarrow PbCl_4 + 2H_2O$$

$$PbO_2 + 4CH_3COOH \rightarrow (CH_3COO)_4Pb + 2H_2O$$

With alkalis it forms plumbates(IV), e.g.,

$$PbO_2 + 2H_2O + 2OH^- \rightarrow Pb(OH)_6{}^{2-}$$

Tin(IV) oxide and lead(IV) oxide both have the rutile structure (p. 82).

Oxoacids and oxo salts

Carbonates

Carbonates contain the discrete ion

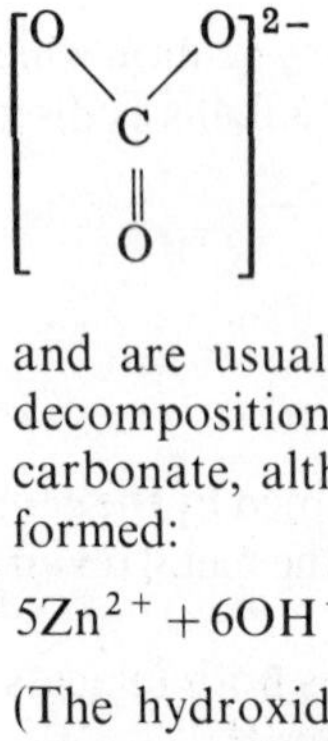

$$\left[\begin{array}{c} O \diagdown \quad \diagup O \\ C \\ \| \\ O \end{array} \right]^{2-}$$

and are usually insoluble; they are therefore often prepared by double decomposition between a soluble salt of the metal and a solution of a soluble carbonate, although this procedure can result in a basic carbonate being formed:

$$5Zn^{2+} + 6OH^- + 2CO_3{}^{2-} \rightarrow 2ZnCO_3.3Zn(OH)_2(s)$$

(The hydroxide ions result from the hydrolysis of the carbonate ions in aqueous solution.)

To prevent the basic carbonate being precipitated, a solution of a hydrogen carbonate is used, the concentration of hydroxide ions produced by hydrolysis of this ion being much less than from the carbonate:

$$Zn^{2+} + HCO_3{}^- \rightarrow ZnCO_3(s) + H^+$$

Although most metals (excluding those of Group 1) can form normal and basic carbonates, there are some, such as copper, which appear to be capable of forming the basic variety only.

The carbonates of the Group 1 metals are soluble and are prepared by first passing carbon dioxide through a solution of the alkali until saturated, and then adding a second quantity of alkali equal in volume and in concentration to the first:

$$OH^- + CO_2 \rightarrow HCO_3^-$$

$$HCO_3^- + OH^- \rightarrow CO_3^{2-} + H_2O$$

Since carbonic acid is a weak acid, it is readily displaced from carbonates by the addition of acids stronger than itself, the more so because it breaks up into water and carbon dioxide and disturbs the equilibrium

$$CO_3^{2-} + 2H^+ \rightleftharpoons H_2CO_3 \rightleftharpoons H_2O + CO_2$$

Carbonates, with the exception of those of Group 1, are also fairly easily decomposed by heat, e.g.,

$$CaCO_3 \rightleftharpoons CaO + CO_2(g)$$

Only the hydrogencarbonates of the alkali metals (and of ammonium) with low polarizing power exist in the solid state; the rest are known only in solution and are decomposed during attempts to isolate them:

$$Mg(HCO_3)_2 \rightleftharpoons MgCO_3(s) + H_2O + CO_2(g)$$

The reversibility of this reaction means that they can be obtained in solution by treatment of a suspension of the carbonate (or of a solution of the hydroxide) with carbon dioxide until the solution is saturated or the suspension has disappeared. Like carbonates, the hydrogencarbonates are decomposed on heating and by the action of most acids.

Silicates

Although silicic acid, H_2SiO_3 (probably better represented as $SiO_2.2H_2O$), is weaker than carbonic acid, silicates are obtained by heating silica with carbonates because of the non-volatility of silica:

$$CO_3^{2-} + SiO_2 \rightarrow SiO_3^{2-} + CO_2(g)$$

With the exception of those of Group 1, silicates, like the acid itself, are insoluble in water. Whilst there are some silicates with discrete ions, e.g., olivine, $Mg_2^{2+}SiO_4^{4-}$, the strength of the Si—O bond is such that most of the naturally occurring silicates have condensed systems (*Figure 13.7*).

Continuation of this trend results in silica, with its three-dimensional lattice, being formed. Replacement of Si^{IV} by Al^{III} can take place, however, because of its similar size and coordination number, giving one negative charge per replacement. The resulting complex oxoanion is then associated with the requisite number of cations, usually Na^+, K^+, Ca^{2+}, Mg^{2+}, Al^{3+}. The choice of resulting structures is very great, accounting for the wide variety of silicates in nature, such as felspars, e.g., orthoclase, $KAlSi_3O_8$, and zeolites, e.g., natrolite, $Na_2Al_2Si_3O_{10}.2H_2O$. In the case of the latter, the alkali cation is capable of ion exchange—an effect now much employed for the softening of water. The lattice is also often capable of allowing the passage of molecules below a certain critical size; this property is employed in 'molecular sieves'.

The fusion of sand with alkali carbonates and basic oxides results in the formation of glasses of fairly variable composition. These materials have a

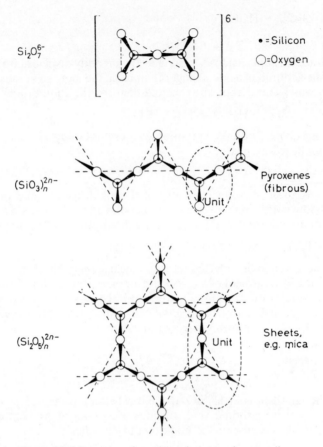

$Si_2O_7^{6-}$

• = Silicon
○ = Oxygen

$(SiO_3)_n^{2n-}$ Pyroxenes (fibrous)

Unit

$(Si_2O_5)_n^{2n-}$ Unit Sheets, e.g. mica

Figure 13.7 Examples of condensed systems in naturally occurring silicates

structure similar to that of a liquid, that is, only short-range ordered regions can be detected. They are in fact sometimes loosely described as supercooled liquids.

Sulphides

All the elements of Group 4, except lead, give a disulphide, MS_2, which can be obtained by direct combination. Characteristically, carbon disulphide exists as small molecules, $S{=}C{=}S$, whilst the rest are macromolecular; for example, SiS_2 consists of the repeating unit

$$\begin{bmatrix} & S & & S & \\ \diagdown & \diagup & \diagdown & \diagup & \diagdown \\ Si & & Si & \\ \diagup & \diagdown & \diagup & \diagdown \\ & S & & S & \end{bmatrix}$$

and typically is hydrolysed by water to the oxide.

Tin gives a monosulphide as well as the disulphide, although this is easily oxidized; for instance, if it is dissolved by ammonium polysulphide to form ammonium trithiostannate(IV), it is the disulphide that is precipitated on acidification

$$SnS \xrightarrow{(NH_4)_2S_n} (NH_4)_2SnS_3 \xrightarrow{H^+} SnS_2(s)$$

Lead(II) sulphide, like tin(II) sulphide, is precipitated by passing hydrogen sulphide through an alkaline or only weakly acidic solution of a suitable salt, but it is not dissolved by ammonium sulphide nor by alkaline solutions, and so the two can easily be distinguished.

Organic compounds

Organic carbon compounds can be made to yield derivatives with the other elements of the Group. Thus, if an alkyl halide is heated to about 300 °C with silicon, in the presence of a copper catalyst, a series of compounds with silicon attached to carbon and halogen is obtained, e.g.,

$$CH_3Cl + Si \xrightarrow{Cu} (CH_3)_3SiCl + (CH_3)_2SiCl_2 + CH_3SiCl_3$$

The mechanism of this reaction involves free radicals:

$$2Cu + CH_3Cl \rightarrow CuCl + CuCH_3$$

$$CuCH_3 \rightarrow Cu + \cdot CH_3$$

$$(4-n)CuCl + Si + nCH_3\cdot \rightarrow (CH_3)_nSiCl_{4-n} + (4-n)Cu \quad (n \not> 3)$$

Hydrolysis of these intermediates gives polymers called *silicones* which, because of their electrical resistance, water repellancy, inertness and elasticity or viscosity, have a wide variety of uses.

$$(CH_3)_2SiCl_2 \rightarrow (CH_3)_2Si(OH)_2 \rightarrow \begin{bmatrix} CH_3 & CH_3 & CH_3 \\ | & | & | \\ -Si-O-Si-O-Si- \\ | & | & | \\ CH_3 & CH_3 & CH_3 \end{bmatrix}_n$$

Because silicon is less electronegative than carbon, the silicon–carbon bond is polar, with silicon electrophilic and carbon, unusually, nucleophilic.

The strength of the bond with carbon falls off as the Group is descended, but tetraethyllead(IV), $(C_2H_5)_4Pb$, obtained by treating lead(II) chloride with ethylmagnesium chloride, is comparatively stable

$$2PbCl_2 + 4C_2H_5MgCl \rightarrow (C_2H_5)_4Pb(g) + 4MgCl_2 + Pb(s)$$

At elevated temperatures, however, it decomposes to give free ethyl radicals, which promote a more regular oxidation of hydrocarbon; it thus finds use as an 'anti-knock' agent in petrol.

Cyanogen, $(CN)_2$

Cyanogen is evolved when certain metallic cyanides are heated. Significantly, these metals are usually not very electropositive and have variable oxidation states: the reaction involves the reduction of the metal by the cyanide ion, e.g.,

$$2Cu(CN)_2 \rightarrow 2CuCN(s) + (CN)_2(g)$$

It is also formed from copper(II) sulphate and potassium cyanide solutions

$$2Cu^{2+} + 4CN^- \rightarrow 2CuCN(s) + (CN)_2(g)$$

Both these reactions can be compared with the reduction of copper(II) by the iodide ion

$$2Cu^{2+} + 4I^- \rightarrow 2CuI(s) + I_2$$

Cyanogen is a very poisonous, colourless gas, smelling of almonds. It readily polymerizes to paracyanogen $(CN)_x$. Many of its reactions earn it the name of a pseudohalogen, e.g.,

$$(CN)_2 + 2K \rightarrow 2KCN$$

$$\underset{\text{cyanide} \quad \text{cyanate}}{(CN)_2 + 2OH^- \rightarrow CN^- + CNO^- + H_2O}$$

Cyanides

Potassium cyanide is made by heating potassium carbonate and carbon in the presence of ammonia:

$$K_2CO_3 + 2C + 2NH_3 \rightarrow 2KCN + CO(g) + 2H_2O(g) + H_2(g)$$

Distillation of potassium cyanide with sulphuric acid results in the evolution of hydrogen cyanide, which can be condensed on cooling to a colourless, very poisonous liquid which behaves as a weak, monobasic acid.

Hydrogen cyanide has a high relative permittivity, and there is evidence for self-ionization of the form

$$HCN \rightleftharpoons H_2CN^+ + CN^-$$

The cyanide ion is a powerful ligand and readily forms complexes with metals. This is the basis for its identification: the cyanide is treated with iron(II) sulphate in acid solution, and as oxidation of some of the iron to iron(III) takes places, so a precipitate of Prussian blue forms:

$$Fe^{2+} + 2CN^- \rightarrow Fe(CN)_2 \xrightarrow{CN^-} Fe(CN)_6^{4-} \xrightarrow{Fe^{2+}} [Fe^{II} Fe^{III} (CN)_6]^-$$

Cyanides occur naturally as glucosides (p. 517), e.g., amygdalin, and can be used in the manufacture of Perspex and propenonitrile (acrylonitrile) (p. 425).

Carbides

Carbides are often made by heating the element or its oxide with carbon:

$Si + C \rightarrow SiC$

$CaO + C \rightarrow CaC_2 + CO(g)$

They can be divided broadly into three classes:

Saline

Most of these contain the C_2^{2-} ion and are consequently decomposed by water with the evolution of ethyne (acetylene). They are known as dicarbides:

$C_2^{2-} + 2H^+ \rightarrow C_2H_2(g)$

A few saline carbides, however, contain discrete carbon atoms or C^{4-} ions; these evolve methane when treated with water and are often called methanides, e.g.,

$Al_4C_3 + 12H_2O \rightarrow 3CH_4(g) + 4Al(OH)_3(s)$

The more covalent dicarbides are insoluble in water and can be prepared by passing ethyne through a suitable solution of the metal, e.g., ammoniacal copper(I) chloride.

Covalent

The covalent carbides are not attacked by acids and are normally very hard. Silicon carbide, for example, is a powerful abrasive, known as carborundum, which approaches diamond in hardness.

Interstitial

The covalent carbides are stoichiometric (or Daltonide). Non-stoichiometric carbides also exist, particularly those of the transition elements. If the radius of the metallic atoms is at least 130 pm, the octahedral holes in the lattice are sufficiently large for carbon atoms to fit in. These carbides are very hard and refractory, inert and able to conduct electricity.

Summary

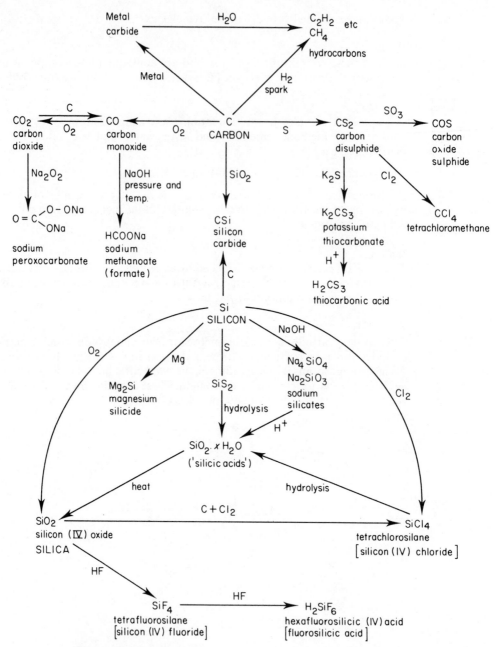

Figure 13.8 Reactions of carbon and silicon

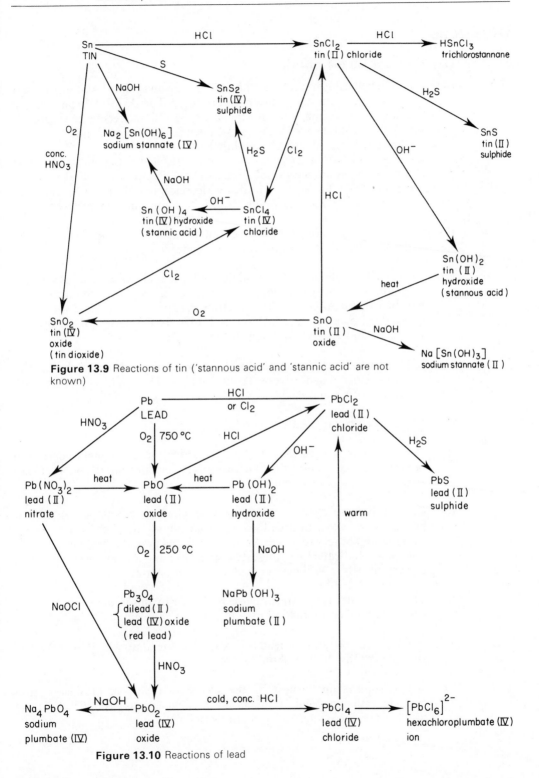

Figure 13.9 Reactions of tin ('stannous acid' and 'stannic acid' are not known)

Figure 13.10 Reactions of lead

Questions

(1)   Comment on the following:

    (a) When hydrogen sulphide is passed through an acidified solution of a salt of a metal, a brown precipitate is produced. This dissolves in ammonium sulphide solution, and on acidification a yellow precipitate is formed.

    (b) A dilute solution of sodium hydrogencarbonate is alkaline to methyl orange but acidic to phenolphthalein.

    (c) Red lead gives a brown deposit with nitric acid; if the filtrate is treated with potassium iodide solution, a yellow precipitate is obtained.

    (d) If sand and fluorspar are heated with concentrated sulphuric acid, a gas is evolved which gives a white deposit on coming into contact with water.

(2)   How far do you think cyanogen can be called a pseudo-halogen and hydrogen cyanide a pseudo-hydrogen halide?

(3)   A light yellow solid heated with dilute sulphuric acid produced a gas smelling of almonds, and with concentrated sulphuric acid, a gas which burnt with a pale blue flame. After this latter reaction, the residue was diluted with water and a dark blue precipitate developed. Identify the nature of the original substance and suggest equations for the reactions.

(4)   It has been said that the acidity of an oxide depends upon the relative amounts of metal and of oxygen contained in it. Discuss this by reference to the oxides of tin and lead.

(5)   Write an essay on silicate minerals.

(6)   'As one descends a group of the Periodic Table, one encounters elements of increasing metallic character.' Discuss this statement from the point of view of the elements of Group 4.

(7)   Carbon, because of the fundamental role which its atoms play in cellular processes, is often regarded as the element of life. Yet several of its compounds are very toxic, carbon monoxide and cyanides being particularly notorious. Indicate how these two poisons function and whether their efficacy reveals a weakness in evolutionary processes.

(8)   The gas industry is in a state of flux. Indicate why this is and summarize the steps being taken by the industry to solve its problems.

(9)   'The commercial future of lead is inextricably bound up with that of the car.' Indicate if and why this should be so.

(10) What products are formed by the successive treatment of silicon tetrachloride with increasing amounts of ehtylmagnesium bromide and what results when these intermediates are hydrolysed?

(11)   (a)   If the solubility of lead chloride is determined in solutions containing an increasing concentration of sodium chloride, it is found that as the concentration of the sodium chloride increases from zero, the solubility of the lead chloride first decreases, passes through a minimum, and then increases. Suggest an explanation for these observations.

(b)   An aqueous solution is 1.00 molar in ammonium chloride, 0.010 molar in manganese(II) sulphate, and also contains some ammonia. Calculate the minimum concentration of ammonia which will just bring about the precipitation of manganese(II) hydroxide, the solubility product of which is 4.00×10^{-14} mol^3 dm^{-9}. The dissociation constant K_b of ammonia is 1.80×10^{-5} mol dm^{-3}, where

$$K_b = [NH_4^+][OH^-]/[NH_3]$$

(Assume that the ammonium ions in the solution derive solely from the ammonium chloride.)

[Oxford 1979]

14 The elements of Groups 1, 2 and 3

Characteristics of the Groups • Occurrence and extraction •
Properties of the elements • Compounds of elements of Groups 1–3 •
Extraction and manufacture of some metals

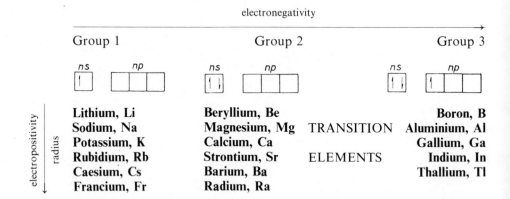

Characteristics of the Groups

The increase in non-metallic character in going along each Period means that of the three Groups, Group 3 is the least metallic. Also, the increase in metallic character on descending each Group results in boron being the least metallic of this selection of elements. However, the ionization energies (*Table 14.1*) of the first element of each Group are sufficiently high for several compounds of lithium, beryllium and boron to be mainly covalent.

Table 14.1 Ionization energies/kJ mol⁻¹ of elements of Groups 1–3

Group 1		*Group 2*			*Group 3*			
	1st		*1st*	*2nd*		*1st*	*2nd*	*3rd*
Li	519	Be	900	1760	B	799	2420	3660
Na	494	Mg	736	1450	Al	577	1820	2740
K	418	Ca	590	1150	Ga	577	1980	2960
Rb	402	Sr	548	1060	In	556	1820	2700
Cs	376	Ba	502	966	Tl	590	1970	2870

Furthermore, the increase in the charge on the ions in proceeding from Group 1 to 3 adds to the effect of the high ionization energy of boron to give this element the properties of a non-metal. For these reasons, much of the chemistry of aqueous solutions of beryllium, aluminium, and particularly boron, is concerned with the element in the form of an anion, such as $BeO_2{}^{2-}$, $Al(OH)_4{}^-$, $H_2BO_3{}^-$, the relevant cations having such considerable polarizing power that under normal conditions they would not be expected to exist.

Although regular gradations in properties may reasonably be expected for the elements of Groups 1 and 2 and their compounds, a less regular change

Table 14.2 Atomic radii of the elements of Groups 1–3

	Group 1 Element	Atomic radius/pm	Group 2 Element	Atomic radius/pm	Group 3 Element	Atomic radius/pm
Progressive increase	Li	123	Be	89	B	80
in atomic size on	Na	157	Mg	136	Al	125
descending Group	K	203	Ca	174	Ga	125*
gives steady	Rb	216	Sr	191	In	150
decrease in	Cs	235	Ba	198	Tl	155
electronegativity						

* Intervening transitional metals, with their contraction in atomic volume, arrest increase in atomic size in Group 3.

must be expected for Group 3. This is the result of the transition elements' producing an irregular increase in size after the second element in the Group (*Table 14.2*).

Valency states

The s^1 electron of the Group 1 elements allows only the univalent state. The elements of Group 2 similarly can only be bivalent because of the s^2 electrons in the outer electron shell. Elements of Group 3 have three electrons in their outer, valency, shell, but these are not all of the same type, being designated s^2p^1 (p. 20). The maximum valency of the Group 3 elements is produced either by ionization of these electrons, giving a trivalent cation or, as always occurs for boron, by excitation of the form

s	p_x	p_y	p_z		s	p_x	p_y	p_z
↑↓	↑			Excitation →	↑	↑	↑	

giving three unpaired electrons and, hence, a covalency of three. Because the s and p electrons can behave independently of each other, it is possible for the p electron to be involved in bonding whilst the pair of s electrons remains inert, so that both univalent and trivalent forms of the Group 3 elements can occur. Indeed, for thallium, because of this 'inert pair' effect, the univalent state is more stable than the trivalent. However, the remaining elements of the

Group are essentially trivalent. In covalent bond formation, the maximum number of electrons which can be accommodated in the second quantum shell is eight (s^2p^6), giving beryllium and boron a possibility of tetrahedral coordination by the further acceptance of two pairs or one pair of electrons, respectively, i.e., by the formation of dative bonds. As a result, beryllium and boron often occur in complexes in which they have acquired the necessary number of electrons, e.g.,

$$BeF_4{}^{2-} \qquad [Be(H_2O)_4]^{2+} \qquad BF_3.NH_3 \qquad BF_4{}^-$$

Occurrence and extraction

Because of their high electropositivity, none of these elements is found native; those of Group 1 occur as their halides, of Group 2 as sulphates and carbonates and of Group 3 as oxides or silicates (often accompanied by the

Table 14.3 Some common naturally-occurring compounds of the elements of Groups 1–3

	Group 1	Group 2	Group 3
Halides	NaCl, rock salt KCl.$MgCl_2$.6H_2O carnallite	CaF_2 fluorspar	Na_3AlF_6, cryolite
Sulphates		$CaSO_4$, anhydrite $CaSO_4$.2H_2O, gypsum $BaSO_4$, barytes $MgSO_4$.7H_2O, Epsom salts	
Oxides			AlO(OH), bauxite $Na_2B_4O_7$.10H_2O, borax
Silicates	$NaAlSi_3O_8$, felspar $LiAl(SiO_3)_2$, spodumene	$Be_3Al_2(SiO_3)_6$, beryl $CaMg_3(SiO_3)_4$, asbestos $Mg_3Si_4O_{10}(OH)_2$, talc	$Al_2Si_2O_5(OH)_4$, kaolin
Carbonates		$MgCO_3$, magnesite $CaCO_3$.$MgCO_3$, dolomite $CaCO_3$ chalk, limestone, marble $BaCO_3$, witherite	

elements of Groups 1 and 2). The general solubility of the salts results in many being found in sea water and salt beds; for example, the approximate concentrations of the more common metallic ions in sea water are Na^+ 1.1 per cent, Mg^{2+} 0.13 per cent, Ca^{2+} and K^+, 0.04 per cent (*Table 14.3*).

Although all the oxides can be reduced to the corresponding element by carbon at suitable temperatures, this method is often not convenient, chiefly because of carbide formation between the element produced and excess carbon, and on account of the high temperatures sometimes needed (*Table 14.4*).

Electrolytic methods

All these elements except boron, barium and, more recently, magnesium are obtained by the electrolytic reduction of the fused salts, with graphite serving as the anode, where halogen and oxygen are evolved, and iron generally as the cathode. Mixed salts are normally employed to lower the melting point of the system (p. 109) and for increasing the conductivity. Examples of the electrolytes used are

(1) Group 1 chlorides fused in the presence of calcium chloride (as impurity).
(2) Fused beryllium chloride made from beryl (with sodium chloride as impurity).
(3) Fused carnallite, the magnesium chloride often being replenished by producing it from magnesite.
(4) Fused calcium chloride, with calcium fluoride or potassium chloride as impurity.
(5) Aluminium oxide dissolved in molten cryolite, Na_3AlF_6; the latter acts as a conducting solvent for the oxide, and is itself significantly consumed.

Thermal reduction

The high temperature required for the electrolysis of fused barium chloride causes too many difficulties, and barium is therefore usually isolated by the reduction of barium oxide (from calcined witherite) with aluminium:

$$3BaO + 2Al \rightarrow 3Ba + Al_2O_3$$

An alternative method for the extraction of magnesium is by the reduction of magnesium oxide, either with ferrosilicon if the oxide is MgO.CaO (from calcined dolomite) or with coke, as in the Pidgeon process (p. 198). In the former case, the ferrosilicon removes the calcium as a slag of calcium silicate:

$$MgO.CaO \xrightarrow{Fe/Si} Mg + CaSiO_3$$

Boron is prepared by reducing boron oxide [itself obtained from disodium tetraborate (borax)] with magnesium or aluminium

$$B_2O_3 + 3Mg \rightarrow 2B + 3MgO$$

Table 14.4 Extraction and uses of the elements of Groups 1–3

Element	Extraction	Uses
Lithium ⎱ Sodium ⎰	electrolysis of fused chloride	scavenger; alloys
		lamps, coolant (atomic reactors); manufacture of 'anti-knock', Na_2O_2 and NaCN; reducing agent, e.g. for titanium extraction
Potassium	electrolysis of fused chloride or hydroxide	reducing agent
Rubidium ⎱ Caesium ⎰	electrolysis of fused cyanide	photoelectric cells
Beryllium	electrolysis of fused chloride	X-ray 'windows'; scavenger; alloys; moderator in fission reactions
Magnesium	electrolysis of fused chloride and thermal reduction of oxide by Fe/Si or C	light alloys; reducing agent, e.g. extraction of uranium; container for nuclear fuel; 'Grignards'
Calcium	electrolysis of fused chloride or reduction of oxide by Al	scavenger, e.g. removal of water from ethanol, air from radio valves
Strontium	electrolysis of fused chloride	⎱ scavengers
Barium	reduction of oxide with Al	⎰
Boron	reduction of the oxide with Mg or Al	hydridoborates
Aluminium	electrolysis of fused Al_2O_3 $+ Na_3AlF_6$	light alloys; electrical conductor; reducing agent (Thermit process)
Gallium ⎱ Indium ⎰ Thallium	electrolysis of aqueous solutions	thermometry—low m.p. (30 °C) and high boiling point (2070 °C)

Properties of the elements

Physical properties

The metals, in contrast to boron, have low melting points (*Table 14.5*) and are good conductors of electricity. With the exception of boron and gallium, the elements crystallize in the typical metallic cubic close-packed, hexagonal close-packed and body-centred cubic structures (*Figure 14.1*). Gallium forms

Table 14.5 Melting points and densities of elements of Groups 1–3

	M.p./°C	Density g cm^{-3}		M.p./°C	Density g cm^{-3}		M.p./°C	Density g cm^{-3}
Li	178	0.53	Be	1285	1.86	B	2300	2.4
Na	97	0.97	Mg	650	1.75	Al	658	2.7
K	63	0.86	Ca	803	1.55	Ga	30	5.9
Rb	39	1.53	Sr	800	2.6	In	156	7.3
Cs	29	1.90	Ba	850	3.6	Tl	449	11.9

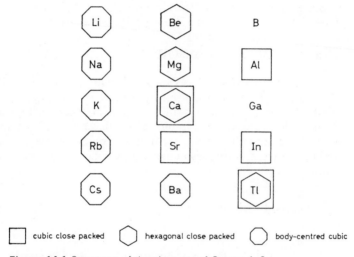

Figure 14.1 Structures of the elements of Groups 1–3

complex orthorhombic crystals, whilst boron has a complex non-metallic lattice of uncertain structure, the atoms being too short of electrons to form an orthodox three-dimensional covalent lattice.

The elements have low densities (*Table 14.5*), except gallium, indium and thallium, and for this reason and because of their tensile strength, magnesium and aluminium find uses in light alloys, e.g., Magnalium (Al 90 per cent, Mg 10 per cent) and Duralumin (Al 95 per cent, Mg 0·5 per cent, Cu 4 per cent, Mn 0·5 per cent); the low electron density of beryllium makes it suitable for 'windows' in *X*-ray work, whilst radium and its salts find applications in radiochemistry.

Chemical properties

The elements of Groups 1 and 2, and aluminium, have the largest negative electrode potentials (*Table 14.6*). Accordingly, they generally displace hydrogen from suitable compounds such as water and acids, their reactivity increasing with increase in atomic number.

Table 14.6 Standard electrode potentials at 25 °C

$E^{\ominus}/V$		$E^{\ominus}/V$		$E^{\ominus}/V$	
Li	−3.02	Be	−1.70		
Na	−2.71	Mg	−2.38	Al	−1.67
K	−2.92	Ca	−2.87		
Rb	−2.99	Sr	−2.89		
Cs	−3.02	Ba	−2.90		

Since the hydroxides of the Group 1 elements are freely soluble in water, these elements react most violently with water, e.g.,

$$2Na \; + \; 2H_2O \; \rightarrow \; 2NaOH \; + \; H_2(g)$$

whilst the hydroxides of the Group 2 elements are not so readily soluble and therefore their initial formation by the above type of reaction hinders the dissolution of the metal in water. Elements of these two Groups also react with alcohols, but with less vigour than with water; such systems, particularly that of sodium and ethanol, are useful as reducing agents. The elements of Group 3, except boron, also react with alcohols, provided that a catalyst is available, e.g.,

$$Al + 3ROH \xrightarrow{\text{HgCl}_2} (RO)_3Al + \tfrac{3}{2}H_2(g)$$

Other reactions of the free elements can be summarized as follows:

Beryllium, aluminium and gallium dissolve in caustic alkalis, although the anions produced are more complex than the usual equations suggest:

$$Be + 2OH^- \rightarrow BeO_2{}^{2-} + H_2(g)$$

$$2Al + 2OH^- + 2H_2O \rightarrow 2AlO_2{}^- + 3H_2(g)$$

The beryllate and aluminate ions are usually hydrated, aluminate occurring mainly as $[Al(OH)_4]^-$.

Heating with hydrogen gives the hydrides of the Group 1 metals as well as those of calcium, strontium and barium in Group 2.

Anhydrous halides are readily obtained by direct union with the free halogen, e.g.,

$$2Al + 3Cl_2 \xrightarrow{\text{heat}} Al_2Cl_6(g)$$

Oxides, peroxides and superoxides are obtained when the elements are heated to only a moderate temperature with oxygen:

$$Li \xrightarrow{O_2} Li_2O \text{ (oxide)}$$

$$Na \xrightarrow{O_2} Na_2O_2 \text{ (peroxide)}$$

$$K \xrightarrow{O_2} KO_2 \text{ (superoxide)}$$

When heated in the air, some nitride is also formed from magnesium. The nitrides of lithium, the remaining Group 2 elements, boron and aluminium are obtained if the relevant element is heated in nitrogen itself. With the exception of boron nitride and aluminium nitride, they contain the nitride ion, N^{3-}, and are readily hydrolysed, e.g.,

$$Mg_3N_2 + 3H_2O \rightarrow 3MgO + 2NH_3(g)$$

(Aluminium nitride is also readily hydrolysed, but boron nitride, one form of which has a covalent structure of the graphite type

reacts only slowly with water.)

Sulphides are obtained by direct combination in a similar manner to the oxides.

Heating with ammonia or dissolution in liquid ammonia results in the formation of amides of the Group 1 metals and those of calcium, strontium and barium. Like the nitrides these are readily hydrolysed:

$$Na + NH_3 \rightarrow \underset{\text{sodamide}}{Na^+NH_2^-} + \tfrac{1}{2}H_2(g)$$

$$Na^+NH_2^- + H_2O \rightarrow Na^+OH^- + NH_3(g)$$

(The name 'amide' is unfortunate, since in organic chemistry an amide has the characteristic group —$CONH_2$).

Compounds of elements of Groups 1–3

Hydrides

The very electropositive metals of Groups 1 and 2 form ionic hydrides on heating in dry hydrogen. As the hydride ion, H^-, is unstable in the presence of protons, H^+, these solids are decomposed by water and by acids, e.g.,

$$Ca^{2+}H_2^- + 2H_2O \rightarrow Ca^{2+}(OH^-)_2 + 2H_2(g)$$

The hydrides of Group 3 are less well characterized; boron gives a series of hydrides, all of which are electron-deficient. The parent member, diborane, B_2H_6, is prepared by adding a boron trifluoride–ether complex slowly to an ethereal suspension of lithium hydride and refluxing:

$$6H^- + 8(C_2H_5)_2O.BF_3 \rightarrow 6BF_4^- + 8(C_2H_5)_2O + B_2H_6(g)$$

Diborane is a gas which is rapidly hydrolysed by water to boric acid and hydrogen. The shape of the molecule is known to be

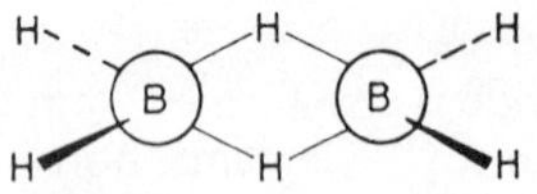

but the method of bonding in the bridge between the boron atoms is uncertain.

Boron, aluminium and gallium form complex hydrides containing the ions BH_4^-, AlH_4^- and GaH_4^-, respectively. Of these, the tetrahydridoaluminate, as a lithium salt, is much used as a reducing agent, especially in organic chemistry.

Aluminium also forms a polymeric hydride, $(AlH_3)_n$.

Halides

The halides can be obtained by the action of halogen acids on the metals, oxides, hydroxides or carbonates:

$$Mg \xrightarrow{HCl} MgCl_2$$

$$KOH \xrightarrow{HI} KI$$

$$Na_2CO_3 \xrightarrow{HBr} NaBr$$

The halides of most of the elements of the three Groups readily form hydrates, and it is these that this method will produce. An alternative method, for preparing the anhydrous salts, is to pass anhydrous hydrogen halide or halogen over the heated metal (or over a mixture of the metal oxide and carbon), e.g.,

$$2Al + 3Cl_2 \rightarrow Al_2Cl_6$$

$$MgO + C + Cl_2 \rightarrow MgCl_2 + CO$$

In each Group, the degree of hydration decreases as the cationic size increases. The most marked difference is between the first and second members, owing to the very small ions of lithium, beryllium and boron. The lithium ion is so small and its charge density so great that it exerts a powerful attraction towards the lone pairs of electrons on the oxygen atom in water or alcohol molecules, with the result that lithium chloride is deliquescent and soluble in alcohol as well as in water, whereas the other halides in this Group are soluble only in water. This characteristic behaviour is more pronounced in Group 2. For example, beryllium halides hydrolyse when their aqueous solutions are heated:

$$BeCl_2 + H_2O \rightarrow BeO(s) + 2HCl$$

Magnesium chloride produces chloride oxides and hydroxides on evaporation of its aqueous solution:

$$MgCl_2 + H_2O \rightarrow Mg(OH)Cl(s) + HCl$$

On the other hand, calcium chloride is used as a drying agent because of the low vapour pressure of its several hydrates, the polarizing power of the cation being insufficient to cause hydrolysis of the salt.

The boron halides have low boiling points and undergo complete hydrolysis in water, e.g.,

$$BCl_3 + 3H_2O \rightarrow B(OH)_3 + 3HCl$$

They are electron-deficient and therefore behave as strong Lewis acids in attempting to complete their octet of electrons, e.g.,

$$BF_3 + F^- \rightarrow [BF_4]^-$$

This is in contrast to aluminium chloride which completes the octet of electrons around the aluminium atom by forming a dimer. Like many other covalent halides, aluminium chloride fumes in moist air and is hydrolysed by small amounts of water:

$$Al_2Cl_6 + 6H_2O \rightarrow 2Al(OH)_3(s) + 6HCl$$

Oxides and hydroxides

The hydroxides, nitrates and carbonates, especially those of the Group 1 metals, are so stable that they do not readily decompose into the oxides on heating. The latter (or peroxides) can be prepared by heating the elements in oxygen or in some cases with other, less stable, oxides; for example, in the Thermit process, aluminium oxide is formed by heating aluminium with the oxide of a transitional metal such as iron or chromium:

$$Fe_2O_3 + 2Al \rightarrow 2Fe + Al_2O_3$$

The oxides of the Group 1 elements are alkaline; being strongly ionic, they react with water to produce hydroxides

$$O^{2-} + H_2O \rightarrow 2OH^-$$

There is an increase in alkalinity of all the oxides and hydroxides as the cation increases in size, so that, whereas magnesium hydroxide is only a weak base, barium hydroxide is sufficiently strong to function as a volumetric alkali. Whereas boron oxide dissolves in hot water to give a weak *acid*

$$B_2O_3 + 3H_2O \rightleftharpoons 2B(OH)_3 \quad \text{(i.e., } H_3BO_3\text{)}$$

$$B(OH)_3 \rightleftharpoons H^+ + H_2BO_3^-$$

aluminium hydroxide is amphoteric (as also is the oxide)

$$Al^{3+}(aq) \underset{H^+}{\overset{OH^-}{\rightleftharpoons}} Al(OH)_3 \underset{H^+}{\overset{OH^-}{\rightleftharpoons}} Al(OH)_4^-$$

The effect of the transition metals intervening before gallium, indium and thallium is to produce comparatively small cations (*see* Chapter 15) for these metals, with the result that their hydroxides, when in oxidation state $+3$, are not very alkaline. However, thallium, because of the inert-pair effect, exists also in the univalent condition, with a correspondingly larger cation; as a result, TlOH rivals the hydroxides of the Group 1 metals in its alkalinity.

The increase in alkalinity of the hydroxides as the cation increases in size can be explained by the fact that, as the cationic size increases, so the positive charge upon it becomes further away from the negative charge on the hydroxide ion. There will thus be a decrease in the coulombic attraction between the two ions, and the hydroxide ion is more free to act as such.

Of the hydroxides, that of sodium is prepared in large quantities, mainly by the electrolysis of sodium chloride solution. One such process, the Kellner–Solvay, has been described earlier (p. 220) when dealing with the preparation of chlorine. To obtain sodium hydroxide, the sodium amalgam obtained is run into large tanks and then decomposed by water in the presence of iron or graphite

$$2Na/Hg + 2H_2O \rightarrow 2NaOH + H_2(g) + 2Hg(l)$$

The resultant solution of the hydroxide can then be evaporated to dryness. It is used as a general alkali, in the manufacture of viscose rayon, soap, textiles and paper, and in oil refining.

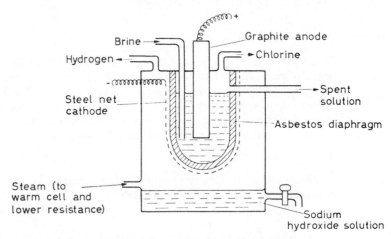

Figure 14.2 Preparation of sodium hydroxide in a diaphragm cell

An alternative method of preparation is by electrolysis of brine in a diaphragm cell, so called because of the porous partition which separates the anode from the cathode. An example of such is the Nelson cell, shown diagrammatically in *Figure 14.2*. The asbestos diaphragm ensures that the chlorine evolved at the anode is not able to react with the solution of sodium hydroxide produced at the cathode. The brine contains the ions Na^+, Cl^-,

H^+ and OH^-. At the anode Cl^- and OH^- are attracted, but only Cl^- is discharged if the solution is concentrated:

$$Cl^- - e^- \rightarrow Cl \rightarrow \tfrac{1}{2}Cl_2(g)$$

At the cathode, Na^+ and H^+ are attracted, but in accordance with the electrochemical series, H^+ is discharged:

$$H^+ + e^- \rightarrow H \rightarrow \tfrac{1}{2}H_2(g)$$

Carbonates

The soluble hydrogencarbonates and carbonates may be obtained by passing carbon dioxide through an aqueous solution of the alkali, the compound produced depending upon the relative amounts of alkali and carbon dioxide used:

$$OH^- + CO_2 \rightarrow HCO_3^-$$

$$2OH^- + CO_2 \rightarrow CO_3^{2-} + H_2O$$

The insoluble carbonates are obtained by double decomposition reactions, e.g.,

$$Ca^{2+} + CO_3^{2-} \rightarrow CaCO_3(s)$$

The carbonate ion is easily polarized by small cations of high charge

breaking down into carbon dioxide and leaving the oxide ion on heating. For this reason, the carbonates of the weakly polarizing ions of Group 1 are the only ones stable to heat and soluble in water. Furthermore, their hydrogencarbonates are the only ones which can be isolated from aqueous solution; hydrogencarbonates of other metals, although existing in solution, cannot be obtained in the solid state because the equilibrium

$$2HCO_3^- \rightleftharpoons CO_3^{2-} + H_2O + CO_2(g)$$

favours formation of the carbonate, particularly if water is lost as vapour and the carbonate is insoluble. In other words, if a solution of such a hydrogencarbonate is heated, or even allowed to evaporate at low temperature, the carbonate is precipitated (e.g., as stalactites). Boiling is, of course, a means of removing the 'temporary' hardness of water caused by the presence of the hydrogencarbonates of calcium and magnesium; since the calcium and magnesium cations are removed from solution as insoluble carbonates, they are no longer able to react with the soap anions, and the water is thus softened.

Aluminium ions have a very large polarizing power, and attempts to prepare aluminium carbonate result in the precipitation of aluminium hydroxide. This result could also be predicted from consideration of the

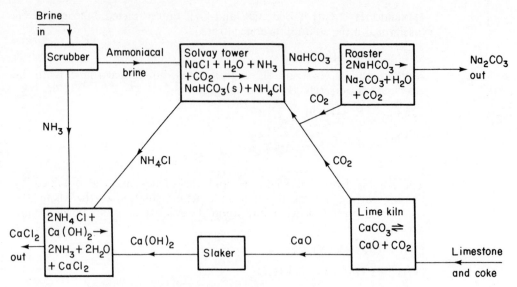

Figure 14.3 Flow sheet for the Solvay process

hydrolysis of aluminium salts in solution. Aluminium hydroxide is only weakly basic, and carbonic acid is a very weak acid: accordingly, complete hydrolysis of aluminium carbonate in water would be expected.

Thallium(I) carbonate *does* exist, showing that the polarizing effect of the univalent cation is much less than that of trivalent ones.

Sodium hydrogencarbonate and sodium carbonate are of commercial value and are manufactured by the Ammonia–Solvay process in which ammonia, carbon dioxide and sodium chloride solution produce an equilibrium which is displaced in favour of the formation of sodium hydrogencarbonate, since this is the least soluble component of the system and is precipitated from the solution

$$Na^+ + H_2O + NH_3 + CO_2 \rightleftharpoons Na^+HCO_3^-(s) + NH_4^+$$

The ammonia also serves to remove hydrogen ions which would otherwise polarize the hydrogencarbonate ions. The flow sheet (*Figure 14.3*) emphasizes the overall economy of the process, the net result of which can be summarized by the equation

$$2NaCl + CaCO_3 \rightarrow CaCl_2 + Na_2CO_3$$

Sodium hydrogencarbonate is used in baking powders, 'health' salts and fire extinguishers, carbon dioxide being produced when hydrogen ions are made available from some suitable source [such as 2,3-dihydroxybutanedioic (tartaric) acid in 'health' salts]. Sodium carbonate finds use as a water softener in washing powders:

$$CO_3^{2-} + Ca^{2+} \rightarrow CaCO_3(s)$$

and as a high-temperature alkali, particularly in glass manufacture:

$$Na_2CO_3 + SiO_2 \rightarrow Na_2SiO_3 + CO_2(g)$$

Calcium carbonate is also of considerable industrial importance, mainly for the manufacture of cement and calcium oxide (quicklime). For the former, a fine slurry of calcium carbonate and clay (which can be regarded as a hydrated double oxide of aluminium and silicon) are reacted together at high temperatures to form an intimate mixture of the oxides of calcium, aluminium and silicon. It is believed that when water is added to this mixture, complex silicates are formed.

Calcium oxide is produced by the decomposition of calcium carbonate (limestone) in a lime kiln

$$CaCO_3 \rightleftharpoons CaO + CO_2$$

The equilibrium is shifted towards the formation of calcium oxide by providing a current of air to remove the carbon dioxide as it is evolved. The calcium oxide is used for improving the texture and reducing the aciditiy of soils and for the production of calcium silicate in glass making.

Salts of oxoanions of strong acids

Although boron oxide reacts with certain oxoacids, the compounds formed are not true salts; for example, the so-called sulphate and phosphate(V) are $B_2O_3.3SO_3$ and $2B_2O_3.P_4O_{10}$, respectively. All the other elements of the three Groups are sufficiently metallic to form salts with nitric, sulphuric and phosphoric(V) acids.

Sulphates

The sulphates of the Group 1 metals are all soluble in water, whilst the solubility of the sulphates of the Group 2 metals decreases in passing from magnesium to barium, presumably owing to a progressive increase in the lattice energies—barium sulphate is one of the least soluble of all sulphates. Typically, the thermal stability depends on the electronegativity of the metal, so that the sulphates of Groups 1 and 2 are very resistant to decomposition by heat; aluminium sulphate, on the other hand, is decomposed to the oxide.

Calcium sulphate occurs as an economic mineral in the form of gypsum, $CaSO_4.2H_2O$, and anhydrite, $CaSO_4$. The former is a source of 'plaster of Paris', which is made by partial dehydration to give $(CaSO_4)_2H_2O$, the setting of the paste being caused by hydration back to interlocking crystals of gypsum. Anhydrite is used as a source of sulphur compounds such as sulphuric acid and ammonium sulphate:

$$CaSO_4 + 4C \rightarrow CaS + 4CO(g)$$
$$CaS + 2H_2O \rightarrow Ca(OH)_2 + H_2S(g)$$

$$2H_2S + 3O_2 \rightarrow 2H_2O + 2SO_2(g)$$

$$2SO_2 + O_2 + 2H_2O \rightarrow 2H_2SO_4$$

$$CaSO_4 + NH_3 + CO_2 \rightarrow CaCO_3(s) + (NH_4)_2SO_4$$

The *alums* are a well known series of double sulphates of general formula $M^IM^{III}(SO_4)_2 12H_2O$, e.g., $KAl(SO_4)_2.12H_2O$, which crystallize as octahedra. They are acidic in solution, because of the hydrolysis which invariably occurs with salts of strong acids and polarizing cations. Aluminium sulphate and sodium alum (M^I = Na, M^{III} = Al) are used as flocculating agents in sewage disposal because of the increased efficiency of a multivalent ion Al^{3+}(aq) in removing the stabilizing charge on colloidal particles (p. 97).

Nitrates
As all the nitrates are soluble in water, they are prepared by reaction of nitric acid with the oxides, hydroxides or carbonates of the metals (although aluminium oxide, like the free metal, becomes passive in concentrated nitric acid), e.g.,

$$NaOH + HNO_3 \rightarrow NaNO_3 + H_2O$$

$$BaO + 2HNO_3 \rightarrow Ba(NO_3)_2 + H_2O$$

The nitrates of the alkali metals yield the corresponding nitrite and oxygen when heated, but the remainder decompose to the oxide, with the evolution of nitrogen dioxide and oxygen:

$$2KNO_3 \rightarrow 2KNO_2 + O_2(g)$$

$$2Sr(NO_3)_2 \rightarrow 2SrO + 4NO_2(g) + O_2(g)$$

Some of the nitrates find commercial application. Potassium nitrate is used in gunpowder, in preference to the cheaper, naturally-occurring sodium nitrate because the latter is hygroscopic. Both sodium and potassium nitrates are important nitrogenous fertilizers. The nitrates of the Group 2 metals are used in pyrotechnics to produce characteristic coloured fires.

Organometallic compounds

Mention has been made of the reaction between these metals and alcohols (p. 330); many other organometallic compounds can also be made. Thus magnesium dissolves in alkyl halides to produce Grignard reagents (p. 445) which react with metals or metallic halides to give *metal alkyls*, e.g.,

$$Na^+CH_3^-$$

$$Mg \xrightarrow{CH_3I} CH_3MgI \nearrow^{Na} \searrow_{BeCl_2}$$

$$Be(CH_3)_2$$

Such compounds hydrolyse in water, e.g.,

$$Be(CH_3)_2 + 2H_2O \rightarrow Be(OH)_2(s) + 2CH_4(g)$$

The simple covalent compounds of the elements beryllium, boron and aluminium are usually unstable because of electron deficiencies, and accordingly they often achieve stability by the acquisition of extra electrons by coordination. Organometallic derivatives of this type, involving two or more bonds from the same organic group, are called *chelate* (Greek chēlē = claw) compounds. Examples of these are 'basic' beryllium ethanoate (acetate), $Be_4O(CH_3COO)_6$, a covalent compound possessing the characteristic properties of high volatility and solubility in organic solvents

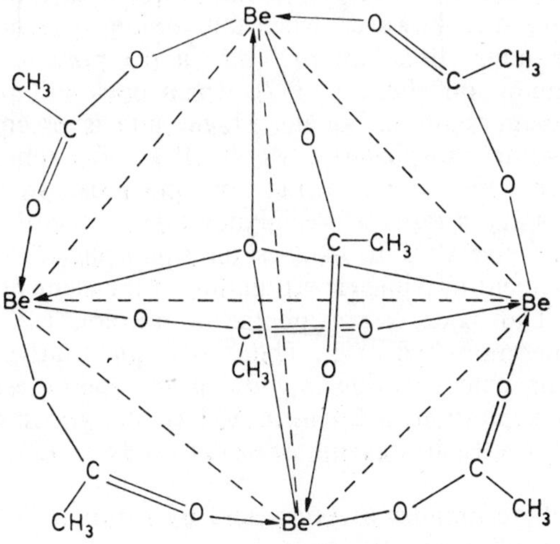

$$Be_4O(CH_3COO)_6$$

[Full name: hexa-μ-ethanoato-(O,O')-μ_4-oxo-tetraberyllium(II)]

and the compound formed between aluminium ions and 8-hydroxyquinoline

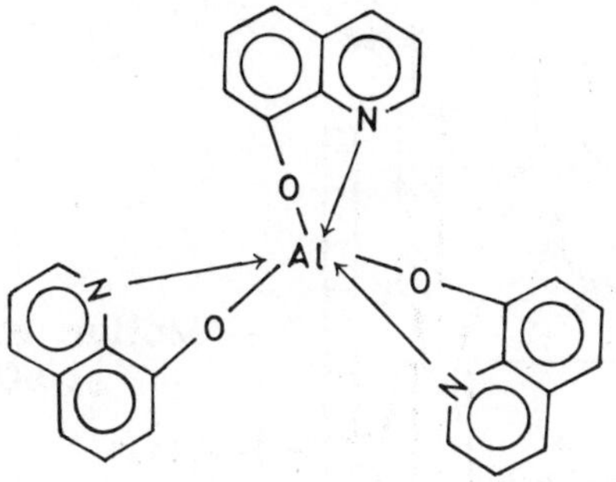

Analysis

The characteristic visible spectra of the elements sodium (yellow), potassium (lilac), calcium (red), strontium (crimson), barium (green), indium (indigo) and thallium (green) can be used both as qualitative and quantitative methods of analysis. Very few insoluble compounds of the alkali metals are known: those which are relatively insoluble and hence afford a means of identification when all other metallic ions have been removed, are potassium chlorate(VII) or perchlorate, $KClO_4$, potassium hexanitrocobaltate(III),

$K_3[Co(NO_2)_6]$, and sodium dizinc diuranyl(VI) ethanoate $NaZn_2(UO_2)_2(CH_3COO)_9$.

The general solubility of the salts of the Group 2 metals results in their being detected towards the end of the conventional scheme of qualitative analysis. The solubility products of the carbonates of calcium, strontium and barium are exceeded, however, when ammonium carbonate is added to solutions containing these cations, even in the presence of ammonium chloride (common ion effect, p. 172), which does still prevent the precipitation of magnesium carbonate. Magnesium is precipitated as magnesium ammonium phosphate(V), $MgNH_4PO_4$, after removal of all other elements except the alkali metals by previous precipitations. The phosphate(V) is quantitatively precipitated and can be filtered off, washed and ignited to convert it into magnesium diphosphate(V), $Mg_2P_2O_7$; this affords a convenient gravimetric estimation of the element.

Aluminium hydroxide is precipitated as a colourless, gelatinous precipitate by ammonium hydroxide. This, too, is quantitative, and ignition to form aluminium oxide provides a suitable gravimetric estimation. Other quantitative precipitations which form the bases for gravimetric estimations of the relevant metals, are calcium as oxalate and strontium and barium as sulphate.

Boron can be estimated as boric acid by titration in the presence of propane-1,2,3-triol (glycerol).

Extraction and manufacture of some metals

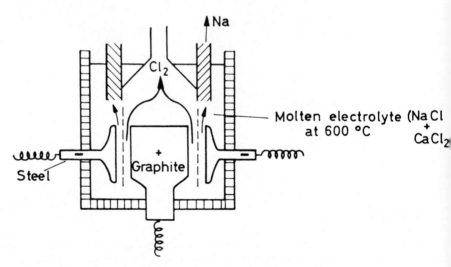

Figure 14.4 The Downs cell for manufacture of sodium

Sodium is manufactured electrolytically in the Downs cell (*Figure 14.4*) from the molten chloride. At the anode, Cl^- is oxidised and at the cathode Na^+ is reduced selectively before Ca^{2+}:

$$Cl^- - e^- \to Cl \to \tfrac{1}{2}Cl_2(g)$$

$$Na^+ + e^- \to Na$$

The molten sodium floats on the electrolyte and can be drawn off. The calcium chloride is present to lower the melting point (p. 109).

Magnesium is extracted either by reduction of calcined dolomite with ferrosilicon and obtained by distillation, or by electrolysis of molten magnesium chloride obtained from sea water (*see Figure 14.8*).

Calcium is also produced electrolytically (*see Figure 8.12*). Aluminium is obtained from bauxite, $Al_2O_3, 3H_2O$. This is treated as outlined in *Figure 14.10* and the aluminium is obtained by electrolysis as shown in *Figure 8.11*. The sources of bauxite and the places where it is converted into aluminium are shown in *Figure 14.11*.

Summary

The typical reactions of sodium, magnesium, and aluminium are summarised in *Figures 14.5, 14.6,* and *14.7* respectively.

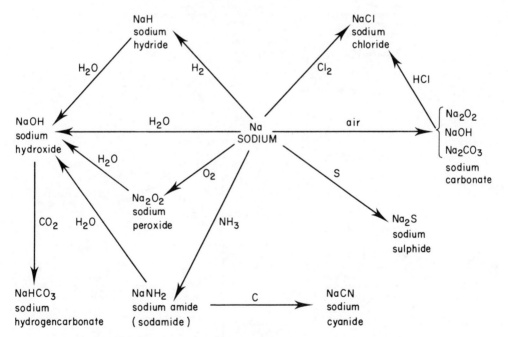

Figure 14.5 Reactions of sodium

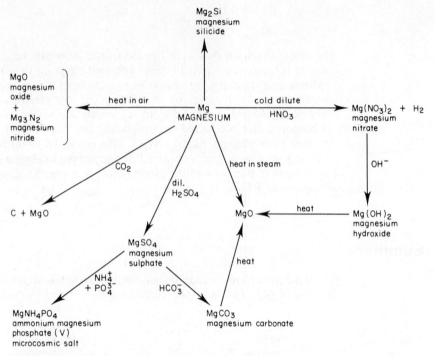

Figure 14.6 Reactions of magnesium

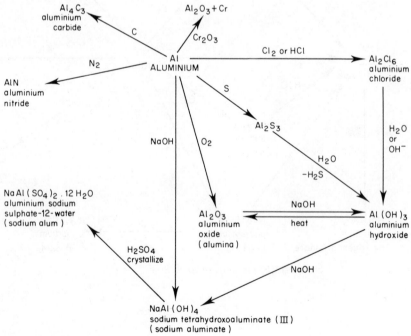

Figure 14.7 Reactions of aluminium

Figure 14.8 An aerial view of a magnesia plant at Hartlepool with an
annual capacity of 250 000 tonnes. The site was chosen for its proximity to
high purity dolomite deposits and because the seawater is of consistent
quality, free from fresh water dilution. Magnesium hydroxide is precipitated
by reaction between seawater and calcined dolomite and drawn to the
centre of the settling tanks prior to filtration and further processing. (By
courtesy of the Steetley Company)

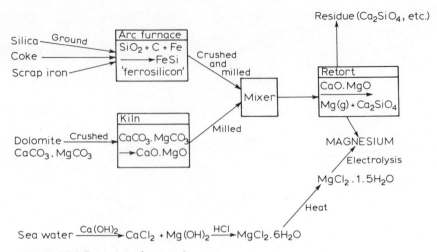

Figure 14.9 Extraction of magnesium

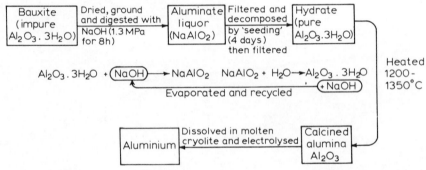

Figure 14.10 Extraction of aluminium

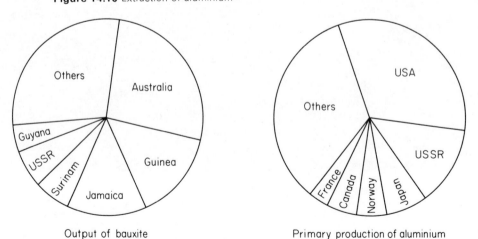

Output of bauxite   Primary production of aluminium

Figure 14.11 Relationship between output of bauxite and production of aluminium

Questions

(1) Show that the oxidation numbers of the elements of Groups 1, 2 and 3 in the following compounds are consistent with their valencies: $Na_2B_4O_7$, $Mg_2P_2O_7$, $Na_3Co(NO_2)_6$, $K_4Fe(CN)_6$ and $NaZn_2(UO_2)_2(CH_3COO)_9$.

(2) Discuss the changes that would occur to a piece of sodium left exposed to the air.

(3) Burning the alkali metals in air gives the following compounds: Li_2O, Na_2O_2, KO_2, RbO_2 and CsO_2. Discuss the structures of these compounds and suggest reasons for their formation.

(4) Explain under what conditions the following substances react together and say why the reactions take place:
calcium hydroxide and sodium carbonate; aluminium chloride and lithium hydride; calcium nitride and water; calcium sulphate and ammonium carbonate.

(5) Discuss the structures of diborane, B_2H_6, and its addition compound with ammonia, $B_2H_6.2NH_3$. On heating the latter, borazole, $B_3N_3H_6$, and hydrogen are formed. Borazole reacts with hydrogen chloride, first giving $B_3N_3H_9Cl_3$ and then $B_3N_3H_6Cl_3$. Because of the similarity in structure, borazole has been called 'inorganic benzene'. On this basis give the structures of the compounds mentioned.

(6) The radii of Rb^+, Ca^{2+}, I^- and O^{2-} are 148, 99, 216 and 140 pm, respectively. What would be the expected crystal structures of rubidium iodide and calcium oxide? (Refer to Chapter 4.)

(7) Why is it possible to use sodium carbonate as an alkali? Why are heavy-metal carbonates precipitated by sodium hydrogencarbonate in preference to sodium carbonate?

(8) How would you titrate (a) disodium tetraborate-10-water (borax) against acid, (b) boric acid against alkali?

(9) How would you analyse calcium, magnesium, sodium and lithium in the presence of each other?

(10) How far can the ammonium ion be regarded as a member of Group 1?

(11) 'There exists a diagonal relationship between lithium and magnesium, beryllium and aluminium, and boron and silicon'. Provide evidence for this statement and suggest a reason for there being such a relationship.

(12) Explain the following:
A little methyl orange was added to some disodium tetraborate solution. The resultant colour was yellow, but it turned pink on the addition of $20 \, cm^3$

of 0.1M hydrochloric acid. Some neutral propane-1,2,3-triol was then added, followed by phenolphthalein, 0.1M sodium hydroxide was subsequently run into the solution until a permanent red colour was obtained; the volume of NaOH required was 10 cm^3.

15 The transitional or *d*-block elements

General properties • Coordination compounds • Occurrence, extraction and reactions • Compounds – halides, oxides and oxocompounds, organometallic compounds • Analysis • Summary– reaction charts for nickel, vanadium, chromium, manganese, iron cobalt, nickel, copper, silver, zinc and mercury

Group $(n-1)d$ ns $(n-1)d$ ns Group

2 ⟶ 3

Ca	Sc	Ti	V	Cr	Mn	Fe	Co	Ni	Cu	Zn	Ga
Sr	Y	Zr	Nb	Mo	Tc	Ru	Rh	Pd	Ag	Cd	In
Ba	La ↑	Hf	Ta	W	Re	Os	Ir	Pt	Au	Hg	Tl

Lanthanoids

General properties

There are three series of so-called 'transitional' elements in which the $3d$, $4d$ and $5d$ orbitals, respectively, are in the process of being filled with electrons. (From the point of view of the Periodic Table, they can be regarded as effecting a *transition* from the electropositive elements of the *s* block, on the one hand, to the more electronegative elements of the *p* block on the other.)

There are ten elements in each series, if one includes zinc, cadmium and mercury in which the final electron needed to complete the *d* orbitals has been added. These last three elements will therefore show departures from those characteristic properties of transitional metals which are the consequence of the *d* orbitals being incomplete.

Table 15.1 shows how the orbitals of the *d* block are filled for the first transition series. In accordance with Hund's rule, electrons occupy the orbitals singly first, before there is any pairing. The increased stability when all five *d* orbitals possess either one or two electrons is indicated by the way in which an electron is transferred from the $4s$ orbital to a *d* orbital in the case of chromium and copper. One thing that quickly emerges from consideration of *Table 15.1* is the large electron availability of most of these metals. They are

Table 15.1 Electronic configurations and oxidation states of first series of transition elements

	3d	4s	Oxidation states (common ones boxed)
Scandium, Sc	[↑ | | | |]	[↑↓]	[3]
Titanium, Ti	[↑|↑| | |]	[↑↓]	−1, +2, 3, [4]
Vanadium, V	[↑|↑|↑| |]	[↑↓]	−1, +1, 2, [3], 4, 5
Chromium, Cr	[↑|↑|↑|↑|↑]	[↑]	−2, −1, +1, 2, [3], 4, 5, 6
Manganese, Mn	[↑|↑|↑|↑|↑]	[↑↓]	−3, −1, +1, [2], 3, 4, 5, 6, 7
Iron, Fe	[↑↓|↑|↑|↑|↑]	[↑↓]	−2, +1, [2],[3], 4, 5, 6
Cobalt, Co	[↑↓|↑↓|↑|↑|↑]	[↑↓]	−1, +1, [2], 3, 4
Nickel, Ni	[↑↓|↑↓|↑↓|↑|↑]	[↑↓]	−1, +1, [2], 3, 4
Copper, Cu	[↑↓|↑↓|↑↓|↑↓|↑↓]	[↑]	+[1],[2], 3
Zinc, Zn	[↑↓|↑↓|↑↓|↑↓|↑↓]	[↑↓]	[2]

therefore very good conductors, and both copper and silver find extensive commercial application on this account. Furthermore, there is powerful metallic bonding, and the elements are, on the whole, strong and with high melting points. These are, in fact, the substances readily recognizable to the non-scientist as metals. They are widely used where great strength and hardness are required; they are largely miscible with each other, and the resultant alloys often have enhanced properties (*Table 15.2*).

Table 15.2 Hardness and tensile strength of some transitional metals and alloys

Metal or alloy	Relative hardness (Brinell scale)	Tensile strength kg cm^{-2}
Copper, wrought	50	2200
Brass, 70/30	160	5500
Iron, wrought	100	3140
Mild steel	130	4710
Nickel–chromium steel	400	13 500
For comparison: wrought aluminium	27	944

Melting points tend to become higher on descending a series and reach a peak with tungsten, which is therefore used in the filament of electric light bulbs. Zinc, cadmium and mercury are exceptional: their melting points not only are very low but decrease from zinc to mercury, which is the only liquid metal at room temperature—at least in temperate climates—and so is used in thermometers, barometers, etc. (*Table 15.3*).

It is also found that, as a particular series is traversed horizontally, the size

Table 15.3 Melting points of *d*-block elements

Element	M.p./°C	Element	M.p./°C	Element	M.p./°C
Sc	1540	Y	1500	La	920
Ti	1675	Zr	1850	Hf	2220
V	1900	Nb	2470	Ta	3000
Cr	1890	Mo	2610	W	3410
Mn	1240	Tc	2200	Re	3180
Fe	1535	Ru	2500	Os	3000
Co	1492	Rh	1970	Ir	2440
Ni	1453	Pd	1550	Pt	1769
Cu	1083	Ag	961	Au	1063
Zn	420	Cd	321	Hg	−38.9

of the atoms steadily contracts (*Figure 15.1, Table 15.4*). The explanation lies in the fact that the diffuse *d* orbitals do not appreciably affect the overall size of the atom, whilst the steadily increasing number of protons exerts a progressively greater attractive force on the outer *s* orbitals, causing them to contract in size. In the case of the third series, where the 5*d* orbitals are being filled, the further effect of the lanthanide contraction must be added. One immediate consequence of this is that the atom of hafnium, coming directly after the lanthanides, is no bigger than that of zirconium: not only does hafnium replace zirconium atoms in minerals, but their chemistry is virtually identical, and their separation is a matter of great difficulty.

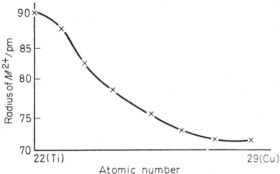

Figure 15.1 Radius of ions M^{2+} of the first series of transitional elements as a function of atomic number

With the atomic volume actually decreasing as nucleons are added, the density must inevitably increase: metals on the right of a series are very dense, those in later series particularly so (*Table 15.4*). Thus osmium, with a relative density of 22.5, is the densest element, closely followed by iridium.

Another consequence of shrinking atomic size and increased nuclear charge is a stronger attraction between nucleus and electrons; that is, the electronegativity increases, together with ionization energies (*Table 15.5*). Therefore, metals towards the end of a series are inactive and resistant to corrosion; in the last two series are found the noble metals which are so valuable because of their resistance to atmospheric oxidation and which are used extensively in jewellery and coinage.

Table 15.4 Atomic sizes (covalent radii) and densities

	Covalent radius/pm	Density g cm^{-3}		Covalent radius/pm	Density g cm^{-3}		Covalent radius/pm	Density g cm^{-3}
Sc	144	3.0	Y	162	4.3	La	169	6.2
Ti	132	4.5	Zr	145	6.5	Hf	144	13.3
V	122	6.0	Nb	134	8.6	Ta	134	16.6
Cr	117	7.2	Mo	129	10.2	W	130	19.4
Mn	117	7.2	Tc	—	11.5	Re	128	20.5
Fe	116	7.9	Ru	124	12.3	Os	126	22.5
Co	116	8.9	Rh	125	12.4	Ir	126	22.5
Ni	115	8.9	Pd	128	12.0	Pt	129	21.4
Cu	117	8.9	Ag	134	10.5	Au	134	19.3
Zn	125	7.1	Cd	141	8.6	Hg	144	13.6

Note: very high density of osmium; atomic radii of zirconium and hafnium; atypical behaviour of zinc, cadmium and mercury.

Table 15.5 Standard electrode potentials $E^{\ominus}$ for the most common ions, and ionization energies, for the first series of transitional elements

Element	$E^{\ominus}/V$ (ion)	Ionization energies/kJ mol^{-1}				
		1st	2nd	3rd	4th	5th
Sc	−2.1 (Sc^{3+})	632	1240	2390	7110	8870
Ti	−1.6 (Ti^{2+})	661	1310	2720	4170	9620*
V	−1.5 (V^{2+})	648	1370	2870	4600	6280
Cr	−0.74 (Cr^{3+})	653	1590	2990	4770	7070
Mn	−1.18 (Mn^{2+})	716	1510	3250	5190	7360
Fe	−0.44 (Fe^{2+})	762	1560	2960	5400	7620
Co	−0.28 (Co^{2+})	757	1640	3230	5100	7910
Ni	−0.25 (Ni^{2+})	736	1750	3390	5400	7620
Cu	+0.34 (Cu^{2+})	745	1960	3550	5690	7990
Zn	−0.76 (Zn^{2+})	908	1730	3828[†]	5980	8260

* Electron removed from inner shell.   † High energy required to remove a *d* electron when all the orbitals are full or half full.

Whilst there is no great change in atomic dimensions throughout the transitional series, nonetheless the differences are sufficient to bring about 'locking' of the crystal planes during alloying, with increase in strength. For instance, alloys of iron and manganese are tough enough to be used in railway lines, whilst alloys of iron with chromium and tungsten have the hardness required in high-speed tools. The transitional metals are also able to accommodate small atoms such as boron, carbon and nitrogen in holes in the lattice, the resultant, non-stoichiometric borides, carbides and nitrides often possessing desirable qualities.

The energy levels of the penultimate *d* orbitals are so close to each other and to the energy level of the outer *s* orbital that it is relatively easy to remove

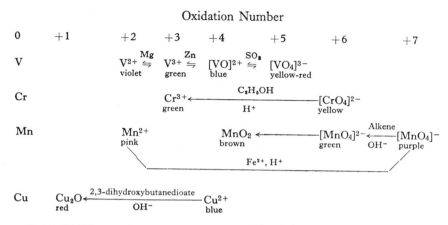

Figure 15.2 Some redox processes among transitional elements

or add varying numbers of electrons and to give different oxidation states (see *Table 15.1*). As is to be expected, this feature is particularly pronounced in the middle of each row (where seven electrons are available for valency purposes) and declines towards the end of the series, so that zinc, cadmium and mercury are always bivalent. Many of the different oxidation states are able to absorb light of different energy (and wavelength) from the visible spectrum and hence appear coloured. Because of the small energy differences involved, conversion of a transition metal from one oxidation state to another is usually a fairly simple operation. Some examples are shown in *Figure 15.2*.

One use of unstable oxidation states is afforded by the NIFE cell in which a higher oxide of nickel (of approximate composition Ni_2O_3) occurs:

$$Fe \text{ (negative electrode)} + 2OH^- \text{ (from electrolyte, KOH)} \rightarrow Fe(OH)_2 + 2e^-$$

$$Ni_2O_3 \text{ (on nickel electrode)} + H_2O + 2H^+ + 2e^- \rightarrow 2Ni(OH)_2$$

Those oxidation states in which the d orbitals are either completely empty or full, or in which each orbital contains one electron, are characterized by greater stability relative to the others, e.g., as shown in *Figure 15.3*.

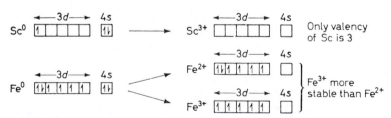

Figure 15.3 Stability of oxidation states

Paramagnetism

Because of its charge, a moving electron, whether the movement is translational or rotational, is associated with a magnetic field. If an atom

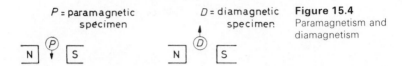

P = paramagnetic
specimen

D = diamagnetic
specimen

Figure 15.4
Paramagnetism and
diamagnetism

contains an unpaired electron, then it gives positive interaction with an external magnetic field; that is, there is attraction and the lines of force become more concentrated. Such a substance is said to be *paramagnetic*. In the case of a *diamagnetic* substance, there are no unpaired electrons, and repulsion occurs when the substance is placed in a magnetic field (*Figure 15.4*). Measurements of magnetic moments can therefore provide significant information about the electronic condition of substances.

Figure 15.5 shows how the paramagnetic moments of the ions of the first transition series vary regularly with the number of unpaired electrons.

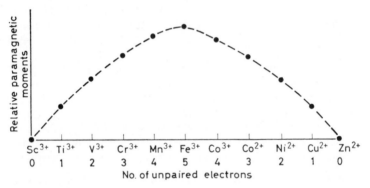

Figure 15.5 Paramagnetism of ions of first series of transitional elements

Coordination compounds

Coordination compounds are formed when ions or molecules (*ligands*) possessing one or more lone pairs of electrons donate these electrons to a central metallic atom or ion to form a complex ion or molecule. Transitional metals, by virtue of the vacant orbitals which they possess and because of the high polarizing power of the small ions of high charge which attract lone pairs towards them, are outstanding in their ability to form coordination compounds. As the water molecule is itself a ligand, it would be naive to expect transitional metal ions to remain unhydrated in aqueous solution. They usually coordinate to four or six molecules of water and it is this water which, persisting in the crystal lattice of the solid, is known as water of crystallization. Not all the water of crystallization, however, need be coordinated to the metallic atom. In the case of copper(II) sulphate-5-water, $CuSO_4.5H_2O$, there are four ligands attached to the copper ion: the remaining water molecule is linked by hydrogen bonding to the sulphate ion.

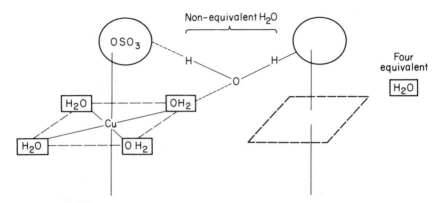

Figure 15.6 Part of the crystal lattice of copper(II) sulphate pentahydrate, showing disposition of water molecules

(That one molecule of water is attached differently to the other four is shown by the fact that the last molecule of water is removed only by prolonged heating at 200 °C; *Figure 15.6*).

Other atoms containing lone pairs capable of being donated to a metallic ion are carbon (especially in the form of the cyanide ion, CN^-), halogen, when present as anion X^-, and nitrogen, particularly as ammonia (when it forms *ammines*) and its organic derivatives such as amines. If the ligand is neutral, then the complex carries the positive charge of the central metallic ion and is known as a cationic complex; if it carries a sufficiently negative charge, then the overall charge of the complex will be negative and it will be anionic. Some examples are shown in *Table 15.6*.

Table 15.6 Coordination complexes of transitional elements

Cationic complexes		Anionic complexes	
$[Cr^{II}(H_2O)_6]^{2+}$	Hexaaquochromium(II)	$[Fe^{II}(CN)_6]^{4-}$	Hexacyanoferrate(II)
$[Cu^{II}(H_2O)_4]^{2+}$	Tetraaquocopper(II)	$[Fe^{III}(CNS)_6]^{3-}$	Hexathiocyanatoferrate(III)
$[Cr^{III}(NH_3)_6]^{3+}$	Hexaamminechromium(III)	$[Ni^{II}(CN)_4]^{2-}$	Tetracyanonickelate(II)
$[Cu^{II}(NH_3)_4]^{2+}$	Tetraamminecopper(II)	$[Co^{II}Cl_4]^{2-}$	Tetrachlorocobaltate(II)
$[Cu^{I}(CO)_2]^+$	Dicarbonylcopper(I)	$[Mn^{VII}O_4]^-$	Tetraoxomanganate(VII)

In some cases, a ligand can donate more than one lone pair and is then known as, e.g., a tridentate ligand. Two well-known and useful examples of ligands of this type are ethane-1,2-diamine and its tetra-acetate derivative (edta) (*Figure 15.7*).

Ligands which give ring formation are often described as *chelating* agents and the ion is said to be *sequestered*.

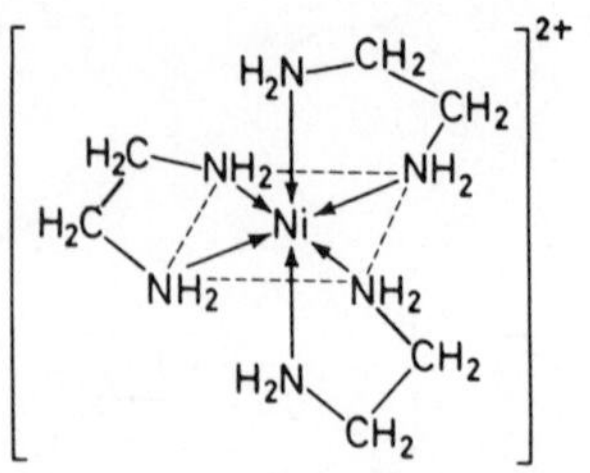

Nickel complex of ethane-1,2-diamine

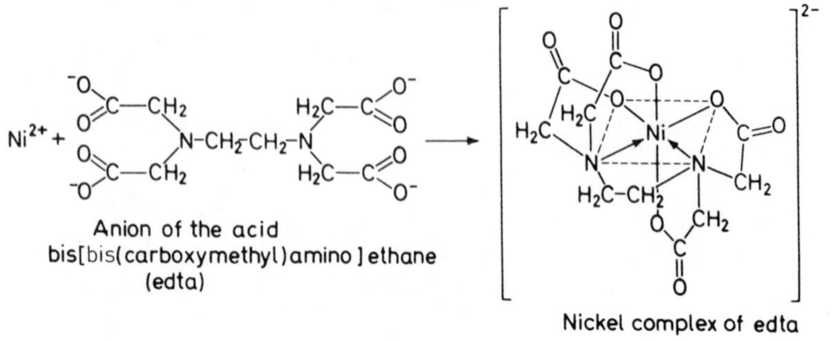

Anion of the acid
bis[bis(carboxymethyl)amino]ethane
(edta)

Nickel complex of edta

Figure 15.7 Complexes of ethane-1,2-diamine and bis[bis(carboxymethyl)amino]ethane ('ethylenediaminetetra-acetic acid, edta')

Ligand reactions

Complete or partial replacement of one ligand by another can take place, e.g.,

$$[Co(NH_3)_6]^{2+} + 6H_3O^+ \rightarrow [Co(H_2O)_6]^{2+} + 6NH_4^+$$

$$[Cu(H_2O)_6]^{2+} + 4NH_3 \rightarrow [Cu(NH_3)_4(H_2O)_2]^{2+} + 4H_2O$$
$$\text{Tetraamminediaquo-}$$
$$\text{copper(II)}$$

The formation of complexes can stabilize a transitional metal ion. For example, simple cobalt(III) ions do not exist, and yet its complex with the nitrite ion is formed readily from the cobalt(II) ion by aerial oxidation:

$$2Co^{2+} + 12NO_2^- + 2H^+ + \tfrac{1}{2}O_2 \rightarrow 2[Co^{III}(NO_2)_6]^{3-} + H_2O$$
$$\text{Hexanitrocobaltate(III)}$$

Those complexes tend to be most stable which contain completed orbitals, and this explains the oxidation of cobalt(II) in the above example and also the stability of hexacyanoferrate(II) (*Figure 15.8*).

Whereas Fe^{3+} is more stable than Fe^{2+}, $[Fe(CN)_6]^{3-}$ is less stable than $[Fe(CN)_6]^{4-}$ and needs one electron to complete the d orbitals, i.e. it is an oxidizing agent.

Stereochemistry of ligands

It can be said very roughly that ligands are so disposed in space about the central atom that there is the maximum distance between the lone pairs. A

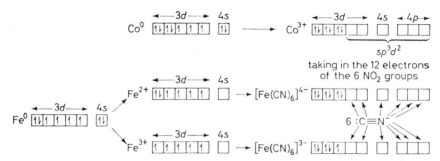

Figure 15.8 Orbitals and stability

Table 15.7 Hybridization and stereochemistry in coordination complexes

Hybridization	No. of ligands	Shape	Typical structure
sp	2	Linear	$[H_3N \rightarrow Ag \leftarrow NH_3]^+$
sp^3	4	Tetrahedral	$\left[\begin{array}{c} Cl \\ Fe-Cl \\ Cl--Cl \end{array}\right]^-$
sp^3d	5	Trigonal pyramid	CO, CO, OC—Fe, CO, CO
sp^3d^2	6	Octahedral	$\left[\begin{array}{c} CN \\ CN \quad CN \\ CN-Fe-CN \\ CN \end{array}\right]^{3-}$
sp^2d	4	Square planar	$\left[\begin{array}{c} H_3N \quad NH_3 \\ Pt \\ H_3N \quad NH_3 \end{array}\right]^{2+}$

more reliable and precise guide is given in terms of hybridization of the relevant orbitals (*Table 15.7*).

Stereoisomerism (p. 402) is possible with many coordinated compounds. Thus, two geometric isomers of the planar diamminedichloroplatinum(II) exist:

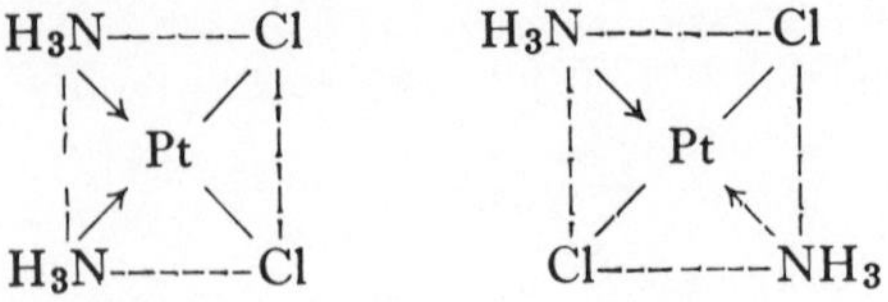

cis-   trans-

For the ion, bis(ethane-1,2-diamine)dichlorocobalt(III), three structures based on an octahedral configuration are possible

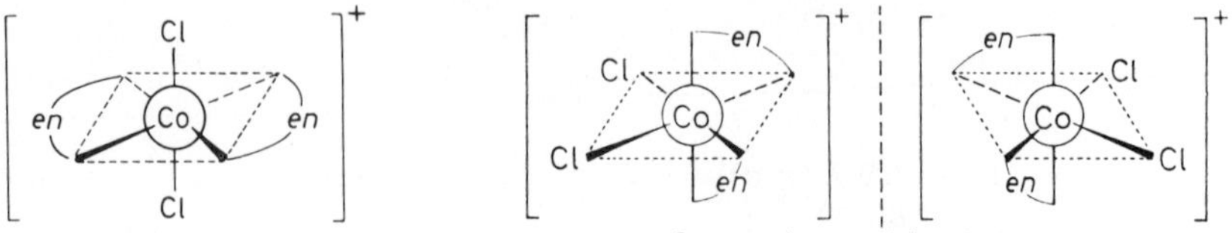

trans form   2 enantiomorphic cis forms

en = ethane-1,2-diamine

Here the two *cis* forms, although having two ligands disposed in the same order around the central atom, cannot be superimposed and, in fact, are mirror images of each other; that is, they are optical isomers (*enantiomorphs*) and rotate the plane of polarized light in different directions.

Occurrence, extraction and reactions

Occurrence

A large number of the transitional elements occur either as their oxides or as the central atom in complex oxoanions (*see Table 15.8*). Notable exceptions are the sulphide ores of molybdenum, iron, cobalt, nickel, copper and the zinc group.

Extraction

The elements at the beginning of each row are quite electropositive, their affinity for oxygen being such that reduction with vigorous reducing agents like aluminium is necessary for extraction of the metal, unless one has recourse to indirect methods. For instance, in the production of titanium, zirconium and hafnium, the oxide is converted into the tetrachloride and this is then reacted with molten magnesium in an inert atmosphere (cf. *Figure 8.10*, p. 199):

$$XCl_4 + 2Mg \rightarrow X + 2MgCl_2$$

Aluminothermal reduction of the oxide is the normal method for the production of vanadium, chromium and manganese:

$$3M_2{}^{n+}O_n + 2nAl \rightarrow 6M + nAl_2O_3$$

Several of the remaining metals can be extracted from their oxides by reduction with carbon; this applies particularly to molybdenum, iron (*see*

Table 15.8 Sources of the transitional elements

Element	Source(s)
Titanium	ilmenite, $FeTiO_3$; rutile, TiO_2
Zirconium	baddeleyite, ZrO_2; zircon, $ZrSiO_4$ (with 1–2 per cent hafnium)
Vanadium	vanadinite, $3Pb_3(VO_4)_2.PbCl_2$
Niobium	niobite, $Fe(NbO_3)_2$ (containing tantalum)
Tantalum	tantalite, $Fe(TaO_3)_2$ (containing niobium)
Chromium	chromite, $Fe(CrO_2)_2$; crocoisite, $PbCrO_4$
Molybdenum	molybdenite, MoS_2; wolfenite, $PbMoO_4$
Tungsten	wolframite, $FeWO_4/MnWO_4$; scheelite, $CaWO_4$; tungstite, WO_3
Manganese	pyrolusite, MnO_2; hausmannite, Mn_3O_4
Technetium	traces only in nature
Rhenium	molybdenite—up to 20 p.p.m.
Iron	magnetite, Fe_3O_4; haematite, Fe_2O_3; pyrites, FeS_2
Cobalt	in sulphides of nickel and copper; smaltite, $CoAs_2$
Nickel	pentlandite, $NiS.2FeS$; garnierite, $MgNi$; silicates
Platinum metals	ruthenium, rhodium, palladium, osmium, iridium; native platinum
Copper	native; copper pyrites, $CuFeS_2$; cuprite, Cu_2O; malachite, $CuCO_3.Cu(OH)_2$
Silver	argentite, Ag_2S; horn silver, $AgCl$; native
Gold	native
Zinc	zinc blende, wurtzite, ZnS; calamine, $ZnCO_3$ (with small amounts of cadmium)
Mercury	cinnabar, HgS

Figures 15.9 to 15.13), tungsten, nickel, zinc and cadmium. Sulphides must first be converted into oxide by roasting in air, e.g., zinc and copper (*see Figures 15.14 and 15.15*),

$$2ZnS + 3O_2 \rightarrow 2ZnO + 2SO_2(g)$$

$$ZnO + C \rightarrow Zn + CO(g)$$

Nickel obtained in like manner is then usually purified by the Mond process which consists of passing carbon monoxide over the crude metal at about 90 °C. Reaction occurs, and gaseous tetracarbonylnickel(0) is formed:

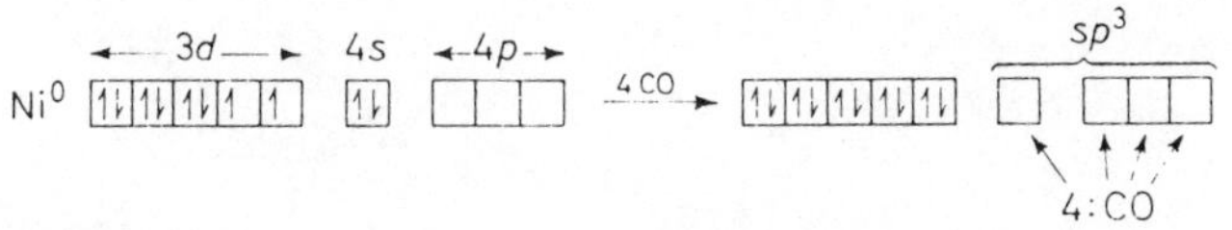

The carbonyl is then heated to about 180 °C, when it decomposes and deposits pure nickel (*see Figure 15.16*)

$$Ni + 4CO \underset{180\,°C}{\overset{90\,°C}{\rightleftharpoons}} Ni(CO)_4$$

Mercury is obtained directly by roasting the sulphide ores (*see Figure 15.17*).

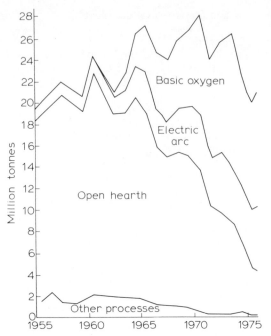

Figure 15.10 Blast furnaces in the Scunthorpe Division of British Steel
Corporation (cf. Figure 15.11). (By courtesy of BSC)

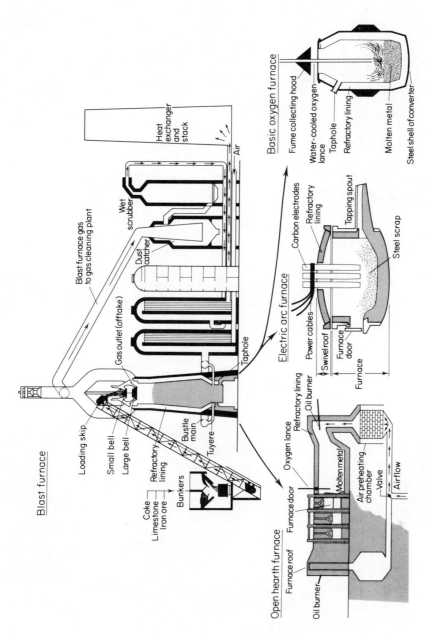

Figure 15.11 The manufacture of iron in the blast furnace and three methods of converting iron into steel

Figure 15.12 Making steel by the open hearth (top left), basic oxygen (bottom left) and electric arc (above) processes (cf. Figure 15.9). (By courtesy of BSC)

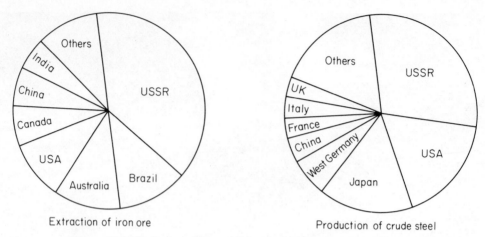

Extraction of iron ore   Production of crude steel

Figure 15.13 Extraction of iron ore and the production of steel; note particularly the extent of Japan's reliance on imported ore

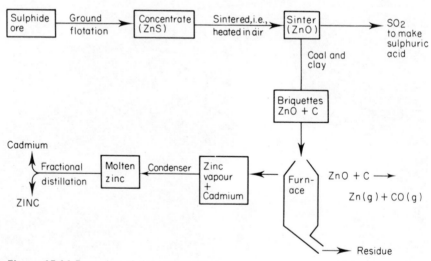

Figure 15.14 Extraction of zinc and cadmium

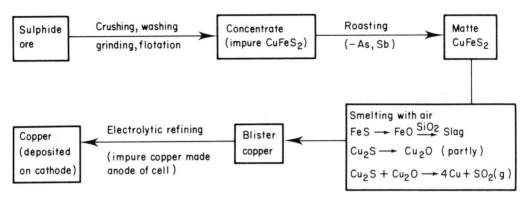

Figure 15.15 Extraction of copper

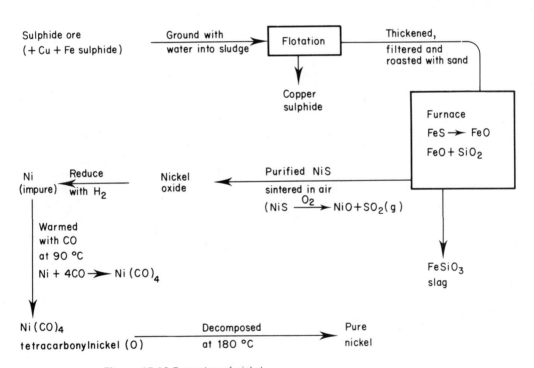

Figure 15.16 Extraction of nickel

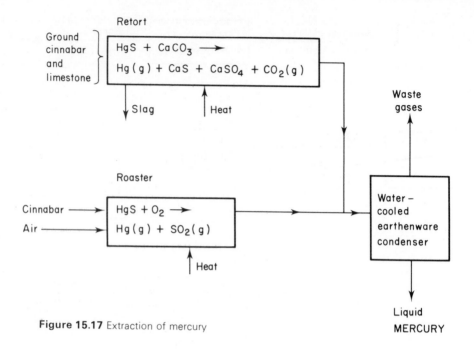

Figure 15.17 Extraction of mercury

Reactions

Many of the transition metals to the right of the series (particularly at bottom right) are noble in character and come below hydrogen in the electrochemical series. They are therefore unattacked by non-oxidizing and non-complexing acids. (Iron, cobalt, nickel and chromium, with negative electrode potentials, are rendered surprisingly passive by concentrated nitric acid.) The tendency of the transitional metals to complex formation is revealed by copper dissolving in concentrated hydrochloric acid, and by gold in aqua regia as well as in potassium cyanide solution:

$$Cu \xrightarrow{\underset{HCl}{conc.}} H_2[CuCl_4] + H_2(g)$$

$$Au \xrightarrow{\underset{O_2}{CN^-}} [Au(CN)_2]^-$$

$$Au \xrightarrow{\underset{HNO_3}{HCl}} [AuCl_4]^-$$

The elements react on heating with halogen, and often on heating with sulphur

$$Ni \xrightarrow{Cl_2} NiCl_2$$

$$Fe \xrightarrow{S} FeS$$

Those to the left of the series will also form nitrides on heating in nitrogen and will decompose steam

$$Ti \xrightarrow[800\,°C]{N_2} TiN$$

$$Cr \xrightarrow{steam} Cr_2O_3$$

Many transitional elements are able to accommodate hydrogen atoms in the interstices of the crystal lattice to give non-stoichiometric hydrides. Palladium is particularly prominent in this respect: at 80 °C and 100 kPa pressure it can absorb 900 times its own volume of hydrogen.

Compounds

Halides

As the oxidation number of the metal increases, so also does the covalent nature of the halide. Thus iron(II) chloride is a simple salt, $Fe^{2+}(Cl^-)_2$, whilst iron(III) chloride can exist as discrete Fe_2Cl_6 molecules:

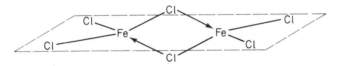

In keeping with its covalent character, anhydrous iron(III) chloride is volatile and soluble in many organic liquids, when it often appears to exist as $FeCl_3$ molecules.

The consequence of increasing oxidation number is increasing polarizing power. The strongly polarizing Fe^{3+} attracts the oxygen of water, so that its halides are readily hydrated. Furthermore, the oxygen-hydrogen bond in the water molecule is weakened, so that a proton can be lost by hydrolysis and the system reacts as an acid:

$$\tfrac{1}{2}Fe_2Cl_6 \xrightarrow{H_2O} [Fe(H_2O)_6]^{3+} + 3Cl^-$$

$$[Fe(H_2O)_6]^{3+} + H_2O \rightleftharpoons [Fe(H_2O)_5(OH)]^{2+} + H_3O^+$$

$$[Fe(H_2O)_5(OH)]^{2+} + H_2O \rightleftharpoons [Fe(H_2O)_4(OH)_2]^+ + H_3O^+$$

Particularly if the solution is sufficiently dilute, hydrolysis continues to give precipitation of $Fe(H_2O)_3(OH)_3$ or the hydrated oxide, $Fe_2O_3.xH_2O$. To prevent such hydrolysis in dilute solution, acidification is necessary to displace the equilibria to the left. Iron(II) chloride also forms a hexahydrate, but because polarization is less, there is a lower tendency to hydrolysis.

Similar considerations apply to compounds of the transitional elements with other non-metals and their oxocompounds.

Which particular oxidation state is the most stable depends to some extent on the halogen involved. The very electronegative fluorine tends to reveal the highest oxidation state of the metal. In the case of tantalum, for example, all the complexes TaF_6^-, TaF_7^{2-}, and TaF_8^{3-}, with tantalum in the highest oxidation state of five, are known. Sometimes, the fluoride of the metal in its highest oxidizing state is not very stable and decomposes fairly easily to give 'nascent' fluorine, so that it can be used as a fluorinating agent, e.g., silver(II) fluoride:

$$AgF_2 \rightarrow AgF + [F]$$

The iodide ion, on the other hand, has a tendency to be associated with the metal in a lower oxidation state (in other words, it is a reducing agent: $I^- \rightarrow \frac{1}{2}I_2 + e^-$). Thus, if copper(II) sulphate solution is treated with potassium iodide solution, copper(I) iodide is precipitated and iodine formed:

$$2Cu^{2+} + 4I^- \rightarrow 2CuI(s) + I_2$$

It should be added, though, that this particular reaction is favoured by the insolubility of copper(I) iodide which offsets the tendency of copper(I) ions to be converted into copper(II) ions in aqueous solution, as indicated by the following data:

$$K = [Cu^{2+}(aq)]/[Cu^+(aq)] \approx 10^6$$

$$2Cu^+(aq) \rightleftharpoons Cu^{2+}(aq) + Cu(s) \qquad E^\ominus = 0.37 \text{ V}$$

Both of these results show how low the concentration of soluble copper(I) ions must be compared with soluble copper(II) ions.

Oxides and oxocompounds

The enthalpies of formation of the oxides of the first transition series (*Table 15.9*) are indicative of the high electropositivity of the metals at the beginning of the period declining as the atomic number increases.

Table 15.9 Enthalpies of formation of oxides

Oxide	TiO_2	$\frac{1}{2}V_2O_5$	CrO_3	MnO	$\frac{1}{2}Fe_2O_3$	CoO	NiO	CuO
ΔH_f/kJ mol^{-1}	-946	-775	-580	-387	-412	-240	-244	-156

With the exception of scandium, the common oxidation number of two is evident from the formation of oxides and hydroxides of general formula MO and $M(OH)_2$, respectively. These are predominantly basic and many of them, for example VO, CrO and MnO, are reducing because of their tendency to acquire higher oxidation states. The hydroxides are especially reactive in this respect, e.g.,

$$Mn(OH)_2 \xrightarrow{\text{air}} MnO(OH) \xrightarrow{\text{air}} MnO_2$$

i.e., $Mn^{2+}Mn^{IV}O_2(OH)_2^{2-}$
manganese(II) dihydroxomanganate(IV)

The preparations of the lower oxides have, in fact, often been carried out under reducing conditions, e.g.,

$$V_2O_5 \xrightarrow[\text{temperature}]{\begin{array}{c}H_2\\ \text{high}\end{array}} VO$$

$$FeC_2O_4 \xrightarrow{\text{heat}} FeO + CO_2 + CO$$

iron(II)
ethanedioate

$$Cu^{2+}(aq) \xrightarrow[\text{alkaline conditions}]{\text{reduction under}} Cu_2O$$

In order to keep the copper ions in solution under alkaline conditions they are complexed with, e.g., ammonia, 2,3-dihydroxybutanedioate (tartrate), or 2-hydroxypropane-1,2,3-tricarboxylate (citrate). The latter two complexes are well known as Fehling's and Benedict's solutions respectively for identifying reducing sugars by use of the above reaction (p. 515).

As the oxygen content rises, so also does the acidity of the oxide as shown in *Figure 15.18*.

Basic	*Basic*	*Amphoteric*	*Acidic*
MnO	Mn_2O_3	MnO_2	Mn_2O_7
$\downarrow H^+$	$\downarrow H^+$	$H^+\swarrow \quad \searrow OH^-$	$H_2O\downarrow$
Mn^{2+}	Mn^{3+}	$Mn^{4+} \qquad MnO_3{}^{2-}$	$H^+[Mn^{VII}O_4]^-$

Figure 15.18 Acidity of oxides

This trend accompanies the decrease in size and increase in cationic charge of the metallic atom, which results in high polarizing power and weakening of the bond between oxygen and hydrogen of any hydroxy group that becomes attached; that is hydrogen ionizes and an oxoanion of the transition metal remains:

$$V_2O_5 + 2OH^- \rightarrow 2VO_3{}^- + H_2O$$
$$CrO_3 + 2OH^- \rightarrow CrO_4{}^{2-} + H_2O$$

The tetraoxochromate(VI) ion on acidification condenses to the μ-oxobistrioxochromate(VI), or dichromate, ion:

$$2CrO_4{}^{2-} + 2H^+ \rightarrow Cr_2O_7{}^{2-} + H_2O$$

and this reaction is reversed on adding alkali.

Manganese(IV) oxide upon alkaline oxidation, usually carried out by fusion with an alkali metal hydroxide in the presence of air, produces a green ion which disproportionates on boiling in water to give the well known purple permanganate [tetraoxomanganate(VII)] ion:

$$MnO_2 + 2OH^- + \tfrac{1}{2}O_2 \longrightarrow Mn^{VI}O_4{}^{2-} \qquad + H_2O \qquad (1)$$
$$\text{tetraoxomanganate(VI)}$$
$$\text{(green)}$$

$$Mn^{VI}O_4{}^{2-} + 4H^+ + 2e^- \rightarrow Mn^{IV}O_2(s) + 2H_2O \tag{2}$$

$$Mn^{VI}O_4{}^{2-} \rightarrow Mn^{VII}O_4{}^- + e^- \tag{3}$$

The addition of equation (2) to twice equation (3) gives the overall (disproportionation) reaction as

$$3MnO_4{}^{2-} + 4H^+ \rightarrow MnO_2(s) + 2MnO_4{}^- + 2H_2O$$
$$\text{purple}$$

'*Mixed*' *oxides* consist of metal atoms in different oxidation states; for example, Fe_3O_4, which is prepared by the action of steam on red-hot iron:

$$3Fe + 4H_2O \rightleftharpoons Fe_3O_4 + 4H_2$$

should be represented as iron(II) diiron(III) oxide:

$$Fe^{2+}Fe_2^{3+}O_4$$

In accordance with this formula, it reacts with acids to produce cations of different charge:

$$Fe^{2+}Fe_2^{3+}O_4 + 8H^+ \rightarrow Fe^{2+} + 2Fe^{3+} + 4H_2O$$

Many oxides of the transition metals are non-stoichiometric. This again is a consequence of variable valency; ions of different charge are present in the crystal lattice, and so electrical neutrality does not require a simple ratio of atoms. For example, the formula for iron(II) oxide approximates to $Fe_{0.95}O$ because the crystal contains some Fe^{3+}.

Organometallic compounds

The ultimate members of the transition periods, zinc, cadmium and mercury, resemble the Group 2 elements in possessing two s electrons outside completed inner orbitals (and therefore, strictly speaking, they are not members of the transition elements). But whereas the elements of Group 2

Table 15.10 Comparison between the last members of the transitional metals series and elements of Group 2

Metals	Atomic radius/pm	Standard electrode potential/V	First ionization energy/kJ mol^{-1}
Mg	136	−2.38	736
Ca	174	−2.76	590
Zn	125	−0.76	910
Sr	191	−2.89	549
Cd	141	−0.40	868
Ba	198	−2.90	507
Hg	144	+0.80	1005

have only s and p orbitals complete in the penultimate shells, zinc, cadmium and mercury have the d orbitals complete as well. The correspondingly higher nuclear charge, the low shielding effect of d electrons, and the contraction of atomic size combine to make the last three elements far less electropositive than their alkaline earth counterparts (*Table 15.10*).

Bearing in mind that elements high in a group are less electropositive than those below, comparisons with magnesium will be sought rather than with calcium, strontium and barium. Such a resemblance is found in organometallic compounds. Whilst the Grignard reagent, $RMgX$, is now far more popular, the first organometallic compounds to be discovered were the zinc alkyls, ZnR_2, obtained by reacting alkyl halides with a zinc–copper couple and decomposing the intermediate, $RZnX$:

$$R\text{---}X \xrightarrow{\text{Zn/Cu}} R\text{---}Zn\text{---}X \xrightarrow{\text{heat}} ZnX_2 + R\text{---}Zn\text{---}R$$

(The actual formulae for these organometallic compounds are more complicated than indicated here, see p. 444).

Cadmium and mercury alkyls also exist; CdR_2 and HgR_2 are formed when halides of the metal are reacted with the appropriate Grignard reagent. (It also seems that, unlike the other metals, mercury *can* form a simple $RHgX$ molecule.)

Mercury shows a great facility for forming links with carbon; for example, 'mercuration' of the aromatic nucleus takes place readily, if it is suitably activated, e.g., as phenol, and reacted with mercury(II) ethanoate (acetate) in ethanoic (acetic) acid:

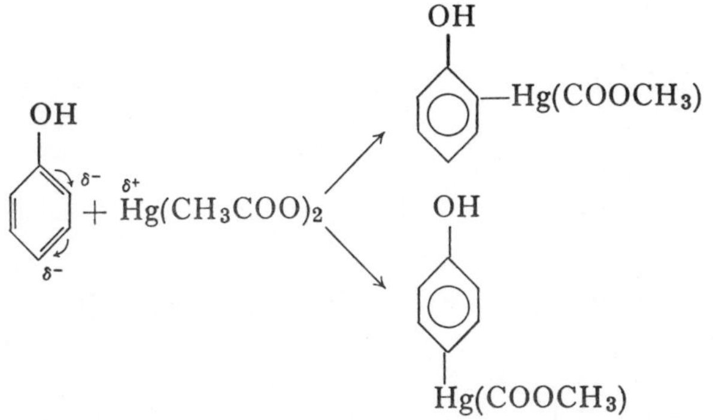

All these organometallic compounds appear to be simple covalent compounds with σ-bonding. There is also the possibility of π-bonding between the available d orbitals of the transitional metals and p electrons of unsaturated organic compounds. Great interest has been aroused in this aspect since the discovery of a very stable compound, ferrocene, by combination of Fe^{II} with the cyclopentadiene ion, $C_5H_5{}^-$. All the $3d$ elements have now been found to give the same type of compound. Other aromatic ring systems, possessing non-localized π-bonding, have also been employed successfully as shown in *Figure 15.19*.

The precipitates of dicarbides, obtained when ethyne is passed through

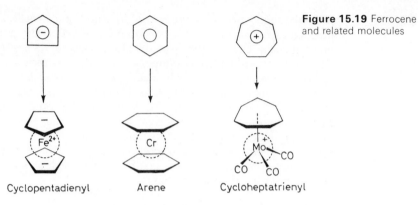

Figure 15.19 Ferrocene and related molecules

Cyclopentadienyl   Arene   Cycloheptatrienyl

ammoniacal solutions of copper(I) and silver, may also involve interaction between the metal and the π-bonding of ethyne (p. 55).

Analysis

The range of coloured compounds produced by alkaline precipitation is shown in *Table 15.11*. The variable oxidation states and ability to form

Table 15.11 Coloured precipitates from solutions of some transitional metal ions with alkali

Ion	Precipitate	Colour
Cr^{3+}	$Cr(OH)_3$	Grey-green
Mn^{2+}	$Mn(OH)_2$	White, rapidly becoming brown owing to oxidation to MnO(OH)
Fe^{2+}	$Fe(OH)_2$	Dirty green
Fe^{3+}	$Fe(OH)_3$	Rust-brown
Co^{2+}	$Co(OH)_2$	Blue (a basic hydroxide) becoming pink on warming
Ni^{2+}	$Ni(OH)_2$	Green
Cu^{2+}	$Cu(OH)_2$	Light blue, becoming black CuO on warming

coordination compounds are also widely used in the identification of the transitional elements.

Chromium

Chromium(III) hydroxide dissolves in sodium peroxide with the formation of yellow chromate(VI):

$$Cr(OH)_3 + H_2O \rightarrow CrO_4^{2-} + 5H^+ + 3e^-$$

$$2H_2O + O_2^{2-} + 2e^- \rightarrow 4OH^-$$

Manganese

Mn^{2+} can be converted into green manganate(VI) by fusion with sodium hydroxide and potassium nitrate; the fused mass on boiling with water forms the purple manganate(VII) solution (p. 361). Alternatively, direct oxidation results by using sodium bismuthate(V)

$$MnO_4^- \xleftarrow{NaBiO_3} Mn^{2+} \xrightarrow{NaOH+KNO_3} MnO_4^{2-}$$

Iron

Fe^{3+} is converted into the blood-red $[Fe(CNS)]^{2+}$ ion by treatment with ammonium or potassium thiocyanate.

Cobalt

Co^{2+} forms a similar blue complex when reacted with the thiocyanate ion

$$Co^{2+} \xrightarrow{CNS^-} [Co(CNS)_4]^{2-}$$

Nickel

Ni^{2+} treated with butanedionedioxime (dimethylglyoxime) in alkaline solution forma s red precipitate:

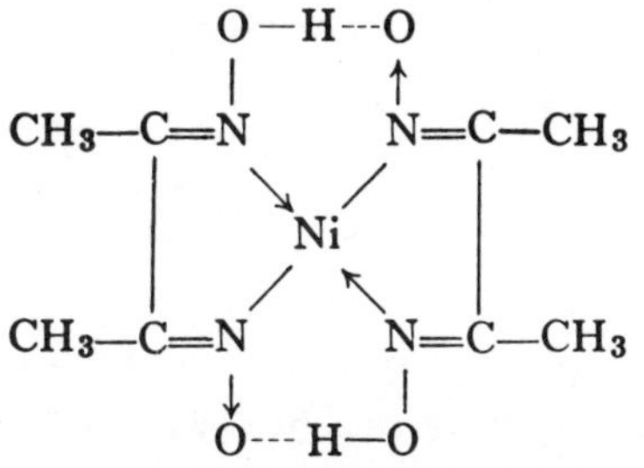

$$\begin{array}{ccc} & O-H\text{---}O & \\ & | \quad\quad \uparrow & \\ CH_3-C{=}N & \quad N{=}C-CH_3 \\ & \searrow \quad \nearrow & \\ | & Ni & | \\ & \nearrow \quad \nwarrow & \\ CH_3-C{=}N & \quad N{=}C-CH_3 \\ & \downarrow \quad\quad | & \\ & O\text{---}H-O & \end{array}$$

This precipitation is sufficiently quantitative to provide the basis for the estimation of nickel.

Copper

The presence of the copper(II) ion is indicated by the development of a deep blue complex when excess of ammonia is added to an aqueous solution:

$$[Cu(H_2O)_4]^{2+} \rightarrow [Cu(NH_3)_4]^{2+}$$

Copper is estimated volumetrically in terms of the iodine liberated when copper(II) is reduced to copper(I) by the addition of potassium iodide:

$$2Cu^{2+} + 4I^- \rightarrow 2CuI(s) + I_2$$

Both copper and cadmium form complexes with the cyanide ion

$$Cu^{2+} + 4CN^- \rightleftharpoons [Cu(CN)_4]^{2-}$$

$$Cd^{2+} + 4CN^- \rightleftharpoons [Cd(CN)_4]^{2-}$$

The cadmium complex dissociates sufficiently to allow cadmium to be precipitated by hydrogen sulphide (in contrast to the far more stable copper complex), and thus cadmium can be identified in the presence of copper.

Silver

Silver chloride is precipitated, along with dimercury(I) chloride and lead chloride, when an aqueous solution of these metal ions is treated with hydrochloric acid. Silver chloride can, however, be distinguished from the other two chlorides by dissolution in excess of ammonia with the formation of the soluble linear complex $[Ag(NH_3)_2]^+$.

Mercury

Dimercury(I) chloride, unlike silver chloride, does not dissolve in excess of ammonia; instead its *s* electron is used to form a covalent bond with the amino group; at the same time, free mercury is liberated and the colour becomes black:

$$Hg_2^{2+} + NH_3 \rightarrow [Hg{-}NH_2]^+ + Hg + H^+$$

The mercury(II) is recognized by being reduced by tin(II) chloride, first to a white precipitate of Hg_2Cl_2 and then to metallic mercury:

$$Hg^{2+} + Sn^{2+} \rightarrow Hg(l) + Sn^{4+}$$

Summary

The transitional metals are characterized by a wide diversity of reactions, many involving ionic equilibria, such as redox, acid–base, precipitation and complex formation. Colour changes are frequent; the name of one element, chromium (Greek: khroma = colour) is an explicit recognition of this. This generally reactive nature is used in many catalysed processes.

The flow-charts (*Figures 15.20–15.29*) for the general chemistry of eight of these elements, should be examined in accordance with the type of reaction, colour change, and structure.

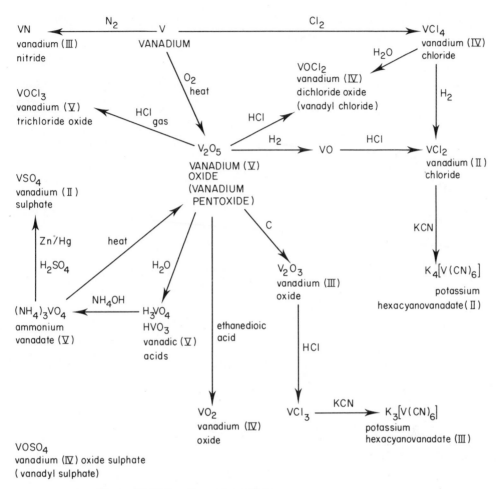

Figure 15.20 Reactions of vanadium

Questions

(1) Starting from the appropriate metal, how could the following compounds be prepared: iron(III) chloride, potassium hexacyanoferrate(III), potassium tetraoxomanganate(VII), potassium tetraoxochromate(VI) and potassium dicyanoargentate(I)?

(2) Suggest how the alums of some of the transition elements might be prepared.

(3) The formulae of the nitrosyl carbonyl and carbonyl of iron are $Fe(NO)_2(CO)_2$ and $Fe(CO)_5$, respectively. Show that these are in agreement with the attainment of a noble gas-type electronic structure.

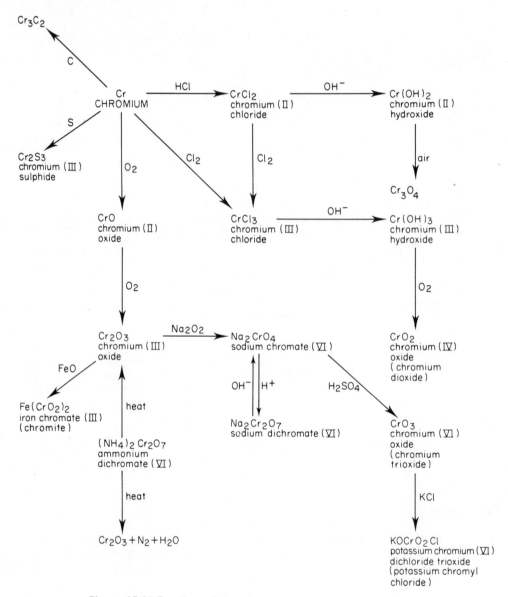

Figure 15.21 Reactions of chromium

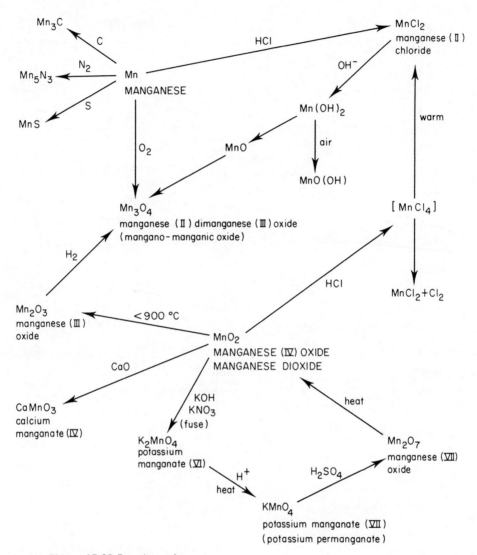

Figure 15.22 Reactions of manganese

Questions (*continued*)

(4) Rationalize the following information by giving the structures of the molecules

$$KCrO_3Cl \xleftarrow{\text{KCl}} CrO_3 \xrightarrow{\text{KOH}} K_2CrO_4$$
$$\downarrow{\text{HCl}}$$
$$CrO_2Cl_2$$

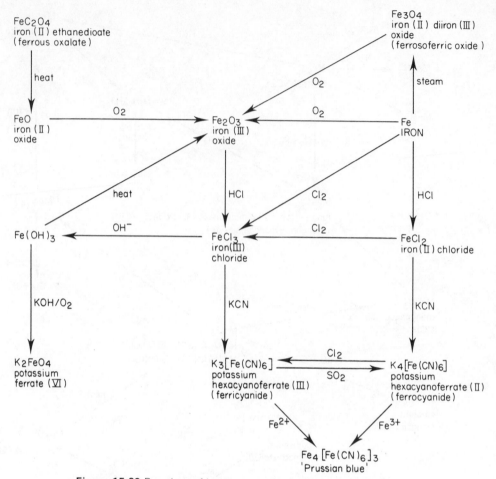

Figure 15.23 Reactions of iron

Questions (*continued*)

(5) How many compounds of the molecular formula $Co(NH_3)_xCl_3$ ($x \not> 4$) could there be? How might they be distinguished by using silver nitrate solution?

(6) Given the following information:

$$[Fe(CN)_6]^{4-} \xrightarrow{HNO_3} [Fe(CN)_5(NO)]^{2-} \xrightarrow[H_2O]{NH_3} [Fe(CN)_5(NH_3)]^{3-} + HNO_2$$

describe the changes in the oxidation number of iron.

(7) Give an account of the organometallic compounds of some of the transitional metals.

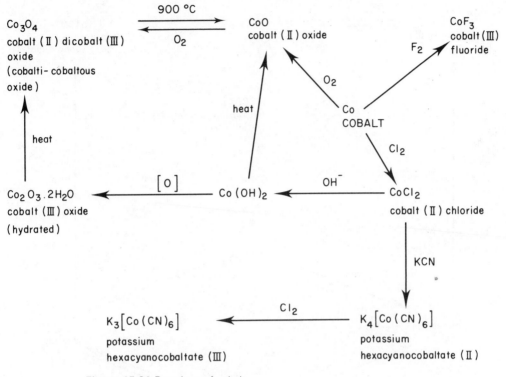

Figure 15.24 Reactions of cobalt

(8)   Give structures for the following compounds so as to accord with the observed molar conductivities/$(\Omega^{-1} \, m^2 \, mol^{-1})$:

$PtCl_4.6NH_3$ 0.0523 $PtCl_4.3NH_3$ 0.0097
$PtCl_4.5NH_3$ 0.0404 $PtCl_4.2NH_3$ 0
$PtCl_4.4NH_3$ 0.0299

(9)   How many isomers with formulae MA_2B_2, MA_2BC and $MABCD$ can exist if M is the central metallic atom and A, B, C and D are the ligands attached (a) tetrahedrally, (b) in a planar form?

(10) Predict what you can of the chemistry of an element with an outer configuration of $d^7 s^2$.

(11) In the presence of excess of mercury, the equilibrium constant for the reaction between Hg(I) and Hg(II) is given by [Hg(II)]/[Hg(I)] and not by $[Hg(II)]/[Hg(I)]^2$. Show that this is in agreement with the existence of mercury in the dimeric form, Hg_2^{2+}, rather than as Hg^+.

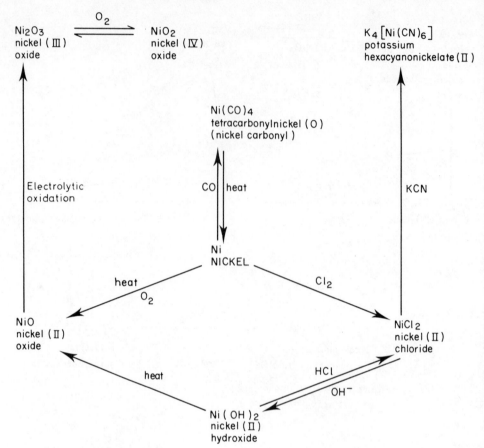

Figure 15.25 Reactions of nickel

Questions (*continued*)

(12) Discuss the coordination compounds of the coinage metals, copper, silver and gold, paying special attention to the stabilization of the silver(II) oxidation state.

(13) Of what use in analysis are the coordination compounds of mercury?

(14) In what respects is the chemistry of mercury unlike that of zinc?

(15) Compare the stabilities of the oxides, chlorides and complex cyanides of five of the transition elements in relation to their oxidation numbers.

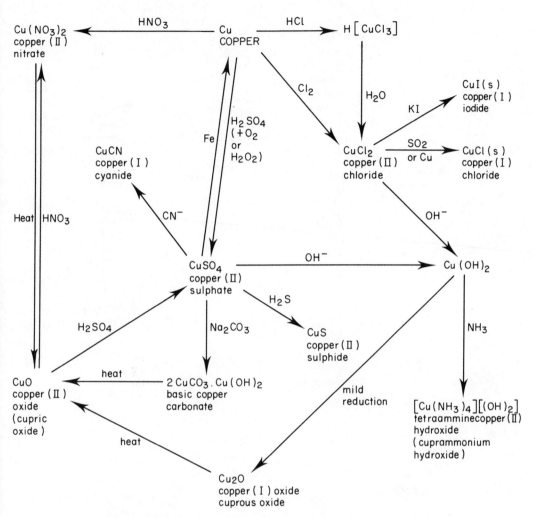

Figure 15.26 Reactions of copper

(16) If the effect of heat on zinc oxide is represented by the equation:

$$Zn^{2+} + O^{2-} \rightarrow Zn_i^{2+} + \tfrac{1}{2}O_2 + 2e^-$$

where Zn_i^{2+} are the interstitial zinc ions formed, show that $[Zn_i^{2+}] \propto [O_2]^{-\frac{1}{6}}$

(17) Some authorities consider zinc, cadmium and mercury to be transitional metals whilst others do not. State and justify your own views on this.

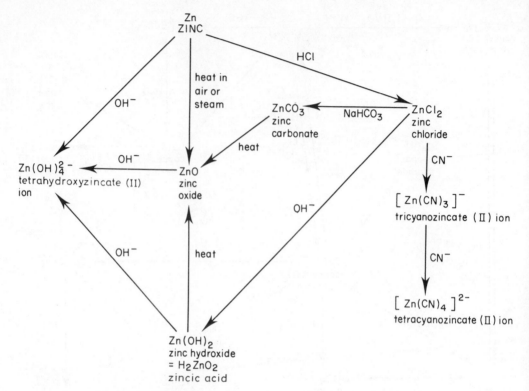

Figure 15.27 Reactions of zinc

Questions (*continued*)

(18)   Explain the following:

(*a*) Black copper oxide dissolves in hot, concentrated hydrochloric acid, to give a greenish solution; this, on boiling with copper is decolorized. When this latter solution is poured into water, a white precipitate is obtained.

(*b*) An aqueous solution of chromium(III) ions gives a green precipitate with ammonia solution; this precipitate dissolves in sodium peroxide to form a yellow solution, which turns orange on the addition of excess acid.

(*c*) On passing sulphur dioxide into ammonium vanadate(V) solution, the colour turns from orange to blue; the addition of zinc powder changes the blue to green, whilst addition of magnesium gives a violet coloration.

(*d*) Black manganese(IV) oxide gives a green mass on fusing with potassium chlorate(V) and potassium hydroxide. Boiling a solution of this in the presence of excess carbon dioxide converts the colour to purple. If the carbon dioxide is replaced by sulphur dioxide, however, the solution is decolorized.

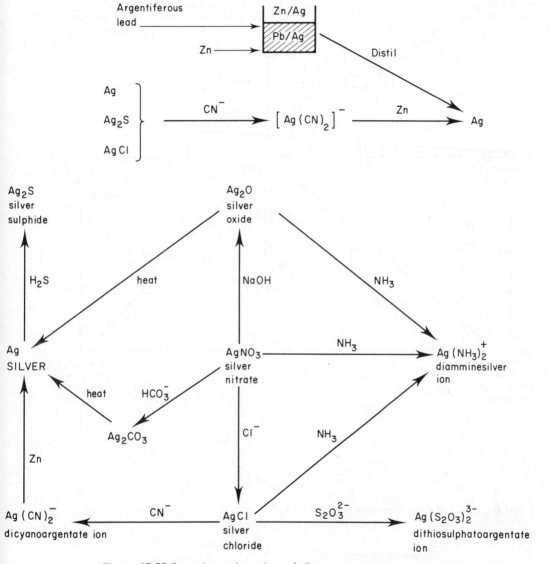

Figure 15.28 Extraction and reactions of silver

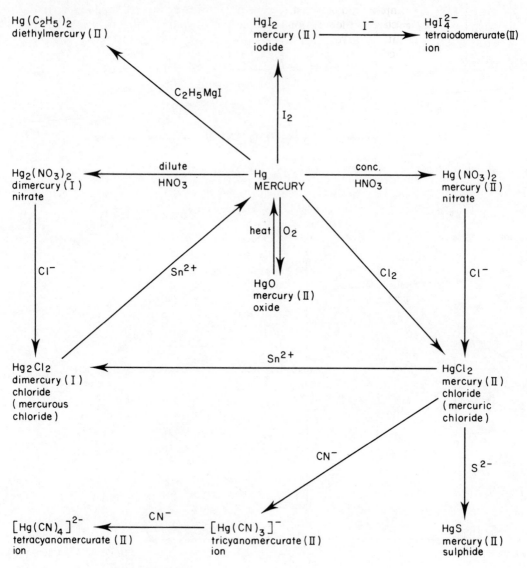

Figure 15.29 Reactions of mercury

(19) List the commonly occurring oxidation states in aqueous solution for each of the first row transition elements from titanium to copper. Indicate any changes in stability of these states across the row. Give two examples (using two different transition elements) to illustrate how a change of ligand can influence the oxidation–reduction behaviour of particular oxidation states.

Compare and contrast the chemistry of copper(II)(aq) and cobalt(II)(aq) by discussing the reactions of aqueous solutions of their sulphates with (a) concentrated hydrochloric acid, (b) concentrated aqueous ammonia in the presence of air.

[JMB 1978]

(20) The following formulae represent some of the carbonyl compounds and ions of the transition metals:

$V(CO)_6^-$, $Cr(CO)_6$, $Cr(CO)_5^{2-}$, $Mn(CO)_5^-$, $Mn(CO)_6^+$, $Fe(CO)_5$, $Fe(CO)_4^{2-}$, $Co(CO)_4^-$, $Ni(CO)_4$, $Mo(CO)_6$

By determining the number of electrons in the valency shell of each metal, suggest which of the following formulae might be correct: $Fe(CO)_6^{2-}$, $Fe(CO)_8^+$, $Fe(CO)_6$, $Fe(CO)_4^+$, $Fe(CO)_6^{2+}$.

(21) (a) Give explanations for the following:
    (i) Concentrated aqueous copper(II) chloride solution is bright green; on dilution with water it changes to a light-blue solution.
    (ii) A solution of copper(I) chloride in hydrochloric acid absorbs carbon monoxide.
    (iii) Copper(I) chloride is soluble in several organic solvents.
  (b) From the following information determine the correct formula of a double salt $Cu(NH_4)_xCl_yzH_2O$ (x, y and z are whole numbers):
    (i) Its molar mass is 277.5 g.
    (ii) The chloride is 1.388 g of the salt is precipitated as silver chloride; after washing and drying the mass of the precipitate is 2.870 g.
    (iii) When 1.388 g of the salt is boiled with excess aqueous sodium hydroxide the ammonia liberated neutralizes 10.0 cm³ of hydrochloric acid of concentration 1.00 mol/dm³.

[A.E.B.]

(22) In what ways are the properties of transition metal ions in aqueous solutions dependent on the nature of the ligands surrounding them?
Suggest explanations of the following observations:
  (a) The compound of formula $Cr(H_2O)_6Cl$ exists in more than one isomeric form.
  (b) Cu_2SO_4 deposits copper metal when treated with water.
  (c) $Fe_2(SO_4)_3$ and KI react in aqueous solution to give iodine but this reaction is suppressed when a large excess of fluoride ion is added.

[C]

16 Lanthanoids and actinoids: the *f*-block elements

General properties • Production of the elements • Reactivity

| La | Ce | Pr | Nd | Pm | Sm | Eu | Gd | Tb | Dy | Ho | Er | Tm | Yb | Lu | | Hf |
| Ac | Th | Pa | U | Np | Pu | Am | Cm | Bk | Cf | Es | Fm | Md | No | Lw | | |

General properties

In these two series of elements, inner *f* orbitals are being filled with electrons. As there are seven such orbitals, there are fourteen lanthanoids and fourteen actinides in which the 4*f* and 5*f* orbitals, respectively, are being filled. *Table 16.1* shows, however, that these orbitals are not filled in a perfectly regular order.

Of these elements, promethium (61), americium (95) and all subsequent ones are unknown in nature but have been prepared artificially. The naturally occurring lanthanoids (and thorium) are usually found together, e.g., in monazite sands, because of their great chemical similarity, which follows from their two outer electron shells having the same configuration for all except gadolinium, lutetium and thorium.

In Chapter 15 it was stated that the *d* orbitals are particularly stable when they are completely empty (d^0), completely full (d^{10}) or half full (d^5). In the same way, the most stable states of *f* orbitals are f^0, f^7 and f^{14}. *Table 16.1* shows how elements tend to adjust their filling of other orbitals so as to attain this configuration. This tendency is also evident in the oxidation states exhibited by the different elements.

The normal oxidation state is $+3$. The reason for this is not immediately obvious from the electronic configurations, which suggest bivalency by the loss of the $6s^2$ or $7s^2$ electrons. However, the energy acquired by the hydration of the tripositive ion of many of the elements is sufficient to compensate for that required to ionize the third electron. An added complication is introduced by the tendency to attain, or at least approach, f^0, f^7 or f^{14} configurations, which leads to cerium, praseodymium,

Table 16.1

Lanthanoids		Actinoids	
Element [Xe]*	Electronic configuration	Element [Rn]*	Electronic configuration
	$4f$		$5f$
Cerium	[↑ ↑]	Thorium	[ ] $6d^2$
Praseodymium	[↑ ↑ ↑]	Protoactinium	[↑ ↑] $6d^17s^2$
Neodymium	[↑ ↑ ↑ ↑] $6s^2$	Uranium	[↑ ↑ ↑] $6d^17s^2$
Promethium	[↑ ↑ ↑ ↑ ↑]	Neptunium	[↑ ↑ ↑ ↑]
Samarium	[↑ ↑ ↑ ↑ ↑ ↑]	Plutonium	[↑ ↑ ↑ ↑ ↑ ↑] $7s^2$
Europium	[↑ ↑ ↑ ↑ ↑ ↑ ↑]	Americium	[↑ ↑ ↑ ↑ ↑ ↑ ↑]
Gadolinium	[↑ ↑ ↑ ↑ ↑ ↑ ↑] $5d^16s^2$	Curium	[↑ ↑ ↑ ↑ ↑ ↑ ↑] $6d^17s^2$
Terbium	[↑↓ ↑↓ ↑ ↑ ↑ ↑ ↑]	Berkelium	[↑↓ ↑↓ ↑ ↑ ↑ ↑ ↑]
Dysprosium	[↑↓ ↑↓ ↑↓ ↑ ↑ ↑ ↑]	Californium	[↑↓ ↑↓ ↑↓ ↑ ↑ ↑ ↑]
Holmium	[↑↓ ↑↓ ↑↓ ↑↓ ↑ ↑ ↑] $6s^2$	Einsteinium	[↑↓ ↑↓ ↑↓ ↑↓ ↑ ↑ ↑] $7s^2$
Erbium	[↑↓ ↑↓ ↑↓ ↑↓ ↑↓ ↑ ↑]	Fermium	[↑↓ ↑↓ ↑↓ ↑↓ ↑↓ ↑ ↑]
Thulium	[↑↓ ↑↓ ↑↓ ↑↓ ↑↓ ↑↓ ↑]	Mendelevium	[↑↓ ↑↓ ↑↓ ↑↓ ↑↓ ↑↓ ↑]
Ytterbium	[↑↓ ↑↓ ↑↓ ↑↓ ↑↓ ↑↓ ↑↓]	Nobelium	[↑↓ ↑↓ ↑↓ ↑↓ ↑↓ ↑↓ ↑↓]
Lutetium	[↑↓ ↑↓ ↑↓ ↑↓ ↑↓ ↑↓ ↑↓] $5d^16s^2$	Lawrencium	[↑↓ ↑↓ ↑↓ ↑↓ ↑↓ ↑↓ ↑↓] $6d^17s^2$

* The elements in each row have an inner core of electrons corresponding to either xenon or radon

neodymium, terbium and dysprosium of the lanthanoid series also having a valency of four, whilst samarium, europium, thulium and ytterbium exhibit bivalency. For the same reason, gadolinium and lutetium are particularly stable when in the trivalent state.

Variation of valency is more pronounced among the actinoids, although the tripositive oxidation state tends to become more common as the series proceeds. *Table 16.2* shows the known oxidation states and corresponding electronic configurations for the two series of elements.

Table 16.2 Oxidation states and electronic configurations

Oxidation state	2	3	4	5		2	3	4	5	6
Ce		f^1	f^0		Th	f^0d^2	f^0d^1	f^0d^0		
Pr		f^2	f^1	f^0	Pa		f^2	f^1	f^0	
Nd		f^3	f^2		U		f^3	f^2	f^1	f^0
Pm		f^4			Np		f^4	f^3	f^2	f^1
Sm	f^6	f^5			Pu		f^5	f^4	f^3	f^2
Eu	f^7	f^6			Am		f^6	f^5	f^4	f^3
Gd		f^7			Cm		f^7	f^6		
Tb		f^8	f^7		Bk		f^8	f^7		
Dy		f^9	f^8		Cf		f^9			
Ho		f^{10}			Es		f^{10}			
Er		f^{11}			Fm		f^{11}			
Tm	f^{13}	f^{12}			Md		f^{12}			
Yb	f^{14}	f^{13}			No		f^{13}			
Lu		f^{14}			Lw		f^{14}			

In both series of elements there is a contraction in atomic and ionic sizes as the atomic number increases. This is because, as successive protons are added, their increased attraction for the outer electrons causes the orbits of the latter to be pulled closer to the nucleus, especially as they are not particularly well shielded by diffuse *f* orbitals. These *lanthanoid and actinoid contractions* are illustrated in *Figure 16.1*.

Like the transitional elements discussed earlier, many ions of these elements exhibit paramagnetism, caused by the presence of unpaired electrons. The magnetic behaviour cannot, however, be as closely correlated with the electronic configurations as was the case with the transition metals, presumably because the *f* orbitals are effectively shielded from external forces by the *s* and *p* orbitals of the outer shells. The greatest paramagnetism in the lanthanoid series is shown by dysprosium, and this has been made use of in the production of very low temperatures. Isothermal magnetization of a dysprosium salt, followed by adiabatic demagnetization, when repeated several times has permitted a temperature as low as 0.09 K to be attained.

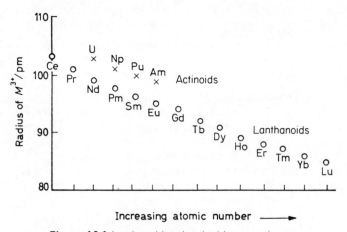

Figure 16.1 Lanthanoid and actinoid contractions

The unpaired electrons give rise to absorption bands in the visible region of the spectrum (as already noted for the *d* block elements), and so the salts of some of these elements are coloured (*Table 16.3*).

The four elements uranium, neptunium, plutonium, americium, and possibly also protactinium, form very stable oxocations of formulae MO_2^+ and MO_2^{2+}, and much of their chemistry revolves around these ions.

Production of the elements

Owing to their very pronounced chemical similarities, the elements of these series usually occur together and are very difficult to separate by common methods of precipitation or crystallization. It is possible to separate some of

Table 16.3 Common ions and their colours

Number of unpaired electrons	Ions		Colour
0		Lu^{3+}	—
1	Ce^{3+}	Yb^{3+}	—
2	Pr^{3+}	Tm^{3+}	green
3	Nd^{3+}	Er^{3+}	pink
4	Pm^{3+}	Ho^{3+}	rose/yellow
5	Sm^{3+}	Dy^{3+}	yellow
6	Eu^{3+}	Tb^{3+}	—
7	Gd^{3+}		—

Table 16.4 pH causing precipitation of hydroxides of lanthanoids

Hydroxide	$La(OH)_3$	$Nd(OH)_3$	$Gd(OH)_3$	$Lu(OH)_3$	$Ce(OH)_4$
Precipitation pH	7.82	7.31	6.83	6.30	2.6

the elements by treatment of solutions with alkali to precipitate the hydroxides which, owing to their differing solubility products, are precipitated at different pH values, e.g., *Table 16.4*.

More recently, *solvent extraction* methods have been utilized in the separation of the lanthanoids. A solution of the elements (in the form of their compounds) is allowed to flow counter to a second solvent. The elements are extracted into the latter in accordance with their respective extraction coefficients. A very satisfactory system consists of a solution of lanthanoids in nitric acid flowing counter to a solution of tributyl phosphate(V) in kerosene.

The most efficient method of separation is *ion exchange*, which depends on differences in the absorbing power of a suitable resin or other absorbent. For example, if a solution of lanthanoids is passed through a 15 m zeolite column, the elements separate into different bands. This method has also been applied very successfully to the artificial actinoid elements, not only for their separation but for their discovery. The accuracy of the work is such that 17 atoms of mendelevium were produced *and* characterized, although its half-life is only 1.26×10^4 s.

Prior to separation by such methods, the artificial elements have first to be produced by suitable radiochemical processes. Americium and curium (in the form of their compounds) are obtained by neutron bombardment of plutonium, which is itself first separated from uranium by solvent extraction:

$$^{239}_{94}Pu \xrightarrow{(n,\gamma)} {}^{240}_{94}Pu \xrightarrow{(n,\gamma)} {}^{241}_{94}Pu \xrightarrow[t_{\frac{1}{2}}=4.2\times10^8\text{ s}]{\beta-} {}^{241}_{95}Am \qquad (t_{\frac{1}{2}}=1.4\times10^{10}\text{ s})$$

where (n,γ) signifies bombardment with a neutron, followed by emission of γ-radiation

$$^{239}_{94}Pu \xrightarrow{(4n,\gamma)} {}^{243}_{94}Pu \xrightarrow[t_{\frac{1}{2}}=1.8\times10^4\text{ s}]{\beta-} {}^{243}_{95}Am \xrightarrow{(n,\gamma)} {}^{244}_{95}Am \qquad (\textit{continued overleaf})$$

$$^{244}_{95}\text{Am} \xrightarrow[t_{\frac{1}{2}} = 1.6 \times 10^3 \text{ s}]{\beta-} {}^{244}_{96}\text{Cm} \qquad (t_{\frac{1}{2}} = 6.0 \times 10^8 \text{ s})$$

Micro and sub-micro amounts of heavier elements have been obtained by bombardment with alpha-particles and stripped-carbon (i.e. carbon nuclei), e.g.,

$$^{241}_{95}\text{Am} \xrightarrow{(\alpha,2n)} {}^{243}_{97}\text{Bk}$$

$$^{238}_{92}\text{U} \xrightarrow{(^{13}\text{C},6n)} {}^{244}_{98}\text{Cf}$$

More recently, bombardment has involved elements other than carbon (*see* *Table 16.5*).

Table 16.5 Production of transuranium elements

$$^{238}_{92}\text{U} + {}^{2}_{1}\text{H} \rightarrow {}^{238}_{93}\text{Np} + 2{}^{1}_{0}\text{n}$$
$$^{238}_{92}\text{U} + {}^{4}_{2}\text{He} \rightarrow {}^{240}_{94}\text{Pu} + 2{}^{1}_{0}\text{n}$$
$$^{239}_{94}\text{Pu} + {}^{4}_{2}\text{He} \rightarrow {}^{241}_{95}\text{Am} + {}^{1}_{1}\text{H} + {}^{1}_{0}\text{n}$$
$$^{239}_{94}\text{Pu} + {}^{4}_{2}\text{He} \rightarrow {}^{240}_{96}\text{Cm} + 3{}^{1}_{0}\text{n}$$
$$^{244}_{96}\text{Cm} + {}^{4}_{2}\text{He} \rightarrow {}^{245}_{97}\text{Bk} + {}^{1}_{1}\text{H} + 2{}^{1}_{0}\text{n}$$
$$^{238}_{92}\text{U} + {}^{12}_{6}\text{C} \rightarrow {}^{245}_{98}\text{Cf} + 5{}^{1}_{0}\text{n}$$
$$^{238}_{92}\text{U} + {}^{14}_{7}\text{N} \rightarrow {}^{247}_{99}\text{Es} + 5{}^{1}_{0}\text{n}$$
$$^{238}_{92}\text{U} + {}^{16}_{8}\text{O} \rightarrow {}^{250}_{100}\text{Fm} + 4{}^{1}_{0}\text{n}$$
$$^{253}_{99}\text{Es} + {}^{4}_{2}\text{He} \rightarrow {}^{256}_{101}\text{Md} + {}^{1}_{0}\text{n}$$
$$^{246}_{96}\text{Cm} + {}^{13}_{6}\text{C} \rightarrow {}^{251}_{102}\text{No} + 8{}^{1}_{0}\text{n}$$
$$^{252}_{98}\text{Cf} + {}^{10}_{5}\text{B} \rightarrow {}^{257}_{103}\text{Lw} + 5{}^{1}_{0}\text{n}$$

Table 16.6 Standard electrode potentials, $M \rightarrow M^{3+}(\text{aq}) + 3e^-$

	$E^{\ominus}/V$		$E^{\ominus}/V$
Ce	-2.52	U	-1.80
Gd	-2.40	Np	-1.83
Lu	-2.25	Pu	-2.03

These elements are all appreciably electropositive (*see Table 16.6*), and thus very reactive. They react readily with hydrogen, oxygen, water and the halogens. In consequence of this reactivity, the metals themselves are obtained by vigorous reductive methods, such as electrolysis of the fused halides, or by metallic reduction of the halides, by use of sodium, magnesium or barium vapour.

Questions

(1)   Write an essay on the extraction of ^{235}U.

(2)   Of what value are the compounds of the lanthanoids?

(3)   Where, in the Periodic Table, would you expect element 104 to be placed? Predict the main chemical features of this element.

(4)   Suggest explanations for the apparent anomalies of atomic sizes in consecutive elements in *Figure 16.2*.

Figure 16.2

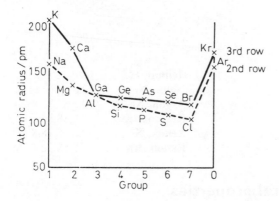

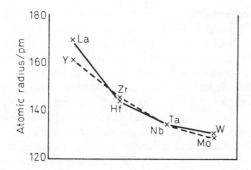

(5)   Write an essay on ion-exchange resins.

17 Group 0: the noble gases

General properties • Occurrence, extraction and uses • Chemical properties

Group 7			Group 1
	ns	*np*	
	Helium, He	2	Li
F	Neon, Ne	2.8	Na
Cl	Argon, Ar	2.8.8	K
Br	Krypton, Kr	2.8.18.8	Rb
I	Xenon, Xe	2.8.18.18.8	Cs
At	Radon, Rn	2.8.18.32.18.8	Fr

Introduction: general properties

The noble gases, helium, neon, argon, krypton, xenon and radon, constitute Group 0 of the Periodic Table (p. 24). Accordingly, the ultimate electron shell has the *s* and three *p* orbitals filled (except helium, which has only an *s* orbital of suitable energy available), and it requires considerable energy either to add or to remove any electrons. Hence they are very unreactive elements, commonly called *inert gases*. This unreactivity is shown by their high ionization energies and negligible electron affinities. However, the effect of the increase in atomic size on descending the Group is that electrons can be lost more readily, i.e., ionization energies decrease (*Table 17.1*) allowing the larger elements to take part in a limited number of chemical reactions.

The almost total lack of interaction between the atoms, consequent upon the complete pairing of electrons in closed shells, results in their existing as

Table 17.1 Atomic size and ionization energies

Element	Atomic radius/pm	First ionization energy/kJ mol^{-1}
He	93	2372
Ne	112	2080
Ar	154	1522
Kr	169	1353
Xe	190	1163
Rn	—	1040

monatomic gases of low boiling point (*see Table 17.2*). Helium has the lowest boiling point of any known substance, 4.2 K. Between 4.2 and 2.2 K liquid helium behaves normally, but below the latter transition temperature it loses its electrical resistance and has such a low viscosity that it is virtually frictionless. It is also able to flow uphill.

Occurrence, extraction and uses

All the gases occur in the atmosphere; their volume percentages are shown in *Table 17.3*. Since helium, like hydrogen, has such a low density, it is lost to

Table 17.2 Volatility of the noble gases

	He	Ne	Ar	Kr	Xe	Rn
M.p./K	0.9 (2.6 MPa)	24.4	83.6	116	161	202
B.p./K	4.2	27.1	87.3	120	166	208

Table 17.3 The composition of the atmosphere

	Volume per cent		Volume per cent
O_2	20.95	CH_4	1.5×10^{-4}
N_2	78.09	Kr	1.1×10^{-4}
Ar	9.3×10^{-1}	N_2O	5×10^{-5}
CO_2	3×10^{-2}	H_2	5×10^{-5}
Ne	1.8×10^{-3}	O_3	4×10^{-5}
He	5.2×10^{-4}	Xe	9×10^{-6}

space fairly quickly but tends to be replenished from radioactive decay products.

Natural gas acts as a reservoir for alpha-particles produced from natural radioactivity, and these readily form helium atoms by removing two electrons from (i.e., by oxidizing) elements of variable valency

$$^4_2He^{2+} + 2e^- \rightarrow {}^4_2He$$

Neon, argon, krypton and xenon are extracted from the atmosphere by liquefaction, followed by fractional distillation. The low vapour density of neon results in its being associated with the nitrogen fraction, whilst the remaining gases come off with the oxygen fraction. Final purification is often effected by selective adsorption on activated charcoal.

All isotopes of radon are radioactive and are themselves products of radioactive decay. The most stable of these is produced in the decay of radium

$$_{88}^{226}\text{Ra} \xrightarrow{(-\alpha)} {}_{86}^{222}\text{Rn} \xrightarrow[t_{\frac{1}{2}} = 3.3 \times 10^5 \text{ s}]{(-\alpha)} {}_{84}^{218}\text{Po} \rightarrow \text{etc.}$$

The chemically unreactive nature of the noble gases, combined with their physical properties, makes them useful as inert atmospheres in arc welding (especially for aluminium, magnesium, titanium and stainless steel) and in electric light bulbs and discharge tubes. Liquid helium is used in low-temperature research, gaseous helium for filling research and weather balloons and also as a diluent for oxygen in diving systems, since it is far less soluble in blood than nitrogen: divers can then surface fairly quickly without ill effect.

Chemical properties

Although the ionization energies of these gases are high, the decrease in these values on descending the Group suggests that those of higher atomic number are more likely to enter into chemical combination than those at the beginning of the Group. Indeed, some compounds have been made, including the tetrafluoride, the difluoride oxide and the trioxide of xenon, and the tetrafluorides of krypton and radon. It is understandable that elements of high electron affinity would be the most likely ones to enter into combination with any of these gases; the lattice energies of the solid products must also be high to compensate for the large ionization energies of the gas (*see* Born–Haber cycle, p. 43).

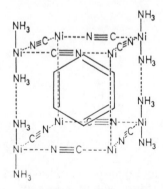

Figure 17.1 A clathrate: the unit cell of benzene amminenickel cyanide

There is also the possibility of the noble gas atoms being trapped in a crystal lattice of suitable dimensions, to give a substance, called a *clathrate* (*Figure 17.1*), that hardly merits the description of a 'compound' in the classical sense. Such clathrates are obtained when, e.g., benzene-1,4-diol (quinol) crystallizes in the presence of the noble gas under high pressure. The lattice is such that three quinol molecules linked by hydrogen bonding comprise a cage large enough to hold one atom of the gas, which can be released on dissolving the crystals in water or by heating. Significantly, helium, the smallest member, is not trapped in this way and so does not form a clathrate with benzene-1,4-diol.

The four heaviest elements form crystalline hydrates of variable composition but approximating to hexahydrates. These also appear to be clathrate-type compounds, the atoms of noble gas being trapped in cages of hydrogen-bonded water molecules. In this case, however, there is a possibility that electrons of the gas atoms are attracted towards the positive end of the polar water molecules.

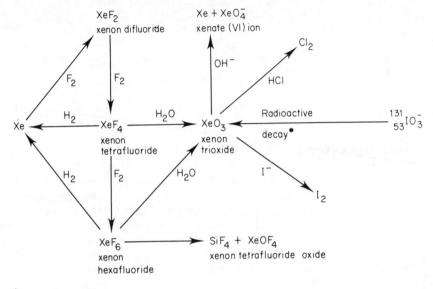

$$^{*}\ ^{131}_{53}I = ^{\ \ 0}_{-1}e + ^{131}_{54}Xe$$

Figure 17.2 Xenon compounds

Questions

(1) In the light of what you have read in this Chapter, discuss the meaning of the word 'compound'.

(2) How far do you think it true to say that the noble gases provide the key to the problem of chemical combination?

(3) The noble gases have also been called 'rare gases' and 'inert gases'. Discuss the merits of these various names.

(4) Do you think the noble gases could form coordination compounds? If you do, what compounds do you think might react in this way and how would you attempt to verify your conclusion?

(5) Write an account of the utilization by industry of the noble gases.

18 Organic chemistry: an introduction

Special properties of carbon • Hydrocarbons • Homologous series • Functional groups • Nomenclature • Isomerism • Organic reactions

Special properties of carbon

The chemistry of the living cell revolves around compounds containing carbon and hydrogen, with or without other elements such as oxygen, nitrogen and halogens; compounds that are therefore conventionally known as *organic* compounds. Over a million such compounds are known already, most of them being based structurally upon a framework of carbon atoms. This profusion bears ample testimony to the versatility of carbon and the stability of carbon bonding, which in turn are partly attributable to the following facts:

(1) The oxidation number of carbon can be between the limits of -4 and $+4$, presumably because carbon lies in the middle of the Second Period;

(2) Being in the Second Period, carbon has only one s and three p orbitals available for bonding, so that the maximum coordination number is four, and four-covalent carbon is therefore in a saturated and stable state, well shielded from attack;

(3) Multiple bonding is at its most conspicuous in Period Two. Thus silicon, in the same group as carbon but in Period 3, does not form such stable multiple bonds as carbon (p. 301).

(4) The power of catenation (Lat. *catena*, chain), i.e., the linking together of identical atoms, reaches a maximum in Group 4 of the Periodic Table. Carbon is pre-eminent in this respect, with silicon exhibiting catenation to a much smaller extent (*see* Chapter 13).

Hydrocarbons

Even the number of compounds containing carbon and hydrogen alone (hydrocarbons) is very large, but fortunately they can be readily classified as

(1) *Aliphatic* hydrocarbons, open-chain structures which may be
(*a*) *saturated*, when they are called *alkanes* or *paraffins*;
(*b*) *unsaturated*, with either double bonds between carbon atoms (the *alkenes* or *olefins*) or triple bonds (*alkynes* or *acetylenes*);

Table 18.1 Types of hydrocarbon

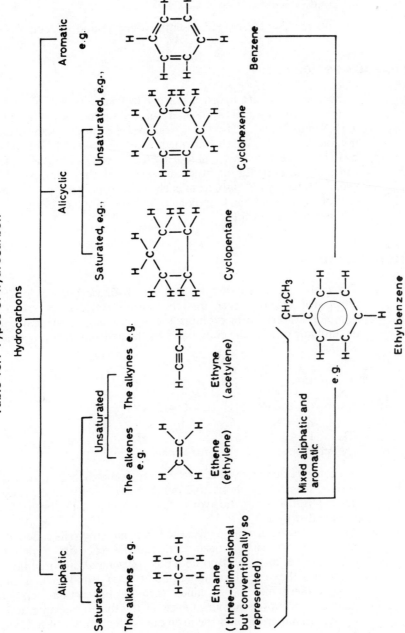

(2) *Alicyclic* hydrocarbons, containing closed rings of carbon atoms;
(3) *Aromatic* hydrocarbons, also containing closed rings but with non-localized orbitals, giving in effect a completely closed, conjugated double bond system (*see* Chapter 3).

Table 18.1 depicts some representative compounds of each of these classes.

Homologous series

The carbon chain of a hydrocarbon can be increased by replacing a terminal hydrogen atom by a methyl ($-CH_3$) group, so that the overall addition is of (CH_3 minus H), i.e., CH_2. The series developed in this manner is termed an homologous series and the members, whose formulae differ from each other by $(CH_2)_n$, are called *homologues*. *Table 18.2* shows the general formulae of some of the aliphatic homologous series. As the members of any series usually differ significantly from each other in physical properties only, the *typical* reactions of one homologue can be regarded as being representative of the series as a whole.

Functional groups

The replacement of hydrogen atoms in hydrocarbons by other atoms or groups of atoms, either directly or indirectly, gives rise to different homologous series. For example, the replacement of one hydrogen atom in an aliphatic or alicyclic compound by the hydroxy group, $-OH$, produces an alcohol:

$$
\begin{array}{cc}
\text{An alkane} & \text{An alcohol}
\end{array}
$$

An alkane   An alcohol

The hydrocarbon residue is referred to as the alkyl group (Alk for short) if derived from an aliphatic hydrocarbon and as the aryl group (or Ar) if derived from an aromatic hydrocarbon. When possible, both Alk and Ar are represented by *R*.

The replacing group is called the *functional group* since, being more labile than the hydrocarbon residue, it confers reactivity upon what would otherwise be fairly inert compounds. In general, all organic compounds containing the same functional group have very similar chemical properties, though these may be modified through interaction between different functional groups in the same molecule. It is due to such interaction between the functional group and the aromatic system to which it is attached that aromatic compounds have modified properties compared with the aliphatic analogues. Nevertheless, it is possible and helpful to consider the typical reactions not only of an homologous series but also of the functional group

Table 18.3 Functional groups

Functional group	Formula	Structure	Systematic name prefix	suffix
Single bond		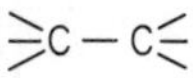 $\geq C - C \leq$		-ane
Double bond		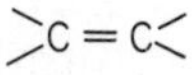 $> C = C <$		-ene
Triple bond		$-C \equiv C-$		-yne
Halide	F, Cl, Br, I		halo-	
Amine	$-NH_2$	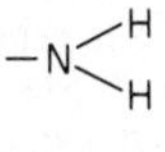 $-N<^H_H$	amino-	amine
Alcohol and phenol	$-OH$	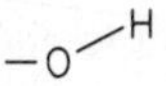 $-O^{\nearrow H}$	hydroxy-	-ol
Ether	$-O-$	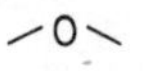 $\diagup O \diagdown$	alkoxy-	
Aldehyde	$-CHO$	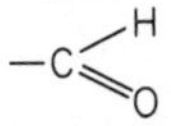 $-C<^H_O$	oxo	-al
Ketone	$> CO$	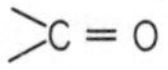 $> C = O$	oxo	-one
Carboxylic acid	$-COOH$	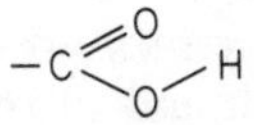 $-C<^O_{O-H}$	carboxy	carboxylic (or -oic)* acid
Acid chloride	$-COCl$	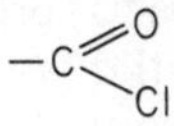 $-C<^O_{Cl}$	chloro-carbonyl	carbonyl (or (-oyl) chloride
Acid anhydride	$> (CO)_2O$	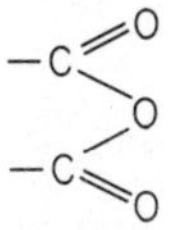 $-C<^O_O$ $-C<^O_O$		-oic anhydride
Ester	$-COO-$	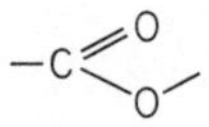 $-C<^O_{O\diagup}$		carboylate (-oate)*
Amide	$-CONH_2$	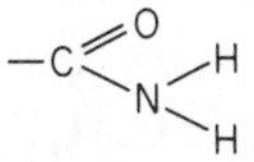 $-C<^O_{N<^H_H}$	carbamoyl	carboxamide (amide)*
Nitro	$-NO_2$	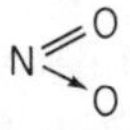 $N<^O_O$	nitro-	
Nitrile (cyanide)	$-CN$	$-C \equiv N$	cyano-	-nitrile
Isonitrile (isocyanide)	$-NC$	$-N \equiv C$	isocyano-	-isonitrile
Sulphonic acid	$-SO_3H$	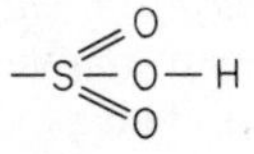 $-S<^O_{O-H}$ (O)		-sulphonic acid

* When the C atom is counted in the carbon chain forming the root.

Table 18.2 Some homologous series

Name of series	General formula
Alkanes	C_nH_{2n+2}
Alkenes	C_nH_{2n}
Alkynes	C_nH_{2n-2}
Monosubstituted derivatives of alkanes	$C_nH_{2n+1}X$ $X = $ halogen, —OH, etc.

present, especially if the latter is regarded from the point of view of the nature of the atoms comprising it. This will be the underlying approach in the following Chapters. *Table 18.3* lists some of the more common functional groups encountered in organic chemistry.

Nomenclature

It is convenient at this juncture to discuss the methods used for naming organic compounds. The systematic naming of molecules should be such that the formula can be constructed unambiguously from the name. However, many common trivial names are so well established that the systematic names are seldom used, and in the course of this book, both trivial and systematic names are used, as appropriate. The systematic name denotes: (*a*) the length of the chain; (*b*) the branching of the chain; (*c*) the position and nature of any functional groups.

Aliphatic compounds

(1) Saturated aliphatic hydrocarbons are given the termination *-ane*, which is prefixed by a suitable contraction of the Greek (occasionally Latin) word for the number of carbon atoms contained in the chain (with the exception of the first four members which retain their original names):

CH_4	methane	C_4H_{10}	butane	C_7H_{16}	heptane
C_2H_6	ethane	C_5H_{12}	pentane	C_8H_{18}	octane
C_3H_8	propane	C_6H_{14}	hexane	C_9H_{20}	nonane

A univalent radical derived from such a hydrocarbon is given the name of the hydrocarbon, except that -ane is replaced by *-yl*. For instance, C_2H_5 is called the *ethyl* radical, from the parent hydrocarbon *ethane*.

The chain length of the molecule is obtained by drawing the structure so that there are *the largest number of carbon atoms in a straight chain*, and the compound is then named as a derivative of the hydrocarbon possessing this chain. The position of any substituent is indicated by prefixing with the number of the carbon atom to which it is attached, the carbon atoms being numbered from that end of the chain which results in the lowest possible number to the group cited by a suffix, then the lowest possible individual numbers to those cited as prefixes. Thus

$$\begin{array}{l} \overset{1}{C}H_3\overset{2}{C}H_2\overset{3}{C}H\overset{4}{C}H_2\overset{5}{C}H_2\overset{6}{C}H_3 \\ \qquad\quad | \\ \qquad\; CH_3 \end{array}$$ **3-(not 4-)methylhexane**

$$\begin{array}{ccccc} & & CH_3 & & \\ & & | & & \\ CH_3 & — & C & — & CH_3 \\ & & | & & \\ & & CH_3 & & \end{array}$$ **2,2-dimethylpropane**

(2) Hydrocarbons containing double bonds are named in a similar fashion except that the termination is *-ene* and the position of the double bond is indicated by the smallest possible number. Similarly, hydrocarbons containing a triple bond end in *-yne*, for example,

$CH_3CH_2CH{=}CH_2$ but-1-ene, from the saturated hydrocarbon butane

$CH_3C{\equiv}CCH_2CH_3$ pent-2-yne, from the saturated hydrocarbon pentane

(3) Halogen derivatives are named by prefixing the name of the parent hydrocarbon by fluoro-, chloro-, bromo- or iodo-:

CH_3Br, bromomethane

$$\begin{array}{l} CH_3CHCH_2CH_2I \\ \qquad | \\ \qquad Cl \end{array}$$ 3-chloro-1-iodobutane

(4) Simple alcohols are named by replacing the final -e of the corresponding hydrocarbon by *-ol*

CH_3CH_2OH ethanol

$$\begin{array}{l} \qquad\quad\;\; OH \\ \qquad\qquad | \\ CH_3CHCHCH_3, \\ \qquad | \\ \qquad CH_3 \end{array}$$ **3-methylbutan-2-ol**

(5) Aldehydes are derived in a similar fashion, the final -e being replaced by *-al*

$CH_3CH_2CH_2CHO$ butanal

(6) In the case of ketones, the final -e is replaced by *-one*. If there is any possible ambiguity, then the position of the carbonyl (ketone) group is indicated numerically in the usual way, e.g.,

$CH_3CH_2COCH_2CH_3$ pentan-3-one

(7) Carboxylic acids are named by replacing the final -e of the parent hydrocarbon by *-oic acid*:

CH_3CH_2COOH propanoic acid

In dibasic acids, the final -e is not removed, e.g.,

COOH
| ethanedioic acid
COOH

(8) Alicyclic compounds are prefixed by *cyclo*, e.g.,

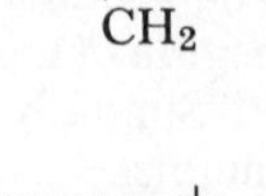

H_2C——CH_2
    \ /   cyclopropane
    CH_2

Aromatic compounds

(9) Aromatic compounds are also named, where convenient, as derivatives of the corresponding hydrocarbon, e.g.,

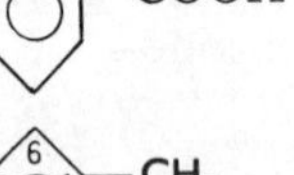

—COOH benzenecarboxylic acid

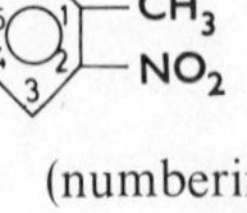

—CH_3
—NO_2 methyl-2-nitrobenzene

(numbering of the ring commences at one of the substituted positions)

Where this type of nomenclature is difficult to use, it is permissible to refer

to the

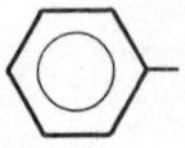

radical as the phenyl radical. This leads (e.g.) to the name triphenylmethanol for the compound having the formula

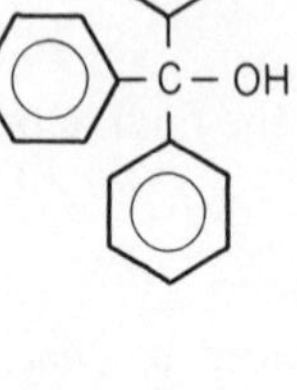

—C – OH

Isomerism

A single molecular formula can represent more than one actual compound. Such substances, having a common molecular formula, are termed *isomers*.

There are two fundamental types of isomerism: structural and stereoisomerism.

Structural isomerism

Structural isomerism is that type of isomerism in which different groups are present in the isomers:

Chain isomerism

Chain isomers differ only in the arrangement of the carbon atoms relative to each other; for example, the molecular formula C_4H_{10} represents the two compounds

$$CH_3CH_2CH_2CH_3 \quad \text{and} \quad \begin{matrix} CH_3 \\ > CH.CH_3 \\ CH_3 \end{matrix}$$

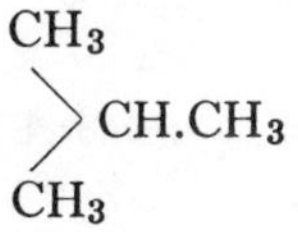

butane   2-methylpropane

Position isomerism

Position isomers possess the same carbon framework but differ in the relative positions of functional groups; for instance, there are three, and only three, straight-chain position isomers with molecular formula $C_6H_{13}Cl$:

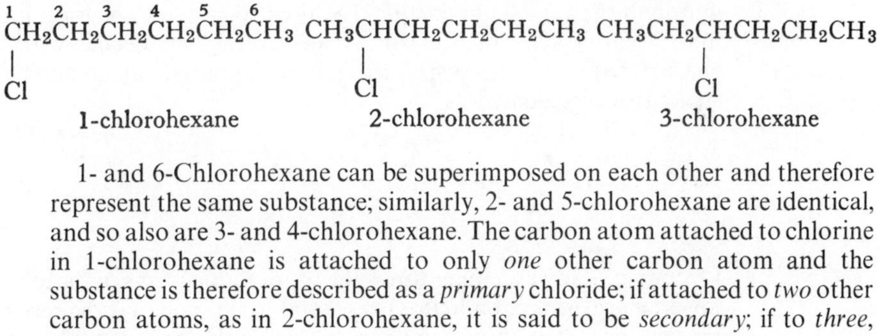

$$\overset{1}{C}H_2\overset{2}{C}H_2\overset{3}{C}H_2\overset{4}{C}H_2\overset{5}{C}H_2\overset{6}{C}H_3 \quad CH_3CHCH_2CH_2CH_2CH_3 \quad CH_3CH_2CHCH_2CH_2CH_3$$

$\quad |$ $\qquad\qquad\qquad\qquad\qquad\qquad |$ $\qquad\qquad\qquad\qquad\quad |$

Cl $\qquad\qquad\qquad\qquad\qquad\qquad$ Cl $\qquad\qquad\qquad\qquad\qquad$ Cl

$\quad$ 1-chlorohexane $\qquad\qquad$ 2-chlorohexane $\qquad\qquad$ 3-chlorohexane

1- and 6-Chlorohexane can be superimposed on each other and therefore represent the same substance; similarly, 2- and 5-chlorohexane are identical, and so also are 3- and 4-chlorohexane. The carbon atom attached to chlorine in 1-chlorohexane is attached to only *one* other carbon atom and the substance is therefore described as a *primary* chloride; if attached to *two* other carbon atoms, as in 2-chlorohexane, it is said to be *secondary*; if to *three*, *tertiary*.

Functional group isomerism

Here the isomers possess atoms arranged in the form of different functional groups; for example, the molecular formula C_2H_6O can be represented as

CH_3CH_2OH and as $\quad CH_3OCH_3$
$\quad$ ethanol $\qquad\qquad\quad$ methoxymethane

Metamerism

Isomers which differ only in the nature of the alkyl groups present are known as metamers, e.g., $CH_3OC_3H_7$ and $C_2H_5OC_2H_5$, both of which are ethers.

Stereoisomerism

Stereoisomers possess the same groups of atoms, but their relative positions in space are different.

Nuclear isomerism
Nuclear isomers are those compounds which differ only in the relative positions of substituent atoms or groups in cyclic compounds; e.g. there are three possible dichlorobenzene molecules

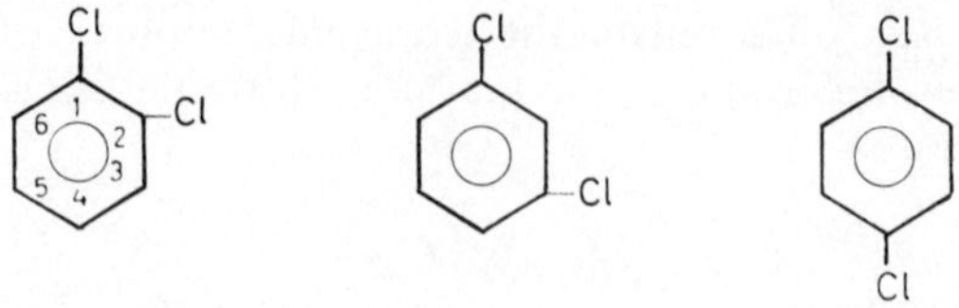

1,2-dichlorobenzene   1,3-dichlorobenzene   1,4-dichlorobenzene
     (ortho-)   (meta-)   (para-)

Geometric isomerism
The possibility of non-aromatic compounds differing only in the disposition of the parts of the molecule relative to each other arises from the directional character of the bonds from a carbon atom (p. 48). It is believed that 'free rotation' occurs about a single bond, and it may therefore be thought that compounds could exist in an infinite variety of ways, each having a slightly different spatial arrangement from the others. In a sense this is correct, although the very slight energies required to convert one form into another mean that a substance exists as an 'average' molecule with uniformly characteristic properties; i.e., there is no apparent difference between the structures represented as

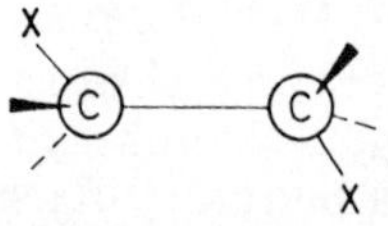

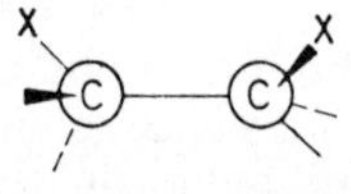

Of course, at lower temperatures the more stable forms of the molecule, with lower energies, predominate. These tend to be such that the repulsive forces present are at a minimum; for example, the more stable form of 1,2-dibromoethane will be (*a*) rather than (*b*), and it will become more predominant as the temperature is decreased

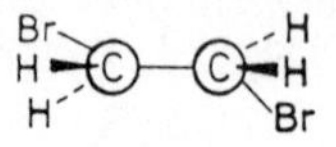

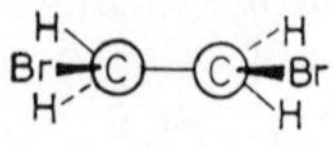

The formation of a double bond between carbon atoms prevents free rotation and enables so-called *geometric isomers* to exist, where different atoms or groups occupy permanently different positions about the double bond. An isomer with two identical groups in proximity, i.e., situated on the

same side of the double bond, is known as the *cis*-isomer and the other as the *trans*-isomer; for example, the 1,2-dichloroethenes

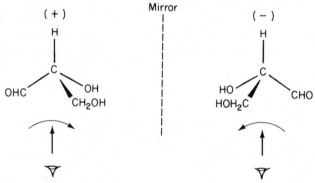

Optical isomerism

If four *different* atoms or groups are attached to a carbon atom, then, by virtue of their tetrahedral distribution, it is possible to represent the resultant structure in two ways, one of which is the mirror image or *enantiomorph* of the other, i.e., the mirror image cannot be superimposed onto the original structure, as shown in *Figure 18.1*. The two enantiomorphs differ only in their ability to rotate the plane of polarized light by equal amounts in opposite directions: that rotating the plane to the right is called dextro- (*d*- or +) and that to the left laevo- (*l*- or −). They are therefore known as optical isomers.

Figure 18.1 Enantiomorphs of 1,2-dihydroxypropanal

Notation of stereochemical forms

Before the absolute configurations were known it became necessary to relate the configurations of the carbohydrates to each other and of the amino acids to each other. In the case of the carbohydrates, those which had the arrangement related at the CH_2OH end (p. 514) to (+)1,2-dihydroxypropanal (glyceraldehyde) were given the prefix D, and those related to (−)1,2-dihydroxypropanal were given the prefix L. Amino-acids were similarly related to (−)serine, and all amino acids stereochemically related to it were (L)amino acids.

It is now known that, in fact, (+)1,2-dihydroxypropanal is as shown in *Figure 18.1*.

The absolute or true stereochemistry at a carbon atom carrying four different groups can be specified by the Cahn–Ingold–Prelog *sequence rule*. First the ligands are placed in an order of preference based on an arbitrary set of rules which in simplified form are as follows:

(1) Higher atomic numbers are preferred to lower
(2) If two atoms are the same, preference is based on the atomic numbers of the next atoms in each chain.

(3) For a double or triple bond, both atoms are duplicated or triplicated
Application of these rules leads to an order $OH > CHO > CH_2OH > H$ since
the atomic number order is $O > C > H$ and CHO counts as C then O, O, H
while CH_2OH counts as C then O, H, H.

Secondly, we now look at the tetrahedron along the line *from* the central
carbon atom *towards* the least preferred ligand, in this case H. If as we go
down the order of preference we see the other groups in a clockwise order, we
designate the configuration as R (*rectus*); if in anticlockwise order, S (*sinister*).

In the case of 1,2-dihydroxypropanal, the R form is the same as the D form,
and is also (+), but this does not always happen since R, D and (+) refer to
different things: the (+) or *d* relates to the sign of optical rotation of a
substance, the D gives a stereochemical relationship of a molecule to 1,2-
dihydroxypropanone, and R, in accordance with the sequence rule, gives the
absolute stereochemistry of a particular central atom.

N.B. Optical activity does not necessitate the presence of an asymmetric
carbon atom, i.e., one with four different groups attached; provided that there
is sufficient absence of symmetry in a molecule to enable two enantiomorphic
forms to exist, the substance will exist in two optically active forms. Such
substances are said to be *chiral* (cf. left and right hands as mirror images); an
achiral molecule has a mirror image identical to the original in all respects.

Organic reactions

Organic reactions involve the breaking of chemical bonds, which in general
are largely covalent. A single covalent bond may break so that either (*a*) both
electrons of the bond are taken by the more electronegative constituent and
two ions result. This is known as *heterolytic fission*

$$A{\mid}B \longrightarrow A^- + B^-$$

(here and elsewhere, the curves arrow indicates the movement of the
electrons)

or (*b*) each constituent of the bond takes an electron by *homolytic fission* and
neutral free radicals are formed:

$$A{\mid}B \longrightarrow A\cdot + B\cdot$$

The kinetics of the reaction will be determined by the slowest stage (*see*
Chapter 6); if fission is the rate-determining step and involves only the
molecule itself, then the reaction is *unimolecular*. If, on the other hand, a
second reactant is involved in the breaking of the bond, the reaction is
bimolecular. The second reactant may be a negative ion or possess a lone pair
of electrons capable of donation; it will then seek a positive site and is called a
nucleophilic reagent. If it is a positive ion or is in any way deficient in
electrons, then it is an *electrophilic* reagent. *Table 18.4* lists some of the more
common reagents.

Table 18.4 Some common reagents

Nucleophilic	Electrophilic
OH^-	NO_2^+
Halide$^-$	H_3O^+
$:O<$	BF_3
$:N\twoheadleftarrow$	Carbocations (formerly carbonium ions), R^+
Carbanions, i.e., R^-	

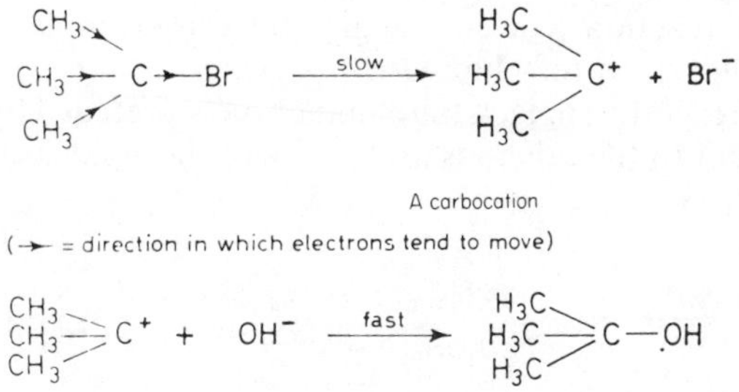

Figure 18.2 Mechanism of hydrolysis of 2-bromo-2-methylpropane (t-butyl bromide)

A carbocation

($\twoheadrightarrow$ = direction in which electrons tend to move)

It is as well to bear in mind that different types of mechanism can operate under different conditions: ionizing solvents, for instance, favour ionic mechanisms, liquids of low relative permittivity non-ionic mechanisms. Furthermore, the mechanism can change in type on ascending an homologous series or with increased branching of the carbon framework, owing to the electron-releasing tendency (inductive effect) of alkyl groups. Thus, bromomethane is hydrolysed chiefly by a bimolecular reaction involving the halide and hydroxide ions:

$$HO^- + CH_3 - Br \rightarrow [HO \cdots CH_3 \cdots Br]^- \rightarrow HO\!-\!CH_3 + Br^-$$
nucleophile

The inductive effects of methyl groups in 2-bromo-2-methylpropane are so marked, however, that the rate-determining step is ionization into a positively charged carbonium ion, which then instantly reacts with a hydroxide ion (*Figure 18.2*).

A complete stoichiometric equation often represents more than one reaction mechanism. In the course of this book, we shall give more or less detail of reactions as we think appropriate.

Reactions can often be classified as

(1) *Displacement* (or *substitution*) reactions:

$$X^- + AY \rightarrow XA + Y^-$$

X^- is said to diaplace Y^- from substance AY.

(2) *Elimination* reactions, where an ion or radical is removed from the system, e.g.,

$$AH^+ \rightarrow A + H^+$$

(3) *Rearrangement* reactions, which involve the migration of an ion from one atom to another in the same molecule:

$$A-B-X \rightarrow X-A-B$$

(4) *Addition* reactions, where unsaturated compounds become saturated (or less unsaturated) by addition of a reagent to a multiple bond

$$X^+ + A{=}B \rightarrow [X-A-B]^+$$

These reactions do not always give rise to a stable compound; in such cases, the product rapidly undergoes another reaction, and sometimes the overall reaction is given a single title. For example, 'condensation' is the name given to the elimination of water or another small molecule from two reactants; in fact, the elimination is preceded by the addition of one reactant to the other, as in the case of the condensation of a ketone with hydroxylamine:

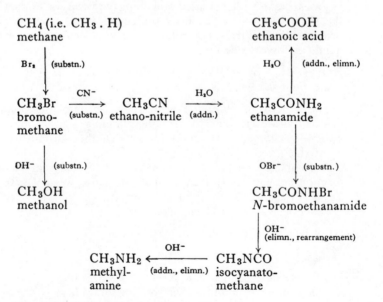

Figure 18.3 consists of a scheme of synthesis which illustrates some of the ways in which functional groups can be changed by the types of reactions listed.

Figure 18.3 Illustration of various reaction paths

Questions

(1) Name the following compounds:

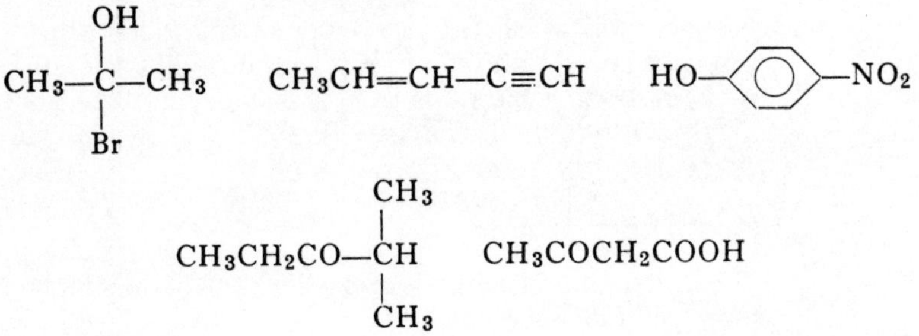

$$
\begin{array}{c}
\text{OH} \\
| \\
\text{CH}_3\text{--C--CH}_3 \\
| \\
\text{Br}
\end{array}
\qquad
\text{CH}_3\text{CH}\text{==}\text{CH}\text{--C}\text{≡}\text{CH}
\qquad
\text{HO---}\langle\bigcirc\rangle\text{---NO}_2
$$

$$
\begin{array}{c}
\text{CH}_3 \\
| \\
\text{CH}_3\text{CH}_2\text{CO--CH} \\
| \\
\text{CH}_3
\end{array}
\qquad
\text{CH}_3\text{COCH}_2\text{COOH}
$$

(2) Write down the formulae of 2-methylbutanal, buta-1.3-diene, 3-amino-benzenecarboxylic acid, cyclopentanone, 1-methoxypropan-2-ol.

(3) Draw and name all possible isomers of C_5H_{12} and of $C_5H_{11}X$.

(4) Draw the structures of all possible compounds called benzenetriol.

(5) Discuss the isomerism of all possible compounds of formula

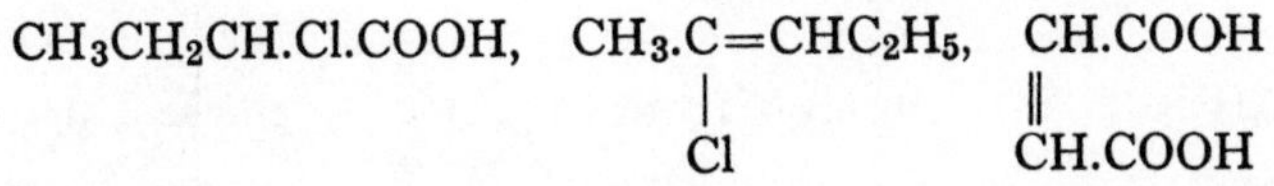

$$
\text{CH}_3\text{CH}_2\text{CH.Cl.COOH}, \quad
\begin{array}{c}
\text{CH}_3\text{.C}\text{==}\text{CHC}_2\text{H}_5 \\
| \\
\text{Cl}
\end{array}
\quad
\begin{array}{c}
\text{CH.COOH} \\
\| \\
\text{CH.COOH}
\end{array}
$$

(6) 'Organic chemistry is the chemistry of the functional group.' Discuss.

19 The hydrocarbons

Aliphatic and alicyclic hydrocarbons • Aromatic hydrocarbons • Occurrence • General methods of preparation • Methods of preparation of alkanes and/or aromatic hydrocarbons • Methods of preparation of alkenes and/or alkynes • Industrial sources • Reactions of hydrocarbons • Metallic derivatives of hydrocarbons • Analysis

The structures and methods of classification of the simpler hydrocarbons are shown in *Tables 19.1* and *19.2*.

Table 19.1 Aliphatic and alicyclic hydrocarbons

(a) *Aliphatic*

Saturated: Alkanes	*Unsaturated:* Alkenes	*Unsaturated:* Alkynes
General formula C_nH_{2n+2}	C_nH_{2n}	C_nH_{2n-2}
methane, CH_4	[methene radical $=CH_2$]	[methyne radical $\equiv CH$]
ethane, C_2H_6 or $CH_3.CH_3$	ethene, C_2H_4 or $CH_2{:}CH_2$	ethyne, C_2H_2 or $CH{\equiv}CH$
propane, C_3H_8 or $\quad CH_3.CH_2.CH_3$	propene, C_3H_6 or $\quad CH_3.CH{:}CH_2$	propyne, C_3H_4 or $\quad CH_3.C{:}CH$
butanes, C_4H_{10}	butenes, C_4H_8	butynes, C_4H_6
$\quad$ butane, $\quad CH_3.CH_2.CH_2.CH_3$	$\quad$ but-1-ene, $\quad CH_2{:}CH.CH_2.CH_3$	$\quad$ but-1-yne, $\quad CH_3.CH_2.C{:}CH$
$\quad$ 2-methylpropane $\quad CH_3.CH.CH_3$ $\qquad\qquad \mid$ $\qquad\qquad CH_3$	$\quad$ but-2-ene, $\quad CH_3.CH{:}CH.CH_3$ $\quad$ 2-methylpropene, $\quad CH_3.C{:}CH_2$ $\qquad\qquad \mid$ $\qquad\qquad CH_3$	$\quad$ but-2-yne, $\quad CH_3.C{:}C.CH_3$

(b) *Alicyclic*

General formula C_nH_{2n}	C_nH_{2n-2}	C_nH_{2n-4}
cyclopropane	cyclohexene	$C\equiv C$
$$\begin{array}{c} CH_2 \\ / \quad \backslash \\ H_2C-CH_2 \end{array}$$	$$\begin{array}{c} HC=CH \\ / \qquad \backslash \\ H_2C \qquad CH_2 \\ \backslash \qquad / \\ H_2C-CH_2 \end{array}$$	$$\begin{array}{c} \mid \quad \mid \\ (CH_2)_n \end{array}$$
cyclohexane $$\begin{array}{c} H_2C-CH_2 \\ / \qquad \backslash \\ H_2C \qquad CH_2 \\ \backslash \qquad / \\ H_2C-CH_2 \end{array}$$		

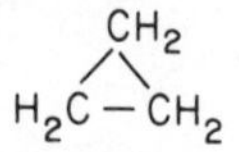

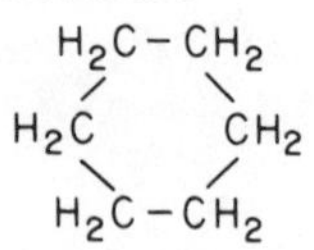

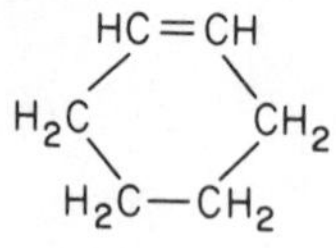

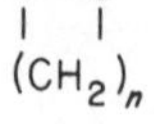

Table 19.2 Aromatic hydrocarbons

(1) *Benzene and its homologues*

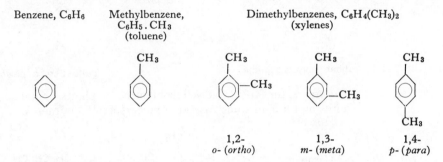

Benzene, C_6H_6 Methylbenzene, $C_6H_5 \cdot CH_3$ (toluene)   Dimethylbenzenes, $C_6H_4(CH_3)_2$ (xylenes)

1,2- o- (*ortho*)   1,3- m- (*meta*)   1,4- p- (*para*)

(2) *Polynuclear*

    (*a*) Rings isolated, e.g.

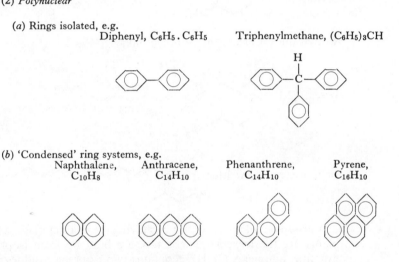

        Diphenyl, $C_6H_5 \cdot C_6H_5$ Triphenylmethane, $(C_6H_5)_3CH$

    (*b*) 'Condensed' ring systems, e.g.

      Naphthalene, $C_{10}H_8$ Anthracene, $C_{14}H_{10}$ Phenanthrene, $C_{14}H_{10}$ Pyrene, $C_{16}H_{10}$

Occurrence

Hydrocarbons are found in concentrated forms in fossilized fuels such as petroleum, peat and coal. A common and important hydrocarbon found in living tissue is 2-methylbuta-1,3-diene (*isoprene*), which is synthesized from acetic acid units present from respiratory and photosynthetic processes (p. 521) and which acts as the precursor of a wide variety of hydrocarbons and related compounds. Polymerization of isoprene into long fibres gives natural *rubber* or *gutta-percha*, depending on the geometric isomerism about the double bonds present in the polymer

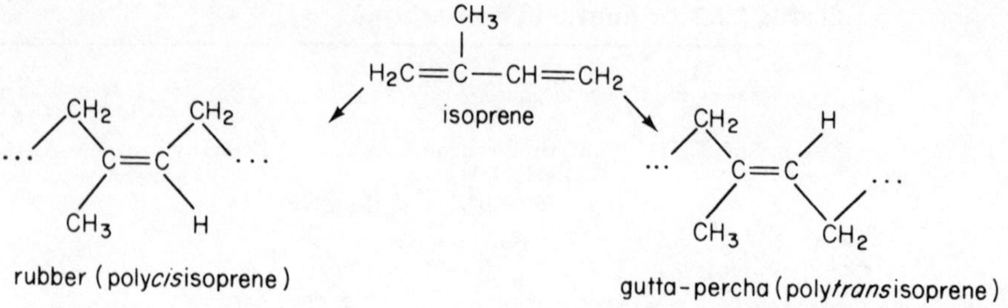

CH3
|
H2C=C—CH=CH2
isoprene

rubber (poly*cis*isoprene)   gutta-percha (poly*trans*isoprene)

Carotenes, the fat-soluble yellow pigments of plants, contain eight isoprene units and are, in some cases, the precursors of Vitamin A:

β-Carotene

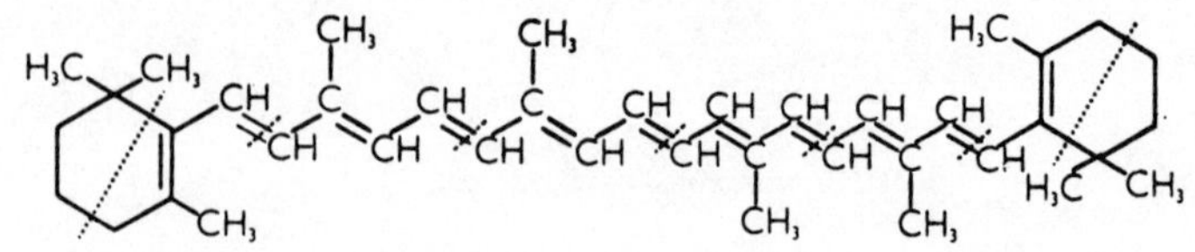

(dotted lines indicate the isoprene links)

Vitamin A

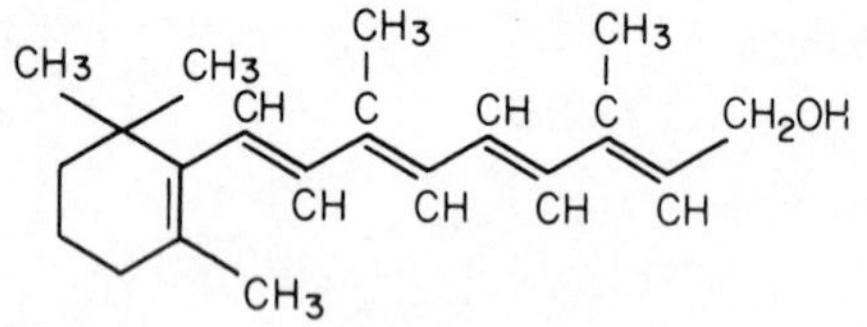

or more shortly written

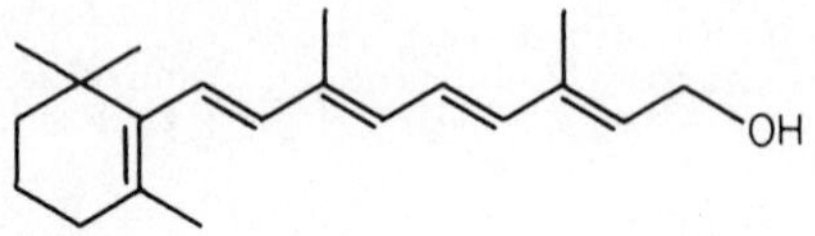

A group of volatile, fragrant oils giving the characteristic odours to many plants are the *terpenes*. These, too, are built up from isoprene units; for example, limonene, $C_{10}H_{16}$, contains two isoprene residues

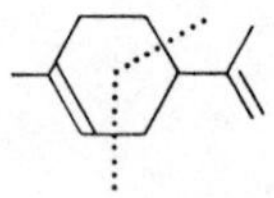

Squalene is a triterpene, i.e., a C_{30} compound, built up from six isoprene units and found particularly in shark liver oil

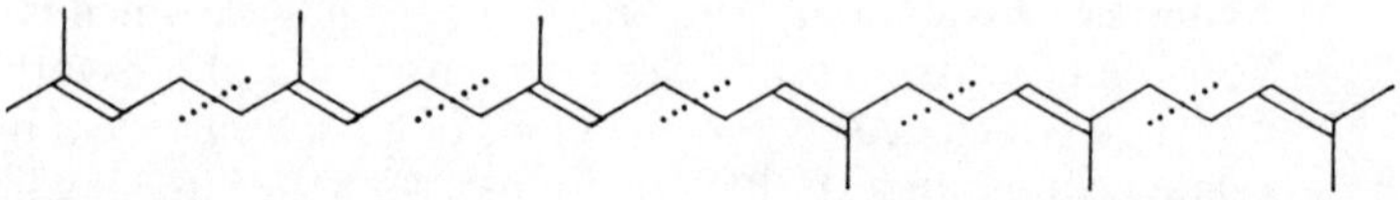

It is of some interest because it is known to be the precursor of cholesterol (p. 476) and thus effects a link between isoprene and the *steroids*, compounds which contain the framework

and which comprise, among other things, bile acids and sex hormones.

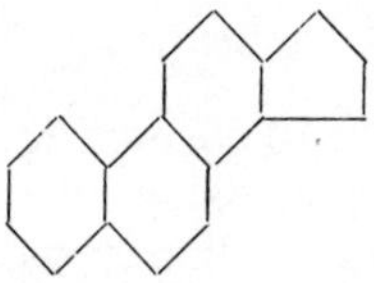

General methods of preparation

Hydrogenation

Unsaturated hydrocarbons can be converted into less unsaturated, or fully saturated, compounds by reaction with hydrogen in the presence of a catalyst such as finely divided nickel at $300\,°C$. More extreme conditions are necessary for the hydrogenation of aromatic hydrocarbons because of the stability conferred by the non-localized π orbitals (p. 56). However, once this is overcome, the system becomes very reactive and no partially unsaturated compounds can be isolated by this method. In general

$$C_nH_{2n-2} \xrightarrow[\text{Ni}]{H_2} C_nH_{2n} \xrightarrow[\text{Ni}]{H_2} C_nH_{2n+2}$$

alkyne   alkene   alkane

e.g.,

$$HC \equiv CH \rightarrow H_2C = CH_2 \rightarrow H_3C.CH_3$$

ethyne   ethene   ethane

and

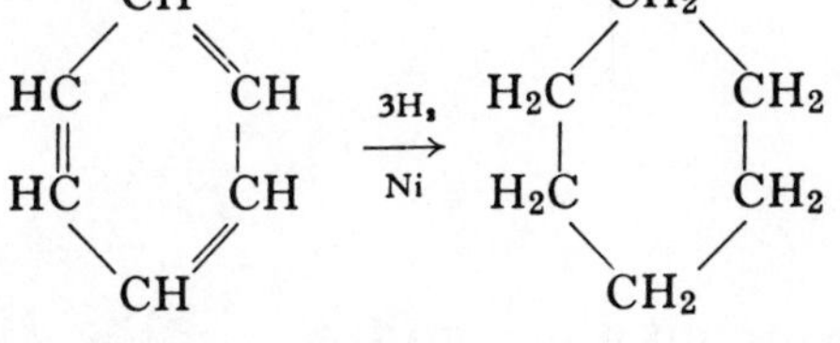

or in abbreviated form

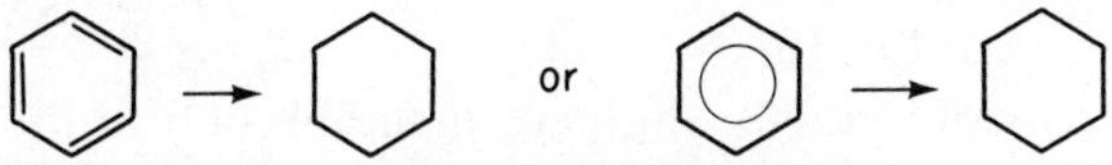

A second method of hydrogenation is by the reduction of halogen derivatives of the hydrocarbons, using reducing agents such as sodium in ethanol:

$$C_2H_5OH + Na \rightarrow C_2H_5O^- Na^+ + [H]$$

$$RI + 2[H] \rightarrow RH + HI$$

The Kolbé process

Electrolysis of the sodium or potassium salts of carboxylic acids in strong aqueous solution produces hydrocarbons possessing an even number of carbon atoms. Monocarboxylic acids give rise to alkanes, dicarboxylic acids to alkenes and unsaturated dicarboxylic acids to alkynes. The reactions at the anode can be summarized by the equations

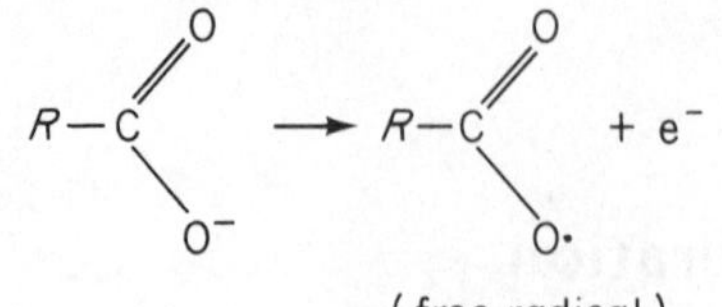

(free radical)

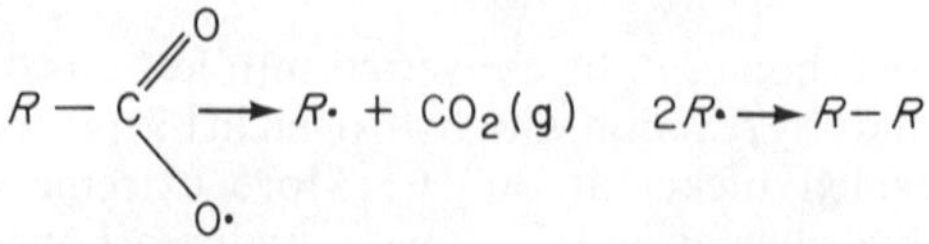

e.g.,

$$2CH_3COO^- \rightarrow C_2H_6(g) + 2CO_2(g) + 2e^-$$
ethanoate   ethane
(acetate)

$$\begin{array}{l} CH_2COO^- \\ | \\ CH_2COO^- \end{array} \longrightarrow \begin{array}{l} CH_2 \\ \| \quad (g) + 2CO_2(g) + 2e^- \\ CH_2 \end{array}$$
butanedioate   ethene
(succinate)

$$\begin{array}{l} CHCOO^- \\ \| \\ CHCOO^- \end{array} \longrightarrow \begin{array}{l} CH \\ \| \| \quad (g) + 2CO_2(g) + 2e^- \\ CH \end{array}$$
cis- or trans-   ethyne
butenedioate
(maleate or fumarate)

Methods for preparation of alkanes and/or aromatic hydrocarbons

There are several methods which are applicable only to the preparation of alkanes and aromatic hydrocarbons, of which the following may be mentioned.

Decarboxylation

Carboxylic acids and their sodium salts, when strongly heated with alkali, eliminate a molecule of carbon dioxide. Poor yields are obtained if the hydrocarbon is unstable, and so the method is only of some value for the preparation of saturated and aromatic hydrocarbons. The alkali normally used is soda lime (calcium oxide slaked with sodium hydroxide), but the aromatic acids will react when heated with calcium oxide alone:

$$RCOO^- + OH^- \rightarrow CO_3^{2-} + RH$$

By this means sodium ethanoate (acetate) will yield methane

$$CH_3COO^- + OH^- \rightarrow CO_3^{2-} + CH_4(g)$$

and benzene can be obtained from sodium benzenecarboxylate (benzoate):

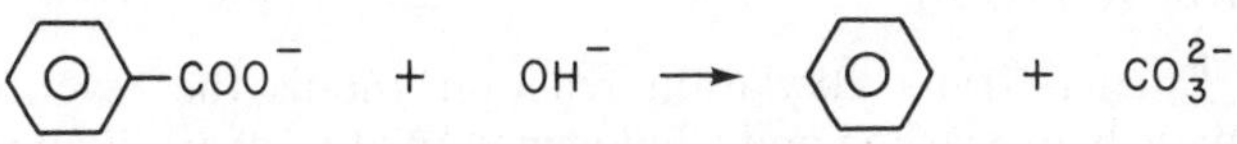

The Wurtz reaction

This is a seldom-used method of obtaining the higher homologues of the alkane series, and depends upon the reaction of sodium with halogen derivatives of hydrocarbons, in ethereal solution where necessary. The reaction mechanism may involve free-radical formation, especially in the vapour phase or in liquids of low relative permittivity:

$$RX + Na \rightarrow R\cdot + Na^+ X^-$$

$$2R\cdot \rightarrow R\text{---}R$$

Alternatively, in more polar solutions the mechanism may involve

$$RX + 2Na \rightarrow R^- Na^+ + Na^+ X^-$$

followed by

$$R^- Na^+ + RX \rightarrow Na^+ X^- + R\text{---}R$$

The Wurtz reaction is satisfactory only for the production of hydrocarbons having an even number of carbon atoms, e.g.,

$$CH_3Br + 2Na + CH_3Br \rightarrow C_2H_6(g) + 2Na^+Br^-$$

and

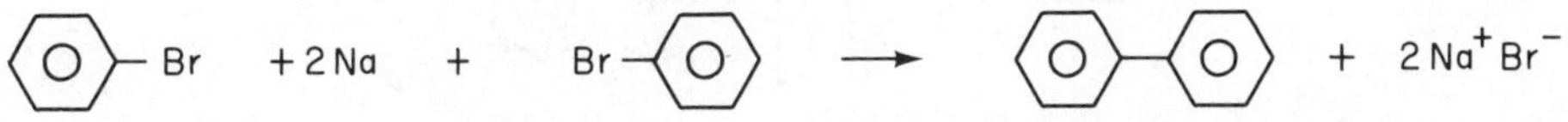

Biphenyl

The use of mixed halides, e.g. $RX + R'X$, results in the formation of $R\text{---}R$, $R\text{---}R'$ and $R'\text{---}R'$ which in general are not easy to separate, as there is little difference in their boiling points.

The reaction is more useful as a method of preparing the alkyl derivatives of aromatic hydrocarbons—a modification due to Fittig—e.g.,

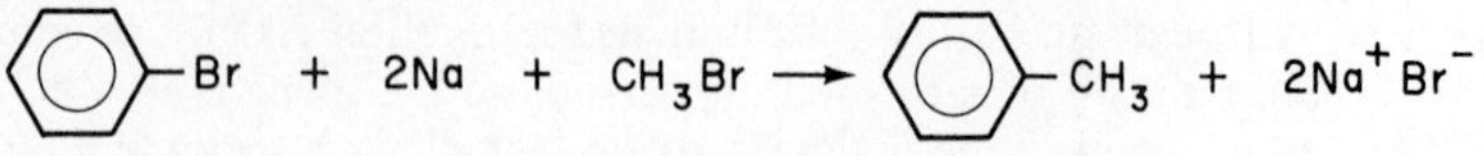

$$\text{Bromobenzene} - Br + 2Na + CH_3Br \longrightarrow \text{Methylbenzene} - CH_3 + 2Na^+Br^-$$

Bromobenzene   **Methylbenzene**
                                                              (toluene)

Although the products obtained in such Wurtz–Fittig reactions may be mixtures, separation is easily effected because of their wide range of boiling points.

The Friedel–Crafts reaction

The Friedel–Crafts alkylation reaction entails the reaction between an aromatic hydrocarbon and a halogen derivative of an alkane in the presence of a catalyst which is normally anhydrous aluminium chloride, although in certain cases boron trifluoride or hydrogen fluoride can be used. It is essential for all the reactants to be free from moisture:

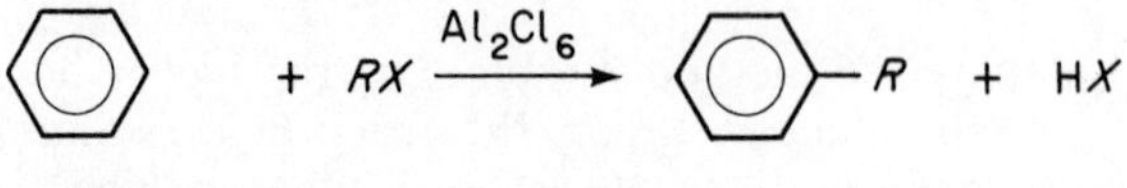

$$\bigcirc \quad + \quad RX \xrightarrow{\text{Al}_2\text{Cl}_6} \bigcirc -R \quad + \quad HX$$

The non-localized π orbitals present in a molecule of benzene (p. 56) will tend to become localized on any part of the molecule which is in proximity to a positive charge. This induced 'polarization' can be represented:

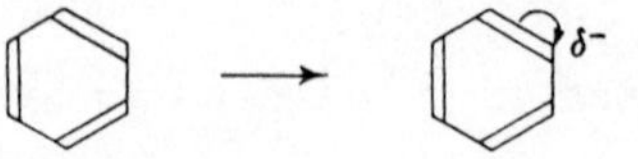

The mechanism of the Friedel–Crafts reaction can then be described in terms of this effect as follows. First, the halogen compound reacts with the anhydrous aluminium chloride:

$$2RX + \text{Al}_2\text{Cl}_6 \rightarrow 2R^+[\text{AlCl}_3X]^-$$

then the ion R^+ attacks the benzene ring

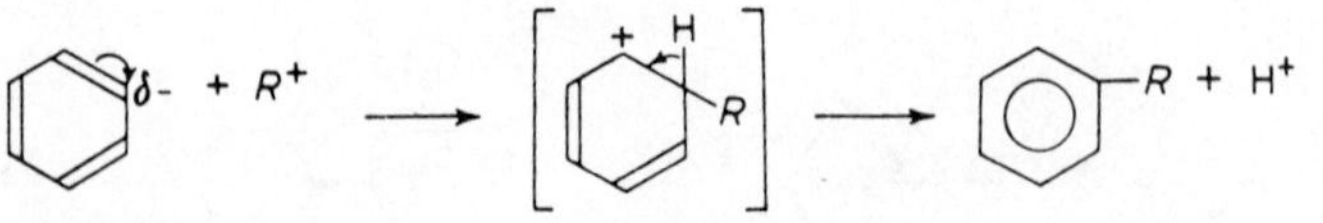

$$\bigcirc^{\delta-} + R^+ \longrightarrow \left[\bigcirc \overset{+}{\underset{R}{\overset{H}{|}}} \right] \longrightarrow \bigcirc -R + H^+$$

                                         Unstable
                                         intermediate

Methylbenzene and ethylbenzene can be prepared by this method but not propylbenzene because the organic ion isomerizes:

$$2CH_3CH_2CH_2X + \text{Al}_2\text{Cl}_6 \rightarrow 2CH_3CH_2\overset{+}{C}H_2[\text{AlCl}_3X]^-$$
$$CH_3CH_2\overset{+}{C}H_2 \rightarrow CH_3\overset{+}{C}HCH_3$$

$$\text{benzene} \quad + \quad CH_3\overset{+}{C}HCH_3 \quad \longrightarrow \quad \text{phenyl}-\overset{\displaystyle CH_3}{\underset{\displaystyle CH_3}{\overset{|}{\underset{|}{C}}}}-H \quad + \quad H^+$$

Similar rearrangements occur with all larger carbon chains.

A further difficulty is that the reaction cannot always be stopped at the monosubstituted stage; e.g., the reaction of benzene with bromomethane yields some 1,2- and 1,4-dimethylbenzenes as well as methylbenzene.

Alcohols and alkenes can also be made to react with aromatic hydrocarbons

$$\text{benzene} \quad + \quad CH_3CH_2CH_2OH \quad \xrightarrow[(-H_2O)]{BF_3} \quad \text{phenyl}-\overset{\displaystyle CH_3}{\underset{\displaystyle CH_3}{\overset{|}{\underset{|}{C}}}}-H$$

propan-1-ol

$$\text{benzene} \quad + \quad CH_3CH=CH_2 \quad \xrightarrow{HF} \quad \text{phenyl}-\overset{\displaystyle CH_3}{\underset{\displaystyle CH_3}{\overset{|}{\underset{|}{C}}}}-H$$

propene   cumene
(2-phenylpropane)

Methods for preparation of alkenes and/or alkynes

Besides the methods described so far, there are others suitable only for unsaturated hydrocarbons.

Dehydrohalogenation

The elements of a hydrogen halide can often be removed from a halogen derivative by refluxing with an alcoholic solution of potassium hydroxide. Of the two possible reactions, hydrolysis ($-X \to -OH$) and dehydrohalogenation, the latter nearly always predominates under these conditions by a first or second order reaction, depending on the concentration of the alkali:

a) 1st order. Rate $\propto$ [halide]

$$-\overset{\displaystyle H}{\underset{\displaystyle H}{\overset{|}{\underset{|}{C}}}}-\overset{\displaystyle H}{\underset{\displaystyle X}{\overset{|}{\underset{|}{C}}}}-H \quad \xrightarrow[-X^-]{Slow} \quad -\overset{\displaystyle H}{\underset{\displaystyle H}{\overset{|}{\underset{|}{C}}}}-\overset{\displaystyle H}{\underset{\displaystyle +}{\overset{|}{\underset{|}{C}}}}-H \quad \xrightarrow[-H^+]{Fast} \quad -\overset{\displaystyle H}{\overset{|}{C}}=\overset{\displaystyle H}{\overset{|}{C}}-H$$

b) 2nd order. Rate $\propto$ [halide] [OH⁻]

$$HO^- \longrightarrow H\overset{\displaystyle H}{\cdots}\overset{|}{\underset{\displaystyle H}{\underset{|}{C}}}-\overset{\displaystyle H}{\underset{\displaystyle X}{\overset{|}{\underset{|}{C}}}}-H \quad \longrightarrow \quad \overset{\displaystyle H}{\overset{|}{C}}=\overset{\displaystyle H}{\underset{\displaystyle H}{\overset{|}{\underset{|}{C}}}}-H \quad + \quad H_2O \quad + \quad X^-$$

The more concentrated the alkali is, the more does the second order reaction predominate.

However, if an attempt is made to dehydrohalogenate a halogenoethane, no more than 1 per cent of ethene is obtained; instead, the main product is ethanol (p. 478).

Dehydrogenation

This is effected by passing the vapour of the saturated hydrocarbon over a heated catalyst, e.g.,

$$CH_3CH_2CH_2CH_3 \xrightarrow[Cr_2O_3]{Al_2O_3} CH_2=CH-CH=CH_2 + 2H_2(g)$$
butane   buta-1,3-diene

Dehydration (for alkenes)

The elements of water can be removed from an alcohol, by heating with excess of either concentrated sulphuric or phosphoric(V) acid:

$$RCH_2CH_2\ddot{O}H \xrightarrow{H^+} RCH_2CH_2\overset{+}{\underset{H}{O}}H \xrightarrow[(-H_2O)]{} [RCH_2CH_2]^+ \xrightarrow[(-H^+)]{} RCH=CH_2$$

or by passing the vapour over heated aluminium oxide

$$RCH_2CH_2OH \xrightarrow{Al_2O_3} RCH=CH_2 + H_2O$$

The latter reaction proceeds because the C—O bond is weakened by adsorption on to the catalyst as shown in *Figure 19.1*.

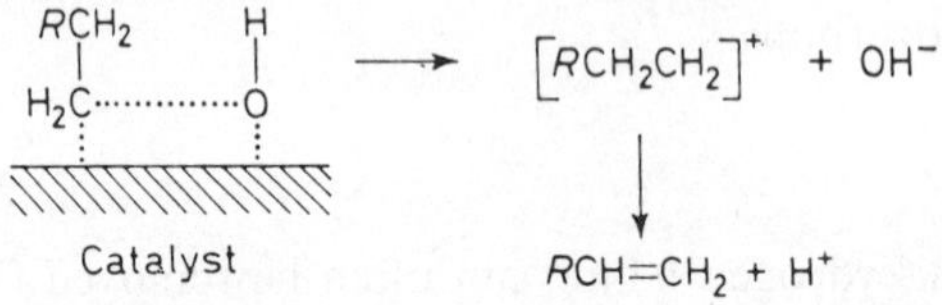

Figure 19.1 Catalytic dehydration of alcohols on alumina

Industrial sources of hydrocarbons

The two major commercial sources of hydrocarbons are coal and petroleum.

Coal

The destructive distillation of coal (p. 304) yields coal gas, coal tar and coke (*Figure 19.2*). The most volatile fraction of coal tar, benzol, is rich in benzene and methylbenzene, which can be either separated by fractional distillation or used together as a means of increasing the calorific and anti-knocking values of petrol. Another constituent of coal tar is the hydrocarbon, naphthalene (*see Table 19.2*) which readily crystallizes out.

The coke remaining in the retort can be heated with calcium oxide in an

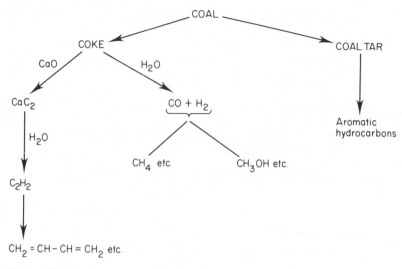

Figure 19.2 Coal as a source of hydrocarbons

Figure 19.3 An aerial view from the east of the Esso oil refinery and chemical manufacturing complex at Fawley. (An Esso Photograph)

electric furnace to give calcium carbide

$$CaO + 3C \rightarrow CaC_2 + CO(g)$$

Water reacts with this product to yield ethyne (acetylene)

$$C_2^{2-} + 2H_2O \rightarrow 2OH^- + C_2H_2(g)$$

which can be used as a fuel (oxyacetylene welding) or converted into other valuable hydrocarbons. For example, if it is passed into mercury(II) sulphate solution, it is hydrated to ethanal (acetaldehyde) which can then be converted into butadiene:

$$CH \equiv CH + H_2O \xrightarrow{HgSO_4} CH_3CHO \xrightarrow{NaOH} CH_3CH(OH)CH_2CHO$$

ethanal

$$\Bigg\downarrow \begin{array}{l} H_2,\ Cu/Cr \\ 30MPa\ 100\,°C \end{array}$$

$$CH_2 = CH—CH = CH_2 \xleftarrow[280\,°C]{Na_3PO_4} CH_3CH(OH)CH_2CH_2OH$$

butadiene

Butadiene, which is also manufactured by dehydrogenation of the C_4 fraction from petroleum 'cracking', is copolymerized with unsaturated compounds such as phenylethene (styrene) and propenonitrile (acrylonitrile) to give various types of synthetic rubber (p. 446).

Alternatively, steam can be passed over white-hot coke to form water gas which, on mixing with more hydrogen and passing, under pressure, over cobalt or nickel catalysts at elevated temperatures (the Fischer–Tropsch reaction), yields various hydrocarbons, e.g.,

$$2CO + 4H_2 \rightarrow 2H_2O + C_2H_4(g)$$

The value of coal as a source of hydrocarbons is summarized in *Figure 19.2* (*see also Figure 13.4*, p. 305).

Petroleum (*Figure 19.3*)

Unlike coal tar, petroleum is chiefly composed of aliphatic hydrocarbons, and primary distillation gives a separation of products mainly in terms of relative molecular mass. Thus hydrocarbons C_1 to C_4 remain in the gaseous condition at the minimum temperature of $30\,°C$ (*Figure 19.4*).

To obtain the maximum utility from petroleum it is necessary at present to break down larger hydrocarbons into smaller molecules, a process known as *cracking*. This is accomplished either by the sole application of heat (thermal cracking) or in conjunction with a catalyst (catalytic cracking), e.g., a mixture of silica and alumina

$$CH_3(CH_2)_8CH_3 \rightarrow CH_3(CH_2)_5CH_3 + CH_2CH = CH_2$$
decane   heptane   propene

When decomposition of this type takes place, one of the products at least is

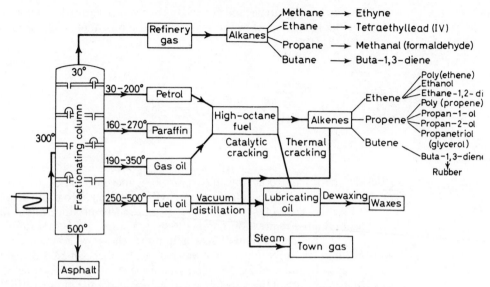

Figure 19.4 Simplified and generalized flow sheet of the petroleum industry (temperatures in °C)

unsaturated. Some of the unsaturated hydrocarbons can be used in the petrochemical industry, e.g.,

$$CH_2=CHCH_3 \xrightarrow[H_2O]{H_2SO_4} CH_3CHCH_3 \xrightarrow[300\,°C]{Cu} CH_3COCH_3(g)+H_2(g)$$

propene $\underset{\displaystyle OH}{|}$ propanone
                                                                                    (acetone)

$$CH_2=CH_2(g)+H_2O(g) \xrightarrow[MPa,\ 300\,°C]{H_3PO_4} C_2H_5OH$$

or *alkylated, polymerized* or *cyclized* (i.e. *re-formed*) to make hydrocarbons suitable for use in petrol. Cyclic compounds so formed can be *dehydrogenated* to give aromatic hydrocarbons; this *reforming* is carried out in the presence of a platinum catalyst and is usually called *platforming*:

Isomerization also can be used as a means of converting a less valuable into a more valuable substance; by this means, ethylcyclopentane is changed into methylbenzene (toluene)

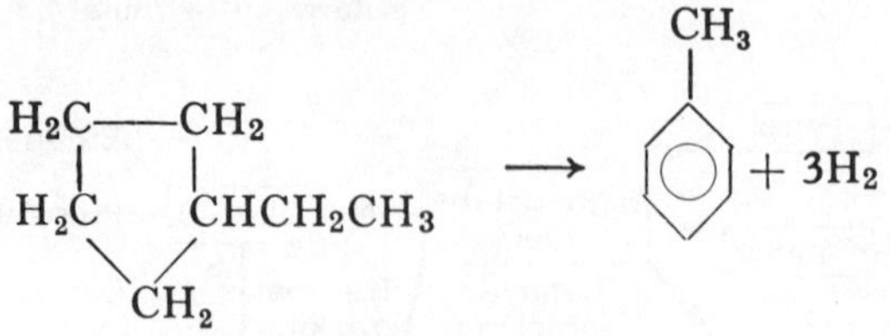

$$H_2C\!-\!CH_2$$
$$H_2C \quad CHCH_2CH_3 \quad \longrightarrow \quad \bigotimes^{CH_3} + 3H_2$$
$$\diagdown CH_2 \diagup$$

Besides petroleum cracking and reforming, methods based on (*a*) the carbonization and gasification of coal; (*b*) hydrogenation of coal and coal tar; and (*c*) Fischer–Tropsch synthesis, have been used to produce liquid fuels suitable for high-compression engines. The tendency of fuels to detonate prematurely ('knock' or 'pink') sets a limit to the power which can be obtained. It is measured in terms of the 'octane number' which is defined as the volume percentage of 2,2,4-trimethylpentane ('iso-octane', octane number 100) in a mixture of this substance with heptane (octane number 0) which equals the fuel in knock intensity when tested in a standard engine under specified conditions. Branched-chain and aromatic hydrocarbons have low knocking properties, whereas straight-chain hydrocarbons readily produce knocking—hence the value of the above processes.

It is often necessary to remove various objectionable constituents at different stages in petroleum refining. Many of these are sulphur-containing compounds which can be removed by treatment with sodium hydroxide:

$$RSH + NaOH \rightarrow RSNa + H_2O$$

or by catalytic hydrogenation, when the hydrogen sulphide formed is burnt to sulphur

$$RSH + H_2 \rightarrow RH + H_2S$$

$$2H_2S + O_2 \rightarrow 2H_2O + 2S$$

The importance of petroleum to the chemical industry is indicated in *Table 19.3*.

Table 19.3 Approximate percentage of substances derived from petroleum, U.S.A.

Substance	Percentage	Substance	Percentage
Propan-2-ol	100	Ethanol	83
Ethene	100	Methylbenzene	78
Methanol	99	Benzene	50
Propanone	98	Propane-1,2,3-triol	42

Reactions of the hydrocarbons

Hydrocarbons are generally of zero or low dipole moment (p. 49); there is therefore little or no interaction between similar molecules. They are consequently of relatively high volatility [although this decreases with increasing relative molecular mass (*Table 19.4*)] and insoluble in polar liquids such as water. The relative densities of the liquid hydrocarbons are less than that of water.

It can be seen from *Table 19.4* that there is a fairly steady increase in boiling point on ascending an homologous series; but the boiling points of isomers differ (*Table 19.5*).

Table 19.4 Volatility of hydrocarbons

Alkanes	B.p./°C	Alkenes	B.p./°C	Alkynes	B.p./°C
CH_4	−160				
C_2H_6	−88	C_2H_4	−104	C_2H_2	−84
C_3H_8	−44	C_3H_6	−48	C_3H_4	−23
C_4H_{10}	−1	$CH_3CH{:}CHCH_3$	4	$CH_3C{\equiv}CCH_3$	27
C_5H_{12}	36	$C_2H_5CH{:}CHCH_3$	37	$C_2H_5C{\equiv}CCH_3$	56
C_6H_{14}	69	$C_2H_5CH{:}CHC_2H_5$	73	$C_3H_7C{\equiv}CCH_3$	84
cyclohexane, C_6H_{12}	81	cyclohexene, C_6H_{10}	83		

Table 19.5 Boiling points of isomeric hydrocarbons

C_5H_{12}	$CH_3(CH_2)_3CH_3$ pentane	$CH_3CH_2CH(CH_3)_2$ 2-methylbutane	$C(CH_3)_4$ 2,2-dimethylpropane
B.p./°C	36	28	9

C_6H_6	$CH_3C{\equiv}CC{\equiv}CCH_3$ hexa-2,4-diyne	benzene
B.p./°C	85	80

Hydrocarbons, when above their respective ignition points, burn in air. If combustion is complete, the flame is non-luminous and the products are carbon dioxide and water. As the molecular complexity increases, progressively more oxygen is required and there is a greater probability of the combustion being incomplete and of resultant free carbon rendering the flame luminous through incandescence. Thus, in the case of methane and naphthalene, for complete combustion:

$$CH_4 + 2O_2 \rightarrow 2H_2O(g) + CO_2(g)$$

$$C_{10}H_8 + 12O_2 \rightarrow 4H_2O(g) + 10CO_2(g)$$

so that, at the same temperature and pressure, six times as much oxygen is required for the complete combustion of a certain volume of naphthalene vapour as for the same volume of methane.

The electron density of the unsaturated hydrocarbons is high (see diagrams on pp. 54, 55), and therefore they are most readily attacked by electrophilic reagents, e.g., reagents having groups with a partial or complete positive charge. In the case of alkenes and alkynes, the result is usually addition, whereas the aromatic systems are generally so stable that addition of the electrophilic reagent is followed by elimination, leaving the delocalized orbitals intact, so that the reaction is essentially one of substitution and the hydrocarbon is reacting as a saturated compound. However, aromatic hydrocarbons may undergo addition reactions in certain cases, behaving then as unsaturated compounds (see Chapter 3 for electron cloud distribution in hydrocarbons).

Addition reactions of unsaturated hydrocarbons

Hydrogenation
This has already been described (p. 411).

Halogenation
This involves the addition of a halogen across a multiple bond. As halogens are symmetrical molecules, there can only be one possible product

$$\text{C}=\text{C} + \text{Br}-\text{Br} \rightarrow -\overset{|}{\underset{|}{\text{C}}}-\overset{|}{\underset{|}{\text{C}}}-$$
$$\qquad\qquad\qquad\quad \text{Br}\ \text{Br}$$

Although the addition is shown as a broadside attack, this is less likely to represent the mechanism than an end-on attack. In the latter case, the induced polarization produced as the two reacting molecules approach one another, allows the transition state to be reached with a lower activation energy than that required for a broadside attack

$$\text{C}=\text{C} + \text{Br}-\text{Br} \rightarrow \left[\text{C}-\text{C}\cdots\text{Br}\cdots\text{Br}\right]$$

$$\text{Br}-\text{C}-\text{C}-\text{Br} \leftarrow \left[\text{C}-\text{C}-\text{Br}\right]^{+} + \text{Br}^{-}$$

Therefore the addition of the two atoms does not take place simultaneously, as is shown by the fact that bromination of ethene in the presence of chloride or nitrate ions produces some $BrCH_2CH_2Cl$ and $BrCH_2CH_2ONO_2$, respectively, e.g.,

$$\text{Br}-\text{C}-\text{C}^{+} + \text{Cl}^{-} \rightarrow \text{Br}-\text{C}-\text{C}-\text{Cl}$$

The addition of a halogen to an aromatic hydrocarbon takes place in the presence of sunlight or ultra-violet radiation. The light activates some of the halogen molecules and produces free radicals which then attack the aromatic ring, e.g.,

$$Cl_2 \xrightarrow[h\nu]{light} Cl\cdot + Cl\cdot$$

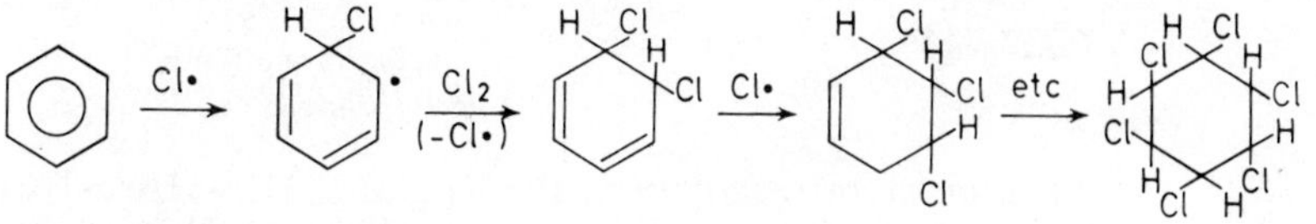

Hexachlorocyclohexane exists in eight isomeric forms, two of which are shown on the left below. One of the isomers is known as 'gammexane', an insecticide, thought to be as shown on the right.

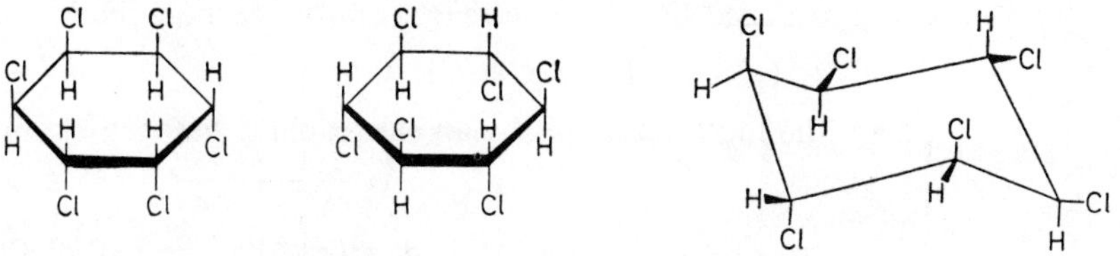

Hydrohalogenation

Addition of a hydrogen halide occurs by the initial attack of the electrophilic hydrogen ion on the more electronegative carbon atom to give a carbocation. Reaction between the halide ion and the carbocation completes the process, in accordance with Markownikoff's rule which states that the hydrogen atom of the hydrogen halide attaches to the carbon which already has the greater number of hydrogen atoms attached to it, e.g.,

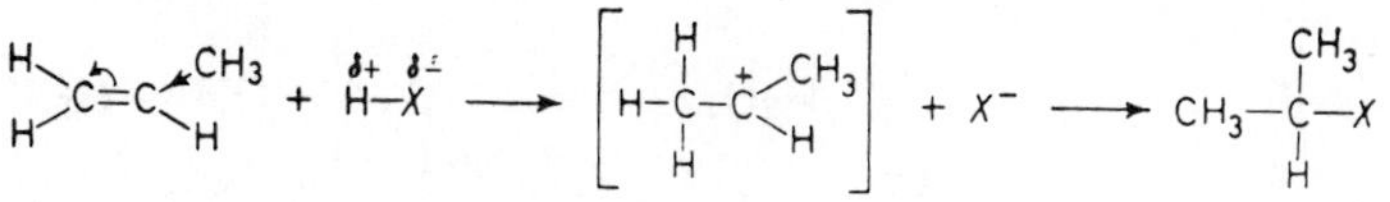

A carbocation

(The methyl group tends to be electron-releasing, so that in a suitable environment polarization occurs as shown.)

Two steps are possible in the hydrohalogenation of an alkyne, e.g.,

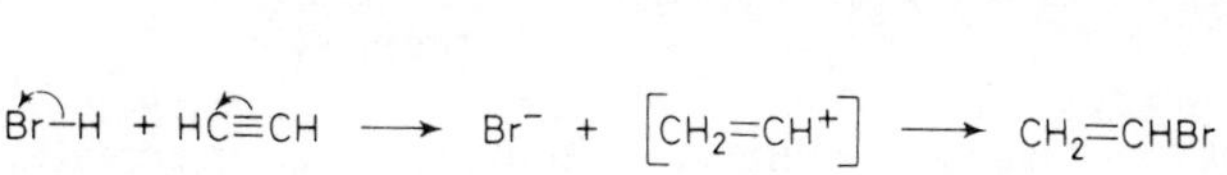

Bromoethene (vinyl bromide)

The bromethene can then react further, polarization of the molecule occurring by the donation of a lone pair of electrons on the bromine atom to give an extension of the π clouds from the unsaturated ethenyl group. Addition again takes place in accordance with Markownikoff's rule:

$$CH_2{-}CH{-}Br \longrightarrow {}^{\delta-}CH_2{-}CH{-}Br^{\delta+} \xrightarrow{\text{HBr}} CH_3CHBr_2$$

1,1-Dibromoethane
(ethylidene bromide)

The order of reactivity of the halides is $HI > HBr > HCl > HF$. This is the same order as that of the respective acid strengths and would be expected from a consideration of the bond energies of the four compounds (p. 217).

Hypohalogenation

Hypohalogen acids, usually chloric(I) (hypochlorous) acid, add across multiple bonds. This is a very useful reaction, as two highly reactive groups are introduced by the same reagent into a carbon chain:

$$HOCl + H_2C{=}CH_2 \rightarrow ClCH_2CH_2OH$$

Again the initial attack is by an electrophilic halogen cation, e.g.,

$$\overset{\delta-}{HO}{-}\overset{\delta+}{Cl} + \overset{n}{\underset{\delta- \; \delta+ \; H}{>C{=}C<^{R}}} \longrightarrow \left[HO^- + \begin{array}{c} Cl \; R \\ | \quad | \\ -C{-}C^+ \\ | \quad | \\ H \end{array} \right] \longrightarrow \begin{array}{c} Cl \; R \\ | \quad | \\ -C{-}C{-}OH \\ | \quad | \\ H \end{array}$$

Miscellaneous addition reactions

(a) Reaction between cold concentrated sulphuric acid and an alkene produces an alkyl hydrogensulphate:

$$>C{=}C< \xrightarrow{H^+} \left[H{-}\overset{|}{C}{-}\overset{|}{C^+} \right] \xrightarrow{HSO_4^-} H{-}\overset{|}{\underset{|}{C}}{-}\overset{|}{\underset{|}{C}}{-}OSO_2OH$$

This can be readily hydrolysed to an alcohol by dilution and distillation:

$$-\overset{|}{\underset{|}{C}}{-}\overset{|}{\underset{|}{C}}{-}OSO_2OH + H_2O \longrightarrow -\overset{|}{\underset{|}{C}}{-}\overset{|}{\underset{|}{C}}{-}OH + H_2SO_4$$

These reactions provide an important route from alkenes to alcohols, e.g.

$$\underset{\text{propene}}{CH_3CH{=}CH_2} \xrightarrow{H_2SO_4} \underset{\underset{\displaystyle OSO_2OH}{|}}{CH_3CH.CH_3} \xrightarrow{H_2O} \underset{\substack{| \\ OH \\ \text{propan-2-ol}}}{CH_3CH.CH_3} + H_2SO_4$$

(b) Alkynes react with dilute acids in the presence of a catalyst, the net result being the addition of the elements of water or some other simple molecule:

$$HC{\equiv}CH \xrightarrow[Hg^{2+}]{H_2SO_4} \left[H_2\overset{\delta-}{C}{=}C{\overset{H}{\underset{\delta+ \; O{-}H}{<}}} \right] \longrightarrow CH_3{-}C{\overset{H}{\underset{O}{<}}}$$

Ethenol   Ethanal
(vinyl alcohol)   (acetaldehyde)

(Ethenol cannot be isolated, as the hydrogen atom of the hydroxyl group is labile because of the polarization shown above.)

Similarly

$$HC \equiv CH \xrightarrow[\text{HCl}]{\text{HgCl}_2} H_2C = CHCl$$
chloroethene
(vinyl chloride)

$$HC \equiv CH \xrightarrow[\text{HCl}]{\text{Ba(CN)}_2} H_2C = CHCN$$
propenonitrile
(vinyl cyanide or acrylonitrile)

The vinyl compounds are important in the manufacture of plastics and paints (p. 446).

(c) Oxidation with alkaline manganate(VII) involves the addition of the elements of hydrogen peroxide, i.e. two hydroxyl groups, e.g.,

$$
\begin{array}{ccc}
CH_2 & & CH_2OH \\
\| & + \; 2OH \longrightarrow & | \\
CH_2 & & CH_2OH
\end{array}
$$
ethane-1,2-diol
(ethylene glycol)

This is a useful diagnostic test for unsaturation because of the appearance of green manganate(VI), but other compounds also reduce alkaline manganate(VII).

The use of ^{18}O in the manganate(VII) has indicated that both oxygen atoms come from this. The following mechanism has been suggested:

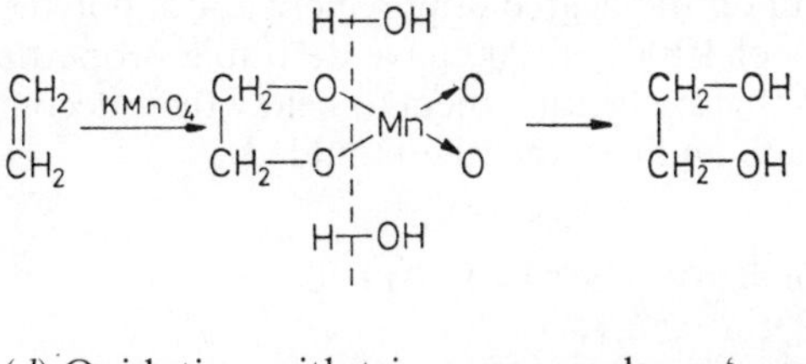

(d) Oxidation with trioxygen produces 'ozonides' which, on hydrolysis with water, yield aldehydes, ketones, carboxylic acids and hydrogen peroxide. The whole process is known as *ozonolysis*. It is useful for determining the position of multiple bonds between carbon atoms by identification of the products:

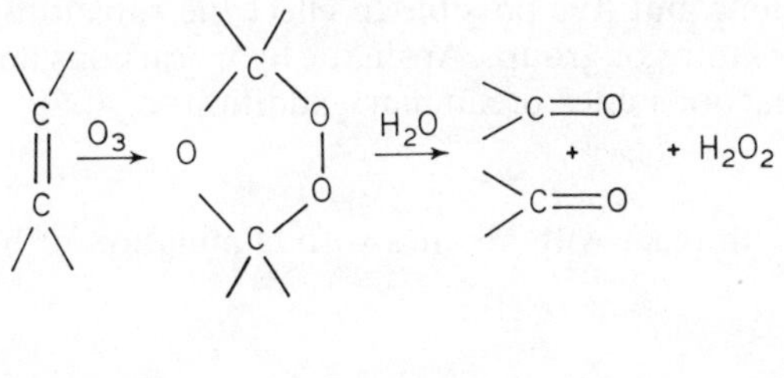

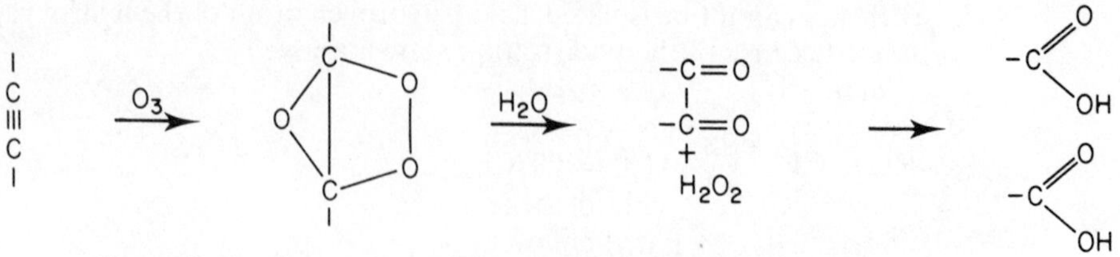

Ozonolysis also occurs with aromatic systems.

(e) The alkene link, when ruptured, gives a molecule with two reactive carbon atoms. Thus *polymerization* can occur, giving a long carbon chain system. Ethene is polymerized on a large scale (1) by using a high pressure and temperature, to give poly(ethene). Propene similarly gives poly(propene) (2) while phenylethene (styrene) forms poly(phenylethene), (polystyrene) (3).

$$2n(CH_2{=}CH_2) \longrightarrow (-CH_2-CH_2-CH_2-CH_2-)_n \qquad (1)$$

$$2n(CH{=}CH_2) \longrightarrow (-CH-CH_2-CH-CH_2-)_n \qquad (2)$$
$$\quad\quad\quad\;\; CH_3 \qquad\qquad\quad CH_3 \qquad CH_3$$

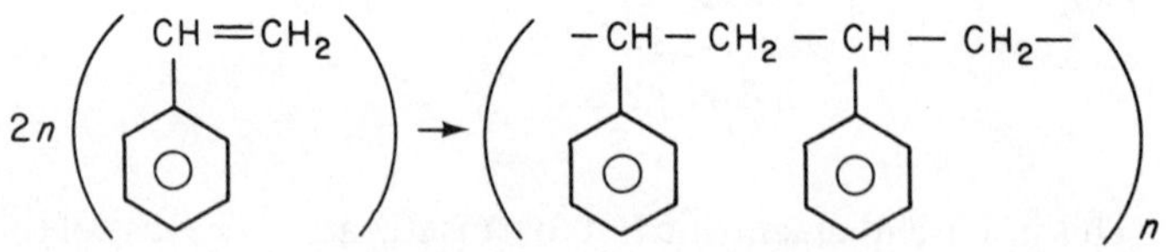

$$\qquad\qquad\qquad\qquad\qquad\qquad\qquad\qquad\qquad\qquad (3)$$

By a similar process, buta-1,3-diene yields synthetic rubber:

$$nCH_2{=}CH{-}CH{=}CH_2 \rightarrow (-CH_2-CH{=}CH-CH_2-)_n$$

The chains are terminated by stray free radicals. The properties of the product depend upon the degree of polymerization; polymers with relative molecular masses of 10 000–40 000 have desirable properties, e.g., they are thermoplastic, so that by heating them to below their decomposition points they become plastic and can then be moulded.

Substitution reactions of alkanes and aromatic hydrocarbons

Alkanes are much less reactive than unsaturated hydrocarbons, and the unsaturation of aromatic hydrocarbons is stabilized by the non-localized nature of the bonding. Because alkanes are saturated, they cannot undergo addition reactions, but it is possible to effect the substitution of hydrogen atoms by other atoms or groups. Aromatic hydrocarbons normally undergo substitution reactions after preliminary addition (p. 406).

Halogenation

The halogens can react with alkanes with elimination of hydrogen halide:

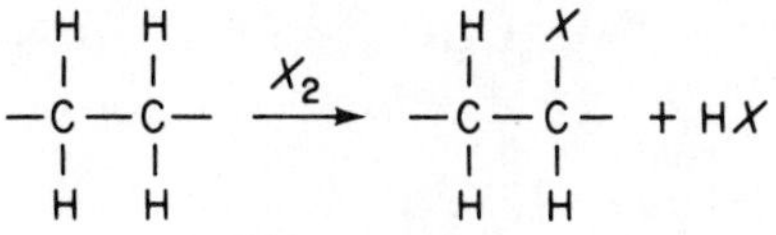

$$\underset{\substack{|\ \ \ | \\ H \ \ H}}{\overset{\substack{H \ \ H \\ |\ \ \ |}}{-C-C-}} \xrightarrow{X_2} \underset{\substack{|\ \ \ | \\ H \ \ H}}{\overset{\substack{H \ \ X \\ |\ \ \ |}}{-C-C-}} + HX$$

A free-radical mechanism operates, e.g.,

$$Cl_2 \xrightarrow[\substack{(h\nu)}]{light} 2Cl\cdot$$

$$CH_4 + Cl\cdot \rightarrow CH_3\cdot + HCl$$

$$CH_3\cdot + Cl_2 \rightarrow CH_3Cl + Cl\cdot, \text{ etc.}$$

The order of reactivity is $F_2 > Cl_2 > Br_2 > I_2$. Fluorine is so reactive that unless the reaction is moderated, tetrafluoromethane is the chief product, irrespective of the hydrocarbon used.

Substitution of hydrogen in the aromatic nucleus is usually effected by an electrophile. Consequently, a substituent group already present in the nucleus assists further substitution if it supplies electrons to the nucleus, and hinders it if it withdraws electrons. The *ortho* and *para*-positions are most vulnerable to this movement, and so *o*- and *p*-substitution predominates if the nucleus is activated. If the nucleus is deactivated, on the other hand, the effect is most marked in the *o*- and *p*-positions, so that substitution takes place very slowly and then chiefly in the *meta*-position. When aromatic hydrocarbons are halogenated at room temperature in the presence of a 'halogen carrier', such as anhydrous aluminium chloride, antimony(III) chloride or iron filings, substitution occurs in the ring, e.g.,

$$2Fe + 5Br_2 \rightarrow 2FeBr_4^- + 2Br^+$$

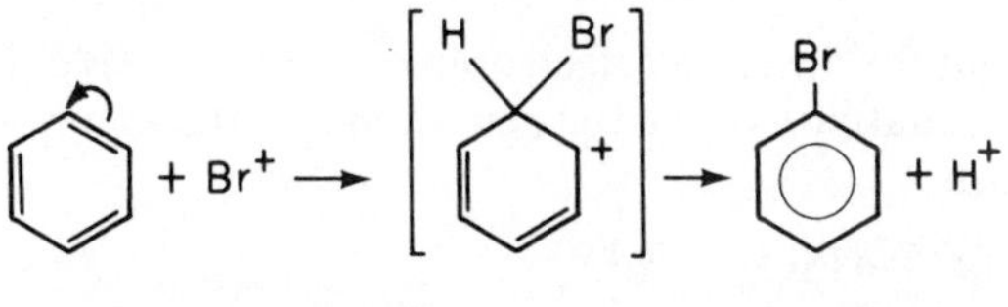

The bromine atom introduced into the benzene nucleus deactivates the ring to electrophilic attack (*see Table 19.6*) because of its high electronegativity which inhibits the sharing of its lone pair with the π-cloud of the benzene ring:

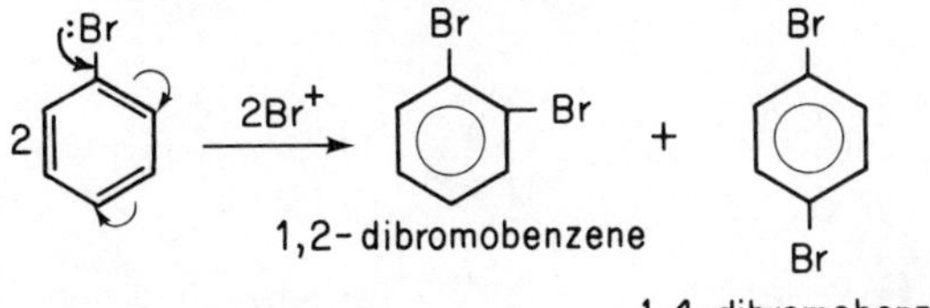

1,2- dibromobenzene

1,4-dibromobenzene

When the reaction is carried out in the absence of a catalyst and at the boiling point of the hydrocarbon, appreciable substitution occurs in any side chain which is present:

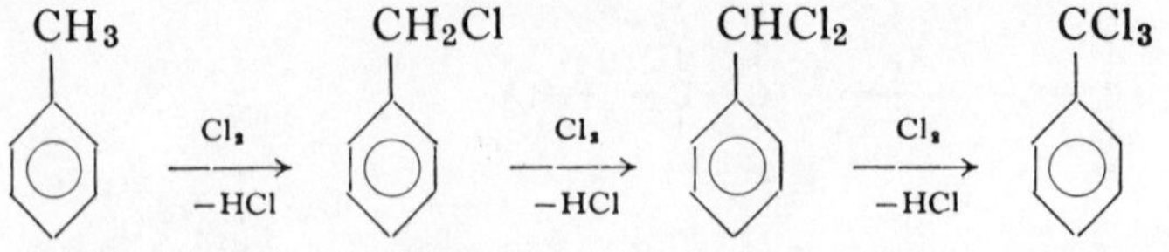
CH3
CH2Cl
CHCl2
CCl3
Cl2
−HCl
Cl2
−HCl
Cl2
−HCl

methylbenzene (toluene)	(chloromethyl)-benzene (benzyl chloride)	(dichloromethyl)-benzene (benzal chloride) (benzylidene chloride)	(trichloromethyl)-benzene (benzotrichloride)

The mechanism involves a chain reaction (see above).

Table 19.6 Directing effects in aromatic substitution

Meta-directing substituents	*ortho- and para-directing substituents*
$—NO_2$	$—CH_3$
—COOH	—Halogen
—CHO	$—NH_2$
—CN	—OH
$—SO_2OH$	

Sulphonation

Concentrated or fuming sulphuric acid fairly readily produces sulphonic acids with aromatic hydrocarbons but only with difficulty with alkanes, e.g.,

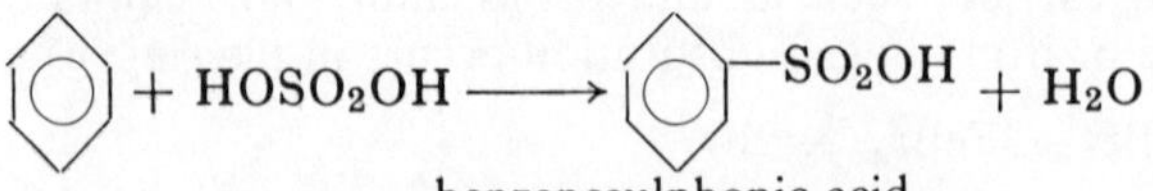
+ HOSO2OH
SO2OH
+ H2O

benzenesulphonic acid

It is thought that the attacking electrophile is sulphur(VI) oxide, in which the greater electronegativity of the oxygen atoms produces a positive charge on the sulphur atom:

$$2H_2SO_4 \rightleftharpoons H_3O^+ + HSO_4^- + SO_3$$

$$H_2S_2O_7 \rightleftharpoons H_2SO_4 + SO_3$$

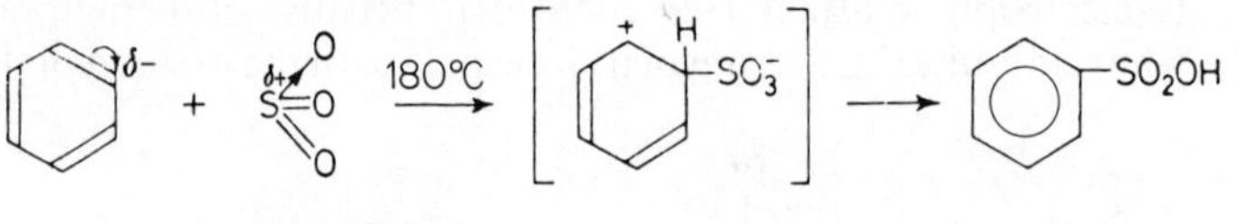
δ-
δ+
180°C
SO3−
SO2OH

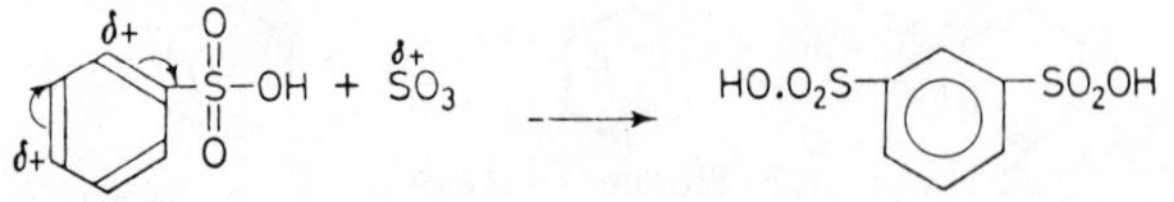
δ+
δ+
S-OH
SO3
HO.O2S
SO2OH

Benzene-1,3-disulphonic acid

Because the sulphonic acid group is deactivating, substitution virtually ends at this stage.

The sulphonic acids obtained from long-chain alkanes (derived from petroleum) are commercially important as detergents:

$$RCH_3 \xrightarrow{\text{H}_2\text{SO}_4/\text{SO}_3} RCH_2\overset{\delta-\ \delta+}{SO_2OH}$$

hydrophobic hydrophilic

Nitration

Nitrating acid consists of a mixture of concentrated or fuming nitric acid and concentrated sulphuric acid. The composition of the mixture has been studied by measurements of the freezing-point depression; the value of the van't Hoff factor (p. 118) is four, in accordance with the ionization:

$$2H_2SO_4 + HNO_3 \rightleftharpoons 2HSO_4^- + H_3O^+ + NO_2^+$$

nitryl (nitronium) cation

The electrophilic nitryl cation is the attacking species and, when aromatic hydrocarbons are treated with the above mixture, nitro derivatives are formed (with activated compounds, this can be a very hazardous procedure):

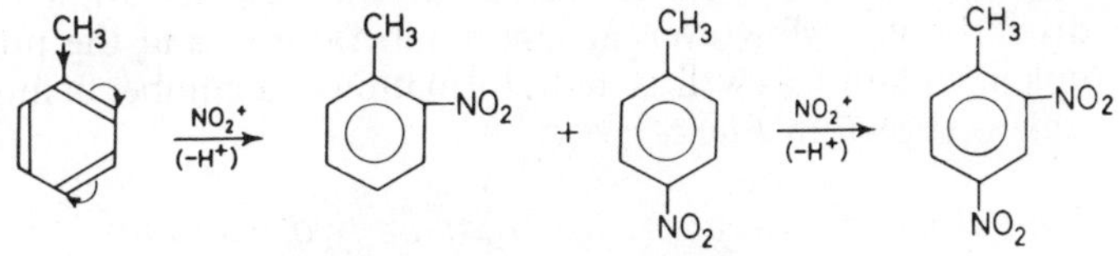

1-Methyl-2-
nitrobenzene

1-Methyl-4-
nitrobenzene

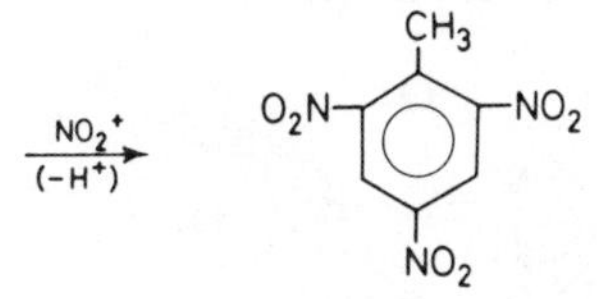

Methyl-2,4,6-
trinitrobenzene
(trinitrotoluene
or T.N.T.)

It will be noticed that the methyl group (or any other alkyl group) activates the *o*- and *p*-positions of the ring towards the electrophilic reagents, as it is an electron-repelling group.

The nitro group, by electron withdrawal, exerts a deactivating influence on the ring towards the usual electrophilic reagents, e.g.,

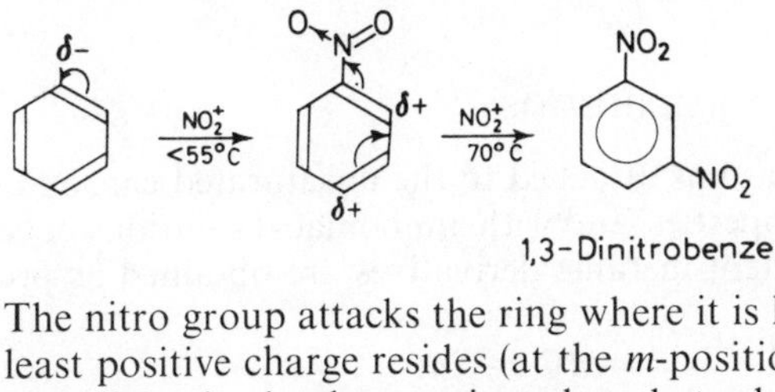

1,3-Dinitrobenzene

The nitro group attacks the ring where it is least deactivated, i.e. where the least positive charge resides (at the *m*-position). It should be noted that no part of the ring has been activated, so that substitution of more than one nitro

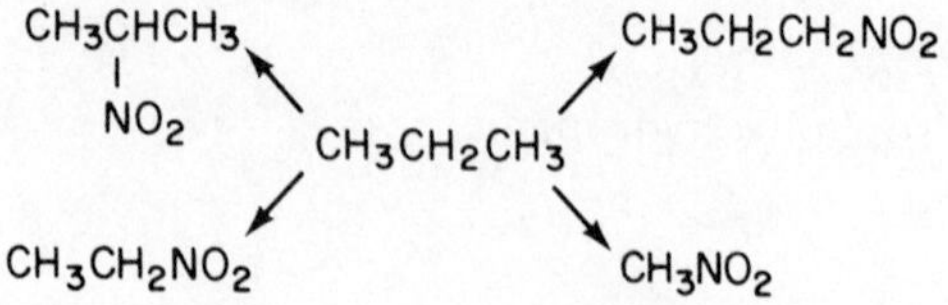

CH₃CHCH₃
|
NO₂   CH₃CH₂CH₃ → CH₃CH₂CH₂NO₂

CH₃CH₂NO₂   CH₃NO₂

The diagram shows: CH_3CHCH_3 (with NO_2), $CH_3CH_2NO_2$, $CH_3CH_2CH_3$ reacting to give $CH_3CH_2CH_2NO_2$ and CH_3NO_2

Figure 19.5 Vapour-phase nitration of propane with nitric acid at 400 °C

group into a benzene ring occurs only at higher temperatures.

In the absence of sulphuric acid, nitric acid normally oxidizes any side chain attached to the ring, producing the corresponding carboxylic acid, e.g.,

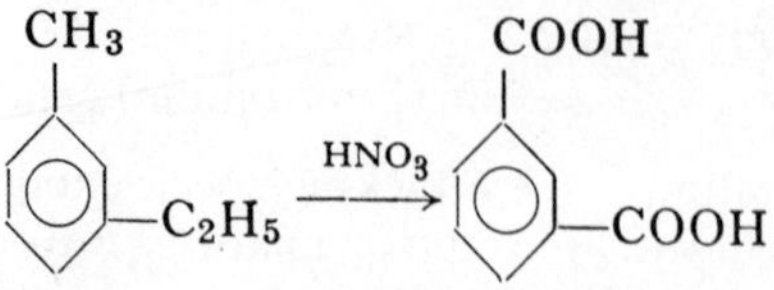

CH_3-C₆H₄-C_2H_5 →(HNO₃) $COOH$-C₆H₄-$COOH$

In contrast to aromatic hydrocarbons, the nitration of aliphatic hydrocarbons is slow even at high temperatures, and the process results in much oxidation as well as in the formation of a number of nitro-compounds, e.g., as shown in *Figure 19.5*.

Friedel–Crafts reaction with aromatic hydrocarbons

The alkylation reaction has already been discussed (p. 414). A useful modification is the acylation reaction, in which an acyl group (RCO—) is introduced into an aromatic ring

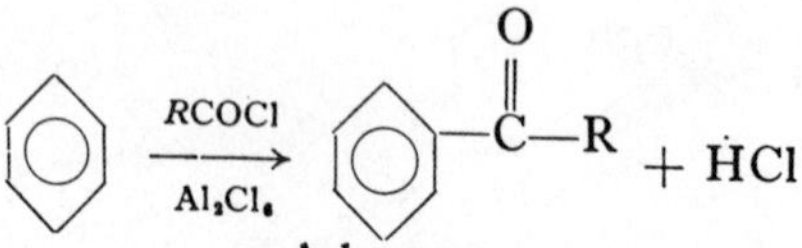

C₆H₆ →(RCOCl / Al₂Cl₆) C₆H₅—C(=O)—R + HCl

A ketone

This reaction is of value for the preparation of aromatic ketones and also, by subsequent reduction of the product, for the introduction of long straight-chain alkyl groups into an aromatic ring:

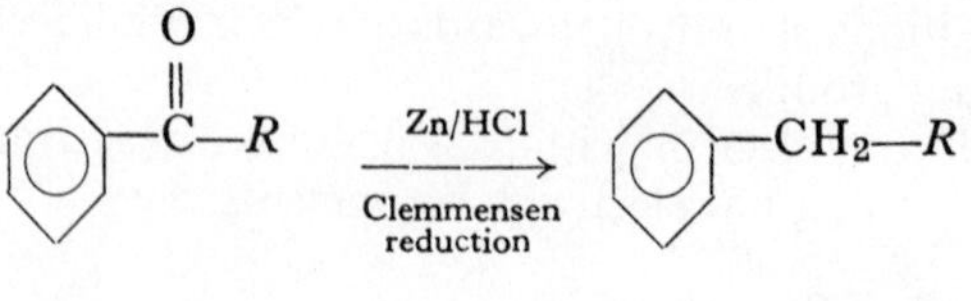

C₆H₅—C(=O)—R →(Zn/HCl, Clemmensen reduction) C₆H₅—CH₂—R

Metallic derivatives of hydrocarbons

The hydrogen atoms attached to the unsaturated carbon atoms in alkynes show acidic properties, and with ammoniacal solutions of copper(I) or silver chlorides, covalent metallic derivatives are obtained as precipitates

$$HC \equiv CH \xrightarrow{M(NH_3)_2{}^+Cl^-} MC \equiv CM$$

Unlike the ionic carbides (cf. calcium carbide), they detonate easily when dry (cf. covalent and ionic azides and nitrates). With metallic sodium dissolved in liquid ammonia, ethyne reacts to give $NaC \equiv CH$ and ultimately $NaC \equiv CNa$, compounds which are of great value in organic syntheses.

Several structurally interesting metallic derivatives of aromatic systems have been prepared. For example, when cyclopentadiene vapour is passed over heated iron(II) oxide, an orange solid (*ferrocene*) can be condensed

$$2C_5H_6 + FeO \rightarrow (C_5H_5)_2Fe + H_2O$$

By the loss of a proton from the methene group of cyclopentadiene, the resulting cyclopentadienyl anion has six *p* electrons and can exhibit aromatic properties. These electrons are believed to combine with the *d* electrons of the iron atom (*see* p. 363).

Analysis

Water and carbon dioxide are liberated when a hydrocarbon is heated with dry copper(II) oxide and can be identified qualitatively in the usual ways. This procedure is also the basis for the quantitative estimation; when a weighed amount of a hydrocarbon is oxidized in this way, the water produced can be absorbed in weighed calcium chloride tubes and the carbon dioxide in weighed potassium hydroxide bulbs

$$C_xH_y + (x + y/4)O_2 \rightarrow xCO_2 + y/2H_2O$$
$$12x + y \qquad\qquad 44x \qquad 9y$$

Example. A hydrocarbon of relative vapour density 29 gave 0.88 g of carbon dioxide and 0.45 g of water on oxidation. The masses of carbon and hydrogen in the compound are thus $0.88 \times 12/44$ g and $0.45 \times 1/9$ g, respectively. Therefore, the atom ratio of carbon and hydrogen is

$$\frac{0.88 \times 12}{12 \times 44} : \frac{0.45 \times 2}{1 \times 18} \qquad \text{i.e. } 2:5$$

so that the empirical formula is C_2H_5, and since the relative molecular mass is 29×2, the molecular formula must be C_4H_{10} (one of the two butanes).

The molecular formula can also be determined volumetrically by the process known as *eudiometry*. A known volume of the hydrocarbon vapour is exploded with a known volume of oxygen (in excess). Cooling results in the elimination of water from the vapour state, and treatment with potassium hydroxide absorbs the carbon dioxide. The final volume represents the volume of unused oxygen. Provided all volumes have been measured at the same temperature and pressure, the molecular formula of the hydrocarbon can be evaluated from the previous equation, i.e. one volume of hydrocarbon vapour required $(x + y/4)$ volumes of oxygen for complete combustion and yields x volumes of carbon dioxide under the same conditions of temperature and pressure.

Example. After exploding 5 cm³ of a hydrocarbon vapour with 50 cm³ of oxygen and cooling, the volume was 40 cm³, of which 15 cm³ were absorbed by potassium hydroxide solution, i.e.,

Volume of carbon dioxide produced $= 15$ cm^3

Volume of oxygen used $= [50 - (40 - 15)] = 25$ cm^3

Therefore

5 cm^3 C$_x$H$_y$ requires 25 cm^3 O$_2$ and produces 15 cm^3 CO$_2$

or

1 mole C$_x$H$_y$ requires 5 mole O$_2$ and produces 3 mole CO$_2$

i.e.,

$x = 3$ $x + y/4 = 5$ and $\therefore$ $y = 4(5 - 3) = 8$

Thus the molecular formula of the hydrocarbon is C$_3$H$_8$ (propane).

To confirm the structural formula of a simple organic compound, after identification of the functional groups present, solid derivatives are usually prepared and their melting points compared with the values of known compounds.

No suitable derivatives of aliphatic hydrocarbons can be prepared for the purpose of identification, either because of the unreactivity of the hydrocarbon or because a number of products result from which a solid substance cannot be readily isolated. Therefore, having classified the hydrocarbon as an alkane, alkene or alkyne, further identification is made by measuring its density and, if a liquid, refractive index. These values are then compared with those of known hydrocarbons.

Useful derivatives of aromatic hydrocarbons are the 'picrates', which are generally readily produced by mixing concentrated alcoholic solutions of 2,4,6-trinitrophenol (picric acid) (p. 480) and the hydrocarbon. 'Picrates' are molecular complexes of general formula (hydrocarbon).n(picric acid), where n is a small integer. Other derivatives of aromatic hydrocarbons suitable for structure determination are the nitro compounds and also the carboxylic acids produced by oxidation of side chains (p. 430).

Summary (*Figure 19.6*)

The characteristic reaction of saturated hydrocarbons where all the chemical bonds are of the single σ type, is substitution of hydrogen by another element or radical. This can equally be regarded as substitution of alkyl, as for example in the photochemical chlorination of methane:

Cl—Cl $\xrightarrow{\text{light}}$ Cl· + ·Cl   dissociation

R—H + ·Cl $\longrightarrow$ R· + H—Cl   substitution

R· + Cl—Cl $\longrightarrow$ R—Cl + ·Cl   substitution, etc.

In the case of unsaturated hydrocarbons, the bond is a source of attraction

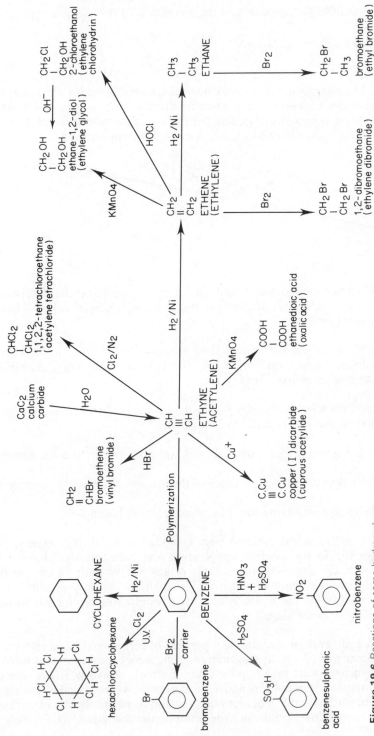

Figure 19.6 Reactions of some hydrocarbons

to electrophilic reagents and the reaction is of the addition type:

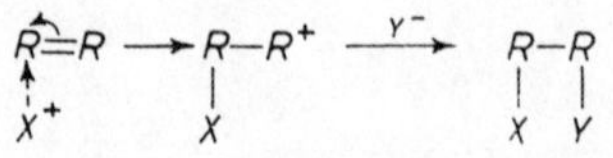

The stability of the aromatic nucleus is such that, whilst the initial attack is by an electrophilic reagent, instead of the reaction proceeding as an orthodox addition type with consequent destruction of the 'resonance' of the system, a proton is expelled and the non-localized system restored.

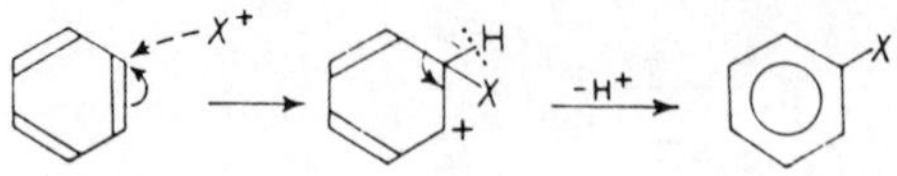

Questions

(1) What are the molecular formulae of rubber, gutta-percha, β-carotene and Vitamin A? What advantages might some synthetic rubbers have over natural rubber?

(2) Why is propan-2-ol, rather than propan-1-ol, obtained from propene via sulphuric acid treatment? Using this reaction, how might propan-1-ol be changed to propan-2-ol?

(3) Devise a scheme whereby you could distinguish between an alkane, alkene, alkyne and an aromatic hydrocarbon.

(4) What bonding forces do you think are operative in 'picrate' compounds?

(5) Give the structures and names of the isomers of molecular formula C_5H_{10}.

(6) What hydrocarbons could be prepared from butan-2-ol?

(7) 10 cm^3 of a hydrocarbon were exploded with 130 cm^3 of oxygen. After cooling to the original temperature and adjusting the pressure to that originally obtaining, the volume of gas was 95 cm^3; 70 cm^3 of this was absorbed in potassium hydroxide solution. Suggest possible structural formulae for the hydrocarbon. How would you distinguish between the different possibilities?

(8) A hydrocarbon contains 88.9 per cent of carbon and has a relative vapour density of 27. On hydrogenation, using a nickel catalyst, two molecules of hydrogen were taken up for every molecule of hydrocarbon. The original compound did not give a metallic derivative when treated with ammoniacal copper(I) chloride, but on careful ozonolysis two molecules of methanal and one molecule of a dialdehyde of molecular formula $C_2H_2O_2$ were formed

from every molecule of hydrocarbon. Deduce the structure of the hydrocarbon.

(9) Suggest a method of preparing 2-bromo(bromomethyl)benzene from methylbenzene. What products are possible when this compound reacts with sodium?

(10) Given that the bond energies for C—C, C=C, and C—H are 348, 620 and 413 kJ mol^{-1} respectively, and that the enthalpy of formation of C(g), H(g) and C_6H_6(g) are $+716$, $+218$ and $+83$ kJ mol^{-1} respectively, determine if these values are in agreement with an alternate single/double bond structure for benzene. (Remember that bond energies conventionally have positive signs; it is the energy required to *break* a mole of bonds.)

(11) An aromatic hydrocarbon, on oxidation with potassium manganate(VII) gave a monobasic organic acid. This, with soda-lime was decarboxylated to a hydrocarbon having a relative vapour density of 39. A Friedel–Crafts reaction using bromomethane gave the original hydrocarbon. Explain these reactions.

(12) Write an essay dealing with the commercial availability of hydrocarbons.

(13) Write an essay on polynuclear hydrocarbons.

(14) Three isomeric hydrocarbons, P, Q, and R contain 88.8 per cent of carbon by mass and have a relative molecular mass of 54. With an aqueous solution containing Cu(I) ions complexed by ammonia, isomer P gave a red precipitate. On vigorous oxidation by aqueous manganate(VII) ions, isomer Q gave only ethanoic acid whereas isomer R gave rise to an unbranched chain, dicarboxylic acid.

(a) Calculate the molecular formula of these isomers.
(b) Write structural formulae for P, Q, and R.
(c) In which of the isomers do the carbon atoms lie in a straight line?

[Cambridge 1978]

(15) But-2-ene can be prepared by the action of an alcoholic solution of potassium hydroxide on 2-bromobutane. Suggest, with the aid of mechanistic equations, the nature of this reaction.

What products are formed when but-2-ene reacts with (a) hydrogen, (b) bromine, and (c) sulphuric acid? In each case give the essential experimental conditions and for either (b) or (c) indicate the mechanism postulated for the reaction.

Explain why but-2-ene exists in two forms but both give the same compound on hydrogenation.

[J.M.B.]

20 Organic halogen compounds

Types of compound • Preparation • Properties and reactions • Industrial aspects • Analysis

The halogenation of hydrocarbons was introduced in Chapter 19, where it was seen that, either by addition to unsaturated hydrocarbons or by substitution of hydrogen in saturated hydrocarbons, one or more carbon–halogen bonds are formed. For more than one such bond in a molecule, two possibilities exist, i.e., polyhalogenated hydrocarbons may be divided into

(1) Those with only one halogen atom attached to any one carbon atom, e.g.,

$ClCH_2.CH_2Cl$ 1,2-Dichloroethane

(2) Compounds in which more than one halogen atom is attached to the same carbon atom, e.g.,

CH_3CHCl_2 1,1-Dichloroethane

The preparation and properties of the former type are similar to those of the monohalogenated hydrocarbons, whereas the latter differ in several respects from compounds containing only one halogen atom.

Preparation

Direct reaction of hydrocarbon with chlorine

This light induced free-radical mechanism (cf. p. 141) involves the following steps:

$$Cl_2 \xrightarrow{h\nu} 2Cl\cdot \qquad\qquad\qquad\qquad\qquad\qquad\qquad \text{initiation}$$

$$\left.\begin{array}{l} Cl\cdot + CH_4 \rightarrow CH_3\cdot + HCl \\ Cl_2 + CH_3\cdot \rightarrow CH_3Cl + Cl \end{array}\right\} \text{ propagation}$$

$$\left.\begin{array}{l} 2CH_3\cdot \rightarrow C_2H_6 \\ 2Cl\cdot \rightarrow Cl_2 \\ CH_3\cdot + Cl\cdot \rightarrow CH_3Cl \end{array}\right\} \text{ termination}$$

By reactions such as these a whole series of chlorinated hydrocarbons can be produced, e.g.,

$$CH_3Cl \longrightarrow CH_2Cl_2 \longrightarrow CHCl_3 \longrightarrow CCl_4$$

chloromethane   dichloromethane   trichloromethane   tetrachloro-

(methyl chloride)   (methylene chloride)   (chloroform)   methane

                                                                           (carbon tetrachloride)

Addition to unsaturated hydrocarbons (*see* also p. 422)

Halogen compounds can be prepared by treating unsaturated hydrocarbons with hydrogen halides or, with the exception of the comparatively unreactive iodine, with the free halogen. The action of fluorine has to be moderated by dilution with nitrogen for the reaction to proceed smoothly, as does also that of chlorine when reacting with an alkyne. Thus

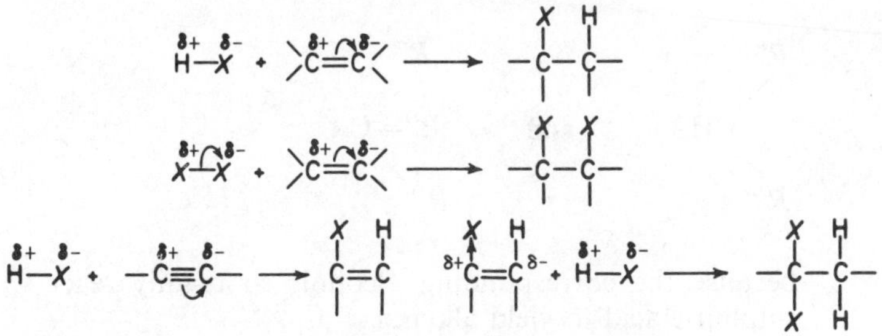

The reagents available are

Hydrogen halides

The mechanism of attack by hydrogen halide involves initial protonation of the alcohol, followed by reaction with halide ion

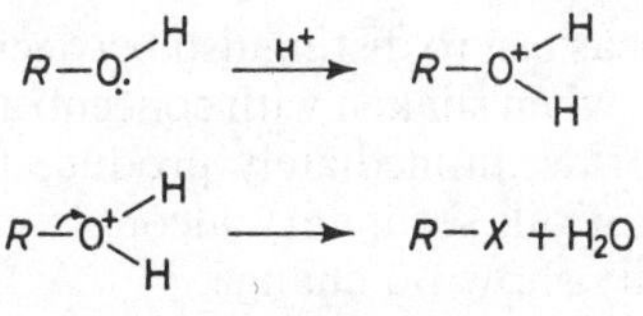

Any factor which can promote ionization of the hydrogen halide will assist the reaction; for example, chloro-compounds can be made by treating an alcohol with dry hydrogen chloride, in the presence of anhydrous zinc chloride, which reacts with hydrogen chloride in accordance with the equation

$$HCl + ZnCl_2 \rightarrow H^+[ZnCl_3]^-$$

The overall reaction can be summarized as, e.g.,

$$C_2H_5OH + HCl \xrightarrow{ZnCl_2} C_2H_5Cl + H_2O$$

ethanol   chloroethane

Bromo- and iodo-derivatives are usually obtained by preparing the appropriate hydrogen halide *in situ*. The alcohol is distilled with a mixture of the potassium halide and concentrated sulphuric acid and the distillate collected under water, which serves to remove halogen acids, free halogen and sulphur dioxide formed in side reactions, e.g.,

$$Br^- + H_2SO_4 \rightarrow HBr + HSO_4^- (+ Br_2 + SO_2)$$

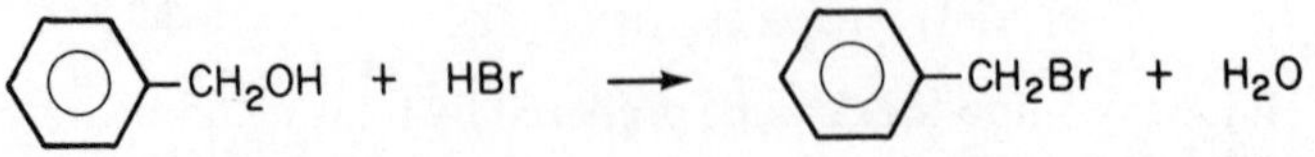

phenylmethanol　　　　　　　　　　(bromomethyl) benzene

This method is not very satisfactory for the preparation of derivatives of general formulae

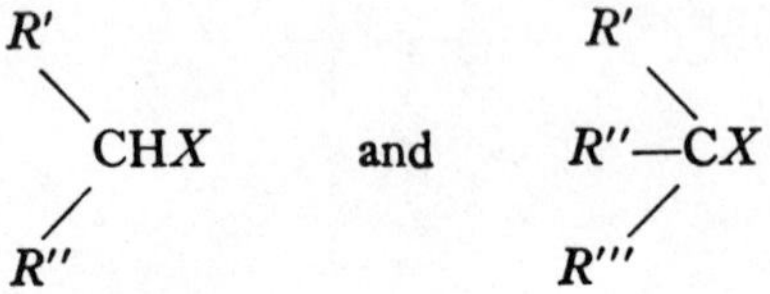

$$\begin{matrix} R' \\ \diagdown \\ \quad CHX \\ \diagup \\ R'' \end{matrix} \quad \text{and} \quad \begin{matrix} R' \\ \diagdown \\ R''-CX \\ \diagup \\ R''' \end{matrix}$$

because the corresponding alcohols so readily react with concentrated sulphuric acid to yield alkenes.

The ease of formation of the halogeno-alkanes is in the order tertiary > secondary > primary alcohols, i.e.,

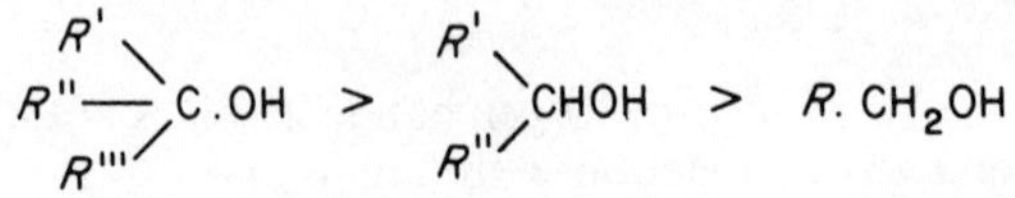

$$\begin{matrix} R' \\ \diagdown \\ R''-C.OH \\ \diagup \\ R''' \end{matrix} > \begin{matrix} R' \\ \diagdown \\ \quad CHOH \\ \diagup \\ R'' \end{matrix} > R.CH_2OH$$

This is the basis of the Lucas test to distinguish between the different alcohol groups. Tertiary alcohols, when shaken with concentrated hydrochloric acid and anhydrous zinc chloride, immediately produce the chloro-derivative, which separates out as an oil; secondary alcohols react only slowly and primary alcohols generally show no change.

Phosphorus halides

These react rapidly with alcohols. The chlorides of phosphorus are readily available but the bromides and iodides are made *in situ* by direct combination of red phosphorus and the halogen. Typical of their reaction with alcohols are

$$3ROH + PCl_3 + 3RCl + P(OH)_3$$

$$ROH + PCl_5 \rightarrow RCl + POCl_3 + HCl(g)$$

Sulphur dichloride oxide (Thionyl chloride)

This reagent, preferably in the presence of pyridine, reacts with alcohols:

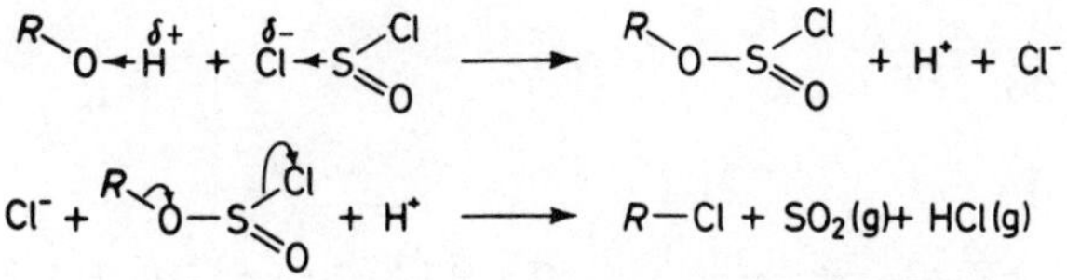

$$R-O \xrightarrow{\delta+} H + \overset{\delta-}{Cl} \xrightarrow{} S \overset{Cl}{\underset{O}{\lessgtr}} \longrightarrow R-O-S \overset{Cl}{\underset{O}{\lessgtr}} + H^+ + Cl^-$$

$$Cl^- + R-O-S\overset{Cl}{\underset{O}{\lessgtr}} + H^+ \longrightarrow R-Cl + SO_2(g) + HCl(g)$$

When pyridine is present, its basic character ensures the ready removal of the acidic by-products.

Substitution of oxygen in a carbonyl group

Halides containing two halogen atoms attached to the same carbon atom are produced by the reaction of phosphorus pentahalide on aldehydes or ketones, e.g.,

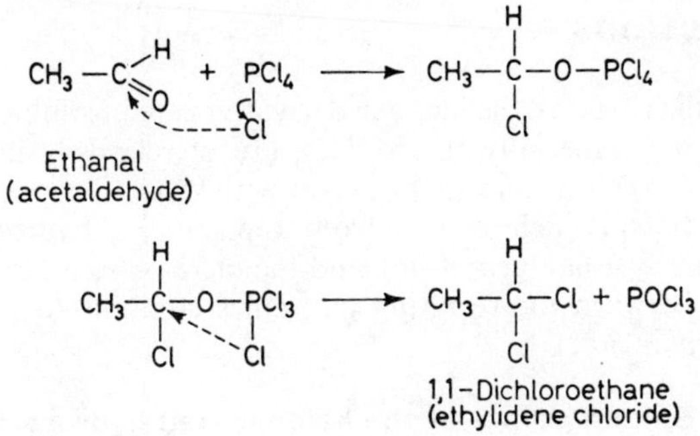

$$CH_3 - C \overset{H}{\underset{O}{\lessgtr}} + PCl_4 \longrightarrow CH_3 - \overset{H}{\underset{Cl}{\overset{|}{C}}} - O - PCl_4$$

Ethanal
(acetaldehyde)

$$CH_3 - \overset{H}{\underset{Cl}{\overset{|}{C}}} - O - PCl_3 \longrightarrow CH_3 - \overset{H}{\underset{Cl}{\overset{|}{C}}} - Cl + POCl_3$$

1,1–Dichloroethane
(ethylidene chloride)

Similarly

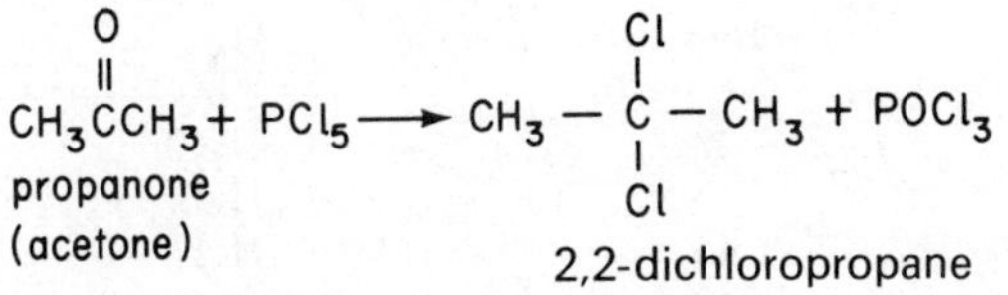

$$\overset{O}{\overset{\|}{CH_3CCH_3}} + PCl_5 \longrightarrow CH_3 - \overset{Cl}{\underset{Cl}{\overset{|}{C}}} - CH_3 + POCl_3$$

propanone
(acetone)   2,2-dichloropropane

The haloform reaction

The compounds CHX_3 can be prepared by the haloform reaction, which involves a compound of structure

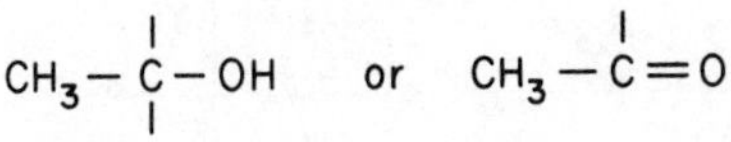

$$CH_3 - \overset{|}{\underset{|}{C}} - OH \quad \text{or} \quad CH_3 - \overset{|}{C} = O$$

(e.g. ethanol, CH_3CH_2OH, or propanone, CH_3COCH_3) undergoing halogenation, followed by hydrolysis. It is usual to perform the reaction with bleaching powder or sodium halate(I), either of which can be regarded both as a source of halogen and alkali, e.g.,

$$\underset{H_3C}{\overset{R}{>}}\!\!\overset{H}{\underset{OH}{\overset{|}{C}}}\xrightarrow[(-2HI)]{I_2}\underset{H_3C}{\overset{R}{>}}C{=}O\xrightarrow[(-H_2O)]{OH^-}\underset{H_2\overset{..}{C}}{\overset{R}{>}}C{=}O\xrightarrow{I_2}\underset{H_2IC}{\overset{R}{>}}C{=}O\xrightarrow{etc.}\underset{I_3C}{\overset{R}{>}}C$$

$$HO^-+\overset{R}{\underset{\overset{|}{\underset{I\ I\ I}{\overset{..}{C}}}}{>}}\overset{\delta+}{C}{=}O\longrightarrow\underset{HO}{\overset{R}{>}}C{=}O+[CI_3]^-\longrightarrow CHI_3+\underset{{}^-O}{\overset{R}{>}}C{=}O$$

Tri-iodo-
methane
(iodoform)

The methylene halides can be prepared by refluxing the appropriate haloform with iron filings and water

$$CHX_3+2[H]\rightarrow CH_2X_2+HX$$

Properties and reactions

Like the hydrocarbons, the halogen derivatives are insoluble in water. Those that are liquid, especially the iodides, are of relatively high density and therefore give a lower, immiscible, layer with water. The increase in relative molecular mass (which results from replacing a hydrogen atom in a hydrocarbon by a halogen atom), and the increase in interaction of dipole-type forces between the resulting molecules combine to produce a fall in volatility (*Table 20.1*).

Table 20.1 Boiling points of the halogenated hydrocarbons

Compound	B.P./°C	Compound	B.p./°C
CH_4	−164	C_6H_6	80
CH_3Cl	−23	C_6H_5Cl	132
CH_8Br	3.5	C_6H_5Br	156
CH_3I	44	C_6H_5I	186
CH_2Cl_2	42		
$CHCl_3$	61		
CCl_4	76		

Table 20.2 Bond energies of carbon–halogen bonds

Bond	C—I	C—Br	C—Cl	C—F
Bond energy/kJ mol^{-1}	188	226	278	408

The characteristic reaction of halides is the substitution of the halogen atom by another atom or radical. Alkyl halides, and those aromatic halides in which the halogen atom is in a side chain, undergo fairly easy replacement of the halogen atom. The ease of replacement is roughly in the order I > Br > Cl > F, i.e., of increasing bond strengths (*Table 20.2*). The order is also dependent upon the proximity of multiple bonds, the number of halogen

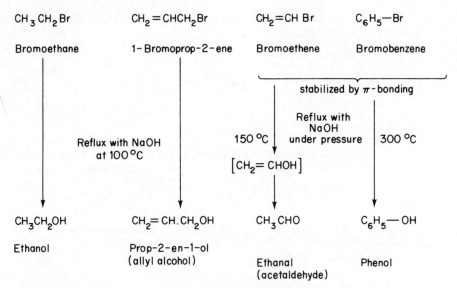

Figure 20.1 Reactivities of alkyl, alkenyl, and aryl bromides towards hydrolysis

atoms attached to the carbon atom and the size of the groups surrounding the halogen atom, i.e., on *steric hindrance*. However, more than one mechanism of replacement is possible, so that experimental conditions are also important.

Aryl halides, in marked contrast to alkyl halides, are very unreactive, because the lone pairs of electrons on the halogen atom can interact with the π bonding of the ring system and so extend the delocalized π clouds, with a resulting increase in stability, e.g.,

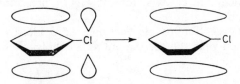

The same considerations apply to unsaturated aliphatic compounds in which the carbon atom to which the halogen is attached is also unsaturated. This change in reactivity between the different halogenated hydrocarbons can be illustrated by considering the necessary conditions required to hydrolyse different compounds shown in *Figure 20.1*.

With the above reservations, the reactions given by the halogenated hydrocarbons are as follows.

Reduction

Halogenated hydrocarbons can be reduced by hydrogen iodide (which is an excellent reducing agent). Normally, moist red phosphorus is also added, as by this means hydrogen iodide is regenerated from the iodine produced in the reduction, e.g.,

$$RI + HI \rightarrow RH + I_2$$

$$P + I_2 \rightarrow PI_3 \xrightarrow{H_2O} P(OH)_3 + HI$$

Formation of amines

When a halogen compound is treated with ammonia under pressure, a mixture of salts of amines results from continued replacement of halogen

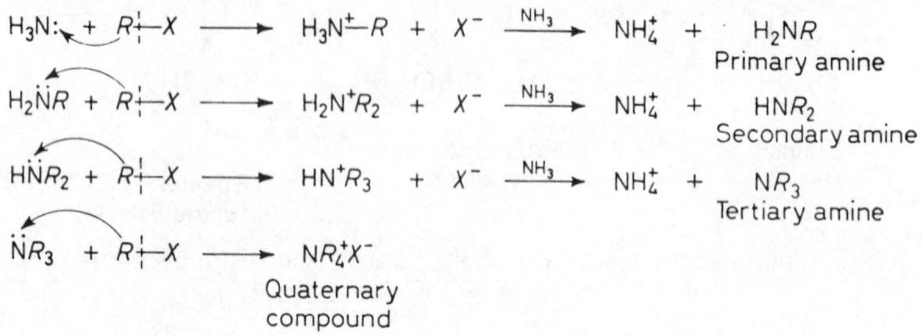

Elimination of hydrogen halide

The elimination of HX to yield an alkene is promoted by boiling a halide with an alcoholic solution of potassium hydroxide:

$$RCH_2CH_2X \xrightarrow[-HX]{alc.KOH} RCH\!=\!CH_2$$

This reaction occurs by a bimolecular mechanism involving attack on the halide by the nucleophilic ethoxide ions, $C_2H_5O^-$, present in the alcoholic hydroxide solution.

Two important exceptions to this reaction are provided by the methyl and ethyl halides, which yield the corresponding alcohols (*see* below).

Formation of ethers

This often accompanies the previous reaction between a halide and alcoholic potassium hydroxide because of the possibility of the alternative mode of attack by the ethoxide ion as shown in *Figure 20.2.*

Hydrolysis to an alcohol

The main reaction when a halide is refluxed with an aqueous solution of an alkali is

$$RX + OH^- \rightarrow ROH + X^-$$

The two possible mechanisms for this reaction have already been described (p. 405).

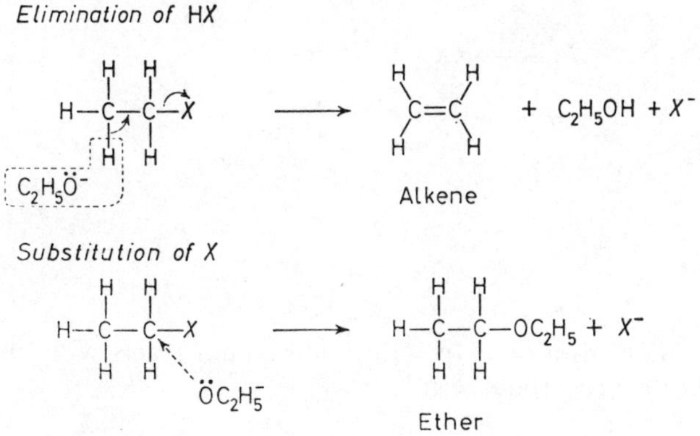

Elimination of HX

Alkene

Substitution of X

Ether

Figure 20.2 Competition between elimination and etherification with alkyl halides

Small amounts of alkene are usually also formed by side reactions.

Hydrolysis of polyhalides initially follows a similar course, but where two or more halogen atoms are attached to the same carbon atom, elimination of water follows upon the first reaction, e.g., in the formation of ethanal (acetaldehyde) from 1,1-dichloroethane:

$$CH_3CHCl_2 \rightarrow [CH_3CH(OH)_2] \rightarrow CH_3CHO + H_2O$$

and in the formation of methanoic acid (formic acid) from trichloromethane:

$$CHCl_3 \rightarrow [CH(OH)_3] \rightarrow H.COOH + H_2O$$

Replacement by the cyano, nitro and nitrite groups

If refluxed with an aqueous-alcoholic solution of potassium cyanide, the cyanide ion replaces the halogen, e.g.,

$$R{-}X + CN^- \rightarrow [\overset{\delta-}{X} \ldots \overset{\delta+}{R} \ldots \overset{\delta-}{CN}] \rightarrow R{-}C{\equiv}N + X^-$$
A nitrile

However, if silver cyanide, which is mainly covalent, is used, the isonitrile is formed:

$$X{-}R + N{\equiv}C{-}Ag \rightarrow [\overset{\delta-}{X} \ldots R \ldots N{\equiv}C \ldots \overset{\delta+}{Ag}] \rightarrow X^- + R{-}N{\equiv}C + Ag^+$$
Isonitrile

Similar reactions occur with solutions of potassium and silver nitrites, nitrites and nitro compounds, respectively, being formed. There is, however, some evidence for the view that both of these reagents yield a 2:1 mixture of nitro compound and nitrite:

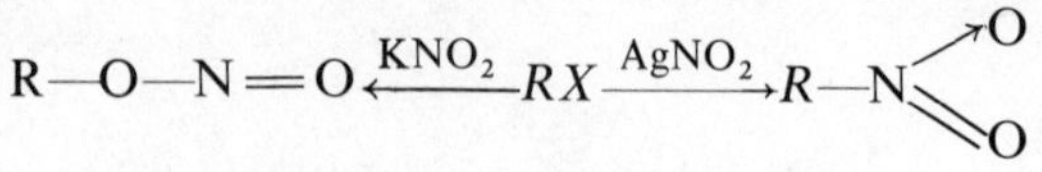

$$R-O-N=O \xleftarrow{KNO_2} RX \xrightarrow{AgNO_2} R-N\big\langle{}^{O}_{O}$$

a nitro
compound

With metals

It has been mentioned (p. 413) that sodium reacts with alkyl halides to produce hydrocarbons, e.g.,

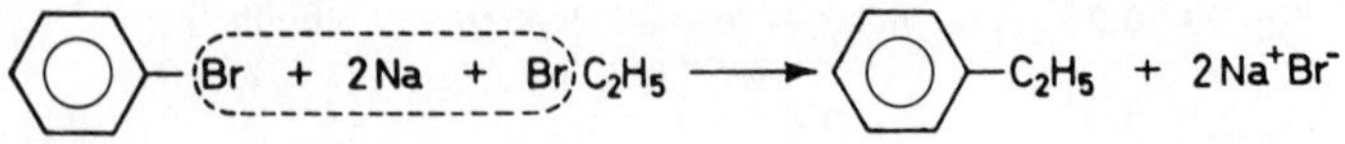

Chloroethane reacts with a lead–sodium alloy to form the important compound tetraethyllead(IV), used as an 'anti-knock' additive in petrol:

$$4C_2H_5Cl + 4Na/Pb \rightarrow (C_2H_5)_4Pb + 3Pb + 4Na^+Cl^-$$

A particularly important reaction of halogen derivatives, including those with the halogen attached directly to an aromatic nucleus, is that with magnesium under anhydrous conditions in ether as solvent. The reaction is normally represented as

$$RX + Mg \rightarrow \overset{\delta-}{R} - \overset{\delta+}{Mg} - \overset{\delta-}{X}$$

The product behaves as $R-Mg-X$ although it appears that it might be an equimolecular mixture of R_2Mg and MgX_2. These so-called *Grignard reagents* are very reactive towards many oxygen-containing compounds as shown in *Figure 20.3*.

In the examples given, an intermediate complex magnesium compound has to be hydrolysed to produce the required product, e.g.,

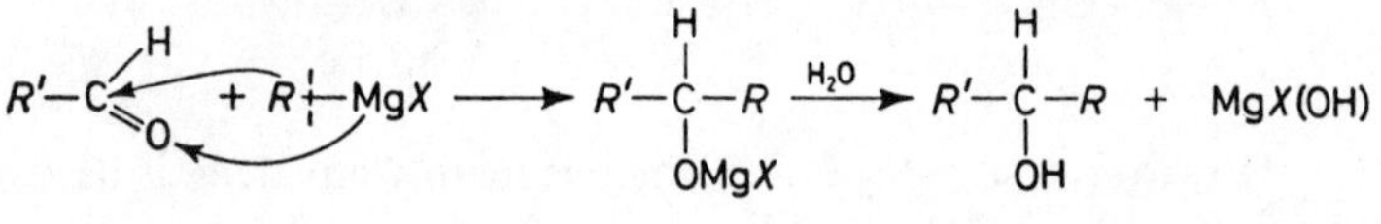

Formation of esters

Reaction of a halide with the silver salts of carboxylic acids leads to esters, e.g., silver ethanoate (acetate):

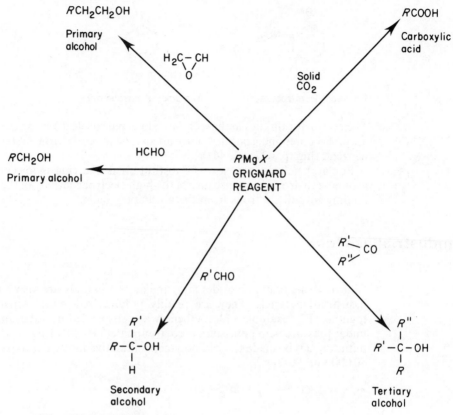

Figure 20.3 Reactions of Grignard reagents

Addition to aromatic systems

Alkyl halides take part in the Friedel–Crafts reaction (p. 414), of which a typical example is:

$$\text{C}_6\text{H}_6 + \text{C}_2\text{H}_5\text{Cl} \xrightarrow{\text{Al}_2\text{Cl}_6} \text{C}_6\text{H}_5\text{-C}_2\text{H}_5$$

The directive influence of halogens in aromatic compounds

A substituent in an aromatic compound directs further attack of the benzene ring. For example, of the three mononitro-derivatives which might be obtained from the nitration of bromobenzene, the *ortho* and *para* compounds predominate:

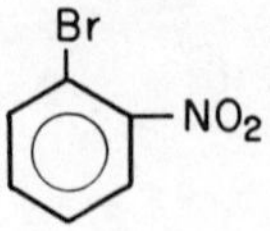

1-bromo-2-nitrobenzene 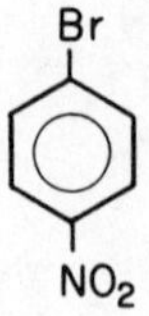   1-bromo-4-nitrobenzene

The reason for this lies in the fact that a lone pair of electrons on the halogen atom becomes incorporated into an extended π electron system of the benzene ring (p. 427 and 441).

For aryl halides, the ease of substitution in the ring is lower than for aromatic hydrocarbons, because of the high electronegativity of the halogen tending to pull electrons from the aromatic system.

Industrial aspects

Polymers

Chloroethene (vinyl chloride) and similar compounds are very important industrial materials. They are readily polymerized, either separately or together. For example, chloroethene when suspended in water and heated under pressure with potassium peroxosulphate(VI) or hydrogen peroxide as initiator, can be made to yield polymers of relative molecular mass between 50 000 and 80 000:

$$n CH_2 = CHCl \longrightarrow \left(-CH_2 - \underset{\underset{Cl}{|}}{CH} - \right)_n$$

The product, commonly referred to as *poly(vinyl chloride)* or PVC, is thermoplastic (i.e. can be moulded when hot), inert and rigid. As vinyl chloride can be made from ethyne by the addition of hydrogen chloride in the presence of mercury(II) ions as catalyst, the main raw materials are coal, salt and limestone.

If the mercury ion catalyst is replaced by copper(I) chloride, in the presence of ammonium chloride, ethyne dimerizes to but-1-ene-3-yne which can then react with hydrogen chloride to produce 2-chlorobuta-1,3-diene (*chloroprene*), in accordance with Markownikoff's rule (p. 423)

$$2HC \equiv CH \rightarrow CH_2 = CH - C \equiv CH$$

$$CH_2 = CH - C \equiv CH + HCl \rightarrow CH_2 = CH - \underset{\underset{Cl}{|}}{C} = CH_2$$

Chloroprene readily polymerizes to *neoprene*, which is an important *elastomer* (synthetic rubber):

$$n CH_2 = \underset{\underset{Cl}{|}}{C} = CH_2 \longrightarrow (-CH_2 - CH = \underset{\underset{Cl}{|}}{C} - CH_2 -)_n$$

This structure should be compared with that of natural rubber (p. 410).

Treatment of 1,2-dichloroethane with a solution of sodium polysulphides results in a condensation polymerization to give an elastomer, *Thiokol*, of comparatively high chemical resistance

$$n\text{ClCH}_2\text{CH}_2\text{Cl} \xrightarrow{\text{Na}_2\text{S}_x} \text{Cl}-(\text{CH}_2-\text{CH}_2-\overset{\overset{\displaystyle S}{\|}}{S}-\overset{\overset{\displaystyle S}{\|}}{S})_n-\text{Cl}$$

A polymer remarkable in view of its stability to a wide range of temperature, its considerable electrical and chemical resistance and also its property of 'self-lubrication', is *poly(tetrafluoroethene)* (PTFE). It is made by the pyrolysis of chlorodifluoromethane at about 700 °C, followed by polymerization:

$$2\text{CHClF}_2 \rightarrow \text{F}_2\text{C}=\text{CF}_2 + 2\text{HCl}$$
$$n\text{F}_2\text{C}=\text{CF}_2 \rightarrow (-\text{F}_2\text{C}-\text{CF}_2-)_n$$

Insecticides

Gammexane, of prime importance in the control of locusts, has been mentioned previously (p. 423).

D.D.T. is made by the reaction between chlorobenzene and 2,2,2-trichloroethanediol (chloral hydrate) (p. 507) in the presence of concentrated sulphuric acid. Because of its stability and fat solubility it tends to persist in animal food chains with harmful results, and its use is restricted.

2,2-bis(4-chlorophenyl)-1,1,1-trichloroethane
(4,4-Dichloro-diphenyl-trichloroethane)

Solvents

Chlorinated hydrocarbons are very popular solvents, particularly in the realms of 'dry cleaning'. The common examples are

Tetrachloromethane, made by the action of chlorine on carbon disulphide at high temperature:

$$\text{CS}_2 + 3\text{Cl}_2 \rightarrow \text{CCl}_4 + \text{S}_2\text{Cl}_2$$

'Westrosol', obtained by passing 1,1,2,2-tetrachloroethane over heated barium chloride:

$$\text{HC}\equiv\text{CH} \xrightarrow{\text{Cl}_2} \text{ClCH}=\text{CHCl} \xrightarrow{\text{Cl}_2} \text{Cl}_2\text{CH}-\text{CHCl}_2 \xrightarrow[(-\text{HCl})]{\text{BaCl}_2} \text{ClCH}=\text{CCl}_2 \text{ 'Westrosol'}$$

More recently on the market, but less toxic than most chlorinated hydrocarbons, is *Genklene*, 1,1,1-trichloroethane.

Refrigerants

Freons, chlorofluoro-derivatives of methane and ethane, are valuable as refrigerants because of their easy liquefaction and low reactivity. They are prepared by reactions such as that between hydrogen fluoride and tetrachloromethane, in the presence of a catalyst and under pressure

$$2CCl_4 + 3HF \rightarrow CFCl_3 + CF_2Cl_2 + 3HCl$$

Analysis

Halides, particularly if the halogen content is high, are not very flammable—tetrachloromethane is even used as a fire extinguisher—and this property is sometimes helpful in their analysis.

Beilstein's test can be used as an indication of the presence of a halogen. In this test, copper wire is heated to give a superficial layer of copper(II) oxide and then reheated with the compound. The appearance of a green flame indicates a halogen forming the comparatively volatile copper(II) halide.

The chlorine, bromine or iodine in alkyl halides is recognized by treatment with alcoholic silver nitrate solution, when a precipitate of the appropriate silver halide is obtained. The reaction with aryl halides is too slow to be worthwhile, but the halogen can be identified by carefully fusing the substance with a little sodium and treating the aqueous extract containing sodium halides with excess of dilute nitric acid and then silver nitrate solution; a characteristic precipitate indicates that halogen is present.

The same basic chemistry is used in the quantitative estimation. A known mass of the substance is heated in a sealed tube with excess of concentrated nitric acid and silver nitrate until reaction is complete. The silver halide formed is washed, dried and weighed. From the results, the percentage by mass of halogen in the substance can be found and, provided that the percentage masses of the other elements are known, the simplest formula can be calculated. A knowledge of the relative molecular mass, e.g., from relative vapour density measurements, then permits the molecular formula to be evaluated.

Example. 1.42 g of an alkyl halide when treated as described above, yielded 2.35 g of silver iodide. The relative vapour density was 71. Calculate the molecular formula.

$$C_nH_{2n+1}I \rightarrow AgI$$
$$(14n+1)+127 \quad 235$$

Mass of iodine in 2.35 g of silver iodide $= \frac{127}{235} \times 2.35$ g $= 1.27$ g

i.e., 1.42 g of iodide contained 1.27 g of iodine

Assume 1 mole of I is contained in 1 mole of the alkyl iodide, then 127 g of iodine are contained in 142 g of the alkyl iodide; therefore

relative molecular mass $= 142$

This is confirmed from the relative vapour density, $(71 \times 2 = 142)$
i.e.,

$(14n + 1) + 127 = 142$

$n = (142 - 128)/14 = 1$

hence

molecular formula: $C_1H_3I_1 = CH_3I$

Summary

Alkyl monohalides undergo substitution of the halogen fairly readily.
Polyhalides, where the halogen atoms are attached to the same carbon atom,

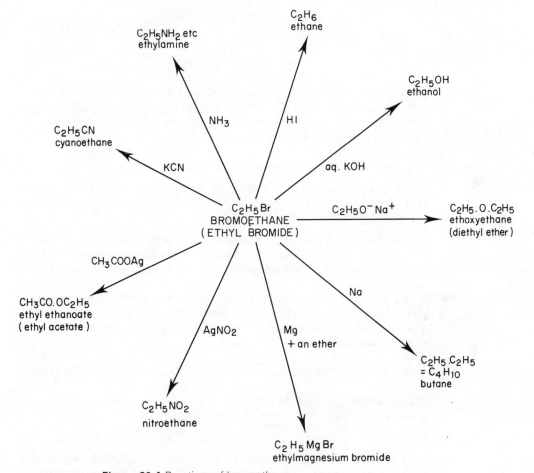

Figure 20.4 Reactions of bromoethane

are less reactive, as are the halides with halogen attached directly to an unsaturated carbon atom (including the benzene nucleus); this is at least partly due to interpenetration of the p orbitals

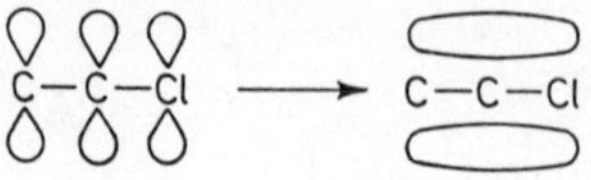

Some common reactions of bromoethane are shown in *Figure 20.4*.

Questions

(1) Name the following:

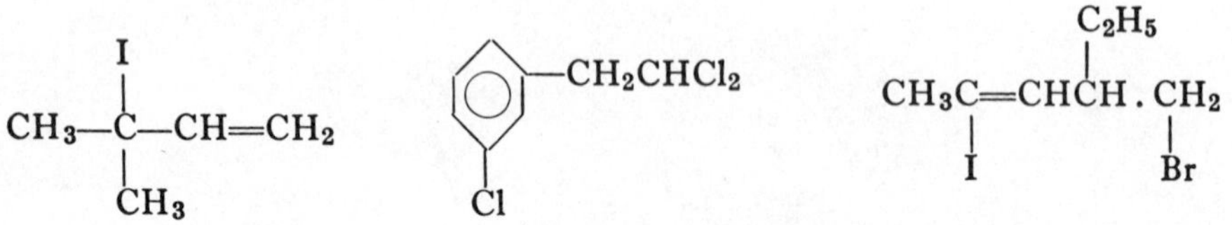

(2) A compound has the following composition by mass: C = 66.4 per cent, H = 5.5 per cent, Cl = 28.1 per cent. Its relative vapour density is 63.25. Write down the structural formulae of substances which accord with this information. Suggest ways of identifying them.

(3) Suggest methods for the following conversions:

(*a*) propanone to methanoic acid; (*b*) methanol to ethane; (*c*) propene to propyne; (*d*) benzene to biphenyl; (*e*) iodoethane to ethoxyethane; (*f*) 1-bromopropane to butanoic acid.

(4) An alcohol, *A*, with sodium carbonate and iodine solution gives a yellow solid, *B*, and the sodium salt of an acid, *C*, which on heating with soda lime gives methane. Identify the substances and explain the reactions taking place.

(5) 0.78 g of an unknown compound on combustion yielded 0.44 g of carbon dioxide and 0.225 g of water. When treated with silver nitrate it yielded 1.175 g of silver iodide. Calculate the empirical formula and suggest a method for determining the molecular formula.

(6) Explain why the halogen derivatives of hydrocarbons are insoluble in water.

(7) It has been stated that the ease with which alcohols react with hydrogen chloride is in the order tertiary > secondary > primary. Explain why this is so.

(8) Comment on the following relative dipole moments: tetrachloromethane, 0.00; chloromethane, 1.83; chlorobenzene, 1.55; 1,4-dichlorobenzene, 0.00.

(9)   What products are likely to be formed when alcoholic potassium hydroxide solution is refluxed with (a) 2-iodopropane and (b) 2-chloro-2-methylbutane?

(10)   Comment on the following data:

Compound	Percentage of alkene formed by dehydrobromination
CH_3CH_2Br	2
$CH_3CHBrCH_3$	80
$(CH_3)_3CBr$	100

(11)   Suggest why the removal of HBr takes place more readily from bromo-fumaric than from bromomaleic acid both of formula
$HOOC.CBr == CH.COOH$.

(12)   Discuss the possible nature of Grignard reagents.

(13)   The reaction between methylbenzene (toluene) and chlorine in the presence of ultraviolet light is said to proceed *via* a free radical mechanism.

(a) State two characteristics of a free radical.
(b) Name one of the products of the reaction between methylbenzene and chlorine under these conditions and write an equation representing its formation.
(c) When methylbenzene reacts with chlorine in the dark in the presence of iron(III) chloride, the reaction proceeds by a different mechanism, and different products are obtained.
(d) Give *one* other example from organic chemistry to illustrate how experimental conditions may alter the products of a reaction between two given reactants.

[A.E.B.]

21 Nitrogen derivatives

Classification • Amines • Diazotization • Identification of
amines • Nitro compounds • Nitrites • Nitriles and
isonitriles • Heterocyclic compounds • DNA

Nitrogen occurs in many organic compounds, some of which, the proteins,
are essential to life. Some of the nitrogen-containing functional groups are
shown in *Figure 21.1*. Nitrogen also occurs in ring systems and amides
(Chapter 26).

Amines

Preparation

Progressive replacement of the hydrogen in ammonia by alkyl or aryl groups
gives rise to *amines*, *primary* for the replacement of *one hydrogen*, *secondary*
for *two*, and *tertiary* when all *three* have been removed:

$$NH_3 \rightarrow RNH_2 \rightarrow R_2NH \rightarrow R_3N$$

As the nitrogen atom possesses a lone pair of electrons, it can be
protonated by acids to give the following types of ions:

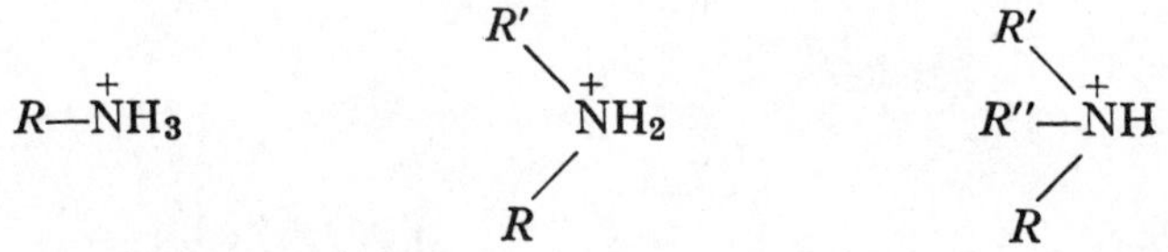

$$R-\overset{+}{N}H_3 \qquad \qquad \overset{R'}{\underset{R}{\diagdown}}\!\!\!\overset{+}{N}H_2 \qquad \qquad \overset{R'}{\underset{R}{\diagdown}}\!\!\!R''-\overset{+}{N}H$$

The hydrogen atom in the last of these can itself be replaced by an alkyl or
aryl group, with the formation of a *quaternary ammonium ion*:

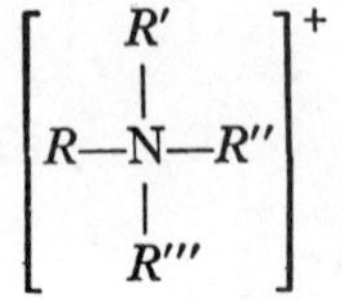

$$\left[\begin{array}{c} R' \\ | \\ R-N-R'' \\ | \\ R''' \end{array} \right]^+$$

All the above types are obtained if ammonia and an alkyl halide are heated
in a sealed tube, e.g.,

$$RBr \xrightarrow{NH_3} \left\{ \begin{array}{l} RNH_2 + HBr \rightarrow RNH_3{}^+Br^- \\ R_2NH + HBr \rightarrow R_2NH_2{}^+Br^- \\ R_3N \ + HBr \rightarrow R_3NH^+Br^- \\ R_3N \ + RBr \rightarrow R_4N^+Br^- \end{array} \right.$$

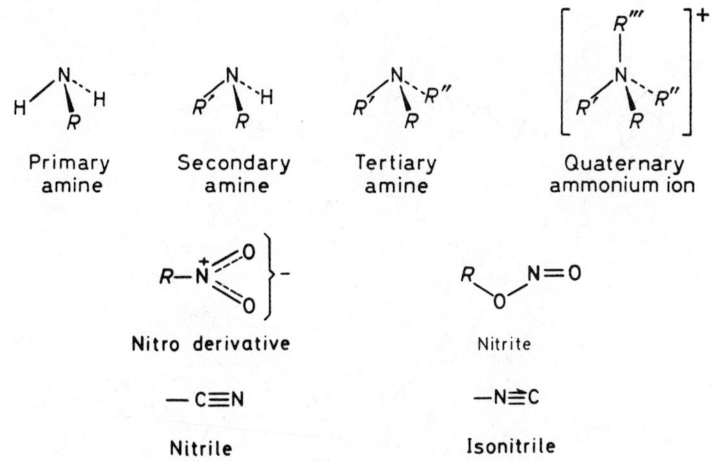

Figure 21.1 Some nitrogen-containing functional groups

Separation of these products can prove tiresome, and this method is seldom employed for the preparation of pure amines.

Aryl halides react only with difficulty, but ammonolysis of chlorobenzene does take place at a pressure of 30 MPa and a temperature of 300 °C in the presence of copper(I) oxide, which removes the chloride ion as copper(I) chloride:

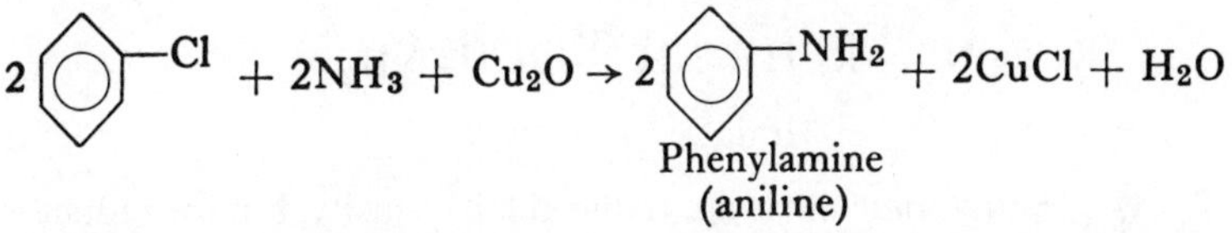

Phenylamine
(aniline)

A modification of the ammonolysis reaction, which yields primary amines uncontaminated by further substituted derivatives, is that in which the ammonia is replaced by benzene-1,2-dicarboximide (*phthalimide*); the sequence of reactions is indicated in *Figure 21.2*.

A method of preparation usually preferred to ammonolysis is the *Hofmann degradation of amides*. Bromination of an amide followed by hydrolysis with hot, concentrated alkali results in decarboxylation *via* a rearrangement reaction:

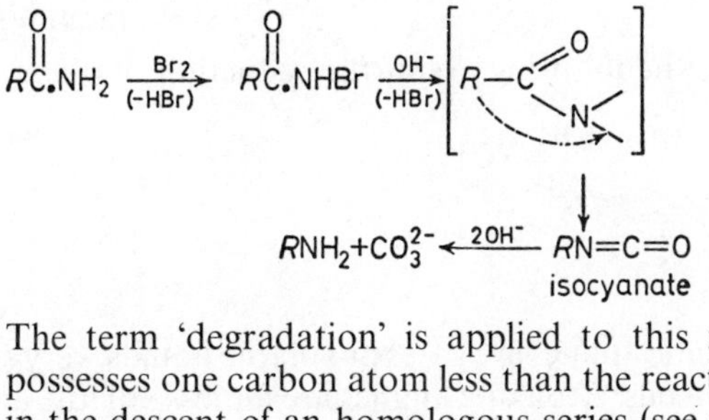

isocyanate

The term 'degradation' is applied to this reaction because the product possesses one carbon atom less than the reactant; it is therefore the key step in the descent of an homologous series (see p. 557).

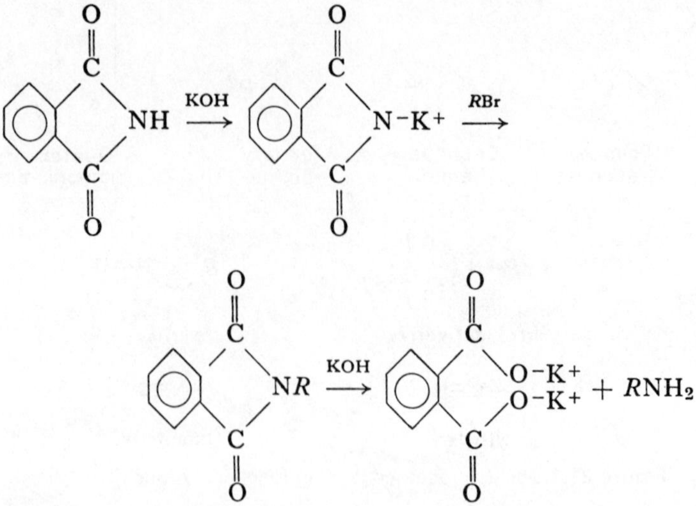

Figure 21.2 Synthesis of primary amines by use of phthalimide

Amines represent the reduction products of many nitrogenous compounds

Nitriles are reduced to amines by dissolving metal systems such as zinc in dilute acid:

$$RC\equiv N \xrightarrow{2H} RCH=NH \xrightarrow{2H} RCH_2.NH_2$$

$$\text{Imine}$$

Nitro compounds are reduced fairly easily, but for reduction to proceed as far as the amine requires the dissolving metal to be in acid media. This method is particularly important in aromatic chemistry as the nitro compound is readily obtained from the parent hydrocarbon:

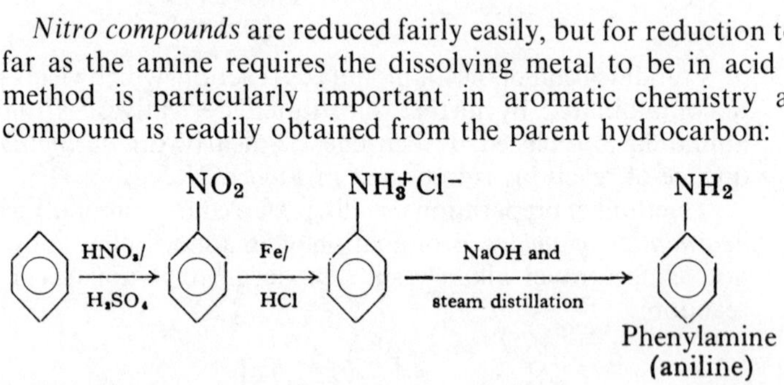

Sodium and ethanol bring about the reduction of *oximes* to amines:

$$RCH=NOH \xrightarrow{Na/C_2H_5OH} RCH_2.NH_2$$

Properties of amines

The lower aliphatic amines are gases or volatile liquids, very soluble in water, and with fishy smells. Aromatic amines are far less volatile and are insoluble in water.

Table 21.1 Dissociation constants of amines

Compound	Formula	K_b/mol dm^{-3}	pK_b
Ammonia	NH_3	1.8×10^{-5}	4.74
Methylamine	CH_3NH_2	4.4×10^{-4}	3.36
Dimethylamine	$(CH_3)_2NH$	5.2×10^{-4}	3.28
Trimethylamine	$(CH_3)_3NH$	5.5×10^{-5}	4.26
Phenylamine	$C_6H_5NH_2$	4.2×10^{-10}	9.38
N-Methylphenylamine	$C_6H_5NHCH_3$	7.1×10^{-10}	9.15
NN-Dimethylphenylamine	$C_6H_5N(CH_3)_2$	1.2×10^{-9}	8.92
2-Methylphenylamine	$2\text{-}CH_3C_6H_4NH_2$	2.5×10^{-10}	9.60
4-Methylphenylamine	$4\text{-}CH_3C_6H_4NH_2$	1.1×10^{-9}	8.96

The presence of a lone pair of electrons on the nitrogen atom ensures basic properties. Alkyl groups exert an electron-releasing inductive effect, and so, when attached to nitrogen, facilitate the acceptance of protons, with the result that aliphatic amines are more basic than ammonia and secondary amines are more basic than primary amines:

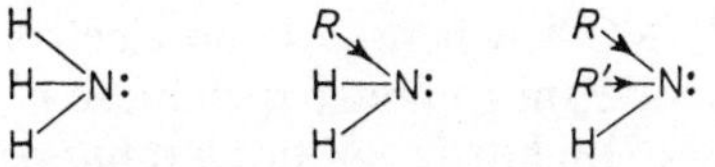

However, this explanation is only partly correct, as tertiary amines are less basic than both primary and secondary amines, a discrepancy which can probably be attributed to the steric hindrance afforded by three large alkyl groups surrounding the nitrogen atom.

The basicity of aromatic amines is much less than that of the aliphatic analogues or of ammonia because the lone pair of electrons on the nitrogen atom is conjugated with the aromatic system and is thus not so readily available for proton acceptance

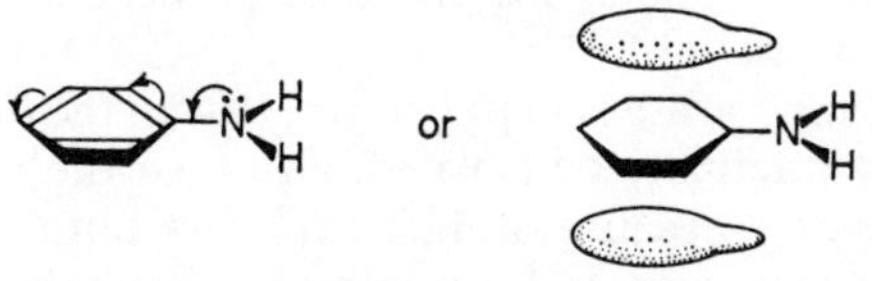

Some dissociation constants K_b for amines, relating to the equations

$$\geqslant N: + H_2O \rightarrow\ \geqslant NH^+ + OH^-$$

$$K_b = \frac{[\geqslant NH^+][OH^-]}{[\geqslant N:]}$$

are given in *Table 21.1*.

This basic character of amines means that they possess nucleophilic properties. Thus *primary and secondary amines are readily acetylated by treatment with an acid chloride*; benzoylation (the Schotten–Baumann reaction) and acetylation are particularly important, e.g.,

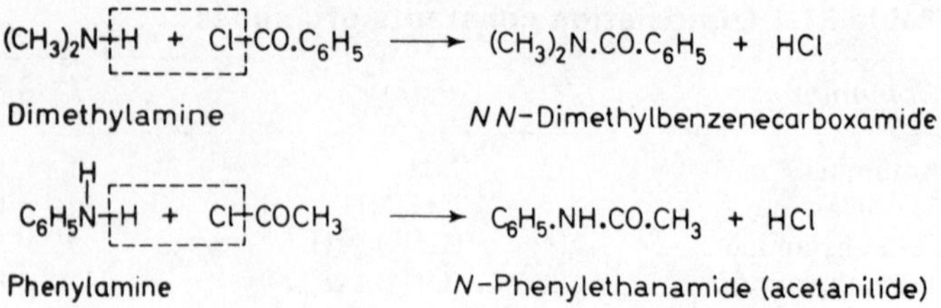

$(CH_3)_2N{+}H$ + $Cl{+}CO.C_6H_5$ $\longrightarrow$ $(CH_3)_2N.CO.C_6H_5$ + HCl

Dimethylamine NN–Dimethylbenzenecarboxamide

$C_6H_5\overset{H}{N}{+}H$ + $Cl{+}COCH_3$ $\longrightarrow$ $C_6H_5.NH.CO.CH_3$ + HCl

Phenylamine N–Phenylethanamide (acetanilide)

The mechanism of the latter reaction is indicated by the sequence of reactions

$$CH_3^{\delta+}\overset{\overset{O^{\delta-}}{\|}}{C}Cl \qquad CH_3.\overset{:O^-}{\underset{|+}{C}Cl} \qquad CH_3\overset{\overset{O}{\|}}{C}$$
$$\uparrow \qquad\qquad\qquad CH_3.\overset{}{C}Cl \qquad\qquad |$$
$$C_6H_5\ddot{N}H_2 \longrightarrow C_6H_5\ddot{N}H_2 \xrightarrow[(-HCl)]{} C_6H_5\ddot{N}H$$

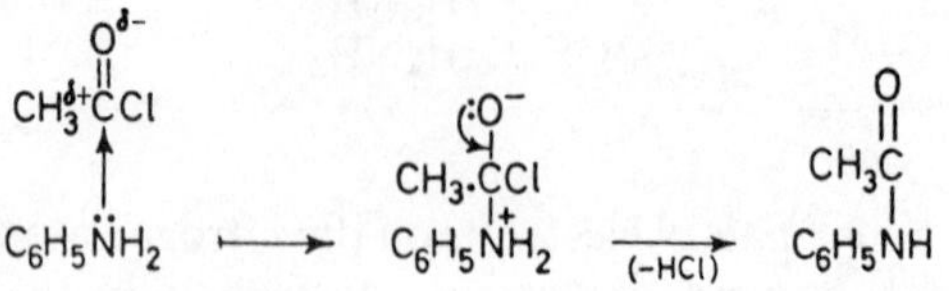

Diazotization

Primary and secondary amines also take part in nucleophilic attack on the nitrogen atom in compounds NOX, where X is an electron-withdrawing group. The classic example involves nitrous acid (produced *in situ* from cold, dilute hydrochloric acid and sodium nitrite solution). It has been shown that oxygen exchange takes place between $H_2^{18}O$ and nitrous acid, in keeping with the equilibrium:

$$2HNO_2 \rightleftharpoons H_2O + N_2O_3$$

The rate of reaction between amine and nitrous acid has been found to be second order with respect to amine and N_2O_3, i.e.,

rate $\propto$ [amine][N_2O_3]

Other substances, like nitrogen chloride oxide (nitrosyl chloride) NOCl, can also bring about similar reactions and the general mechanism outlined in *Figure 21.3* has been proposed.

Provided that the temperature is kept below $-10\,^\circ$C, the diazonium salts of aromatic amines can actually be isolated, although the resultant solids must be handled with care because of their explosive nature. However, as these compounds serve as very useful intermediates, there is no need to isolate them; instead, the solution obtained can be used directly. The sequence of reactions involved in making benzenediazonium chloride may be shown as:

$C_6H_5NH_2$ + $H^+ + Cl^-$ $\rightarrow$ $C_6H_5NH_3^+ + Cl^-$
phenylamine   hydrochloric   phenylammonium chloride
                        acid

NO_2^- $+H^+$ $\rightarrow$ HONO
nitrite   nitrous acid

$C_6H_5NH_3^+$ $+HONO$ $\rightarrow C_6H_5N_2^+ + 2H_2O$

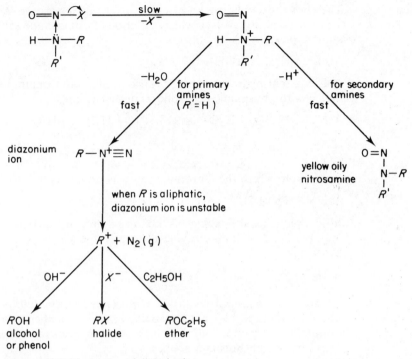

Figure 21.3 Mechanism of diazotization

and because hydrochloric acid was used, benzenediazonium chloride, $C_6H_5N_2^+Cl$, is present in solution.

The reactions of the aromatic diazonium compounds can be conveniently classified into three types.

(a) Nitrogen is evolved
If the diazonium salt is refluxed with potassium iodide solution, the iodo derivative is formed

$$C_6H_5N_2^+ + I^- \rightarrow \quad C_6H_5I \quad + N_2(g)$$
$$\text{iodobenzene}$$

Similar reactions do not take place with potassium bromide or chloride; to introduce either of these halogens into the aromatic nucleus, it is necessary to have copper or copper(I) halide present. Complex copper ions are believed to be the active agents, e.g.,

$$[Cl—Cu—Cl]^- + Ar—N^+ \equiv N \rightarrow Ar—Cl + CuCl + N_2(g)$$

$$CuCl + Cl^- \rightarrow [CuCl_2]^-$$

The diazo group can be replaced by a cyano group on treatment with copper(I) cyanide in the presence of potassium cyanide

$$ArN_2^+ + CuCN \xrightarrow{KCN} ArCN + Cu^+ + N_2(g)$$

The above reactions are known collectively as *Sandmeyer reactions.*

Nitrogen is also evolved when a solution of a diazonium salt is boiled, the organic product being a phenol

$$ArN_2^+ + H_2O \rightarrow ArOH + H^+ + N_2(g)$$

The parent hydrocarbon is produced when a diazonium salt solution is boiled with phosphinic (hypophosphorous) acid

$$ArN_2^+ + H_3PO_2 + H_2O \rightarrow Ar.H + N_2(g) + H_3PO_3 + H^+$$

(b) The multiple bond between the nitrogen atoms is reduced
Vigorous reducing agents, e.g., zinc in acid, produce the salt of the aromatic amine from which the diazonium salt was originally derived

$$ArN_2^+ + 3Zn + 7H^+ \rightarrow ArNH_3^+ + NH_4^+ + 3Zn^{2+}$$

Milder reducing agents, such as sodium sulphite or tin(II) in acid, reduce the multiple bond without cleaving it, to give the salt of an aryl-substituted hydrazine

$$ArN_2^+ + 4H^+ + 2Sn^{2+} \rightarrow ArNH.NH_3^+ + 2Sn^{4+}$$

(c) The diazonium ion 'couples' with another organic molecule
The diazonium ion is an electrophilic reagent and therefore combines or 'couples' with compounds in which there are centres activated to such reagents, e.g., phenols and amines:

The phenol can be further activated by dissolving in alkali to form the strongly nucleophilic ion:

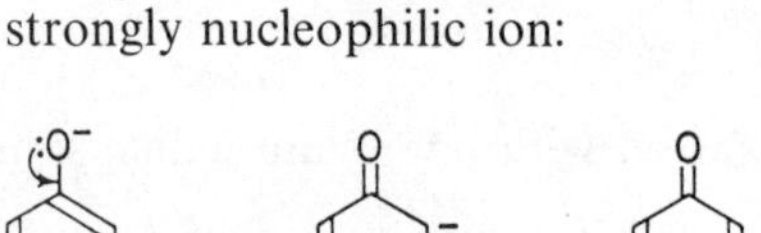

It also follows that the presence of electron-withdrawing groups in appropriate positions on the diazonium ion will further assist the progress of the reaction, as for example a sulphonic acid group in positions 2 or 4:

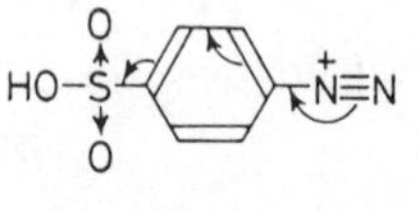

An important example of the use of this coupling reaction is the preparation of the indicator, *methyl orange*, outlined in *Figure 21.4.* The reactions of aromatic diazonium ions are summarized in *Figure 21.5.*

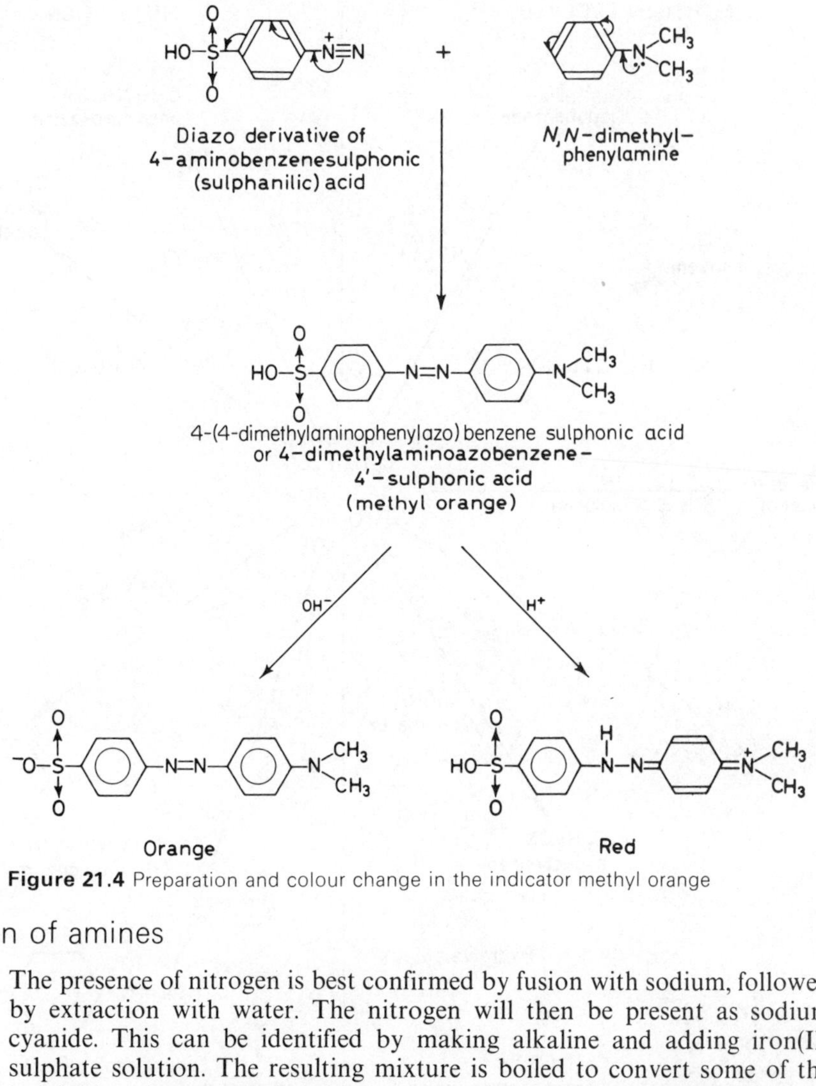

Diazo derivative of
4-aminobenzenesulphonic
(sulphanilic) acid

N,N-dimethyl-
phenylamine

4-(4-dimethylaminophenylazo)benzene sulphonic acid
or 4-dimethylaminoazobenzene-
4'-sulphonic acid
(methyl orange)

Orange

Red

Figure 21.4 Preparation and colour change in the indicator methyl orange

Identification of amines

The presence of nitrogen is best confirmed by fusion with sodium, followed by extraction with water. The nitrogen will then be present as sodium cyanide. This can be identified by making alkaline and adding iron(II) sulphate solution. The resulting mixture is boiled to convert some of the iron(II) ions into iron(III). Acidification then causes the iron(II) ions to react with the cyanide ions to give the complex hexacyanoferrate(II) ion which, in turn, reacts with the sodium and iron(III) ions to give the characteristic 'Prussian blue' colour of iron(III) sodium hexacyanoferrate(II)

$$Fe^{2+} + 6CN^- \rightarrow [Fe(CN)_6]^{4-}$$

$$Na^+ + Fe^{3+} + [Fe(CN)_6]^{4-} \rightarrow Na^+[Fe^{II}Fe^{III}(CN)_6]^-(s)$$

All primary amines react with trichloromethane in alcoholic potassium hydroxide to give evil-smelling isonitriles (the carbylamine test):

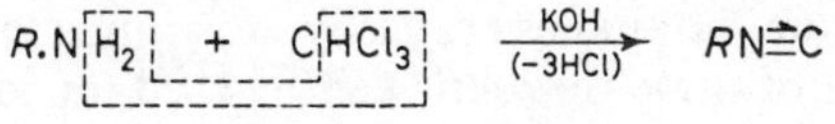

$$R.NH_2 + CHCl_3 \xrightarrow[(-3HCl)]{KOH} RN{\equiv}C$$

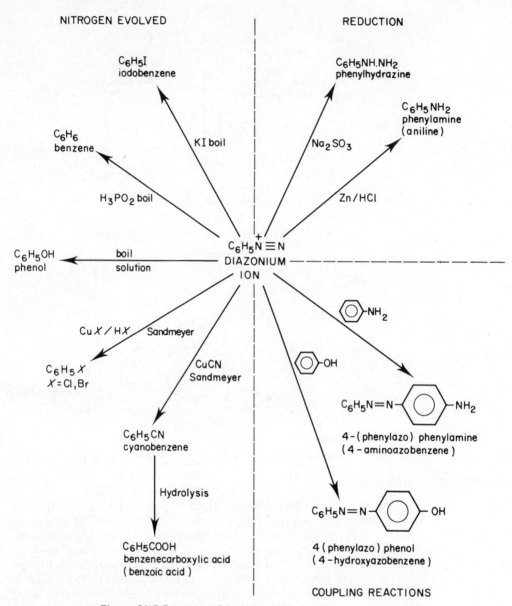

Figure 21.5 Reactions of the benzenediazonium ion

Nitrogen is evolved if a primary amine is treated with nitrous acid solution (p. 456). The formation of a yellow oil without the evolution of nitrogen is indicative of a secondary amine (p. 457), whilst tertiary amines give no visible reaction with nitrous acid.

The activating effect of the amino group when attached to the benzene nucleus means that aromatic amines are easily substituted in the 2-, 4- and 6-positions, and several of these reactions can be used for identification; for example, a white precipitate of 2,4,6-tribromophenylamine is instantly formed when phenylamine is treated with bromine water:

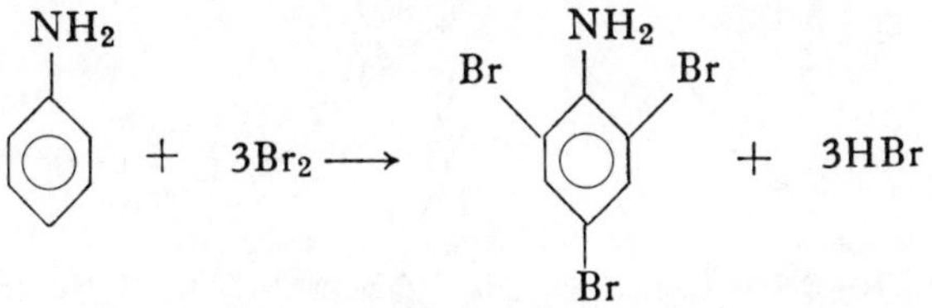

The more usual derivative for amines, however, is the benzenecarboxamide, prepared from benzoyl chloride (the Schotten–Baumann reaction)

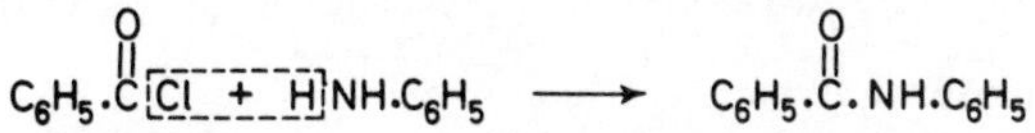

N–Phenylbenzenecarboxamide

The ability of the lone pair on the nitrogen atom to undergo protonation is the basis for the purification of amines. If an impure ethereal solution of an amine is shaken with an acid, the amine forms a salt and dissolves in the aqueous phase. Separation of this phase, followed by treatment with alkali, leads to regeneration of the amine:

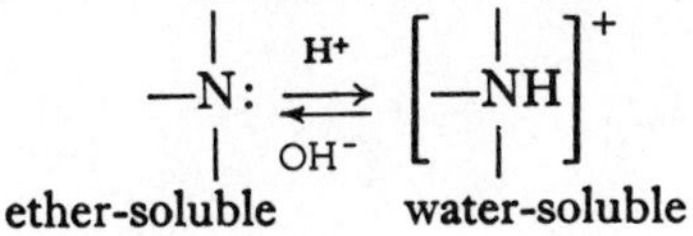

ether-soluble   water-soluble

Nitro compounds

Nitro compounds can be prepared by direct nitration of a hydrocarbon using nitric acid, but the ease with which the reaction takes place depends upon whether there are any π electrons present in the hydrocarbon and upon the susceptibility to oxidation. Aromatic compounds, for example, possess the electron clouds attractive to the nitryl ion and so reaction takes place readily, particularly in the presence of substances such as concentrated sulphuric acid which favour the formation of this ion (p. 429).

$$HNO_3 + 2H_2SO_4 \rightarrow H_3O^+ + 2HSO_4^- + NO_2^+$$

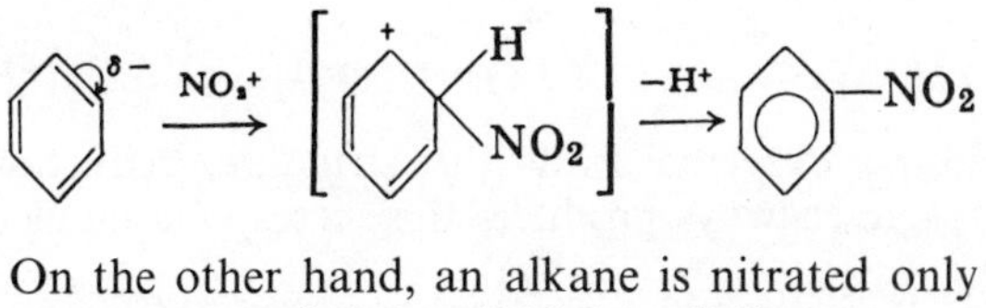

On the other hand, an alkane is nitrated only in the gaseous phase at a temperature of about 400 °C (p. 430), e.g.,

$$CH_4 + HNO_3 \rightarrow CH_3NO_2 + H_2O$$

All nitro compounds are fairly easily reduced, the extent of reduction

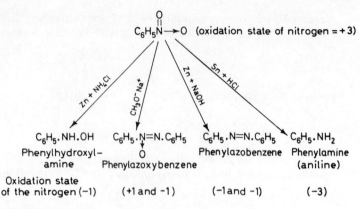

Figure 21.6 Reduction of nitrobenzene

depending on the conditions employed. Thus, nitrobenzene is reduced to phenylhydroxylamine in neutral solution with zinc and ammonium chloride, to phenylazoxybenzene with sodium methoxide, to phenylazobenzene with zinc in caustic alkali and to phenylamine with metals in acid solution as shown in *Figure 21.6.* Primary and secondary nitro compounds, by virtue of the hydrogen atom on carbon-1, exhibit dynamic isomerism or *tautomerism*, in which the two forms differ only in the position of the hydrogen atom:

$$-\overset{\displaystyle |}{\underset{\displaystyle H}{C}}-\overset{\displaystyle ||}{\underset{\displaystyle O}{N}}\rightarrow O \quad \rightleftharpoons \quad -\overset{\displaystyle |}{C}=\underset{\displaystyle OH}{N}\rightarrow O$$

As a result of this equilibrium, these compounds exhibit weakly acidic properties. Tertiary (including aromatic) nitro compounds are without the requisite hydrogen atom and are accordingly neutral.

Nitrites

Nitrites, $RO.N{=}O$, are isomeric with nitro compounds; the difference between them is that, whereas carbon is linked with nitrogen in the nitro compounds, in the nitrites carbon is attached to oxygen. Nitrites can, in fact, be regarded as esters of nitrous acid and can be prepared by reaction between nitrous acid (i.e. sodium nitrite and hydrochloric acid) and the alcohol

$$ROH + HO.N{=}O \rightarrow RO.N == O + H_2O$$

As would be expected from the structure, reduction of nitrites is unambiguous and always produces the corresponding alcohol

$$RO.N{=}O \xrightarrow{[H]} ROH$$

Consideration of the structures also suggests that nitro compounds are more polar than the nitrites; in accordance with this, there is more molecular

Table 21.2 Boiling points of isomeric nitro compounds and nitrites

Nitro compound	B.p./°C	Nitrite	B.p./°C
$CH_3N(:O \to O$	101	$CH_3ON=O$	-12
$C_2H_5N(:O) \to O$	113	$C_2H_5ON=O$	16

interaction and nitro compounds are found to be less volatile than the nitrites, as shown in *Table 21.2*.

Nitriles, $RC{\equiv}N$, and isonitriles, $RN{\equiv}C$

Nitriles can be obtained in the following ways:

(1) Dehydration of an amide with phosphorus(V) oxide:

$$RC.NH_2 \xrightarrow{P_4O_{10}} RC{\equiv}N + H_2O$$

(the O is double-bonded to the first C: R–C(=O).NH₂)

(2) The action of cyanide on alkyl halides:

$$RX + CN^- \to RC{\equiv}N + X^-$$

Some isonitrile is also formed in this reaction; in fact, by using the covalent silver cyanide instead of the electrovalent potassium cyanide, isonitrile predominates:

$$AgCN + RX \to AgX + RN{\equiv}C$$

Isonitriles are the evil-smelling substances referred to previously in the reaction between primary amines, trichloromethane and alcoholic potassium hydroxide:

$$RNH_2 + CHCl_3 + 3OH^- \to RN{\equiv}C + 3Cl^- + 3H_2O$$

The bond between nitrogen and carbon in both compounds is multiple and the difference in electronegativity between the two elements results in some polarization, so that the molecule can undergo *nucleophilic attack at the carbon* atom and *electrophilic attack at nitrogen*. Thus, both nitriles and isonitriles can be hydrolysed by acid, the former giving rise to carboxylic acids and the latter to amines. In the case of nitriles, the initial attack is by a proton on the nitrogen

$$R-C{\equiv}N \ (+H^+) \longrightarrow R-\overset{+}{C}=NH$$

The overall reactions can be summarized as follows:

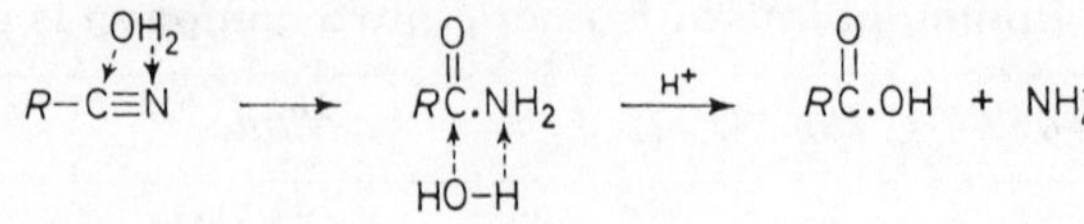

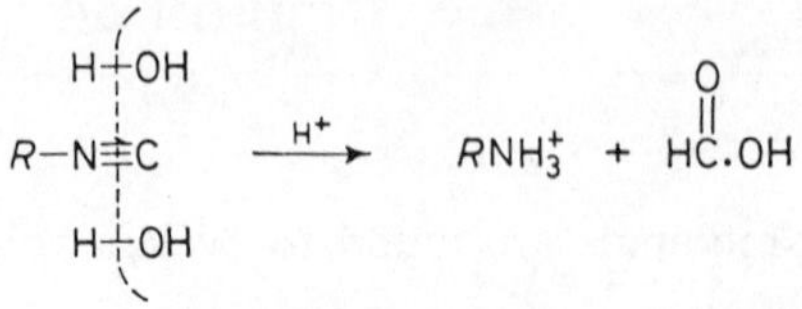

$$R-C\equiv N \xrightarrow[\substack{\uparrow\\ HO-H}]{OH_2} \underset{HO-H}{R\overset{O}{\overset{\|}{C}}.NH_2} \xrightarrow{H^+} R\overset{O}{\overset{\|}{C}}.OH + NH_4^+$$

$$R-N\overset{\substack{H+OH}}{\overset{\|}{\underset{H+OH}{C}}} \xrightarrow{H^+} RNH_3^+ + H\overset{O}{\overset{\|}{C}}.OH$$

Reduction of nitriles gives primary amines, but isonitriles are converted into secondary amines. This is a natural consequence of the fact that, in the case of the latter, nitrogen is already attached to two carbon atoms:

$$RC\equiv N + 4[H] \rightarrow RCH_2NH_2$$

$$RN\equiv C + 4[H] \rightarrow RNHCH_3$$

Heterocyclic compounds

Many naturally occurring compounds of great importance contain nitrogen as one or more members of a closed ring system, i.e., as part of a *heterocyclic* system. For example, they play a fundamental part in processes of biological inheritance and in the transfer of energy in the cell, as well as comprising the *alkaloids*, a group of physiologically active, basic substances derived from plants. Any book attempting a broad survey of the field of chemistry would be incomplete without some reference to such compounds.

Pyridine

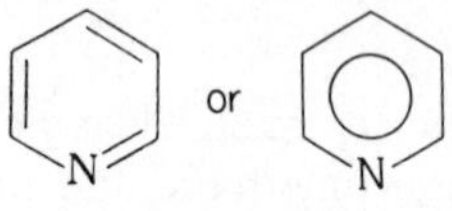

Pyridine is an unpleasant-smelling liquid present in coal tar. It is basic by virtue of the lone pair on the nitrogen but it is weaker than ammonia and comparable with tertiary amines. The *p* orbitals of nitrogen allow it to play a full part in the non-localized system of the hexagon, so that pyridine is truly aromatic.

The pyridine residue is present in the alkaloid nicotine and in vitamin B_6, the anti-pellegra factor.

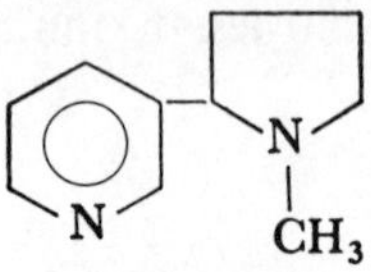

Nicotine 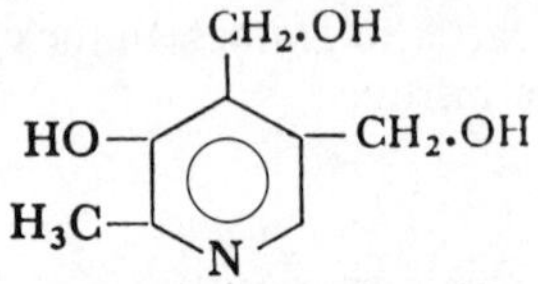   Vitamin B_6

Compounds containing two quaternary ammonium groups present as pyridine residues are found to interfere with the photosynthetic process in grasses and are finding application as non-emergent weed killers, e.g., as seed dressing such as:

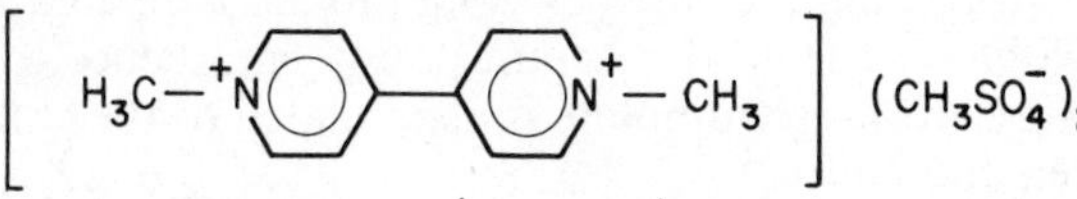

$$\left[H_3C-{}^{+}N \bigcirc - \bigcirc N^{+}- CH_3 \right] (CH_3SO_4^{-})_2$$

'Paraquat'

Pyrrole

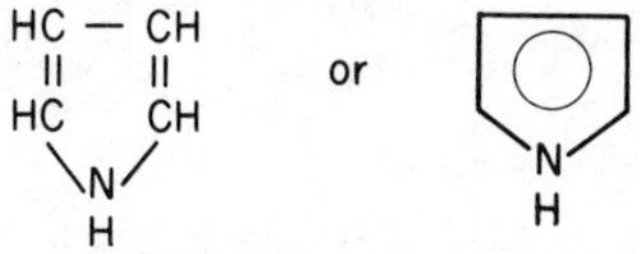

HC — CH
‖   ‖   or
HC   CH
  \N/
    H

This compound has a five-membered aromatic ring with the lone pair on the nitrogen contributing to the resonance of the system; it is therefore not available for protonation, so that pyrrole is not basic: it is, in fact, slightly acidic.

Four pyrrole rings joined together by carbon 'bridges' comprise the interesting *porphyrin* framework; *haem* contains the iron(II) ion at the centre of the structure and *chlorophyll*, magnesium. Both haem and chlorophyll are *prosthetic* groups, i.e., they are attached to protein in the normal, active condition. In the case of haem, the complete substance, known as *haemoglobin*, plays a vital part in respiration, transporting oxygen in the form of *oxyhaemoglobin* from the lungs to the cells. Chlorophyll is of prime importance as a catalyst in photosynthesis (p. 519).

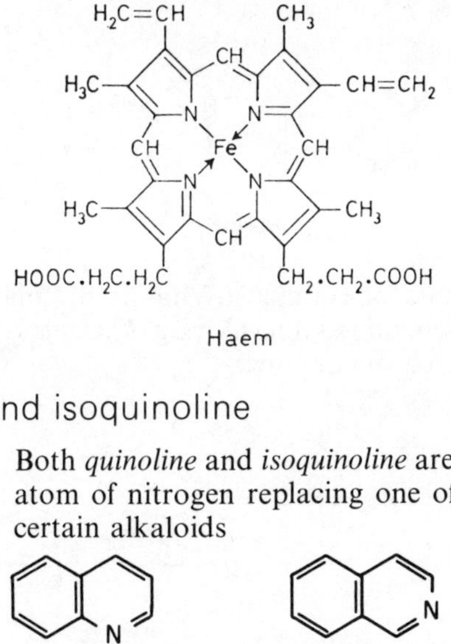

Haem 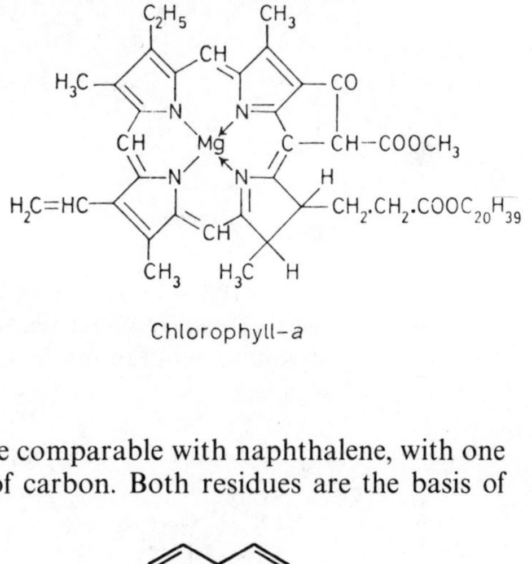   Chlorophyll–a

Quinoline and isoquinoline

Both *quinoline* and *isoquinoline* are comparable with naphthalene, with one atom of nitrogen replacing one of carbon. Both residues are the basis of certain alkaloids

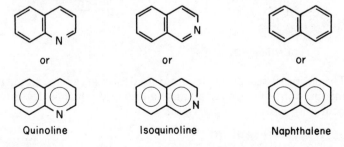

Quinoline   Isoquinoline   Naphthalene

Quinoline is related to the poisonous alkaloid strychnine and also to the antimalarial drug quinine, whereas isoquinoline is related to the alkaloid papaverine, obtained from the opium poppy, and hence to its derivatives morphine and heroin. Isoquinoline is also related to the stimulant caffeine, present in tea and coffee.

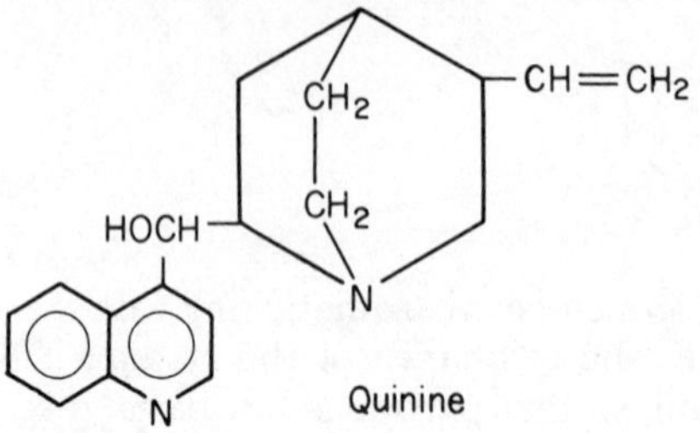

Quinine 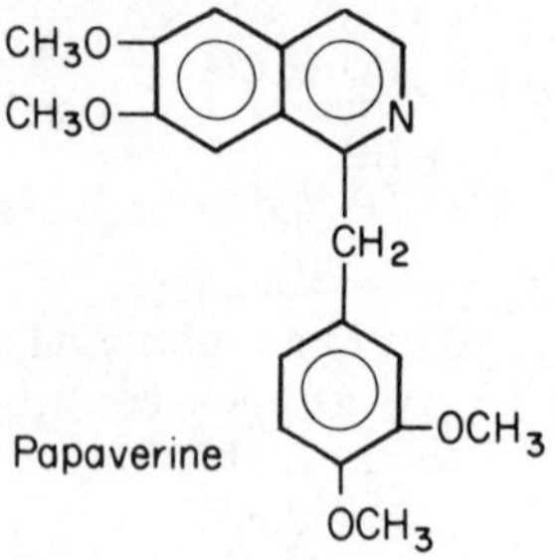   Papaverine

Pyrimidine and purine; nucleotides

Pyrimidine and purine bases are of especial importance nowadays, with the present emphasis on research into the chemistry of genetics. Pyrimidines contain two nitrogen atoms replacing carbon in the benzene hexagon, whilst purines are pyrimidines attached to a 5-membered closed system containing two nitrogen atoms

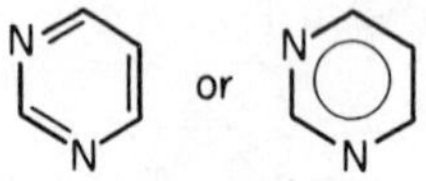

Pyrimidine 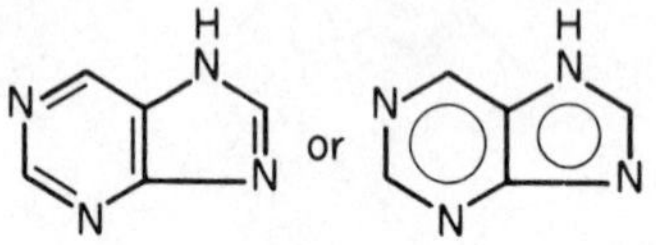   Purine

The pyrimidine ring occurs for example in Vitamin B_1 (thiamin), the anti-beriberi factor, which also contains a thiazole ring. Thiazole is an example of a sulphur-containing heteroaromatic ring.

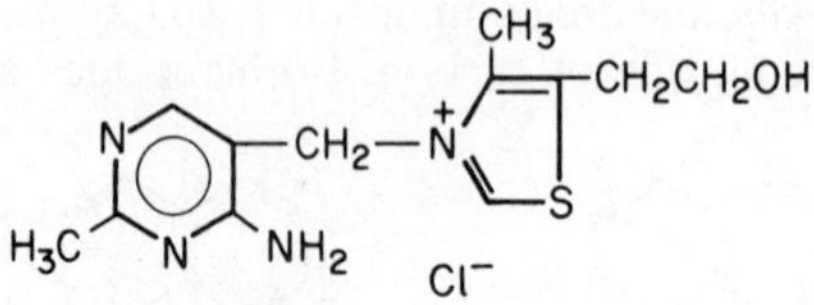

Vitamin B_1 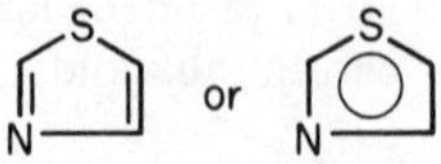   Thiazole

The purine system occurs in an oxidized form in caffeine and in the stimulant theobromine (in cocoa), and also in uric acid which is deposited in cases of gout.

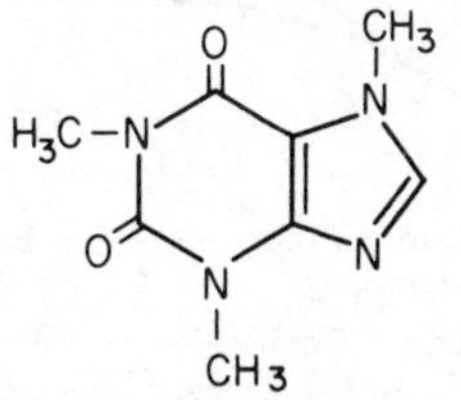

Caffeine 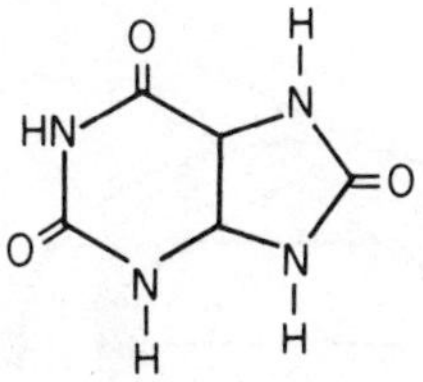   Uric acid

The genetic material DNA (deoxyribonucleic acid) contains the two pyrimidines *thymine* and *cytosine*, and the two purines *adenine* and *guanine*.

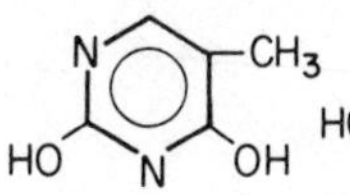

Thymine 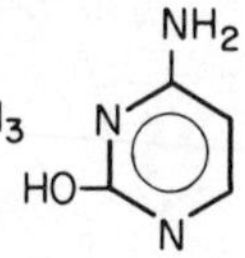   Cytosine 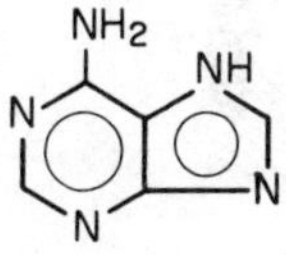   Adenine 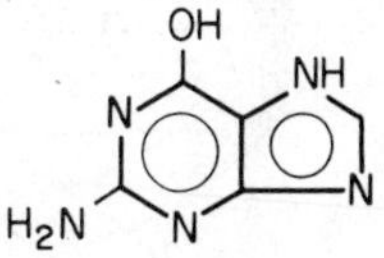   Guanine

These bases are linked to deoxyribose, which is also linked to a phosphoric(V) acid residue, the whole unit being called a *nucleotide*.

An example of a nucleotide is adenylic acid

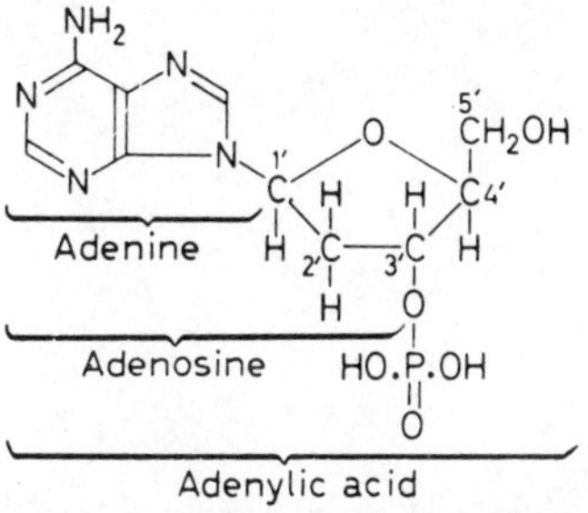

Adenylic acid

Condensation of this particular nucleotide with one molecule of phosphoric(V) acid gives ADP and, with two molecules, ATP (p. 518).

Repeating of units like the adenylic acid shown above through position 5' of the sugar and through a hydroxyl group of the phosphoric(V) acid residue gives one-half of a DNA molecule, which in some respects can be compared to a spiral staircase. The other half is constructed in a similar way and the 'rungs' are established by hydrogen bonding between adenine and thymine and between guanine and cytosine (*Figure 21.7*). Only if one base is a purine and one a pyrimidine is there the 'right' geometric fit for the 'rungs' to be formed.

The affinity between the corresponding pairs of bases means that when a molecule splits lengthwise during cell division, two identical molecules, one for each new cell, can be constructed from materials available in the protoplasm; that is, a mechanism for replication is established, as in *Figure 21.8*.

The structure of the DNA molecule determines the protein produced, i.e.,

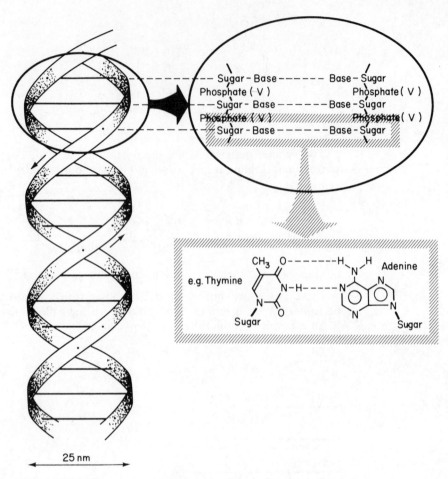

Figure 21.7 Details of part of a DNA molecule showing the double helical structure

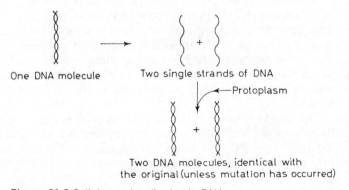

Figure 21.8 Splitting and replication in DNA

it provides the code for protein synthesis, but the intermediary, RNA (ribonucleic acid), is required to transport the information from the cell nucleus to the site of the protein synthesis. This 'transfer' RNA differs from DNA in that the sugar unit is ribose instead of deoxyribose and uracil replaces thymine

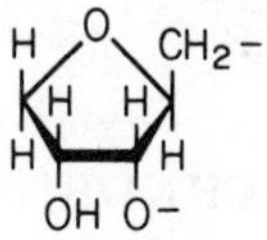

| Ribose unit | Deoxyribose unit | Uracil |

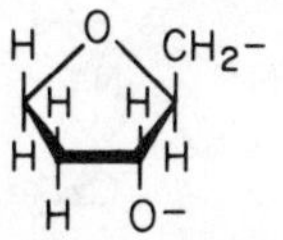

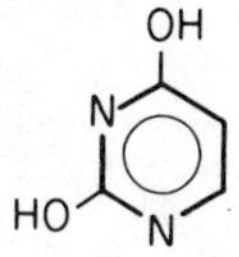

Uracil has an affinity for adenine and so it can serve a purpose similar to thymine. This means that when a DNA molecule divides, a single strand, instead of making a new molecule of DNA, can synthesize an RNA molecule of complementary structure. Every 'transfer' RNA molecule is specific for a particular amino acid, and so the particular DNA, by controlling the nature of the 'transfer' RNA molecules made from it, also controls the nature of the subsequent protein that is synthesized. It can be said to carry the genetic information in terms of a four-symbol code, according to the way in which the four bases are repeated and arranged along the length of the molecule.

The production of 'transfer' RNA may be shown as in *Figure 21.9*. (For details of the role played by RNA in protein synthesis, see p. 559).

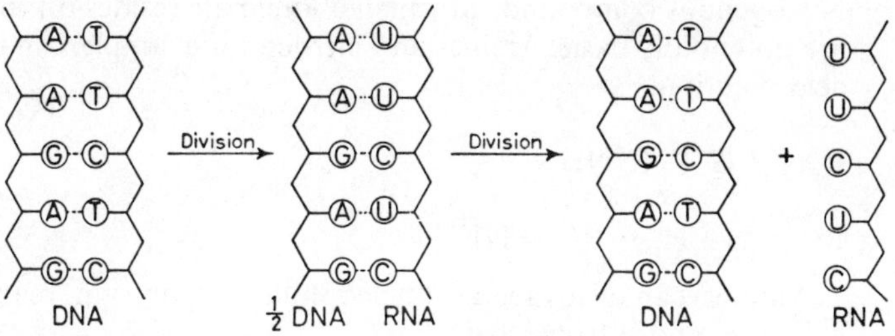

Figure 21.9 Production of transfer RNA

Pteridines

Pteridines resemble purines in that they too contain the pyrimidine residue attached to a ring containing two nitrogen atoms, but this latter ring is 6- instead of 5-membered. Pteridines (Greek *pteron* = wing) are present on the wings of insects, e.g., white *leucopterin* in cabbage white butterflies and yellow *xanthopterin* in brimstones. Apart from adornment, these substances appear

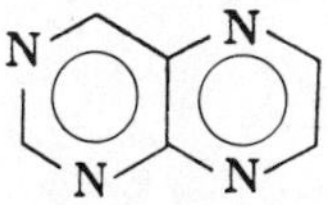

| Pteridine | Leucopterin | Xanthopterin |

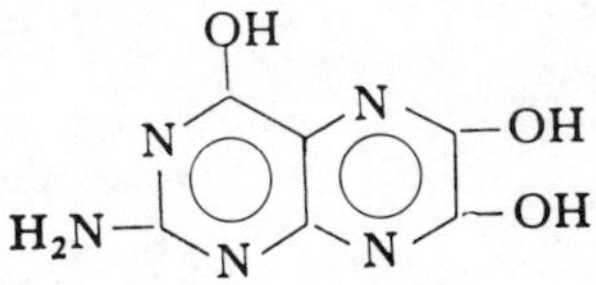

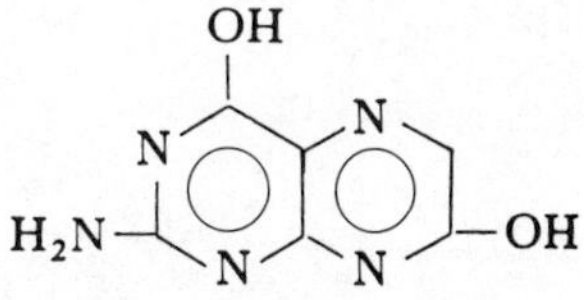

to have growth-controlling properties and indeed bear a close resemblance to two vitamins, *riboflavin* (Vitamin B$_2$) and folic acid.

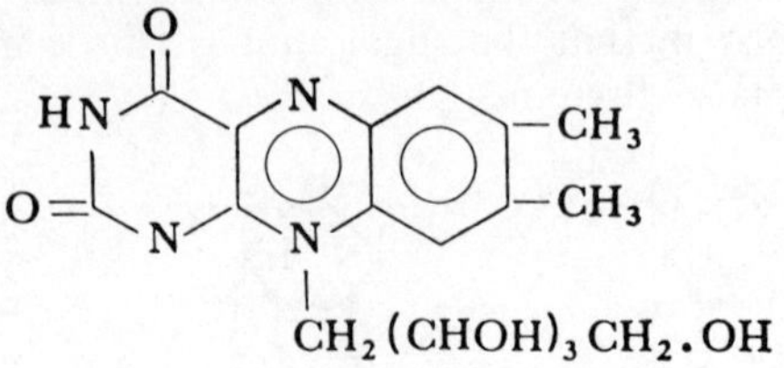

Riboflavin

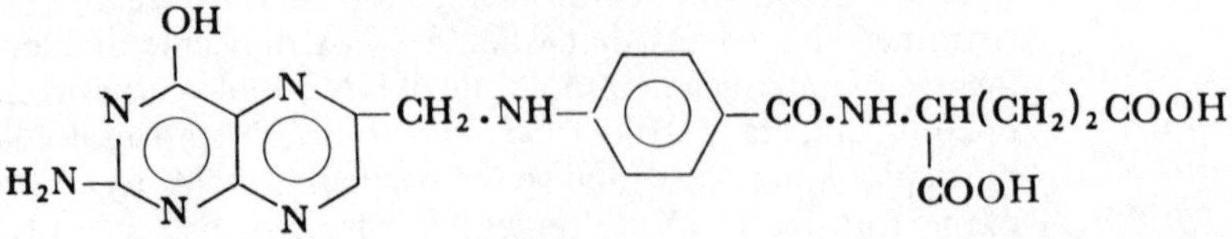

Folic acid

Summary

Nitrogenous compounds in which a lone pair resides on the nitrogen atom are potentially basic. Amines and nitriles, for example, can be protonated in acid solution:

$$R_3N: + H^+ \rightarrow R_3NH^+$$

$$RC \equiv N + H^+ \rightarrow RC \overset{+}{=} NH$$

Amines can also act as nucleophilic reagents in reactions with acid chlorides and nitrous acid.

If the nitrogen atom is attached directly to an aromatic nucleus, then the lone pair forms part of the non-localized system. By activating the nucleus, it also encourages 2,4-substitution:

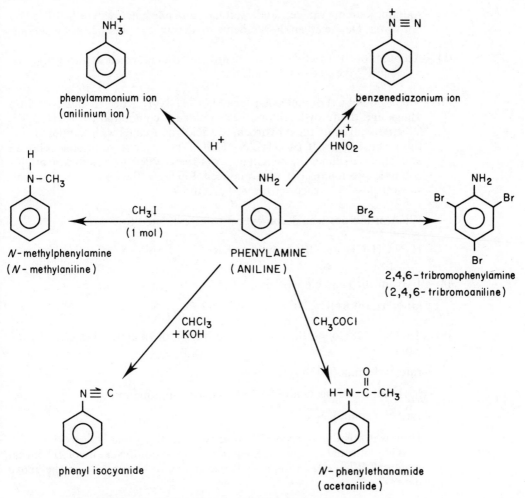

Figure 21.10 Some reactions of phenylamine

Questions

(1) Starting from ethene, how could (*a*) N-methylaminoethane, (*b*) butane-1,4-dioic acid, (*c*) butane-1,4-diamine, (*d*) propanoic acid, (*e*) 2-hydroxy-1-(*NN*-dimethyl)aminoethane be obtained?

(2) Explain mechanistically the difference in the action of potassium cyanide and silver cyanide on alkyl halides.

(3) Give a qualitative explanation of the differences in the dissociation constants of the amines listed in *Table 21.1*.

(4) In the method referred to as Lassaigne's sodium test, if an organic substance containing nitrogen, a halogen, sulphur and phosphorus is fused with

sodium, sodium cyanide, halide, sulphide and phosphide can be extracted by hot water. Devise an analysis scheme to identify the presence of the anions.

(5) Account for the reactivity of the hydrogen atom attached to nitrogen in the benzene-1,2-dicarboximide molecule.

(6) A compound (A) of molecular formula $C_7H_7NO_2$, when treated with iron filings and dilute hydrochloric acid, yielded a compound (B), $C_7H_{10}ClN$. When dissolved in dilute hydrochloric acid and treated with sodium nitrite, this reacted to give (C), $C_7H_7ClN_2$. Half of the resultant solution was boiled and made alkaline; the remainder was then added to this and a bright-coloured precipitate, (D), was produced. Elucidate the above reactions and suggest possible structures for A, B, C and D.

(7) What properties would you expect to be exhibited by

$H_2N.CH_2CH == CHCl$ and by H_2N—⟨○⟩—NO_2 ?

(8) Explain the characteristic properties of nitrogen when present in:

(a) NH_3, (b) $RNHR'$, (c) $C_6H_5.NH_2$, (d) $RC≡N$

(9) Taylor (1928) found the rate of diazotization of an amine was given by the equation

rate $\propto$ [amine][HNO_2]^2

Hammett subsequently (1933) modified this equation to

rate $\propto$ [amine][N_2O_3]

How is it possible to reconcile these two rate equations?

What information can a rate equation give about the nature of a chemical reaction? What, for example, is it reasonable to conclude from the above data?

(10) Write an essay on heterocyclic compounds.

(11) Why should the acid concentration not be too high during diazotization, especially when the amine concerned is aliphatic?

(12) Replication of a DNA molecule has been dealt with in this chapter from the point of view of the human being. How do the basic constituents of the DNA (or similar genetic material) vary throughout the animal world? How reasonable do you think is the view that these constituents emerged by chance in the geological past and since then have controlled the course of organic development?

(13) Nitriles undergo hydrolysis in the presence of acid or alkali, whilst isonitriles only hydrolyse in the presence of acid. Suggest a mechanistic reason for this difference.

(14)   [1-^{14}C]Chlorobenzene treated with potassium amide in liquid ammonia gives an equimolar mixture of phenyl-1-amine and phenyl-2-amine. Suggest a mechanism to account for this observation.

22 Hydroxy compounds: alcohols and phenols

Alcohols and phenols • Occurrence • Preparation • Properties • Industrial aspects • Analysis

Alcohols and phenols

Organic compounds containing the hydroxy group (—OH) are normally classified as alcohols or phenols. The latter are those compounds in which the hydroxy group is directly linked to an aromatic ring. It has been seen in Chapter 20 how the aromatic ring can modify the properties of a halogen atom, and its effect on an amino group has been discussed in Chapter 21. In similar fashion it can considerably alter the properties of a hydroxy group.

The number of hydroxy groups in a compound is denoted by use of the terms monohydric, dihydric, etc. Monohydric alcohols can be further divided into three types, depending on the number of hydrogen atoms attached to the carbon atom carrying the hydroxy group

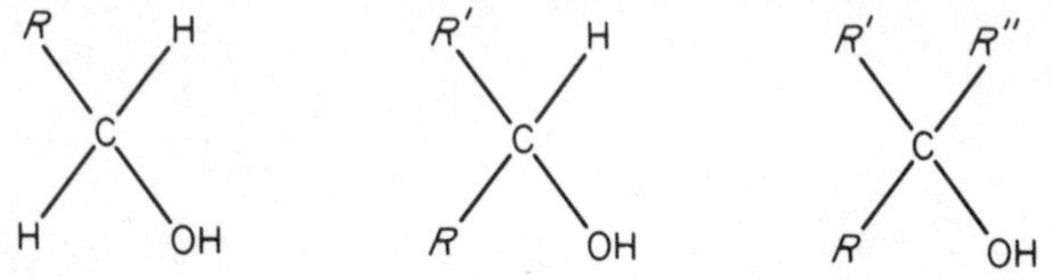

or	$RCH_2.OH$	$R'RCH.OH$	$R'R''RC.OH$
	primary	secondary	tertiary

Alcohols and phenols are named by replacing the final -e of the corresponding hydrocarbon by -ol (monohydric), or adding -diol (dihydric), etc. to the full name of the hydrocarbon root, together with the relevant numbers of carbon atoms where necessary. An exception is C_6H_5OH which retains the name *phenol*.

Some typical hydroxy compounds are shown in *Table 22.1*.

Occurrence

Many alcohols occur as terpene derivatives in plants, e.g., *phytol*, $C_{20}H_{40}O$, containing four isoprene units, is produced in the hydrolysis of chlorophyll (p. 465):

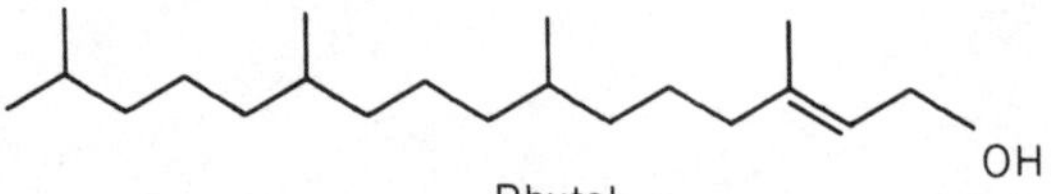

Phytol

Table 22.1 Representative hydroxy compounds

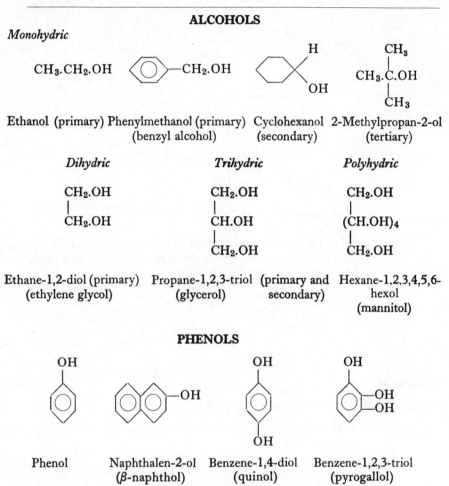

ALCOHOLS

Monohydric

CH₃.CH₂.OH

Ethanol (primary)

Phenylmethanol (primary)
(benzyl alcohol)

Cyclohexanol
(secondary)

2-Methylpropan-2-ol
(tertiary)

Dihydric

CH₂.OH
|
CH₂.OH

Ethane-1,2-diol (primary)
(ethylene glycol)

Trihydric

CH₂.OH
|
CH.OH
|
CH₂.OH

Propane-1,2,3-triol (primary and
(glycerol)   secondary)

Polyhydric

CH₂.OH
|
(CH.OH)₄
|
CH₂.OH

Hexane-1,2,3,4,5,6-
hexol
(mannitol)

PHENOLS

Phenol

Naphthalen-2-ol
(β-naphthol)

Benzene-1,4-diol
(quinol)

Benzene-1,2,3-triol
(pyrogallol)

and *menthol* is a reduced hydroxy derivative of limonene

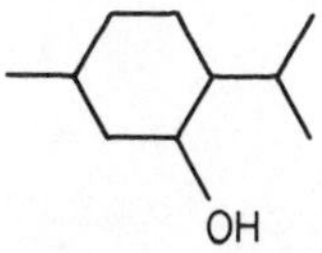

The formation of 3β-hydroxycholest-5(6)-ene *cholesterol* (a precursor of Vitamin D) from the skeleton of squalene (p. 410) can be seen by reorientating the structure shown earlier:

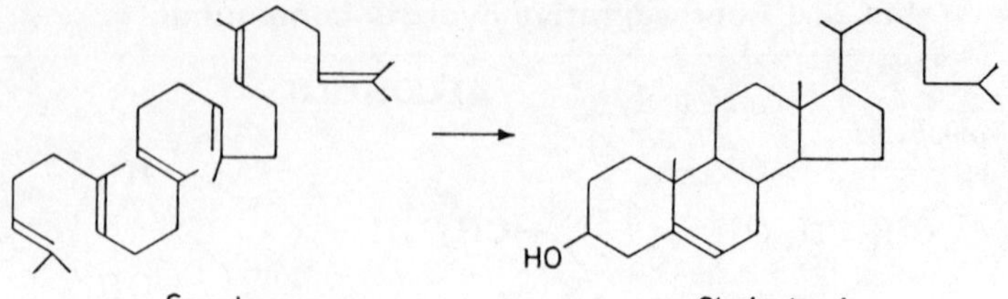

Squalene   Cholesterol

Vitamins D_1–D_4 are closely related to this compound, e.g., Vitamin D_2:

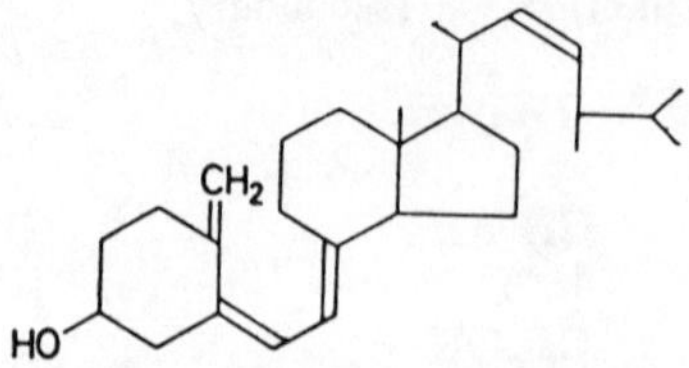

In some biochemical processes the benzene ring is readily hydroxylated, giving rise to many naturally occurring phenols such as the tannins, some phenolic ethers and Vitamin E:

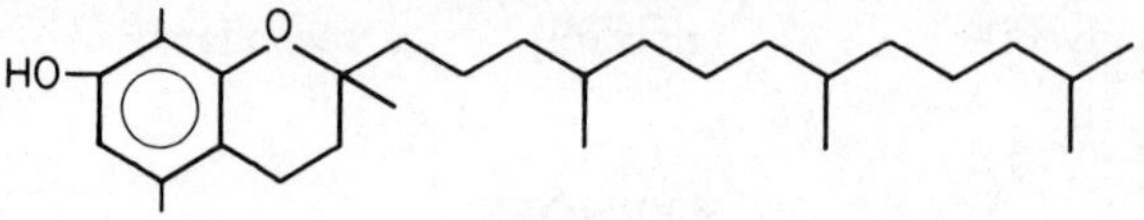

Vitamin E

Preparation

Preparation from the hydrocarbon

(a) If an alkene is treated with concentrated sulphuric acid, an alkyl hydrogen-sulphate results:

$$R\text{CH}=\text{CH}_2 + \text{HO}-\overset{\displaystyle\text{O}}{\underset{\displaystyle\text{OH}}{\overset{\|}{\underset{|}{\text{S}}}}}=\text{O} \longrightarrow R\text{CH}.\text{CH}_3$$
$$\text{O.SO}_2.\text{OH}$$

Addition of water converts the alkyl hydrogensulphate into the alcohol, which can be removed by distillation:

$$\begin{array}{c} R\text{CH}.\text{CH}_3 \\ | \\ \text{O.SO}_2.\text{OH} \end{array} + \text{H}_2\text{O} \longrightarrow \begin{array}{c} R\text{CH}.\text{CH}_3 \\ | \\ \text{OH} \end{array} + \text{H}_2\text{SO}_4$$

Hydration of the alkene can also be achieved in the vapour phase under suitable conditions:

$$R-\text{CH}=\text{CH}_2 + \text{H}_2\text{O(g)} \xrightarrow[\substack{\text{pumice} \\ 300\,°\text{C}}]{\text{H}_3\text{PO}_4 \text{ on}} R\text{CH}_2\text{CH}_2\text{OH}$$

(b) Aromatic hydrocarbons are slowly sulphonated by hot, concentrated sulphuric acid. As the reaction is believed to involve sulphur(VI) oxide, the velocity is increased by using 'oleum' (p. 428), e.g.,

$$C_6H_6 + H_2SO_4 \rightarrow C_6H_5.SO_2.OH \qquad + H_2O$$
$$\text{Benzenesulphonic acid}$$

Fusion of the sodium salt of benzenesulphonic acid with sodium hydroxide, followed by acidification, results in the formation of phenol. This method was of economic importance but now only accounts for about 5 per cent of world production:

$$C_6H_5.SO_2.O^- + OH^- \rightarrow C_6H_5.O^- + HSO_3^-$$
$$C_6H_5.O^- + H^+ \rightarrow C_6H_5.OH$$

Preparation from the amine

Primary amines react with nitrous acid (i.e., cold dilute hydrochloric acid and cold dilute sodium nitrite solution) to give diazonium compounds (p. 456):

$$RNH_2 + HNO_2 \rightarrow RN_2^+ + H_2O + OH^-$$

Aliphatic diazonium compounds decompose immediately, so that the overall reaction is often simply represented by the equation:

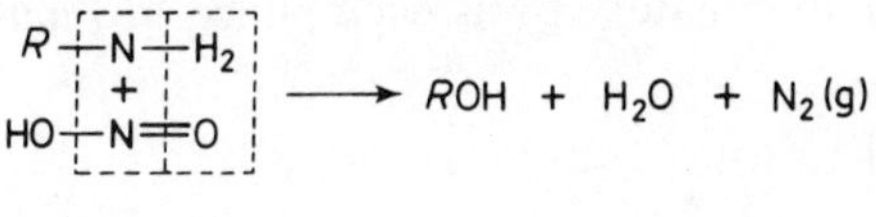

Aromatic diazonium compounds are, however, stable at low temperatures ($< 10\,°C$) and are converted into phenols on warming, e.g.,

$$C_6H_5.\overset{+}{N} \equiv N \rightarrow C_6H_5^+ + N_2(g)$$
$$C_6H_5^+ + H_2O \rightarrow C_6H_5.OH + H^+$$

Preparation from the halides

Halogen derivatives can be hydrolysed by refluxing with alkali. Alkyl halides are easily hydrolysed, e.g.,

$$\text{CH}_3\text{I} \qquad + OH^- \rightarrow CH_3.OH + I^-$$
$$\text{Iodomethane} \qquad\quad \text{Methanol}$$

whereas aryl halides must be subjected to a pressure of 20 MPa and a temperature of about 300 °C (p. 441). Tertiary halides tend to give alkenes under these conditions, so that in these cases other methods are preferred.

Preparation from carbonyl compounds, etc.

Esters, ketones and aldehydes can be reduced to alcohols and phenols by a variety of methods, including treatment with a Grignard reagent or with

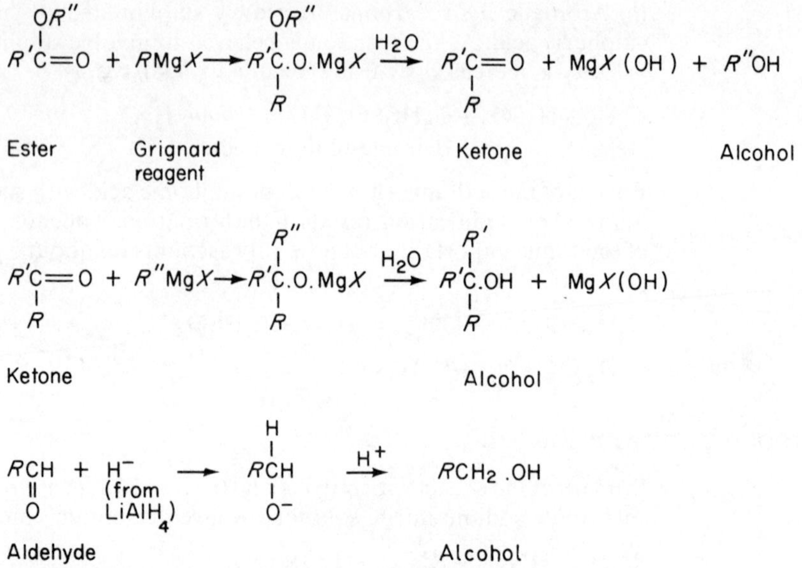

$$\underset{\text{Ester}}{R'\overset{\displaystyle OR''}{\underset{\displaystyle R}{C}}=O} + \underset{\substack{\text{Grignard}\\\text{reagent}}}{RMgX} \longrightarrow R'\overset{\displaystyle OR''}{\underset{\displaystyle R}{C}}.O.MgX \xrightarrow{H_2O} \underset{\text{Ketone}}{R'\underset{\displaystyle R}{C}=O} + MgX(OH) + \underset{\text{Alcohol}}{R''OH}$$

$$\underset{\text{Ketone}}{R'\underset{\displaystyle R}{C}=O} + R''MgX \longrightarrow R'\overset{\displaystyle R''}{\underset{\displaystyle R}{C}}.O.MgX \xrightarrow{H_2O} \underset{\text{Alcohol}}{R'\overset{\displaystyle R'}{\underset{\displaystyle R}{C}}.OH} + MgX(OH)$$

$$\underset{\substack{\text{Aldehyde}}}{R\underset{\displaystyle O}{\overset{\displaystyle\|}{C}}H} + \underset{\substack{\text{(from}\\\text{LiAlH}_4)}}{H^-} \longrightarrow R\overset{\displaystyle H}{\underset{\displaystyle O^-}{C}}H \xrightarrow{H^+} \underset{\text{Alcohol}}{RCH_2.OH}$$

Figure 22.1 Use of Grignard reagents and lithium tetrahydridoaluminate in production of alcohols from carbonyl compounds

lithium tetrahydridoaluminate, e.g., as exemplified in *Figure 22.1*.

Preparation by hydrolysis of esters

Esters can be hydrolysed to acids, together with alcohols or phenols, by acid or base catalysis, e.g.,

$$R\underset{O}{\overset{\|}{C}}.OR' \longrightarrow R\underset{O}{\overset{\|}{C}}.OH + R'OH$$

The base-catalysed hydrolysis of fats (see saponification, p. 551) is important as a means of manufacturing soaps and glycerol

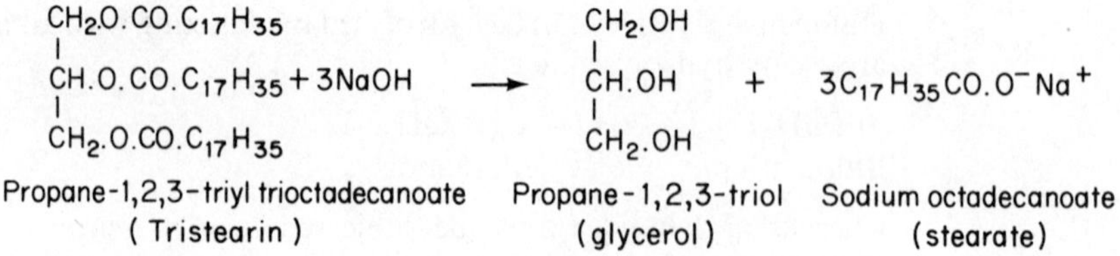

$$\begin{array}{l} CH_2O.CO.C_{17}H_{35} \\ | \\ CH.O.CO.C_{17}H_{35} + 3NaOH \\ | \\ CH_2.O.CO.C_{17}H_{35} \end{array} \longrightarrow \begin{array}{l} CH_2.OH \\ | \\ CH.OH \\ | \\ CH_2.OH \end{array} + \quad 3C_{17}H_{35}CO.O^-Na^+$$

Propane-1,2,3-triyl trioctadecanoate   Propane-1,2,3-triol   Sodium octadecanoate
(Tristearin)   (glycerol)   (stearate)

Biochemical processes

The anaerobic respiration of micro-organisms living in suitable media can produce a variety of compounds, including several alcohols (see 'fermentation', p. 523).

Table 22.2 Boiling points of hydrocarbons and hydroxy compounds of comparable relative molecular mass

Hydrocarbons	Relative molecular mass	B.p./°C	Hydroxy compounds	Relative molecular mass	B.p./°C
Ethane	30	−88	Methanol	32	65
Propane	44	−42	Ethanol	46	78
Butane	58	0	Propanol	60	98
Methylbenzene	92	110	Phenol	94	182
1,4-Dimethyl-benzene	106	138	Benzene-1,4-diol	110	286

Properties

The high electronegativity of oxygen in the —OH *group results in the hydroxy compounds being polarized* (cf. water) and allows the lower homologues of the alcohols to be completely miscible with water, i.e., —OH is a hydrophilic group. As the hydrocarbon chain increases in size, the solubility of the alcohols decreases. An increase in the number of hydroxy groups also results in an increase in hydrophilic character, so that propane-1,2,3-triol is hygroscopic and completely miscible with water but immiscible with ether. (The equilibrium between phenol and water is discussed on p. 103.)

The polarization of the hydroxy compounds is sometimes so marked that hydrogen bonding is present, giving rise to aggregates of molecules of relatively low volatility (*Table 22.2*):

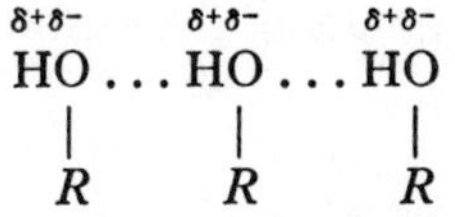

$$\overset{\delta+\delta-}{HO} \ldots \overset{\delta+\delta-}{HO} \ldots \overset{\delta+\delta-}{HO}$$
$$\underset{R}{|} \qquad \underset{R}{|} \qquad \underset{R}{|}$$

In the case of phenols, polarization is so pronounced that there is a measurable amount of ionization: they are particularly non-volatile and even acidic:

$$C_6H_5.OH \rightleftharpoons C_6H_5.O^- + H^+$$

The forward reaction is favoured by the fact that lone pairs of electrons on the oxygen atom can form part of the delocalized π bonding of the benzene ring

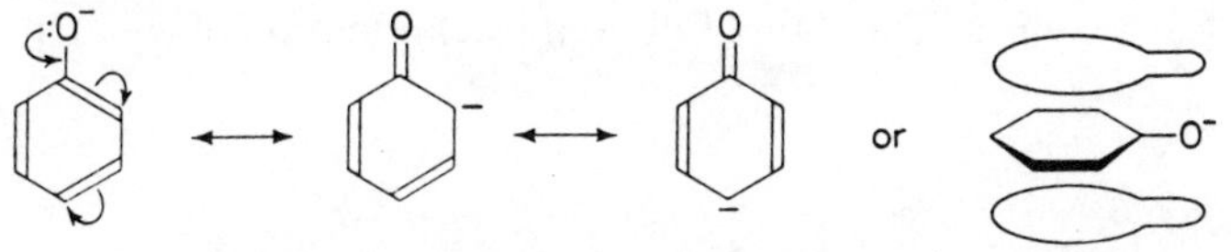

Because of their acidity, phenols, unlike alcohols, are soluble in sodium hydroxide solution, but generally their acidity is not great enough for them to

dissolve in the low concentration of hydroxide ions provided by sodium carbonate solution. Their acidity is enhanced by the presence of electron-attracting groups in the 2- and 4-positions. For example, 4-nitrophenol exhibits stronger acid properties than does phenol itself

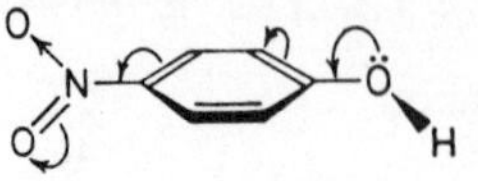

and 2,4,6-trinitrophenol (picric acid) will even decompose carbonates.

Although alcohols are not acidic, they will react with metallic sodium to form the corresponding *alkoxide*:

$$ROH + Na \rightarrow RO^- Na^+ + \tfrac{1}{2}H_2(g)$$

The oxygen atom of alcohols, by virtue of its lone pairs of electrons, *can be protonated*:

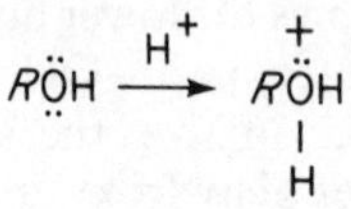

Protonation of this type is often the precursor to reactions given by alcohols and is responsible for the several differences between the chemistry of alcohols and phenols.

In the case of phenols, the lone pairs on the oxygen are less available, as they form part of the delocalized electron system. However, they do activate the ring towards electrophilic attack, especially at positions 2, 4 and 6. Thus, bromine water gives an immediate precipitate of 2,4,6-tribromophenol with phenol:

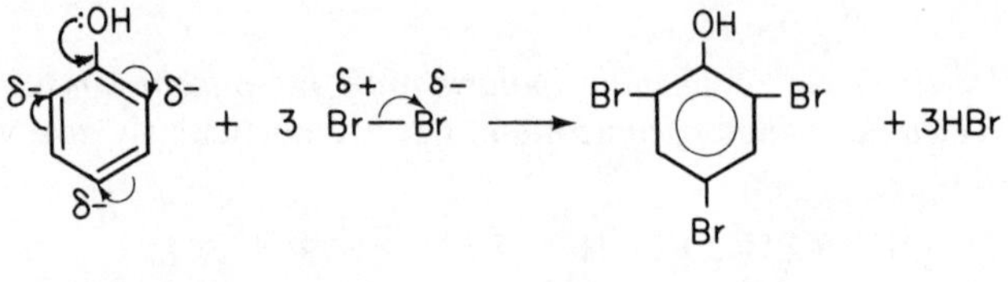

Ester formation (esterification)

The lone pair on the alcohol is able to attack the carbon of a protonated carboxylic acid and this, followed by elimination of water, produces an ester:

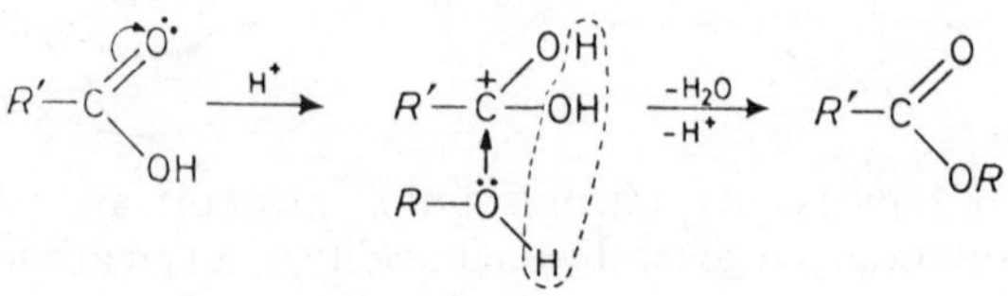

that is

$$\overset{O}{\underset{\parallel}{R'C}}.OH + ROH \longrightarrow \overset{O}{\underset{\parallel}{R'C}}.OR + H_2O$$

That it is an oxygen of the carboxylic acid which is eliminated was shown to be the case by using an ^{18}O-enriched alcohol; after esterification, no enrichment of the oxygen in the water produced was found, so that the reaction is

$$RCO.\overline{|OH} + \overline{H|}OR' \longrightarrow RCO.OR' + H_2O$$

in accordance with nucleophilic attack by the alcohol on protonated acid, and *not*

$$RCO.O\overline{|H} + HO|R' \longrightarrow RCO.OR' + H_2O$$

The reaction is reversible and the equilibrium can be favourably displaced for ester formation by using concentrated sulphuric acid, anhydrous zinc chloride or dry hydrogen chloride under reflux.

Many alcohols, particularly the secondary and tertiary ones, which readily form carbonium ions, easily eliminate the elements of water (p. 416) to produce alkenes at moderate temperatures in the presence of concentrated sulphuric acid or anhydrous zinc chloride, e.g.,

$$CH_3 \rightarrow \underset{\overset{|}{CH_3}}{\overset{CH_3}{\overset{+}{C.OH}}} \xrightarrow{-H_2O} CH_3 - \underset{\overset{|}{H_2CH}}{\overset{CH_3}{\overset{|}{C^+}}} \xrightarrow{-H^+} CH_3 - \underset{\overset{\parallel}{H_2C}}{\overset{CH_3}{\overset{|}{C}}}$$

so that dry hydrogen chloride must be used as catalyst.

Because of their acidic nature, phenols do not readily form esters using the methods just described for alcohols. It is generally necessary to reflux the phenol with an acid halide, although benzoylation of phenols can be effected by merely shaking with benzenecarbonyl (benzoyl) chloride

$$C_6H_5.OH + C_6H_5.COCl \longrightarrow \overset{O}{\underset{\parallel}{C_6H_5.C}}.O.C_6H_5 \qquad + HCl$$

Phenyl benzenecarboxylate
(phenyl benzoate)

The product is less stable than an ester produced from an alcohol and is more readily hydrolysed into its components.

Esters can also be formed with mineral acids such as sulphuric and hydrochloric acids, e.g.,

$$ROH + H_2SO_4 \rightarrow RO.SO_2.OH + H_2O$$

$$ROH + HCl \rightarrow RCl + H_2O$$

The readiness with which halides are formed from alcohols and hydrogen halides (p. 438) forms a method for distinguishing between primary, secondary and tertiary alcohols. The order of reactivity can be explained in terms of the electron-repelling effect exerted by alkyl groups; in primary alcohols, there is only one such effect tending to weaken the C—O bond; in a secondary alcohol there are two such effects, and in a tertiary three:

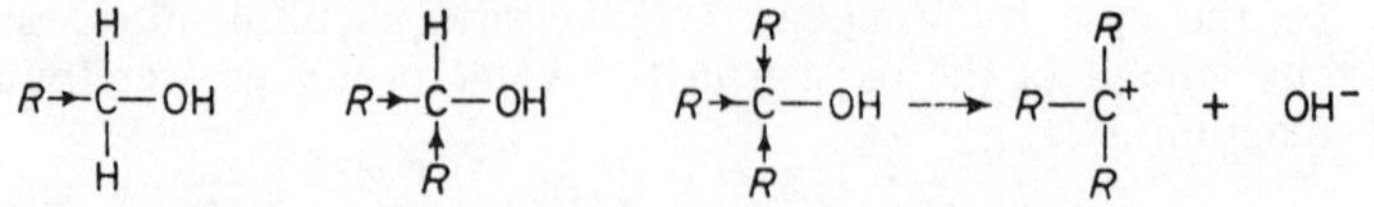

The mechanism of halide formation must involve reaction between the hydroxy group of the alcohol and the hydrogen of the hydrogen halide, i.e., the C—O bond of the alcohol is broken: a mechanism different from that in which a carboxylic acid is esterified.

Ether formation

Alcohols are protonated in the presence of cold concentrated sulphuric acid, and if there is insufficient acid to give complete protonation, then the following reaction can take place:

Ether formation

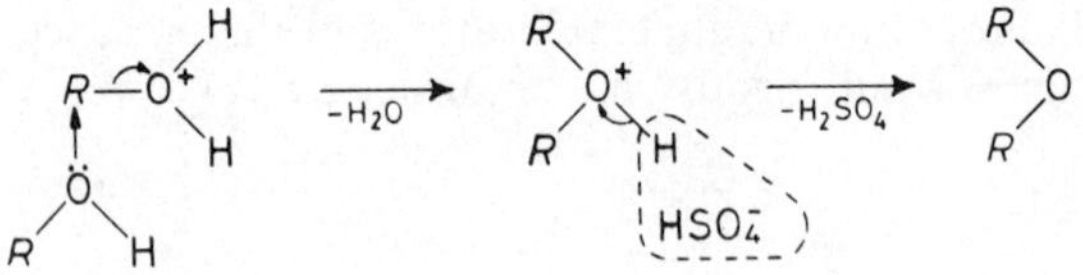

Alkene formation

High temperature and excess of concentrated sulphuric acid result in an elimination reaction of the protonated alcohol (p. 416).

Alkene formation

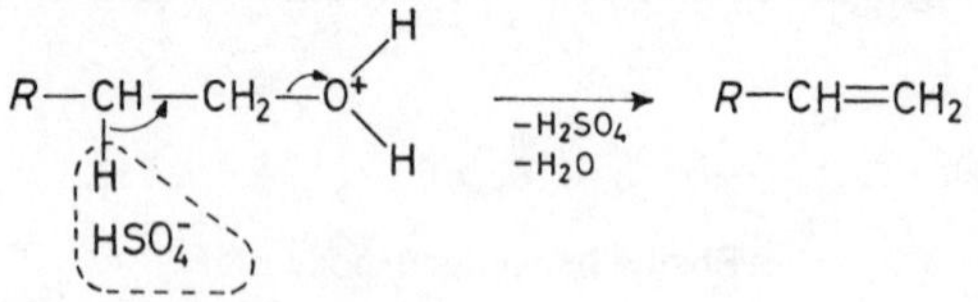

Reduction and oxidation

Alcohols are very resistant to reduction, but if phenol vapour mixed with hydrogen is passed over a heated nickel catalyst, it is reduced to cyclohexanol:

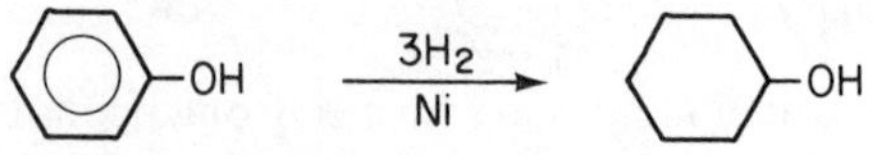

Phenols are oxidized only with great difficulty if rupture of the aromatic ring is involved. However, monohydric phenols are oxidized with peroxosulphates(VI) in alkaline solution to dihydric phenols, e.g.,

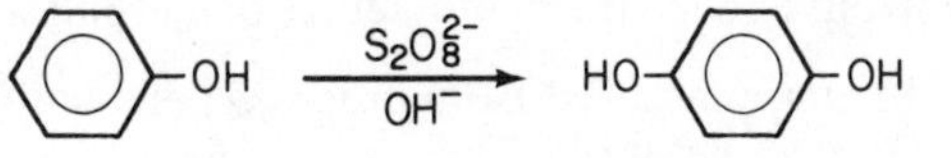

$$\text{C}_6\text{H}_5\text{-OH} \xrightarrow[\text{OH}^-]{\text{S}_2\text{O}_8^{2-}} \text{HO-C}_6\text{H}_4\text{-OH}$$

Benzene-1,4-diol
(quinol)

Quinol and the related *N*-methyl-4-aminophenol (metol) are readily oxidized to cyclohexadiene-1,4-dione (quinone) and so are strong reducing agents, used, for example, in photographic developers:

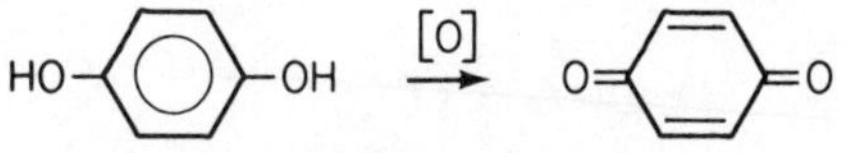

$$\text{HO-C}_6\text{H}_4\text{-OH} \xrightarrow{[\text{O}]} \text{O=C}_6\text{H}_4\text{=O}$$

Primary and secondary alcohols are easily oxidized using, for example, sodium dichromate and sulphuric acid in the liquid phase or catalytic dehydrogenation in the vapour phase, e.g.,

$$\underset{\text{Ethanol}}{\text{CH}_3.\text{CH}_2\text{OH}} \xrightarrow[+\text{H}_2\text{SO}_4]{\text{Na}_2\text{Cr}_2\text{O}_7} \underset{\substack{\text{Ethanal} \\ \text{(acetaldehyde)}}}{\text{CH}_3.\text{CHO}} \xrightarrow[+\text{H}_2\text{SO}_4]{\text{Na}_2\text{Cr}_2\text{O}_7} \underset{\substack{\text{Ethanoic (acetic)} \\ \text{acid}}}{\text{CH}_3.\text{COOH}}$$

The oxidation products afford a means of distinguishing between the three types of alcohols as shown in *Figure 22.2*.

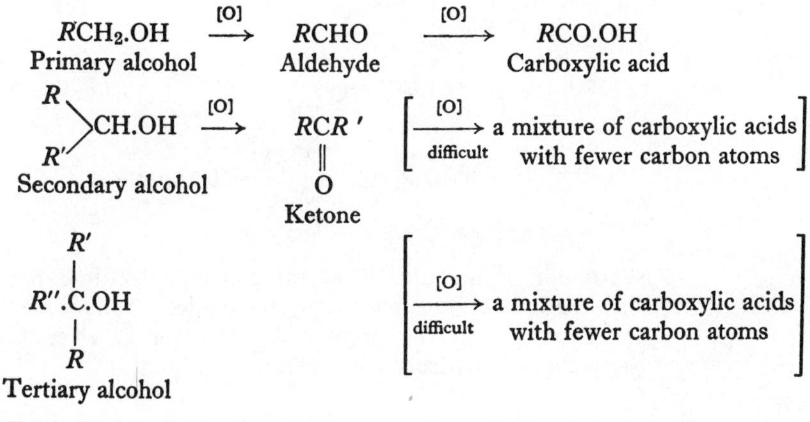

$$\underset{\text{Primary alcohol}}{R\text{CH}_2.\text{OH}} \xrightarrow{[\text{O}]} \underset{\text{Aldehyde}}{R\text{CHO}} \xrightarrow{[\text{O}]} \underset{\text{Carboxylic acid}}{R\text{CO.OH}}$$

$$\underset{\text{Secondary alcohol}}{\overset{R}{\underset{R'}{>}}\text{CH.OH}} \xrightarrow{[\text{O}]} \underset{\underset{\text{Ketone}}{\overset{\parallel}{\text{O}}}}{R\text{C}R'} \quad \left[\xrightarrow[\text{difficult}]{[\text{O}]} \begin{array}{l} \text{a mixture of carboxylic acids} \\ \text{with fewer carbon atoms} \end{array}\right]$$

$$\underset{\text{Tertiary alcohol}}{R''.\overset{\overset{\textstyle R'}{\mid}}{\underset{\underset{\textstyle R}{\mid}}{\text{C}}}.\text{OH}} \quad\quad\quad \left[\xrightarrow[\text{difficult}]{[\text{O}]} \begin{array}{l} \text{a mixture of carboxylic acids} \\ \text{with fewer carbon atoms} \end{array}\right]$$

Figure 22.2 Oxidation of alcohols

Industrial aspects

Monohydric alcohols

Methanol can be manufactured from the products of petroleum cracking

$$3CH_4 + CO_2 + 2H_2O \xrightarrow[800\,°C]{Ni} 4CO + 8H_2 \xrightarrow[30\ MPa]{ZnO/Cr_2O_3} 4CH_3OH$$

(*see* Fischer–Tropsch process, p. 418).

It is used as a solvent and for the manufacture of methanal (formaldehyde).

Ethanol can also be manufactured from cracked petroleum

$$CH_2{=}CH_2 + H_2O \xrightarrow[\substack{pumice\\300\,°C}]{H_3PO_4\ on} CH_3CH_2OH$$

As ethanol is the by-product of the anaerobic respiration of yeast with carbohydrates as substrate (p. 521), it has been known to man for centuries. The 'fermentation' process can be summarized as

$$C_6H_{12}O_6 \xrightarrow[\text{(in yeast cells)}]{\text{enzymes}} 2C_2H_5OH + 2CO_2(g)$$

At a concentration approaching 15 per cent, development of yeast cells is inhibited and the fermentation ceases. Increase of the alcoholic content above this concentration requires distillation, leading to the production of 'spirits'. Ethanol, like methanol, is used widely as a solvent and increasingly as a fuel.

Polyhydric alcohols

Ethane-1,2-diol (*ethylene glycol*) is in demand as an 'antifreeze' (a substance which depresses the freezing point of water, p. 113), as a solvent and, because of its difunctional nature, for making the polyester 'Terylene' (*ter*ephthalic acid and eth*ylene* glycol):

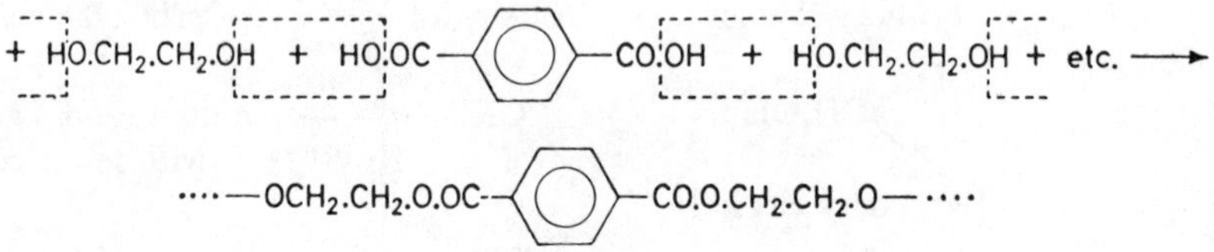

$$\cdots {-}OCH_2.CH_2.O.OC{-}\bigcirc{-}CO.O.CH_2.CH_2.O{-}\cdots$$

'Terylene'

Nearly all of the diol is at present manufactured from ethene (from petroleum) *via* epoxyethane (ethylene oxide); ethene is reacted directly with air or oxygen in the presence of a silver catalyst and the resultant epoxyethane is hydrolysed to ethane-1,2-diol:

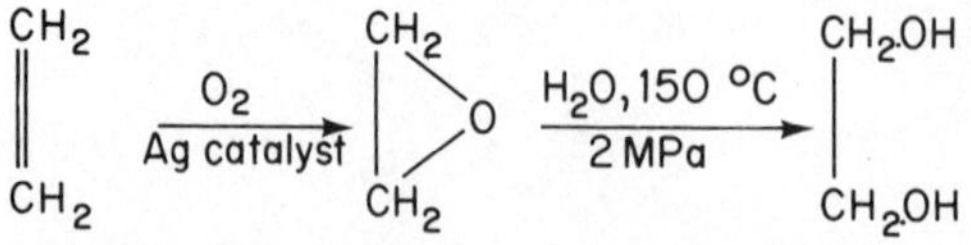

Important derivatives of ethane-1,2-diol are the 'cellosolves' (pp. 496, 498).
Propane-1,2,3-triol (*glycerol*) is used widely for the preparation of the

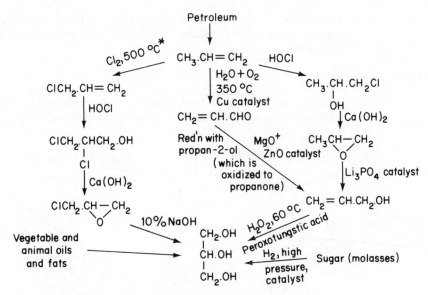

Figure 22.3 Routes to propane-1,2,3-triol.* At lower temperatures the 'expected' addition of chlorine to propene occurs

explosive propane-1,2,3-triyl trinitrate (commonly, and incorrectly, called nitroglycerine):

$$\begin{array}{ccc}
CH_2.OH & & CH_2.O.NO_2 \\
| & & | \\
CH.OH & +3HONO_2 \xrightarrow{H_2SO_4} & CH.O.NO_2 + 3H_2O \\
| & & | \\
CH_2.OH & & CH_2.O.NO_2
\end{array}$$

This substance is too dangerous to handle on its own and so is either adsorbed on to sawdust (which, with ammonium nitrate, is one form of *dynamite*), mixed with 'gun cotton' (*gelignite*) or with gun cotton and vaseline (*cordite*).

Glycerol is also a constituent of the *alkyd resins* (p. 548).

The esters of glycerol with the fatty acids are the main constituents of vegetable oils (p. 550) and animal fats, and so are one important source of glycerol as a by-product in the manufacture of soap. Other sources are available, chiefly from propene, a product of petroleum cracking (p. 418). The reactions involved in the manufacture of glycerol are summarized in *Figure 22.3*.

Tetra(hydroxymethyl)methane (*pentaerythritol*), a polyhydric alcohol is prepared by the reaction of methanal (formaldehyde) with ethanal (acetaldehyde):

$$4HCHO + CH_3.CHO + H_2O \xrightarrow[50°C]{Ca(OH)_2} HO.CH_2\!-\!\underset{\underset{CH_2.OH}{|}}{\overset{\overset{CH_2.OH}{|}}{C}}\!-\!CH_2.OH + HCO.OH$$

The alcohol is an important constituent of alkyd resins, because of the number of functional groups present, and its tetranitrate is a powerful explosive.

Phenol

Phenol is in such demand industrially that it is essential to obtain it from a wide variety of sources. Some of the methods used are shown in *Figure 22.4.*

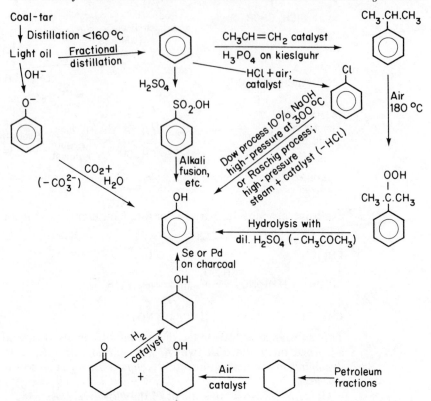

Figure 22.4 Routes to phenol

The activating effect of the hydroxy group for positions 2, 4 and 6 of the benzene ring allows many useful derivatives of phenol to be made. Among them are:

2,4,6-Trichlorophenol (T.C.P.)
This is prepared from chlorine water and phenol (cf. tribromophenol) and is used as a germicide.

2-Ethanoyloxybenzenecarboxylic acid
This substance, acetylsalicylic acid or aspirin, an analgesic, is made as shown in *Figure 22.5.* Trichloromethane is reacted with phenol in the presence of alkali; after acidification, when 2-hydroxybenzenecarbaldehyde is formed, the product is oxidized and acetylated.

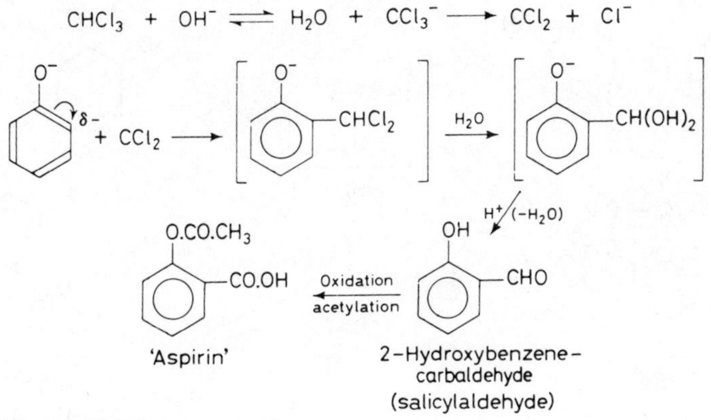

$$CHCl_3 + OH^- \rightleftharpoons H_2O + CCl_3^- \longrightarrow CCl_2 + Cl^-$$

Figure 22.5 Route to aspirin

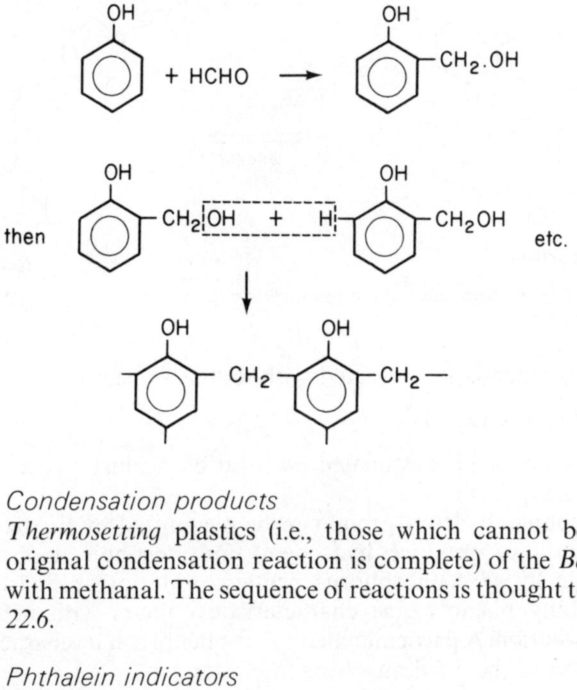

'Aspirin'

2-Hydroxybenzene-carbaldehyde (salicylaldehyde)

Figure 22.6 Thermosetting reactions in Bakelite-type plastics

Condensation products

Thermosetting plastics (i.e., those which cannot be remoulded after the original condensation reaction is complete) of the *Bakelite* type are formed with methanal. The sequence of reactions is thought to be as shown in *Figure 22.6*.

Phthalein indicators

Phenolphthalein results when phenol is heated with benzene-1,2-dicarboxylic (phthalic) anhydride in the presence of concentrated sulphuric acid (*Figure 22.7*).

Analysis

The amount of oxygen present in an organic compound is determined by heating the substance in a stream of nitrogen and passing the oxides of carbon so formed over heated carbon. This reduces any carbon dioxide to the

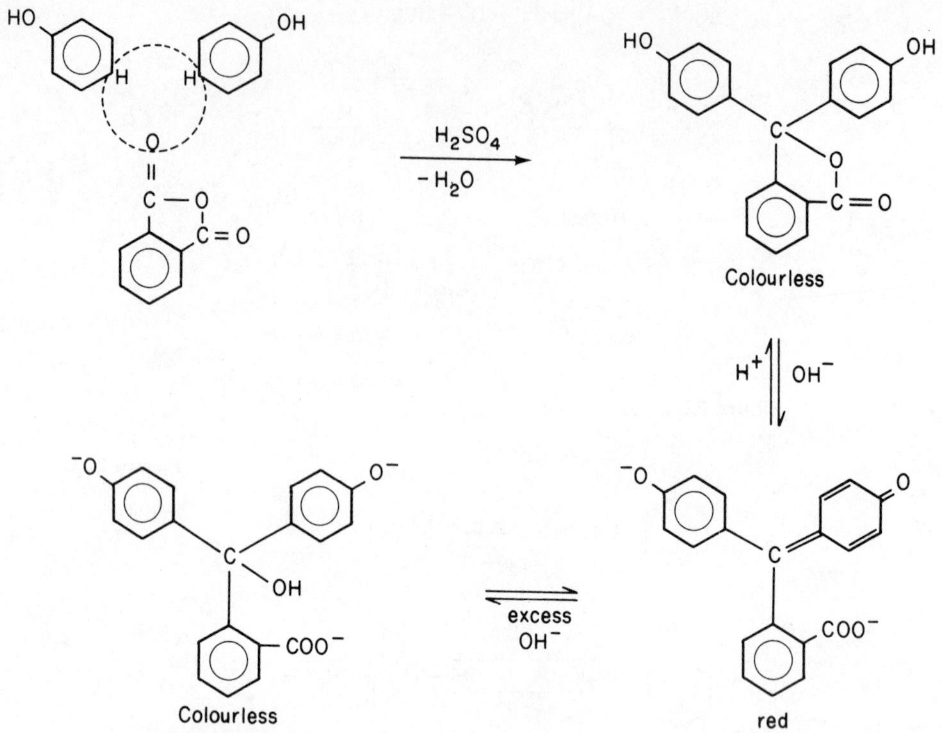

Figure 22.7 Formation and colour changes in the indicator phenolphthalein

monoxide, which is then reacted with iodine(V) pentoxide

$$5CO + I_2O_5 \rightarrow 5CO_2 + I_2$$

The iodine liberated is estimated by titration against a solution of sodium thiosulphate (p. 258).

The alcoholic hydroxy group can be recognized by the reaction it gives with sodium to produce hydrogen. Both alcohols and phenols react exothermically with phosphorus pentachloride giving hydrogen chloride fumes. Many phenols give characteristic colours with neutral iron(III) chloride solution. A particular alcohol or phenol can be characterized by the preparation of the 3,5-dinitrobenzoate ester, e.g.,

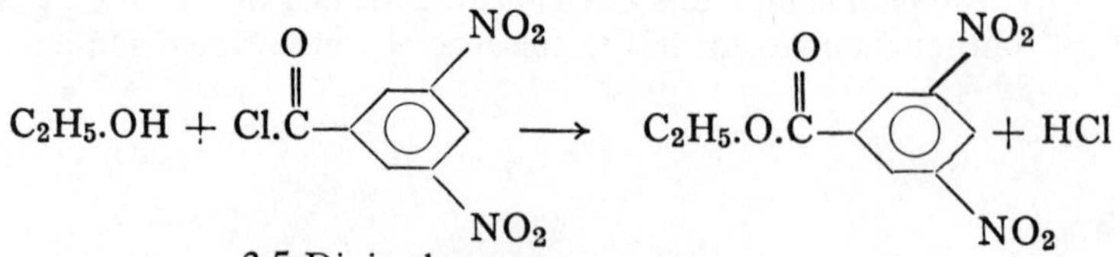

3,5-Dinitrobenzene-
carbonyl chloride

The number of hydroxy groups in a polyhydric alcohol or phenol can be determined by completely acetylating the compound, using ethanoic anhydride or ethanoyl chloride:

$$-\overset{|}{\underset{|}{C}}.OH + CH_3.COCl \rightarrow -\overset{|}{\underset{|}{C}}.O.CO.CH_3 + HCl$$

The ester produced, after careful washing, may be hydrolysed with a known amount of standard alkali and the quantity of unused alkali determined by titration with standard acid. From these values the number of ethanoyl groups (and, therefore, the number of hydroxy groups in the original compound) can be found:

$$-\overset{|}{\underset{|}{C}}.O.CO.CH_3 + OH^- \rightarrow -\overset{|}{\underset{|}{C}}.OH + CH_3.CO.O^-$$

Summary

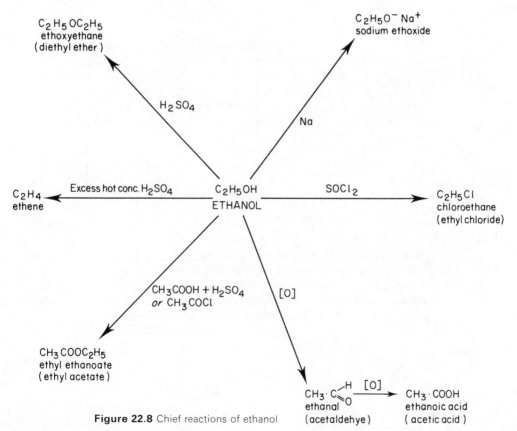

Figure 22.8 Chief reactions of ethanol

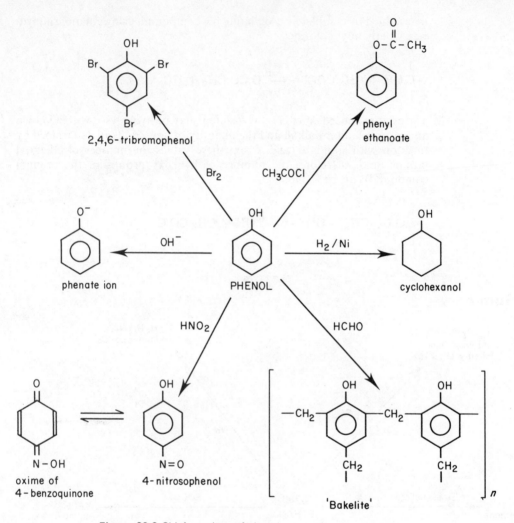

Figure 22.9 Chief reactions of phenol

Many of the reactions of alcohols can be attributed to the availability of a lone pair on the oxygen atom. The fact that this lone pair is, because of interaction with the benzene nucleus, far less available in phenols than in alcohols leads to marked differences between them. The main reactions of ethanol and phenol are summarized in *Figures 22.8* and *22.9*.

Questions

(1) Give the five possible oxidation products of ethane-1,2-diol in which the two carbon atoms remain attached.

(2) Draw the various structures of the alcohols with the molecular formula $C_4H_{10}O$.

(3) Suggest reasons for the following:

(a) Phenol is soluble in sodium hydroxide solution but insoluble in sodium carbonate solution;

(b) Of the three mononitrophenols, the strongest acidity is shown by 4-nitrophenol;

(c) The melting point of 2-nitrophenol is only 43 °C, whereas 3- and 4-nitrophenol melt at the much higher temperatures of 97 °C and 114 °C, respectively;

(d) 2,4,6-Trinitrophenol is a stronger acid than phenol, whilst ethanol is neutral.

(4) An organic compound has a relative molecular mass of 92. 0.46 g of the substance was completely ethanoylated and the product treated with 200 cm^3 of 0.1M sodium hydroxide solution. The resultant solution required 25 cm^3 of 0.2M hydrochloric acid for neutralization. Calculate the number of hydroxy groups in a molecule of the original compound and suggest a structural formula.

(5) A substance A , of molecular formula C_3H_8O, with excess of concentrated sulphuric acid yields B, C_3H_6, and with a little concentrated sulphuric acid C, $C_6H_{14}O$. B reacts with bromine to give D, $C_3H_6Br_2$, which with aqueous potassium hydroxide produces E, $C_3H_8O_2$, and with alcoholic potash, F, C_3H_4. Identify compounds $A–F$ and illustrate the reactions by suitable equations.

(6) A compound of molecular formula $C_{10}H_{15}NO$ gives benzoic acid on oxidation; when treated with nitrous acid, a nitroso compound is formed. The substance forms a dibenzoyl derivative with benzoyl chloride and also, on heating with hydrochloric acid, methylamine and 1-phenylpropan-2-one are formed. Deduce structures which would account for these properties.

(7) Compare the properties of ethanol and phenol.

(8) The ease of esterification of an alcohol with a carboxylic acid is in the order: primary 〉 secondary 〉 tertiary, whereas the reverse order holds for the esterification using hydrogen halides. Suggest reasons for these observations.

(9) Many phenols and alcohols are nowadays in great commercial demand and their production illustrates the changing balance of industrial supply. Review the major sources of organic materials, with particular reference to hydroxy compounds.

(10)   A compound of formula C_3H_7Br is converted into C_3H_9N by reaction with
       alcoholic ammonia. This compound, in turn, is changed into C_3H_8O by
       treatment with nitrous acid solution, which then reacts with sodium
       carbonate solution and iodine to give a yellow precipitate. Explain all these
       reactions.

(11)   A compound B, molecular formula C_6H_6O, is soluble in sodium hydroxide
       solution and when its vapour is passed with hydrogen over heated nickel, a
       compound, C, molecular formula $C_6H_{12}O$, is formed. The vapour of C when
       passed over heated alumina forms D, of molecular formula C_6H_{10} which can
       be reduced to E, C_6H_{12}. C on oxidation forms F, $C_6H_{10}O$, which is
       unreactive towards ammoniacal silver nitrate, but forms a precipitate with
       2,4-dinitrophenylhydrazine. Identify the compounds B–F, giving all
       equations.

(12)   When pentan-2-ol, $CH_3CH_2CH_2(OH)CH_3$, is heated with acidified pot-
       assium dichromate(VI) solution, the colour of the solution changes from
       orange to green and the alcohol is oxidized to the corresponding ketone.

       (a) What is the oxidation state of the chromium responsible for the green
           colour?
       (b) Name one other reagent which could have been used instead of the
           potassium dichromate(VI) to oxidize the alcohol to the ketone.
       (c) What is the structural formula of the ketone formed by oxidizing pentan-
           2-ol?
       (d) Show, by means of equations, how this ketone reacts with: (i) 2,4-
           dinitrophenylhydrazine and (ii) hydroxylamine.
       (e) How might you use the 2,4-dinitrophenylhydrazine derivatives of two
           isomeric ketones in order to distinguish between them?

                                                                          [J.M.B.]

23 Ethers

Preparation and properties • Cyclic ethers • Sulphur analogues • Industrial aspects • Analysis

Ethers are derived from alcohols, the hydrogen of the hydroxyl group being replaced by alkyl or aryl. Alcohols and ethers both possess saturated oxygen atoms, unlike carbonyl and carboxyl compounds where the p electrons of the oxygen atoms are involved in 'double' bonds:

		$\overset{\displaystyle R'}{\underset{\displaystyle \vert}{}}$	$\overset{\displaystyle OR'}{\underset{\displaystyle \vert}{}}$
ROH	ROR'	$R-C{=}O$	$RC{=}O$
Alcohol	Ether	Carbonyl compound	Carboxyl compound

Preparation and properties of ethers

Preparation

It has already been stated (p. 480) that alcohols are protonated by strong acids

$$ROH \xrightarrow{\;H^+\;} R\overset{+}{O}H_2$$

In the presence of excess of concentrated sulphuric acid and at elevated temperatures, a proton is removed from the β-carbon atom, and simultaneous elimination of water gives rise to an alkene. At lower temperatures and in the presence of excess of alcohol, so that not all the alcohol is protonated, a different reaction predominates by which an ether is formed (see p. 482).

The latter method is suitable for the preparation of the lower homologues only, because the decrease in volatility resulting from increased relative molecular mass means that the temperature of distillation is so high that alkene formation predominates.

Diols give, by the above reaction, cyclic ethers or *epoxy compounds*

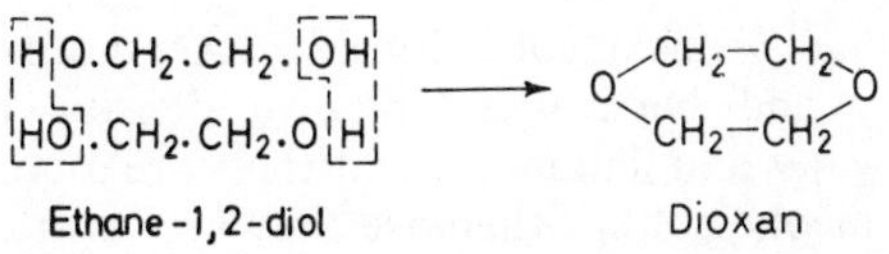

Ethane-1,2-diol   Dioxan

A general method of preparation of ethers involves the action of heat on a mixture of alkyl halide and sodium alkoxide or phenate (Williamson's synthesis):

$$R'X + Na^+OR^- \rightarrow R'OR + Na^+X^-$$

The reaction is bimolecular, the alkoxide or phenate ion constituting a powerful nucleophile and attacking the carbon atom attached to the halogen. This carbon atom comes more and more under the influence of the nucleophile, and the halogen atom steadily recedes, taking the electron pair of the original covalent bond with it; ultimately, a new bond is formed, as the old one is broken:

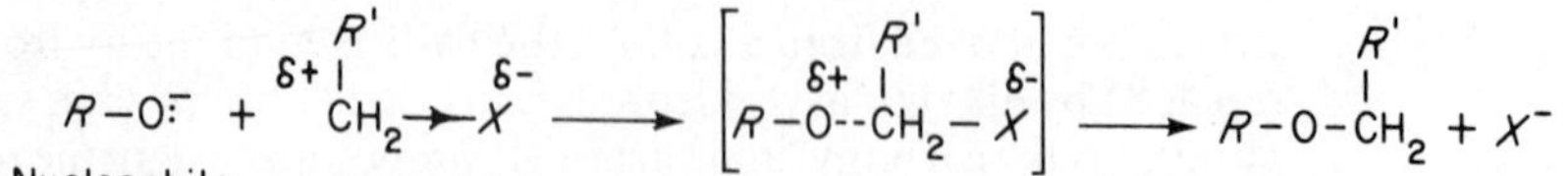

Aryl halides cannot be used in this reaction; conjugation of the 'lone pair' on the halogen atom with the aromatic system leads to the latter being negatively rather than positively charged, and thus unattractive to the nucleophile.

Nomenclature

Ethers can be regarded as alkoxy derivatives of hydrocarbons, and this forms the basis for nomenclature. For example, diethyl ether is ethoxyethane and both anethole, a constituent of aniseed, and eugenol, found in oil of cloves, are methoxy derivatives of propenylbenzenes

$C_2H_5.O.C_2H_5$

Ethoxyethane

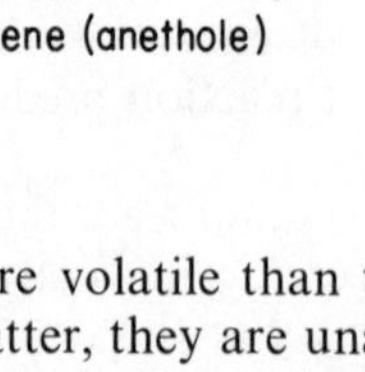

4-Methoxyprop-1'-enyl-benzene (anethole)

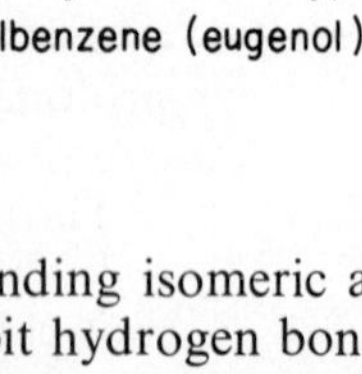

4-Hydroxy-3-methoxyprop-2'-enylbenzene (eugenol)

Properties of ethers

Ethers are much more volatile than the corresponding isomeric alcohols because, unlike the latter, they are unable to exhibit hydrogen bonding (p. 57). *Table 23.1* gives the boiling points of some common isomeric alcohols and ethers, together with hydrocarbons of comparable relative molecular mass.

It will be seen that, in terms of volatility, ethers are more like alkanes. Nor does the similarity end here; it extends to density, refractive index, immiscibility with water and flammability. Ethers are more flammable than the corresponding alcohols, but otherwise they are less reactive. Although

Table 23.1 Comparison of boiling points of ethers and alcohols

Molecular formula	Alcohol	B.p./°C	Ether	B.p./°C	Hydrocarbon	B.p./°C
C_2H_6O	ethanol, C_2H_6OH	78	methoxymethane, $CH_3.O.CH_3$	-24	propane, C_3H_8	-42
C_3H_6O	propan-1-ol, C_3H_7OH	97	methoxyethane, $CH_3.O.C_2H_5$	8	butane, C_4H_{10}	0
	propan-2-ol, $(CH_3)_2.CHOH$	82				

the oxygen atom is capable of protonation by strong acids, the complexes, unlike those of alcohols, are incapable of eliminating water. Thus, although ethers dissolve in concentrated sulphuric acid, they are regenerated on dilution:

$$\underset{\cdot\cdot}{RO}R' \xrightarrow{H_2SO_4} [ROR']^+ HSO_4^- \xrightarrow{H_2O} \underset{\cdot\cdot}{RO}R' + H_2SO_4.H_2O$$
$$|$$
$$H$$

Donation of the electron pair on the oxygen atom of an ether to other molecules is also common:

$$R_2O \rightarrow BF_3 \qquad \overset{\displaystyle R}{\underset{\displaystyle \underset{O(C_2H_5)_2}{\uparrow}}{RMg \leftarrow O(C_2H_5)_2}} \qquad (C_2H_5)_2O \rightarrow CrO_5$$

However, fission *does* result from attack by the powerful electrophile formed when acyl chlorides (p. 545) react with anhydrous zinc chloride; subsequent elimination of alkyl chloride from the intermediate gives an ester:

$$RC\overset{\frown}{=}\overset{\cdot\cdot}{O}: \leftarrow\text{-}-ZnCl_2 \longrightarrow R\overset{+}{C}.O.\bar{Z}nCl_2$$
$$| \qquad\qquad\qquad\qquad |$$
$$Cl \qquad\qquad\qquad\qquad Cl$$

$$\underset{\underset{\overset{|}{R''}}{\cdots-:OR'}}{\overset{\displaystyle Cl}{\overset{|}{RC}\overset{+}{}.O.\bar{Z}nCl_2}} \longrightarrow \underset{R'\overset{+}{-}OR''}{\overset{\displaystyle Cl}{\overset{|}{RC}.O\overset{=}{.}\bar{Z}nCl_2}} \longrightarrow \underset{\overset{|}{OR'}\quad\overset{|}{OR''}}{RC=O\;(+RC\overset{\cdots}{=}O) + R''Cl\,(+R'Cl) + ZnCl_2}$$

Ethers are also attacked by constant-boiling-point hydriodic acid; this time, the strongly nucleophilic iodide ion is the active agent

$$\underset{\underset{\overset{|}{H^+}}{\overset{|}{\uparrow}}}{ROR'} \longrightarrow \overset{+}{RO}R'; \quad \underset{I^-}{R\overset{\frown}{\overset{+}{-}O}} \underset{H}{-R'} \longrightarrow RI + R'OH$$
$$\qquad\qquad H$$

The hydroxy compound formed, if aliphatic, can be converted by more hydrogen iodide into the iodoalkane but phenols resist further attack. The iodoalkane formed can be absorbed in alcoholic silver nitrate solution, and the resulting silver iodide can be weighed. This is the basis for the determination of methoxy groups by the Zeisel method:

$$C_6H_5.O.CH_3 \xrightarrow{HI} C_6H_5.OH + CH_3I$$

$$CH_3I + AgNO_3 \xrightarrow{H_2O} AgI(s) + CH_3OH + HNO_3$$
142 g   235 g

Cyclic ethers

Epoxyethane (ethylene oxide)

This is prepared industrially by the catalytic oxidation of ethene or by treating 2-chloroethanol (ethylene chlorohydrin) with alkali:

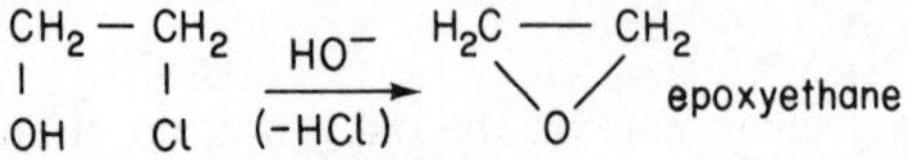

This compound can be compared with cyclopropane in that the strain imposed upon the bonds by excessive departure from the tetrahedral angle makes it unduly reactive. In this case, cleavage can readily take place in the presence of acid to give ethane-1,2-diol or, in the presence of acid and alcohol, a 'cellosolve':

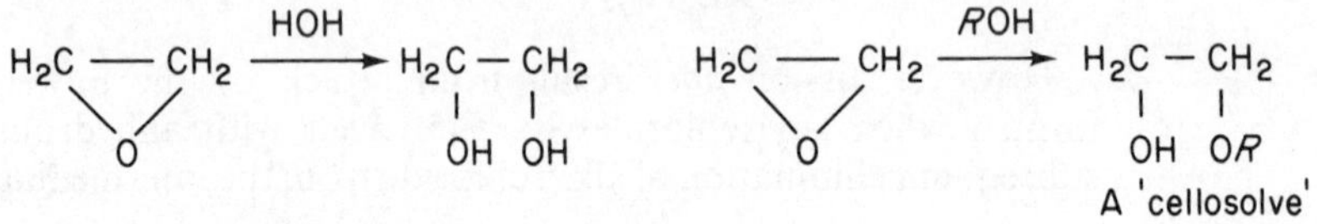

Furan

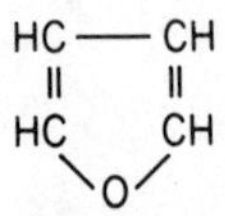

Furan occurs in the distillates of most woody materials and is also obtained by heating 2,3,4,5-tetrahydroxyhexane-1,6-dioic (mucic) acid

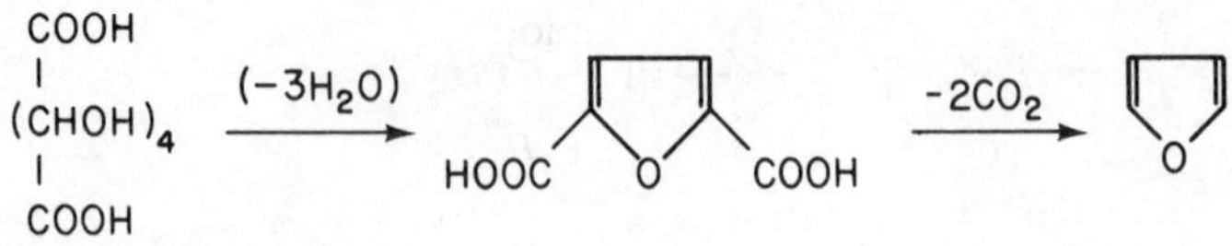

It reveals its conjugated double-bond character by taking part in what is known as the Diels–Alder reaction with, for instance, cis-butenedioic (maleic) anhydride:

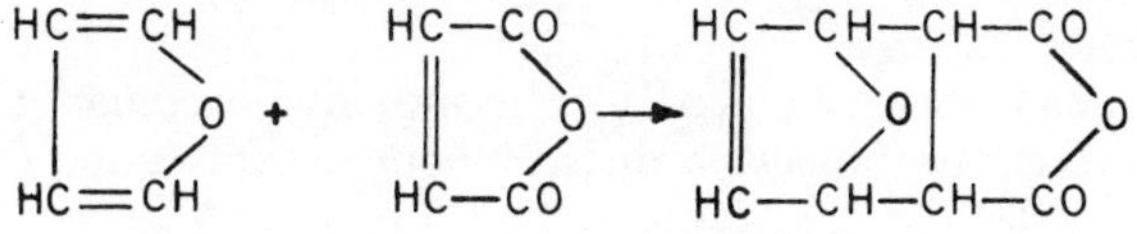

Pyran

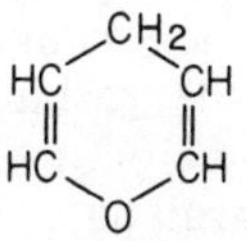

Pyran itself is not known but the removal of an electron from the oxygen enables it to participate in a non-localized π electron system and so form the pyrilium compounds that are so widespread as plant pigments, e.g.,

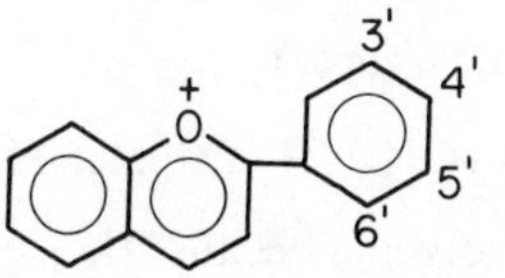

4′—OH pelargonidin (from pelargonium)
3′,4′—OH cyanidin (from cornflower)
3′,4′.5′—OH delphinidin (from delphinium)

Tetrahydrofuran and tetrahydropyran provide the basis for the furanose and pyranose ring structures present in carbohydrates (pp. 515, 516).

Sulphur analogues

Sulphur can replace oxygen in many types of organic compound. Thus the thiols are analogues of alcohols, C_2H_5SH being ethanethiol, the thioketones are analogues of ketones, $C_2H_5C(:S)CH_3$ being butane-2-thione. One or both oxygen atoms may be replaced in carboxylic acids, leading to, e.g., dithiobenzoic acid C_6H_5CSSH. The ether analogues are sometimes called thioethers or sulphides: $C_2H_5.S.C_2H_5$ is ethylthioethane.

Thiophen

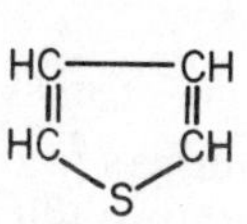

Thiophen is similar to furan but contains an ether link through sulphur instead of oxygen. This —S— link is present in the penicillins, e.g., penicillin G

$$C_6H_5.CH_2 - CONH - \underset{\underset{CO-N}{|}}{CH} - \underset{\underset{CH-COOH}{|}}{CH} \quad \overset{S}{\underset{}{\diagup}} \quad \overset{CH_3}{\underset{|}{C-CH_3}}$$

These penicillins, produced by the mould *Penicillium*, are examples of *antibiotics*, i.e., substances produced by one organism and toxic to another.

Industrial aspects

The lower aliphatic homologues are of considerable importance as solvents (ethoxyethane is also used as an anaesthetic) and are manufactured on the large scale by treating alkenes (by-products from the cracking of petroleum) with cold, concentrated sulphuric acid. Addition of water in sufficient quantity converts some of the alkyl hydrogensulphate formed by this reaction into alcohol; the ether results from interaction between this and more alkyl hydrogensulphate, e.g.,

$$CH_2 = CH_2 \xrightarrow{H_2SO_4} CH_3CH_2.O.SO_2.OH$$

$$CH_3.CH_2.O.SO_2.OH + H_2O \rightarrow CH_3CH_2OH + H_2SO_4$$

$$CH_3CH_2.O.SO_2.OH + CH_3CH_2OH \rightarrow CH_3CH_2.O.CH_2CH_3 + H_2SO_4$$

'Cellosolves' are particularly good solvents because they contain both the hydrophilic alcohol group and the hydrophobic ethereal linkage, e.g.,

$$HO.CH_2CH_2.O.CH_3$$

Cyclic ethers can be polymerized to epoxy resins by treatment with a little water in the presence of acid or base as catalyst, e.g.,

$$H_2C \overline{\qquad} CH_2 \rightarrow \ldots -O.CH_2CH_2.O.CH_2 - \ldots$$

Copolymerization is effected by mixing the ether with a polyhydric alcohol: the use of propane-1,2,3-triol, with its three hydroxy groups, establishes cross-linkages throughout the structure to give a three-dimensional resin. Epoxy resins are becoming increasingly important as cements and adhesives of high strength.

Compounds containing the ether link are also playing an expanding role in horticulture, e.g. as selective weedkillers such as 2,4-D.C.P.A. (see p. 536).

Analysis

Advantage is taken of the presence of oxygen in ethers to distinguish them from the comparably inert alkanes; they dissolve, through protonation, in concentrated sulphuric acid but are regenerated on dilution.

Suitable derivatives of ethers are not easy to find, and so characterization is often effected by means of physical properties such as refractive index.

Questions

(1)   Give the structural formulae and the names of compounds of molecular formula $C_4H_{10}O$. Briefly say how you would distinguish between them chemically.

(2)   What products may be obtained from propan-1-ol and concentrated sulphuric acid, with and without treatment with water?

(3)   What ether may be prepared from 4-bromomethoxybenzene and 4-methyl-phenol? What other ether would be produced by treating the product with potassium manganate(VII) under reflux, followed by hydriodic acid?

(4)   How could you prepare $(CH_3)_2CH.O.C_2H_5$, starting from ethanol and propan-1-ol? What is the name of this compound?

(5)   Suggest a synthesis of phenoxybenzene from benzene.

(6)   How would you separate a mixture of benzene, methoxybenzene and phenylamine?

(7)   A pure substance was found to contain carbon, hydrogen and oxygen only. It dissolved in cold, concentrated sulphuric acid, but the addition of excess of water served to regenerate it. 0.03 g of this compound displaced 13.2 cm^3 of air in a Victor Meyer apparatus, measured dry at 27 °C and at a pressure which supported a column of mercury 750 mm high. Deduce what you can from this information.

(8)   Write an essay on the role of ethers in either agriculture or adhesives.

24 Carbonyl compounds

Aldehydes and ketones

The carbonyl group contains carbon attached to oxygen by a double bond:

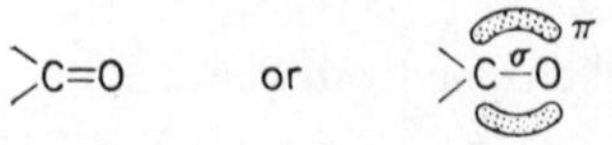

$$>C=O \quad \text{or} \quad >C \overset{\sigma}{\underset{}{=}} O \; \pi$$

If the carbon atom of this unsaturated group is attached to two other carbon atoms, the compound is called a *ketone*; if to one carbon and one hydrogen, an *aldehyde*. Aliphatic aldehydes are named by replacing the final -*e* of the corresponding hydrocarbon by -*al*, aliphatic ketones by replacing it by -*one* and numbering the relevant carbon atom if necessary

$$CH_3-C=O \qquad\qquad CH_3-C=O$$
$$\qquad | \qquad\qquad\qquad\qquad |$$
$$\qquad H \qquad\qquad\qquad\qquad CH_3$$

Ethanal   Propanone

Preparation

By oxidation

Aldehydes bear a close relationship to primary alcohols, and ketones to secondary alcohols

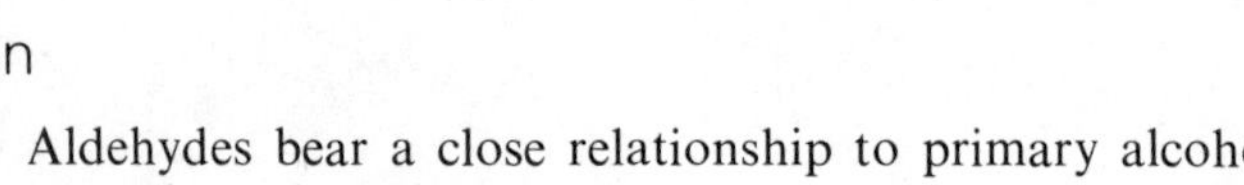

$$\underset{H}{\overset{R}{>}}C\underset{OH}{\overset{H}{<}} \longrightarrow \underset{H}{\overset{R}{>}}C=O \qquad \underset{R'}{\overset{R}{>}}C\underset{OH}{\overset{H}{<}} \longrightarrow \underset{R'}{\overset{R}{>}}C=O$$

Consequently, an aldehyde is obtained by oxidation of a primary alcohol with acidified sodium dichromate(VI). It should be distilled off immediately, without refluxing, to minimize further oxidation to the corresponding carboxylic acid, e.g.,

$$CH_3CH_2CH_2OH \xrightarrow{[O]} CH_3CH_2CHO$$
Propan-1-ol   Propanal

Alternatively, dehydrogenation of the primary alcohol can be effected catalytically by passing the vapour over heated copper (in fact, the name 'aldehyde' is derived from '*al*cohol *dehyd*rogenatus'), e.g.,

$$CH_3CH_2OH \xrightarrow[(-2H)]{} CH_3CHO$$

Ethanol   Ethanal
                   (acetaldehyde)

The latter method has the advantage that no aldehyde is lost by conversion into acid.

Ketones are produced from secondary alcohols in a similar manner, e.g.,

$$(CH_3)_2CHOH \xrightarrow[(-2H)]{} (CH_3)_2CO$$

Propan-2-ol   Propanone
                   (acetone)

By hydrolysis

Hydrolysis of compounds in which two halogen atoms are attached to the same carbon atom yields the corresponding carbonyl compound:

$$RCHCl_2 \xrightarrow[(-2HCl)]{2H_2O} [RCH(OH)_2] \rightarrow RCHO + H_2O$$

This is of particular value in the case of benzenecarbaldehyde (benzaldehyde); methylbenzene (toluene) can be chlorinated to the required state and the product hydrolysed (p. 428).

By Friedel–Crafts reaction

Aromatic and arylalkyl ketones are readily prepared by the Friedel–Crafts reaction, i.e., by reacting aromatic hydrocarbons with an acid chloride in the presence of anhydrous aluminium chloride (p. 430)

$$C_6H_6 + RCOCl \xrightarrow[(-HCl)]{Al_2Cl_6} C_6H_5.COR$$

By reduction

Acid chlorides can be selectively reduced to aldehydes by treatment with hydrogen in the presence of a palladium catalyst, partially deactivated with, e.g., barium sulphate (Rosenmund reaction). The deactivation is essential to prevent further reduction of the aldehyde to alcohol

$$\begin{array}{ccc} RC=O & \xrightarrow[(-HCl)]{H_2} & RC=O \\ | & & | \\ Cl & & H \end{array}$$

Properties and reactions

Because oxygen is more electronegative than carbon, there is a drift of electrons (the inductive effect) towards the oxygen in the carbonyl bond. That is, the bond is *polarized*:

$$\overset{\delta+}{C}\!\!-\!\!\overset{\delta-}{O}$$

This condition may be represented as an increase in the probability of the valency electrons becoming associated with the oxygen atom, i.e., there is an increased electron density around the oxygen

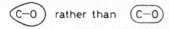   rather than

The lower aliphatic carbonyl compounds are freely soluble in water. However, as there is no prospect of hydrogen bonding with the solvent, aldehydes and ketones are not as a rule as soluble as the corresponding alcohols. For the same reason they are more volatile. There is, however, sufficient molecular interaction arising from polarization to ensure that the carbonyl compounds are considerably less volatile than the corresponding hydrocarbons.

The characteristic reaction of carbonyl compounds involves nucleophilic attack by a base at the carbon atom of the carbonyl group

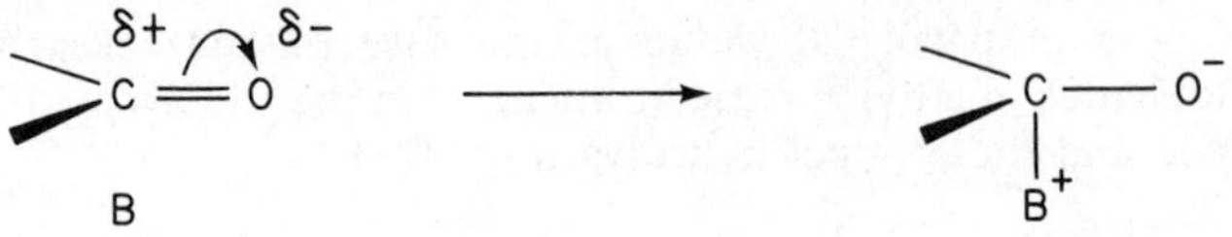

(B = base)

The reaction is sometimes reversible, and attempts to isolate the product may result in the reappearance of the original compound. In other cases, the preliminary addition is followed by the elimination of water or another small molecule; the combined reaction is then known as a *condensation*.

Generally, reactivity decreases as the molecular complexity increases, owing at least partly to an increase in *steric hindrance*, i.e., to the carbonyl group being shielded from attack by large neighbouring groups, e.g., as in an aromatic ketone:

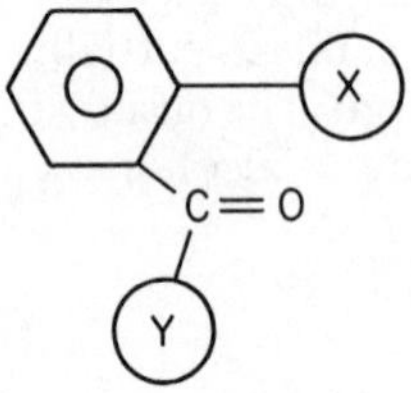

Steric hindrance will be more marked with ketones than with aldehydes where one adjacent group is always a small hydrogen atom. *It is the presence of this hydrogen atom that is responsible for some fundamental differences between aldehydes and ketones.* For example, aldehydes are readily oxidized to acids (with the same number of carbon atoms) and are therefore good reducing agents; those soluble in aqueous solution reduce Fehling's solution (which contains copper(II) ions (kept in alkaline solution as a 2,3-dihydroxy-butanedioate complex) to copper(I) oxide or even, in the case of methanal, to metallic copper:

$$RCHO + 3OH^- \rightarrow RCOO^- + 2H_2O + 2e^-$$

$$2Cu^{2+} + 2OH^- + 2e^- \rightarrow Cu_2O(s) + H_2O$$

Aliphatic aldehydes will also reduce a solution containing diamminesilver(I) ions (ammoniacal silver nitrate) to give a silver mirror:

$$RCHO + 3OH^- \rightarrow RCOO^- + 2H_2O + 2e^-$$

$$2[Ag(NH_3)_2]^+ + 2e^- \rightarrow 2Ag(s) + 4NH_3$$

Ketones, on the other hand, are oxidized with difficulty under more extreme conditions, to give acids with fewer carbon atoms than the original compound. They do not, therefore, give the reducing reactions of aldehydes, and this affords a means of distinguishing between these two classes of compound.

Nucleophilic attack by nitrogen

Carbonyl compounds undergo many reactions with nitrogenous compounds. Initially, the lone pair on nitrogen is attracted to the carbon of the carbonyl group:

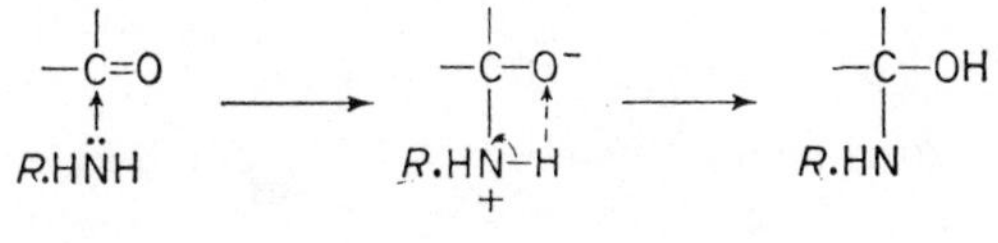

Ammonia
When $R = $ hydrogen, the attack is by ammonia and the initial product is a 1-amino-substituted alcohol (an 'aldehyde ammonia'), e.g.,

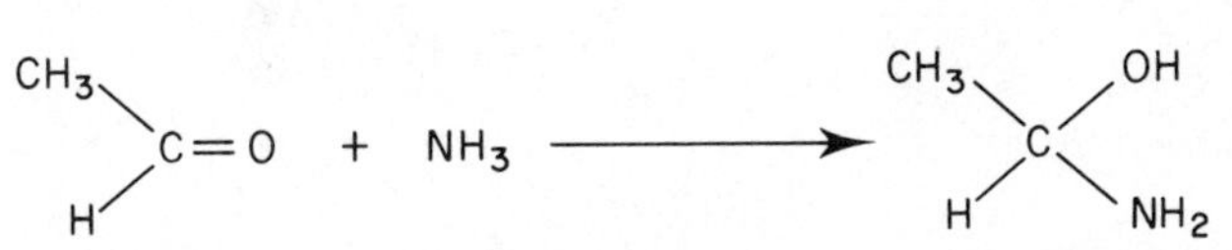

This is seldom isolable, usually either reverting to the original substance or undergoing *condensation polymerization*. For example, methanal (formaldehyde) yields 'hexamethylenetetramine' as shown in *Figure 24.1*.

Amines
A primary amine is involved if $R = $ alkyl or aryl. Under acidic conditions, the product condenses to give an aldimine or Schiff's base, e.g.,

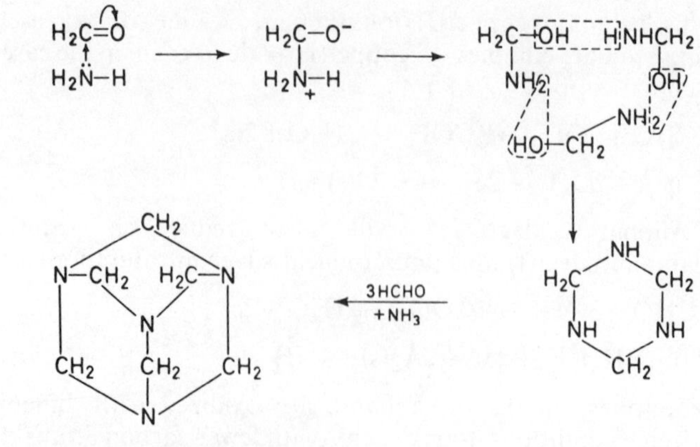

'Hexamethylenetetramine'

Figure 24.1 Polymerization of methanal

$$CH_3CH_2CHO + CH_3NH_2 \longrightarrow \left[\begin{array}{c} CH_3CH_2CH.OH \\ | \\ NH.CH_3 \end{array} \right] \xrightarrow{-H_2O} CH_3CH_2CH=NCH_3$$

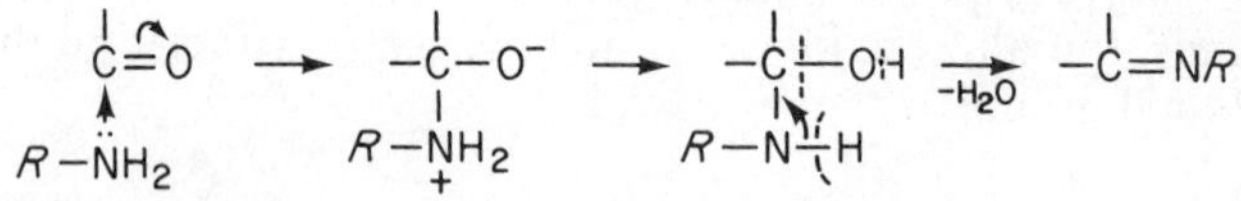

Hydrazines
In the case of hydrazine, i.e., when $R = NH_2$, condensation leads first to hydrazones and then to azines, e.g.,

$$(CH_3)_2 C=O + H_2N.NH_2 \xrightarrow{-H_2O} (CH_3)_2C=N.NH_2 \xrightarrow[{-H_2O}]{(CH_3)_2CO} (CH_3)_2C=N.N=C(CH_3)_2$$

A hydrazone   An azine

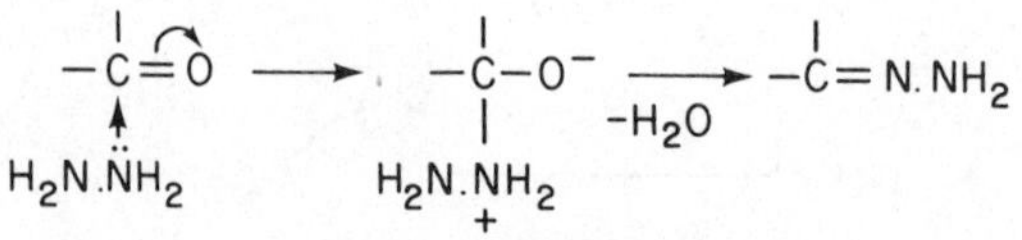

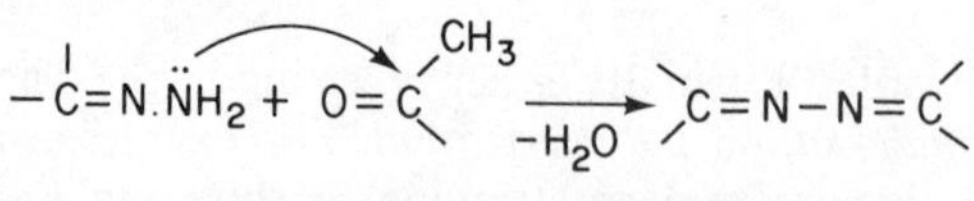

However, if phenylhydrazine is used, reaction stops at the hydrazone stage, and the product can be used to identify the carbonyl compound involved. Better crystalline products for melting-point determination are obtained if 2,4-dinitrophenylhydrazine (Brady's reagent) is used, e.g.,

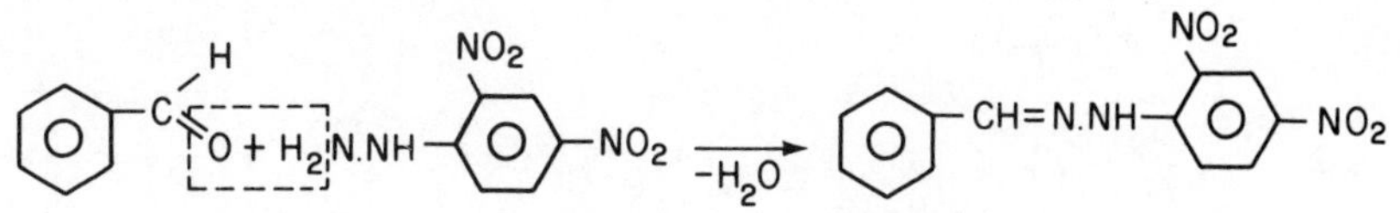

'Benzaldehyde
2,4-dinitrophenylhydrazone'

Hydroxylamine

When $R = OH$, oximes are formed which can also be used for the characterization of aldehydes and ketones:

$$>C=O + H_2N.OH \xrightarrow[-H_2O]{} >C=N.OH$$

$$(CH_3)_2C=O + H_2NOH \xrightarrow[-H_2O]{} (CH_3)_2C=N.OH$$

Nucleophilic attack by carbon

Cyanide

An aqueous solution of cyanide ions, in the presence of dilute acid, converts carbonyl compounds into hydroxynitriles (cyanohydrins). Typically, the initial attack is by the nucleophile, CN^-:

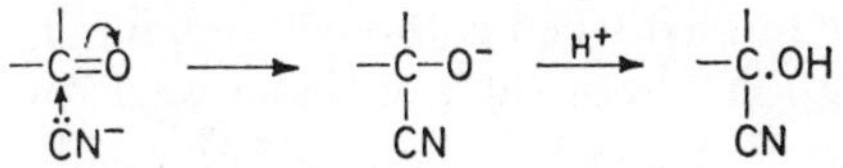

Cyanohydrin

As the nitrile group, —CH, can be hydrolysed to the corresponding carboxylic acid, the cyanohydrin synthesis is an important method of making hydroxy acids (p. 529).

In the case of aromatic aldehydes, the cyanohydrin can release a proton to form a *carbanion*, which then attacks a second molecule of aldehyde. In the case of benzenecarbaldehyde (benzaldehyde), ultimate loss of the elements of hydrogen cyanide gives 2-hydroxy-1,2-diphenylethanone (benzoin), and this reaction is therefore known as the *benzoin condensation (Figure 24.2)*.

Aldol condensation

Hydroxide ions are capable of removing a proton from the carbon atom adjacent to the carbonyl group. A carbanion is thereby formed and, being a nucleophile, can then attack a second carbonyl group, e.g. for ethanal:

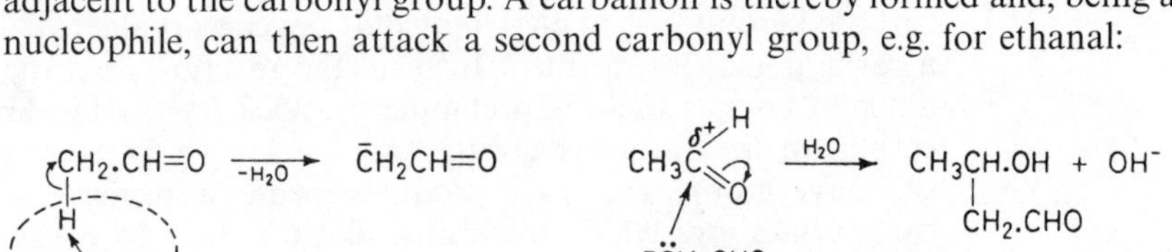

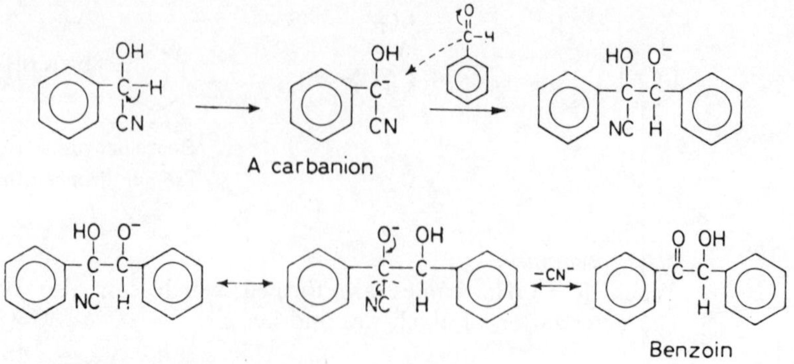

A carbanion

Benzoin

Figure 24.2 The benzoin condensation

This addition reaction, paradoxically known as the aldol condensation, is a useful method of ascending an homologous series. Hydroxide ions are provided by an aqueous solution of potassium carbonate. In the presence of a large concentration of hydroxide ions, e.g., sodium hydroxide solution, a genuine condensation reaction (*the Claisen condensation*) develops with the formation of an unsaturated carbonyl compound

$$RCH\underset{\overset{|}{OH}\ H}{-}CH.CHO \longrightarrow RCH{=}CH.CHO + H_2O$$

The carbonyl group in the product often perpetuates the process, so that resinous polymers are formed when carbonyl compounds are boiled with sodium hydroxide solution. The possible reactions occurring with hydroxide ions are then, e.g.

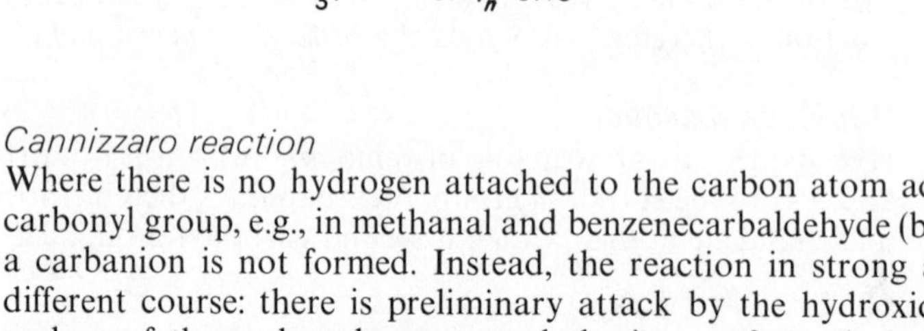

$$CH_3CHO \xrightarrow{K_2CO_3} CH_3CH(OH).CH_2.CHO \xrightarrow{dil.\ NaOH} CH_3.CH{=}CH.CHO$$

conc. NaOH   conc. NaOH   Heat

$$CH_3(CH{=}CH)_n\ CHO$$

Cannizzaro reaction

Where there is no hydrogen attached to the carbon atom adjacent to the carbonyl group, e.g., in methanal and benzenecarbaldehyde (benzaldehyde), a carbanion is not formed. Instead, the reaction in strong alkali takes a different course: there is preliminary attack by the hydroxide ion on the carbon of the carbonyl group, and the ion so formed attacks a second carbonyl group, the final products being a primary alcohol and the carboxylate ion, which mineral acid converts into carboxylic acid. This *Cannizzaro reaction* is of wide application in aromatic chemistry, e.g.,

$$2C_6H_5.CHO + OH^- \rightarrow C_6H_5.CH_2.OH + C_6H_5.CO.O^-$$

Benzaldehyde Phenylmethanol Benzoate

or, in general terms of reaction mechanisms

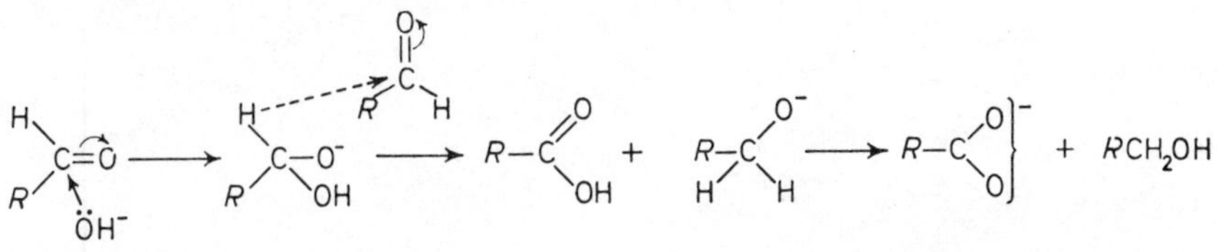

Perkin reaction

In this reaction, the carbanion $CH_3.CO.O.CO.CH_2^-$ is produced by the removal of a proton from ethanoic anhydride by the ethanoate ion. This carbanion then attacks the carbonyl carbon of an aromatic aldehyde.

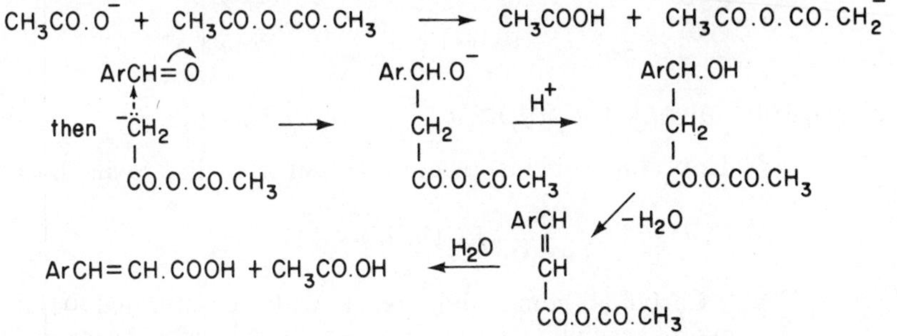

Figure 24.3 The Perkin reaction

Subsequent dehydration and hydrolysis gives rise to an unsaturated acid (*Figure 24.3*). The overall reaction of *benzenecarbaldehyde* with ethanoic (acetic) anhydride and sodium ethanoate (acetate) is

$$C_6H_5.CH\boxed{O + H_2}CH.CO.O.CO.CH_3 \xrightarrow{CH_3.CO.ONa} C_6H_5.CH{:}CH.CO.OH + CH_3.CO.OH$$

Benzenecarbaldehyde Ethanoic anhydride 3-phenylpropenoic acid (Cinnamic acid) Ethanoic acid

Nucleophilic attack by oxygen

It is believed that aqueous solutions of the lower aliphatic aldehydes contain hydrates, the result of nucleophilic attack by the oxygen of the water molecule

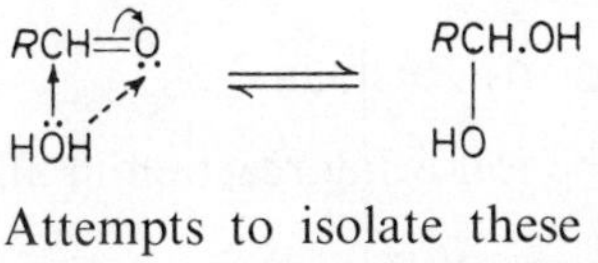

Attempts to isolate these hydrates result in their reconversion into the original aldehyde. The hydrate of trichloroethanal (chloral hydrate), however, is so stable that it is not even decomposed by moderate heat; the two

hydroxyl groups in this compound are stabilized by the inductive effect of the chlorine atoms, and possibly by hydrogen bonding, so that the normal sequence of elimination of the two hydroxyl groups attached to the same carbon atom is prevented:

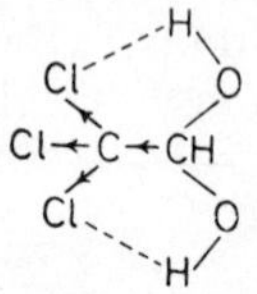

If the nucleophile is an alcohol instead of water, 'acetals' are formed (with aldehydes)

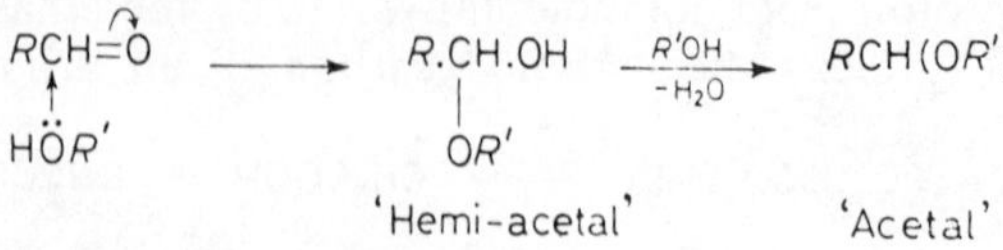

Nucleophilic attack by sulphur

Thiols (mercaptans) react in a manner similar to alcohols

$$RCH = O \xrightarrow[(-H_2O)]{2R'SH} RCH(SR')_2$$

Carbonyl compounds react with a saturated solution of sodium hydrogensulphite to yield products that can often be readily crystallized. The nucleophile here is the hydrogensulphite ion

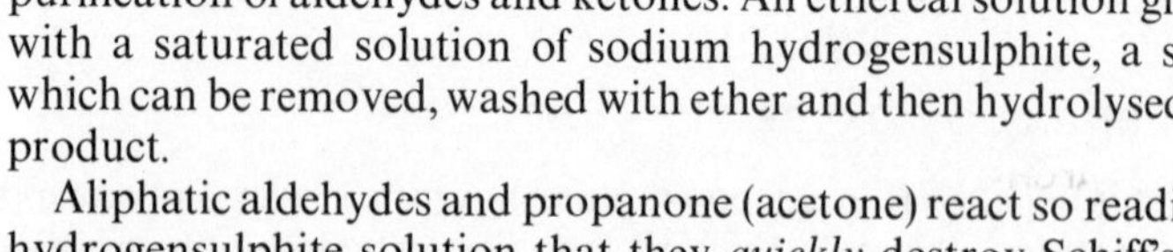

The original substance is easily liberated from the 'bisulphite' compound by treatment with acid or alkali. This reaction is therefore useful for the purification of aldehydes and ketones. An ethereal solution gives, on shaking with a saturated solution of sodium hydrogensulphite, a solid derivative which can be removed, washed with ether and then hydrolysed to yield a pure product.

Aliphatic aldehydes and propanone (acetone) react so readily with sodium hydrogensulphite solution that they *quickly* destroy Schiff's reagent (magenta bleached with sulphur dioxide) to give a red coloration.

Reaction with halogens and halogen compounds

Carbonyl compounds undergo the following reaction in alkaline solution:

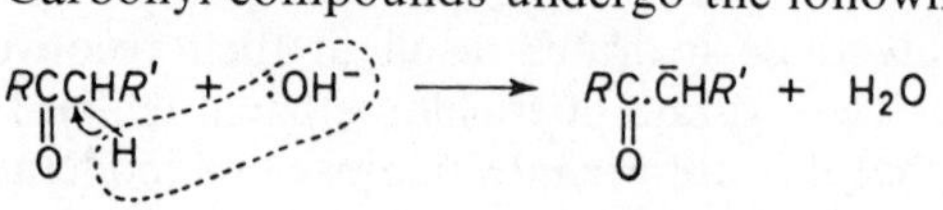

The resultant anion undergoes reaction with halogens

$$RC.\bar{C}HR' + X\!-\!X \longrightarrow RC.CHR' + X^-$$
$$\underset{O}{\|} \qquad\qquad \underset{O\ X}{\|\ \ |}$$

In the case of ketones containing the methyl group, i.e., when $R' = H$, the process can continue until all three hydrogen atoms are replaced by halogen. Hydrolysis then yields the 'haloform', e.g., triiodomethane (iodoform) (p. 439):

$$RCO.CH_3 \xrightarrow[I_2]{OH^-} RCO.CH_2I \xrightarrow[I_2]{OH^-} RCO.CHI_2 \xrightarrow[I_2]{OH^-} RCO.CI_3 \xrightarrow{OH^-} CHI_3$$
$$+ RCO.O^-$$

Phosphorus pentachloride replaces the oxygen atom of the carbonyl group by chlorine, giving a dichloride

$$RCHO + PCl_5 \rightarrow RCHCl_2 + POCl_3$$

Reduction of aldehydes and ketones

The carbonyl group can be reduced to different extents by various systems; most dissolving metal systems, e.g., sodium in ethanol, reduce the $>\!C=O$ to $>\!CHOH$, thus producing primary alcohols from aldehydes and secondary alcohols from ketones. Zinc amalgam and concentrated hydrochloric acid, however, can effect the complete reduction of $>\!C=O$ to $>\!CH_2$ (the Clemmensen reduction).

The hydride ion, as provided by lithium tetrahydridoaluminate, selectively reduces the carbonyl group, leaving other unsaturated bonds intact, e.g.,

$$CH_2\!=\!CHCH_2.\overset{H}{\underset{\uparrow}{C}}\!=\!O \longrightarrow \left[CH_2\!=\!CHCH_2.\overset{H}{\underset{H}{C}}\!-\!O\right]^- \xrightarrow{H^+} CH_2\!=\!CHCH_2.CH_2OH$$
$$H\!-$$

Complexed with Al

Reduction to alcohols is also effected by treatment with Grignard reagents, e.g.,

$$CH_3.\underset{O}{\overset{\|}{CH}} + \underset{MgBr}{C_2H_5} \longrightarrow CH_3.\underset{O.MgBr}{\overset{|}{CH}}.C_2H_5 \xrightarrow[-Mg(OH)Br]{Hydrolysis} CH_3.\underset{OH}{\overset{|}{CH}}.C_2H_5$$

Magnesium, activated by amalgamation, can initiate the reduction of ketones to 'pinacols'

$$\underset{O}{\overset{R'}{\underset{\|}{R\!-\!C}}} + \ddot{M}g + \underset{O}{\overset{R'}{\underset{\|}{C\!-\!R}}} \longrightarrow \underset{O\!-\!Mg\!-\!O}{\overset{R' \qquad R'}{R\!-\!C\!-\!\!-\!\!-\!C\!-\!R}} \xrightarrow{Hydrolysis} \underset{OH \quad OH}{\overset{R' \quad R'}{R\!-\!C\!-\!\!-\!C\!-\!R}}$$

The directive influence of the carbonyl group

The inductive effect of the oxygen atom in the carbonyl group, when attached directly to the benzene ring, results in the withdrawal of electrons from the benzene nucleus. This effect, as usual, is most apparent in positions 2, 4 and 6. Therefore, the benzene ring is deactivated, and substitution by the customary electrophiles takes place slowly, chiefly in positions 3 and 5: i.e., the carbonyl group is *meta*-directing to electrophilic reagents

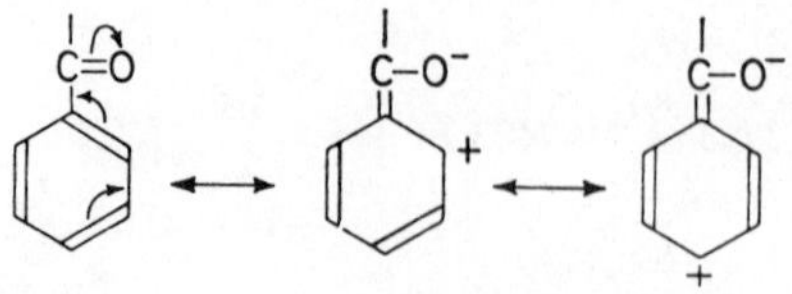

Carbonyl compounds in industry

It is clear from the foregoing that carbonyl compounds are reactive materials, and consequently they have many commercial uses.

Methanal

Methanal (formaldehyde) is required in large quantities

(1) For polymerization with, for example, phenol and urea to give thermosetting resins which are useful for making moulded articles and adhesives (pp. 487, 560).
(2) For the manufacture of dyes.
(3) As a disinfectant and preservative (an aqueous solution used for this purpose is known as Formalin).
(4) For hardening gelatine.

Methanal is manufactured by the catalytic dehydrogenation or oxidation, in the vapour phase, of methanol, which is itself prepared from water gas (p. 484):

$$CO + H_2 \rightarrow CH_3.OH \xrightarrow[-H_2]{} HCHO$$

Ethanal

Ethanal (acetaldehyde), $CH_3.CHO$, is prepared commercially from the other plentiful two-carbon compounds, ethyne (acetylene) and ethene (ethylene).

The elements of water are added catalytically to ethyne by treating it with a dilute solution of mercury(II) sulphate acidified with sulphuric acid and kept at about 75 °C. The yield is about 95 per cent of the theoretical, but as it employs the highly endothermic ethyne, it is comparatively costly (p. 425):

$$CH \equiv CH + H_2O \rightarrow CH_3.CHO$$

A cheaper route to ethanol involves direct oxidation of ethene in the Wacker process, the catalyst being a solution of palladium(II) chloride and copper(II) chloride. The reaction is very complex, but the net result can be summarized as

$$CH_2 = CH_2 + O \rightarrow CH_3.CHO$$

In more detail:

$$PdCl_2 + 2Cl^- \rightleftharpoons [PdCl_4]^{2-} \xrightarrow{C_2H_4} [PdCl_3C_2H_4]^-$$

$$\downarrow H_2O$$

$$[PdCl_2(OH)C_2H_4]^- \xleftarrow{-H^+} [PdCl_2(H_2O)C_2H_4]$$

$$-Cl^- \downarrow$$

$$Cl-Pd-CH_2CH_2OH \rightarrow Pd + HCl + CH_3CHO$$

Alternatively, ethene can be absorbed in cold, concentrated sulphuric acid and the intermediate hydrolysed to ethanol with water (p. 484); the ethanol, like methanol, can be dehydrogenated to the corresponding aldehyde, using a copper catalyst

$$\begin{matrix} CH_2 \\ \| \\ CH_2 \end{matrix} \xrightarrow{H^+} \begin{matrix} CH_3 \\ | \\ _+CH_2 \end{matrix} \xrightarrow{OH^-} \begin{matrix} CH_3 \\ | \\ CH_2OH \end{matrix} \xrightarrow{-H_2} \begin{matrix} CH_3 \\ | \\ CHO \end{matrix}$$

Ethanal is used in the manufacture of ethanoic (acetic) acid, D.D.T. and various drugs.

Propanone

Propanone (acetone) is now largely manufactured from propene, which results from the cracking of petroleum, by a series of reactions described above for ethanal:

$$\begin{matrix} CH_3.CH \\ \| \\ CH_2 \end{matrix} \xrightarrow{H_2O} \begin{matrix} CH_3.CH.OH \\ | \\ CH_3 \end{matrix} \xrightarrow{-H_2} \begin{matrix} CH_3.C=O \\ | \\ CH_3 \end{matrix}$$

It is also obtained in considerable quantities as a by-product in the cumene process for phenol (p. 486).

Propanone is used as a solvent, e.g., for paints and ethyne, for the manufacture of trichloromethane (chloroform) (p. 439) and for making Perspex (p. 446).

Propenal

Propenal (acrolein) manufacture is also a by-product of the petroleum industry; propene can now be oxidized directly to propenal:

$$CH_2 = CH.CH_3 + O_2 \xrightarrow{Catalyst} CH_2 = CH.CHO + H_2O$$

Propenal, possessing a carbonyl group and the alkenic double bond, is very reactive. It is used in the manufacture of ring systems by the Diels–Alder reaction:

$$
\begin{array}{ccc}
CH_2 & & CH_2 \\
\parallel & & \diagup \diagdown \\
CH & CH_2 & HC \quad\; CH_2 \\
| \quad + \quad \parallel & \xrightarrow{\text{Heat}} & \parallel \qquad | \\
CH & CH.CHO & HC \quad\; CH.CHO \\
\diagdown & \text{Propenal} & \diagdown \diagup \\
CH_2 & & CH_2
\end{array}
$$

Buta-1,3-diene

Reaction with phenylamine (aniline) gives rise to pyridine derivatives, e.g.,

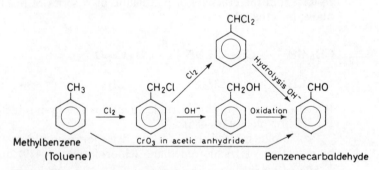

Quinoline

Other commercial applications include conversion to glycerol (p. 485), amino acids and plastics.

Benzenecarbaldehyde

Benzenecarbaldehyde (benzaldehyde) is in demand for perfumes, flavouring essences and dyes. The main source is coal tar. Methylbenzene, extracted from the benzol fraction, can be converted into benzenecarbaldehyde in several ways (*Figure 24.4*).

Figure 24.4 Routes from methylbenzene to benzenecarbaldehyde

Analysis of carbonyl compounds

The presence of an unmodified carbonyl bond is indicated by the eventual formation of a solid 'bisulphite' compound on shaking with a saturated solution of sodium hydrogensulphite. Those aldehydes which are soluble in water quickly produce a colour in Schiff's reagent.

The derivative most often prepared is the condensation product with 2,4-dinitrophenylhydrazine.

The molecular formula can be determined by qualitative analysis, followed by quantitative analysis and calculation of the relative molecular mass, for instance from vapour density measurements. The presence of oxygen in the formula might suggest an alcohol or ether. However, the oxygen atom in a carbonyl group is equivalent to two hydrogen atoms, and so results in the molecule having two fewer hydrogen atoms than the corresponding saturated hydrocarbon, e.g.,

Propane, C_3H_8

C_3H_8O $\begin{cases} \text{Propan-1-ol, } CH_3CH_2.CH_2.OH \\ \text{Methoxyethane, } CH_3CH_2.O.CH_3 \end{cases}$

C_3H_6O $\begin{cases} \text{Propanal, } CH_3CH_2.CHO \\ \text{Propanone, } CH_3CO.CH_3 \end{cases}$

(If the alkyl group is unsaturated, C_3H_6O also represents the molecular formula for $CH_2 = CH.CH_2.OH$, prop-2-en-1-ol.)

The remaining obstacle in the way of elucidation of the structural formula is the determination of the nature of the alkyl or aryl groups. This is usually overcome by degradation to simpler molecules capable of ready recognition and by final unambiguous synthesis of the substance.

Carbohydrates

Carbohydrates make up the major part of the vegetable kingdom and contain potential carbonyl groups. They have the general formula $C_xH_{2y}O_y$—it was the ratio of hydrogen to oxygen that led the French to give them the unfortunate name 'hydrates de carbone'—and may be classified as

(1) *Monosaccharides* or simple sugars, the most common of which are called hexoses and have the molecular formula $C_6H_{12}O_6$, e.g., glucose. They are soluble in water and have a sweet taste.

(2) *Disaccharides*, which are usually the condensation products of two hexose units

$$2C_6H_{12}O_6 - H_2O \rightarrow C_{12}H_{22}O_{11}$$

Sucrose (cane sugar) is one such example. They resemble the monosaccharides in sweetness and solubility.

(3) *Polysaccharides*, representing the condensation of hundreds and possibly thousands of hexose units. Not surprisingly, in view of the size of their molecules, they are either completely insoluble in water, e.g., cellulose, or present in the colloidal form only, e.g., starch.

It is important to realize that carbohydrates have a common origin (see Photosynthesis, p. 519) and are interconvertible to a large extent, simple hexoses being transformed to complex molecules by the process of condensation, and polysaccharides converted to simple sugars by hydrolysis, e.g.

$$C_{12}H_{22}O_{11} + H_2O \xrightarrow[\text{or enzymes}]{H^+} 2C_6H_{12}O_6$$

Structures of carbohydrates

Hexoses

Hydroxyl groups are attached to five of the six carbon atoms, and the sixth forms a carbonyl group; if the sixth is a terminal atom, then the carbonyl group is part of the aldehyde function, and the sugar is known as an *aldose*:

$$CH_2\overset{\times}{-}CH\overset{\times}{-}CH\overset{\times}{-}CH\overset{\times}{-}CH-CH{=\!=}O$$
$$\ \ |\ \ \ \ \ |\ \ \ \ \ |\ \ \ \ \ |\ \ \ \ \ |$$
$$OH\ \ OH\ \ OH\ \ OH\ \ OH$$

There are four asymmetric carbon atoms present (marked $\times$), and therefore 2^4 or 16 enantiomorphs exist. Two of these are $(+)$- and $(-)$-glucose; the $(+)$ and $(-)$ refer to the optical activity (dextro and laevo), but the D and L refer to the configuration at carbon atom 5. Any hexose where the configuration at C(5) is the same as it is in $(+)$glyceraldehyde [$(+)$-2,3-dihydroxypropanal] is a D-hexose, whatever its own rotation.

$$
\begin{array}{ll}
\text{^{1}CHO} & \text{CHO} \\
| & | \\
\text{H—^{2}C—OH} & \text{HO—C—H} \\
| & | \\
\text{HO—^{3}C—H} & \text{H—C—OH} \\
| & | \\
\text{H—^{4}C—OH} & \text{HO—C—H} \\
| & | \\
\text{H—^{5}C—OH} & \text{HO—C—H} \\
| & | \\
\text{^{6}CH$_2$OH} & \text{CH$_2$OH}
\end{array}
$$

$(+)$ (or D)-Glucose $\quad|\quad$ $(-)$(or L-)-Glucose

It is now known that most of the hexose unit is in the form of a ring, usually 6- but sometimes 5-membered, brought about by addition to the double bond of the carbonyl group:

$$
\begin{array}{l}
| \\
C{=}O \\
| \\
H{-}C{-}OH \\
| \\
H{-}C{-}OH \\
| \\
H{-}C{-}OH \\
|
\end{array}
\quad\longrightarrow\quad
\begin{array}{l}
\diagup OH \longleftarrow \text{ Potential} \\
C \qquad\qquad\quad \text{carbonyl} \\
| \qquad\qquad\quad \text{group} \\
H{-}C{-}OH \\
| \qquad O \\
H{-}C{-}OH \\
| \\
C \\
|\ \diagdown H
\end{array}
$$

A further asymmetric carbon atom is therefore introduced and the number of optical isomers doubled. In the case of glucose, the possible forms are $\alpha(+)$, $\beta(+)$, $\alpha(-)$ and $\beta(-)$. The α and β forms are readily interconvertible, by the opening of the ring of one form and its reclosure into the other form. Thus, whereas ordinary crystalline glucose consists of the $\alpha(+)$ form, the solution soon contains an equilibrium mixture of this and the less rotatory $\beta(+)$ form.

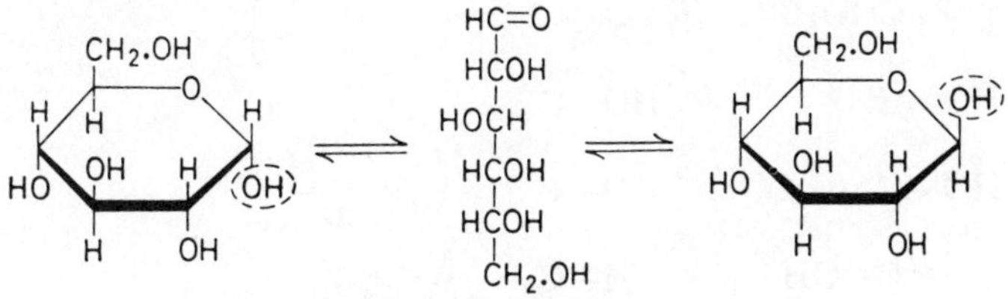

$\alpha(+)$-Glucose   (+)-Glucose $\beta(+)$-Glucose

Figure 24.5 Mutarotation in (+)-glucose

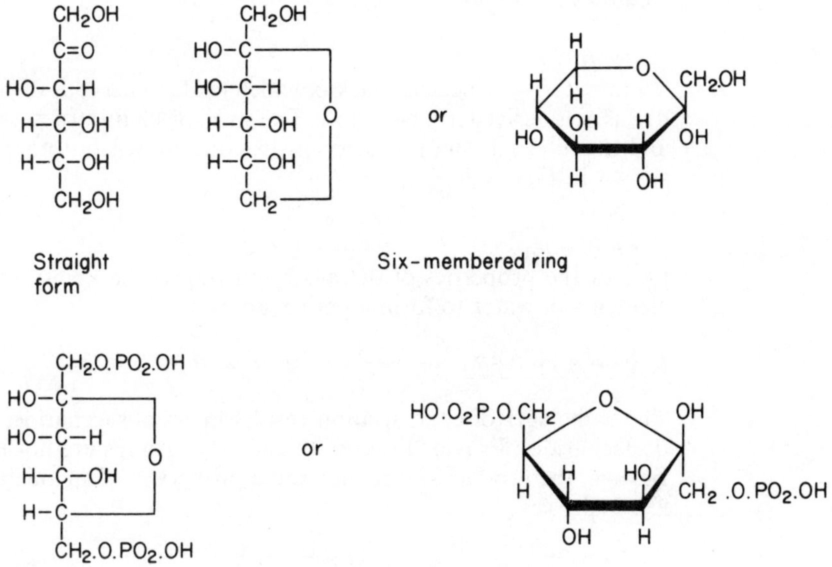

Straight form   or   Six-membered ring

or

Fructose-1,6-diphosphate (V)

Figure 24.6 Structures of fructose

When a solution is made, its optical activity diminishes to an equilibrium value, a process called *mutarotation* (*Figure 24.5*). Some of the open form is always present in solution, and so hexoses have the properties of carbonyl compounds, although the possibility of modification by the hydroxyl groups must be borne in mind.

Fructose is a reducing sugar, similar to glucose but containing a keto group instead of an aldehyde group, so being a ketose instead of an aldose. $(-)$-Fructose forms the same osazone (p. 519) as (+)-glucose, showing that the arrangement about the carbon atoms 3–6 is identical with that of (+)-glucose. The straight-chain structure is shown on the left of *Figure 24.6*, but in the free state it exists also in the 6-membered ring structure, two formulations of which are shown. In compounds it often is in a 5-membered ring form, as shown for the 1,6-diphosphate(V).

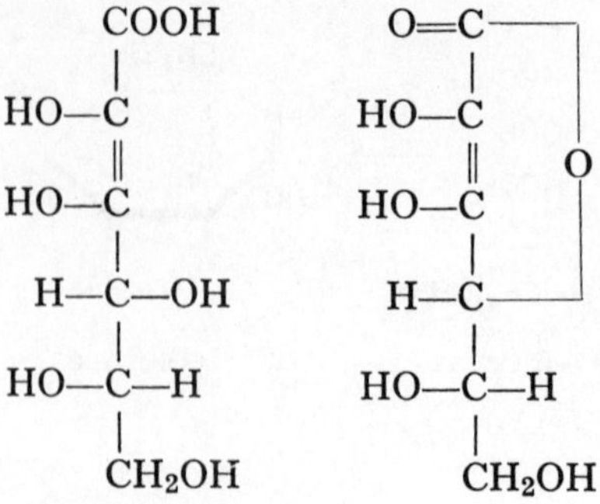

$$
\begin{array}{cc}
\text{COOH} & \text{O=C}\text{---} \\
| & | \quad\quad\quad \\
\text{HO---C} & \text{HO---C} \\
\| & \| \quad\quad\quad \text{O} \\
\text{HO---C} & \text{HO---C} \\
| & | \quad\quad\quad \\
\text{H---C---OH} & \text{H---C}\text{---} \\
| & | \\
\text{HO---C---H} & \text{HO---C---H} \\
| & | \\
\text{CH}_2\text{OH} & \text{CH}_2\text{OH}
\end{array}
$$

Figure 24.7 Straight-chain and lactone forms of ascorbic acid

Vitamin C
Vitamin C (ascorbic acid), lack of which leads to nervous disorders and skin diseases, is derived from a hexose. The straight-chain structure of the vitamin and the internal ester (or lactone) structure in which it normally exists are shown in *Figure 24.7*.

Disaccharides
One of the properties of the alcohol group is the ability to eliminate the elements of water to form an ether link:

$$R{-}O{+}H \; + \; H{-}O{+}R' \longrightarrow R{-}O{-}R' \; + \; H_2O$$

The same sort of condensation results in monosaccharides giving rise to disaccharides. Provided that both carbonyl groups are not involved in the linkage, the product still retains some properties of an aldehyde or ketone, e.g. maltose:

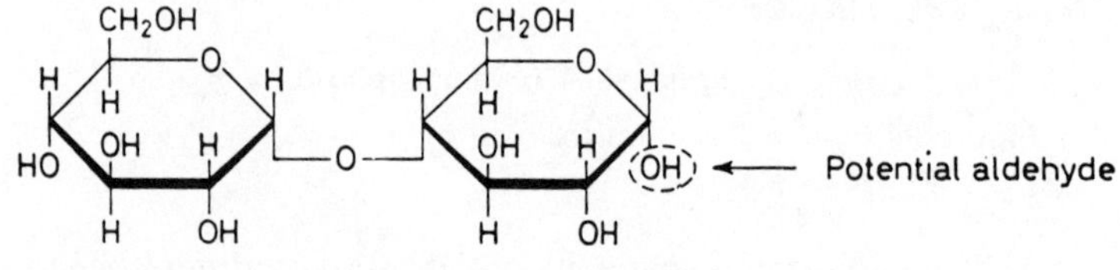

$\alpha(\text{D})$-Glucose $\alpha(\text{D})$-Glucose   Potential aldehyde

If, on the other hand, the linkage involves both potential carbonyl groups, then the sugar is non-reducing, e.g., sucrose:

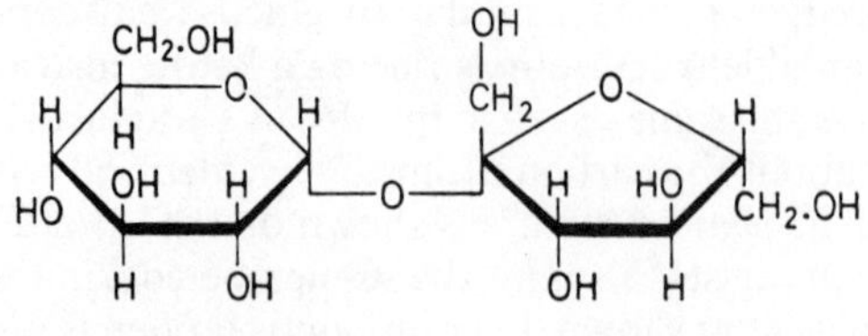

$\alpha(\text{D})$-Glucose $\beta(\text{D})$-Fructose

Polysaccharides

Cellulose, so called because it is the chief component of the walls of the plant cell, is made up solely of $\beta(\text{D})$-glucose units, the number involved being perhaps as many as 4000

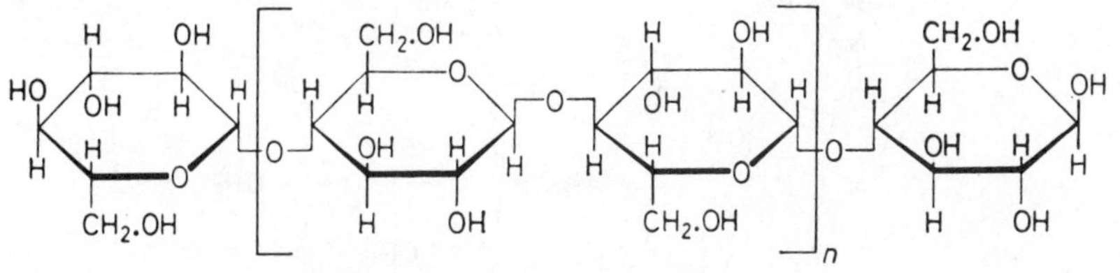

Starch, on the other hand, contains $\alpha(\text{D})$-glucose units, partly at least in the form of a linear polymer

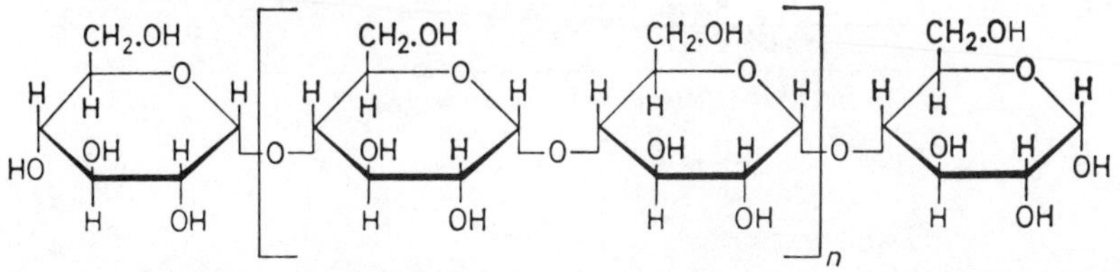

Starch represents the main energy reservoir in plants and animals; prior to its utilization in the respiratory processes of the cells (p. 520), it is hydrolysed first to maltose and then to glucose. Cellulose, however, cannot be digested in those animals which are not hosts to the intestinal bacteria that provide the cellulase enzymes; the cow is thus able to extract far more than the human being from grass. This point illustrates the specificity of enzyme action, when it is realized that the main difference between starch and cellulose is merely the configuration of one of the six carbon atoms of the hexose unit.

Glycosides, which occur in plants, are another important source of carbohydrates. On hydrolysis they split into saccharides and other sub-stances (called aglycones):if the saccharide is glucose then the glycoside is called a glucoside, e.g.,

$$\text{simple tannins} \xrightarrow{\text{Hydrolysis}} \text{glucose units} + 2,3,4\text{-trihydroxybenzenecarboxylic acid}$$
$$\text{(gallic acid)}$$

Properties of carbohydrates

Esterification

Carbohydrates possess the characteristic alcoholic capacity for esterifi-cation. Ethanoic anhydride produces ethanoyl derivatives which are useful as a means of locating the positions of hydroxyl groups, and hence the carbon atoms, involved in the ether-type link of the ring system. Esters with phosphoric(V) acid are fundamental in the processes of respiration and photosynthesis (pp. 519–523). For instance, adenosine represents the condensation products of (+)-ribose (a pentose) with the base adenine;

further condensation of the sugar with phosphoric(V) acid gives rise to the substances adenosine di- and tri-phosphate(V) (ADP and ATP) which play such a vital part in the energy transformations in the living cell:

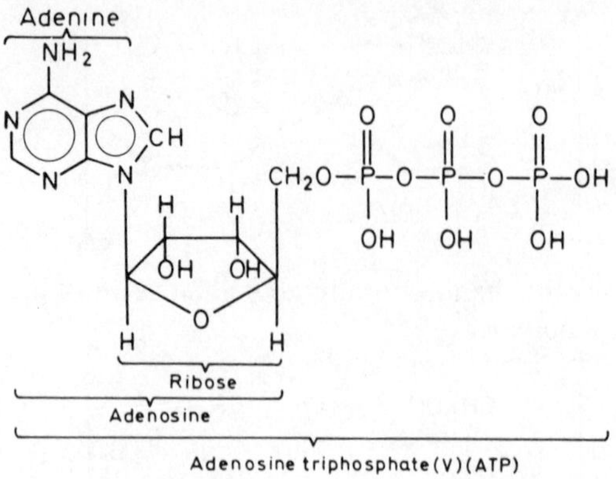

Adenosine triphosphate(V)(ATP)

Oxidation
Strong oxidation converts the terminal groups into carboxylic acids; e.g. glucose gives saccharic acid

$$\begin{array}{ccc}
\text{CHO} & & \text{CO.OH} \\
| & & | \\
(\text{CH.OH})_4 & \xrightarrow{\text{HNO}_3} & (\text{CH.OH})_4 \\
| & & | \\
\text{CH}_2\text{.OH} & & \text{CO.OH}
\end{array}$$

whilst mild oxidation affects the carbonyl groups only; e.g., glucose gives 2,3,4,5,6-pentahydroxyhexanoic (gluconic) acid

$$\begin{array}{ccc}
\text{CHO} & & \text{CO.OH} \\
| & & | \\
(\text{CH.OH})_4 & \xrightarrow{\text{Bromine water}} & (\text{CH.OH})_4 \\
| & & | \\
\text{CH}_2\text{.OH} & & \text{CH}_2\text{.OH}
\end{array}$$

Carbohydrates which possess latent carbonyl groups and which are soluble in aqueous media, give a silver mirror with diamminesilver(I) hydroxide solution

$$RCHO + [Ag(NH_3)_2]^+OH^- \rightarrow RCOOH + Ag(s)$$

Reduction
Powerful reducing agents convert the carbonyl group into an alcohol group:

$$
\begin{array}{ccc}
\text{CHO} & & \text{CH}_2\text{.OH} \\
| & & | \\
\text{HO—C—H} & & \text{HO—C—H} \\
| & \xrightarrow{\text{P+HI}} & | \\
\text{HO—C—H} & & \text{HO—C—H} \\
| & & | \\
\text{H—C—OH} & & \text{H—C—OH} \\
| & & | \\
\text{H—C—OH} & & \text{H—C—OH} \\
| & & | \\
\text{CH}_2\text{.OH} & & \text{CH}_2\text{.OH} \\
\text{Mannose} & & \text{Hexane-1,2,3,4,5,6-hexol (Mannitol)}
\end{array}
$$

Identification

The presence of the carbonyl group is revealed by formation of oximes (p. 505) as well as by the reactions mentioned above.

Phenylhydrazine forms phenylhydrazones by condensation, but these are not the ultimate products; an adjacent alcohol group is oxidized to carbonyl, which then condenses with more phenylhydrazine to give a double phenylhydrazone or *osazone*. These derivatives are often of characteristic crystalline shape and can be useful in identifying a particular sugar:

$$
\begin{array}{c}
\text{HC}{=}\text{O} \\
| \\
\text{CH.OH} \\
|
\end{array}
\qquad\qquad\qquad
\begin{array}{c}
\text{CH}{=}\text{N.NH.C}_6\text{H}_5 \\
| \qquad\qquad \text{Osazone} \\
\text{C}{=}\text{N.NH.C}_6\text{H}_5 \\
|
\end{array}
$$

$H_2 N.NH.C_6H_5 \Big| (-H_2O)$ $\qquad\qquad\qquad\qquad\qquad -H_2O \Big| C_6H_5 NHNH_2$

$$
\begin{array}{c}
\text{CH}{=}\text{N.NH.C}_6\text{H}_5 \\
| \\
\text{CH. OH} \\
|
\end{array}
\xrightarrow[\text{(-2H)}]{\text{H}_2\text{N.NH.C}_6\text{H}_5}
\begin{array}{c}
\text{CH}{=}\text{N.NH.C}_6\text{H}_5 \\
| \\
\text{C}{=}\text{O} \\
|
\end{array}
$$

Phenylhydrazone $\qquad\qquad (+C_6H_5.NH_2+NH_3)$

Photosynthesis

Carbohydrates are produced naturally *via* photosynthesis. Basically, this process consists of the reduction, involving photochemical activation with chlorophyll as catalyst, of carbon dioxide by water. In the simplest terms it can be represented as

$$6CO_2 + 6H_2O \xrightarrow[\text{Energy}]{\text{Light}} C_6H_{12}O_6 + 6O_2 \qquad\qquad \Delta H = 2810 \text{ kJ}$$

Efficiency ~ 1 per cent

This equation is misleading in its simplicity, since all of the oxygen evolved comes from the water and not from the carbon dioxide. In recent years, however, the use of radioactive tracers and other new techniques has permitted a much more detailed picture of the mechanism of photosynthesis to be drawn.

It is a significant fact that photosynthesis involves a relatively minute part of the electromagnetic spectrum. Chlorophyll appears green because it absorbs from white light the radiation from the red end of the spectrum. It is now believed that the energy of this radiation activates an electron from the chlorophyll, so that it can combine with a proton from water to give a hydrogen atom. The hydroxide ion remaining from the water molecule loses a low-energy electron to the electron-deficient chlorophyll molecule and then decomposes to water and oxygen. This part of the light-activated process can be summarized as

$$H_2O \rightarrow H^+ + OH^-$$

$$H^+ + e^- \rightarrow H\cdot$$

$$OH^- - e^- \rightarrow OH$$

$$4OH \rightarrow 2H_2O + O_2$$

The hydrogen atom, in combination with enzyme systems, reduces 3-phosphoglyceric acid to 3-phosphoglyceraldehyde, which is converted to fructose-1,6-diphosphate(V). This either leaves the cycle as the precursor of carbohydrates and other essential substances or combines with phosphoglyceraldehydr to give, by various routes, a 5-carbon compound [ribulose-5-phosphate(V)] which reacts with carbon dioxide to produce 3-phosphoglyceric acid and so complete the cycle (*Figure 24.8*). Clearly one molecule of fructose-1,6-diphosphate(V) is formed for six molecules of carbon dioxide entering the cycle, i.e., one molecule of monosaccharide is manufactured from six revolutions of the cycle. The role of ATP and enzymes as catalysts in these processes must not be overlooked; for example, ATP activates molecules by phosphorylating them.

Respiration

Photosynthesis is essentially the means of manufacturing foods, and appears important only when viewed against that process whereby the foods are consumed and energy (originally sunlight) is liberated. This process is known as respiration and can be of two types: *aerobic*, where oxygen is absorbed and utilized, and *anaerobic*, which takes place in the absence of oxygen. Aerobic respiration can be summarized as

$$C_6H_{12}O_6 + 6O_2 \rightarrow 6CO_2 + 6H_2O \qquad \Delta H = -2810 \text{ kJ}$$

In this form, aerobic respiration is seen to be the reverse of photosynthesis, although this does not imply that the entire mechanism is reversed.

Figure 24.9 shows that starch and cellulose are hydrolysed to glucose, which is subsequently converted into 2-oxopropanoic (pyruvic) acid. In anaerobic respiration in the human body, 2-oxopropanoic acid becomes 2-hydroxypropanoic (lactic) acid; in the case of the micro-organism, yeast, it

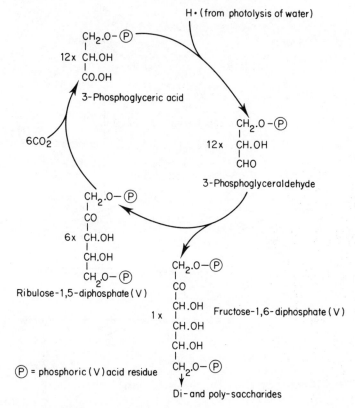

Figure 24.8 Photosynthesis (much simplified)

it transformed into ethanol, a process known as *fermentation*. In aerobic respiration, 2-oxopropanoic acid loses carbon dioxide and interacts with coenzyme A to give 'acetyl coenzyme A' which then enters the 'citric acid cycle'. In this, carbon dioxide is eliminated, together with hydrogen, which after passage through a series of enzymes is oxidized to water. The energy released is used for synthesizing high-energy ATP from ADP. As a result, one molecule of glucose is oxidized aerobically to carbon dioxide and water by two 'revolutions' of the 'citric acid' cycle; oxidation of hydrogen in the course of this results in the manufacture of

$$\frac{5 \times 4}{2} \times 3 = 30$$

molecules of ATP. As eight molecules are synthesized by glycolysis, the total number of molecules of ATP produced is 38, i.e.,

$$C_6H_{12}O_6 + 6O_2 \rightarrow 6CO_2 + 6H_2O + \text{energy stored in 38 ATP}$$

For the conversion of 1 mole ATP into ADP, 50 kJ of energy is evolved. Therefore

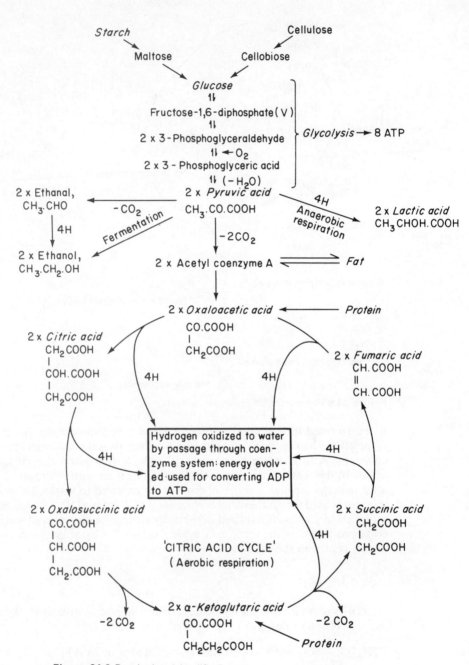

Figure 24.9 Respiration (simplified)

Energy obtained from glucose $= 38 \times 50 = 190$ kJ mol^{-1}

≈ 10.5 kJ/g

Protein can also be used in respiration, the amino acids resulting from hydrolysis being eventually absorbed into the 'citric acid cycle'.

It should be realized that, because many of the processes mentioned are reversible, they can be part of both respiratory and photosynthetic processes. For example, fructose-1,6-diphosphate(V) can appear as the result of photosynthesis from 3-phosphoglyceraldehyde or from the respiratory degradation of starch.

Commercial applications of carbohydrates

It is clear from what has been said that carbohydrates are an essential part of animal diet and that they are provided, in the first place, by plants. Starch is the reserve food of plants and is particularly concentrated in tubers, from some of which, e.g., potatoes, it is extracted. Sucrose, the most common sugar, is extracted from sugar cane and beet.

Cellulose is the main component of paper and for this purpose is obtained from wood by removal of the lignin, a complex polysaccharide, with solutions containing the hydrogensulphite ion. The resultant slurry is converted into paper by the controlled removal of water.

Cellulose can be dissolved in solutions containing tetraamminecopper(II) ions. If the solution is then forced through a narrow jet into a bath of acid, the tetraamminecopper(II) complex is decomposed and the cellulose regenerated. This process therefore results in the production of cellulose in the form of filaments, suitable for use in the textile industry, where it is known as *cuprammonium rayon*. *Viscose rayon* is the product obtained by treating cellulose first with concentrated alkali and then with carbon disulphide when 'cellulose xanthates', $RO.CS.SNa$, are formed; these too are reconverted into cellulose by subsequent treatment.

Modified, as well as regenerated, cellulose is of wide application. Nitration to varying extents gives products ranging from *pyroxylin* (with up to two nitrate groups per hexose unit) to *gun cotton* (up to three nitrate groups per hexose unit). The former is used as the base of cellulose paints and the latter as an explosive

$$\overset{|}{-\text{C.OH}} + \text{HO.NO}_2 \xrightarrow[-\text{H}_2\text{O}]{} \overset{|}{-\text{C.O.NO}_2}$$

Acetates, of greater stability than the nitrates, are manufactured by esterifying cellulose with ethanoic acid and are used in the manufacture of photographic films.

Fermentation processes can yield a wide variety of products depending on the conditions, e.g., pH and temperature, and the carbohydrate and microorganism employed. The conversion of starch to ethanol has been mentioned already under the heading 'Respiration'; in the Weizmann process, *Clostridium acetobutylicum* converts starch into butan-1-ol and propanone.

Perhaps the chief potential of fermentation processes lies in the fact that the fermentation is roughly contemporaneous with the growth of the plant producing the carbohydrate, in sharp contrast with the coal and petroleum industries, where the interval might be 200 million years: the long-term implications are clear.

Summary

Many reactions of the carbonyl group involve the addition of a nucleophile at the carbon atom; addition is often followed by elimination and the overall process is then known as condensation.

The more common reactions of ethanal are summarized in *Figure 24.10*.

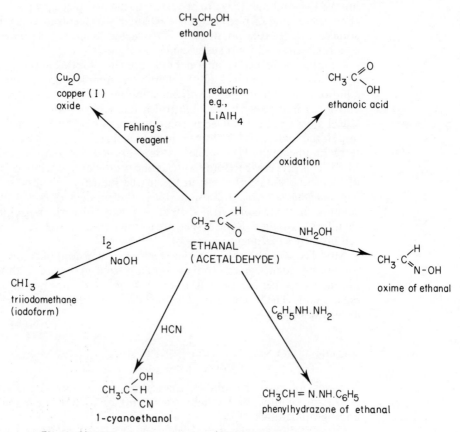

Figure 24.10 The more common reactions of ethanal

Questions

(1) Give the equations and state the conditions for each of the processes shown in the flow chart:

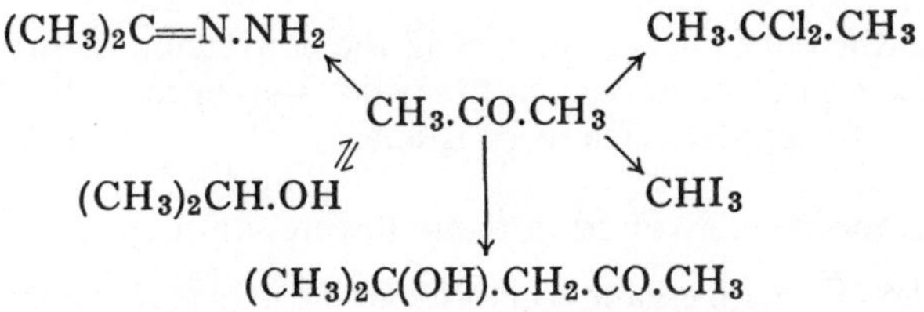

$(CH_3)_2C{=}N.NH_2$ $CH_3.CCl_2.CH_3$

$CH_3.CO.CH_3$

$(CH_3)_2CH.OH$ CHI_3

$(CH_3)_2C(OH).CH_2.CO.CH_3$

(2) How might propanone be converteʹoʹo: (a) 2-hydroxy-3-methylbutanoic acid, (b) propene, (c) 2-methylpropylamine?

(3) Give the structures of the isomeric aldehydes and ketones of molecular formula C_4H_8O and suggest chemical methods for distinguishing between them.

(4) Glycerol, on heating with anhydrous magnesium sulphate, yields an acrid-smelling distillate of molecular formula C_3H_4O, which immediately decolorizes bromine water and also restores the colour to Schiff's reagent. Indicate a possible structure for the compound and suggest its mode of formation from glycerol.

(5) When propanone is distilled with concentrated sulphuric acid, a hydrocarbon of molecular formula C_9H_{12} is formed, which on oxidation gives $C_9H_6O_6$, an acid, of which 0.85 g required 12.2 cm³ of sodium hydroxide solution of concentration 4.0 g dm⁻³ for neutralization. Suggest a possible structure for, and mode of formation of, the hydrocarbon.

(6) Ethanal, on treatment with concentrated sulphuric acid, gives $(CH_3.CHO)_3$, a trimer, and $(CH_3.CHO)_4$, a tetramer. Methanal also gives a trimer on standing. Suggest cyclic structures for these polymers.

(7) When the vapour of an acid A, of empirical formula CH_2O, was passed over heated aluminium oxide, compound B, of empirical formula C_3H_6O, was produced. This had a vapour density of 29; it reacted with 2,4-dinitrophenylhydrazine to give an orange precipitate of C. Compound B did not reduce diamminesilver(I) ions to metallic silver. Deduce the nature of A, B, and C and write equations to illustrate the above reactions.

(8) The compound A, C_7H_9N, when reacted with chlorine, yielded B, $C_7H_7Cl_2N$; this was readily hydrolysed to C, C_7H_7NO. C oxidized easily on standing to D, $C_7H_7NO_2$, which exhibited a few properties characteristic of a carboxylic acid. C also reacted with nitrous acid to give a product which, after boiling with ethanol, gave a compound E, C_7H_6O. In the presence of traces of the cyanide ion this reacted to produce F, having the same empirical formula as E and a vapour density of 106.

Elucidate the reactions and name the compounds A–F.

(9)   A naturally occurring compound has the molecular formula $C_{14}H_{18}O_8$. On hydrolysis, it yields a hexose and an ether $C_8H_8O_3$. Treatment of the latter with hydriodic acid and subsequent oxidation gives 3,4-dihydroxybenzene-carboxylic acid. Suggest a possible structure for the original compound.

(10) Trace the chief energy changes in the conversion of sunlight to muscular contraction. How far is it true to say that the concentration of carbon dioxide in the atmosphere remains constant?

(11) Trace the steps involved in the following syntheses:

   (a) Benzene with ethanoyl chloride in the presence of anhydrous aluminium chloride, produces substance A. A, treated with methylmagnesium iodide gives B, which on hydrolysis forms C. C, on heating, gives D, which on reduction gives E.

   (b) Ethene produces substance F on treatment with chloric(I) acid. F, with potassium cyanide forms G, which with alkali gives H.

   (c) Cyclohexanol treated with concentrated sulphuric acid gives J, which with bromine forms K (molecular formula $C_6H_{10}Br_2$). K reacts with an excess of dimethylamine to form L (molecular formula $C_8H_{15}N$). L reacts with iodomethane to give a salt M, which reacts with silver 'hydroxide' to form N (molecular formula $C_9H_{19}NO$). N on heating forms P, a cyclohexadiene. By repeating the total process show how benzene may ultimately be prepared.

(12) Discuss some of the evidence available for the mechanisms of respiration and photosynthesis suggested in this Chapter.

(13) Explain why acetals are produced in the presence of dry hydrogen chloride but decomposed by dilute hydrochloric acid.

(14) Write an essay on nucleophilic substitution.

(15) Using structural formulae account for the following properties of vitamin C (p. 516): (a) it forms a monosodium salt, (b) it decolourises bromine, (c) it gives no Schiff's reaction but forms a phenylhydrazone, (d) with iron(III) chloride solution, a violet colour results. On boiling with dilute hydrochloric acid, vitamin C forms water, carbon dioxide and an aldehyde of furan (p. 496). Deduce the position of the aldehyde group in the furan ring.

Carboxylic acids are difunctional compounds, containing both the *carbonyl* and hydr*oxyl* groups, hence the name. Furthermore, both of these groups are attached to the same carbon atom; the inductive effect of the two oxygen atoms is sufficiently powerful for the hydrogen of the hydroxyl group to be released as a proton, so that these substances are acidic, although very weak in comparison with the common mineral acids:

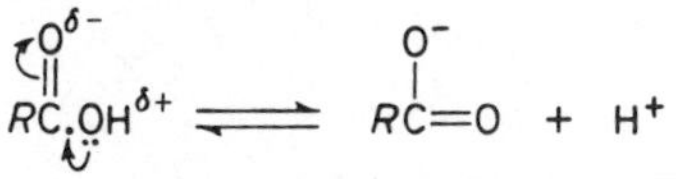

$$RC.OH^{\delta+} \rightleftharpoons R\overset{O^-}{\underset{}{C}}=O \ + \ H^+$$

Other functional groups are often also present, for example, $-NH_2$ in amino acids and a second hydroxy group in hydroxy acids.

Carboxylic acids are named by replacing the final *-e* of the hydrocarbon by *-oic acid*, or in some cases adding -carboxylic acid to a hydrocarbon name.

Occurrence and preparation

Carboxylic acids are of widespread occurrence. Thus methanoic (formic) acid is a common irritant in both the plant and animal kingdoms (Lat. *formica* = ant); ethanoic (acetic) acid represents the end product of one type of fermentation of carbohydrates (when it is known as vinegar); ethanedioic (oxalic) acid is present in the *Oxalis* group of plants, e.g. sorrel; hydroxy acids, e.g. citric and tartaric acids, are the characteristic acids of citrus fruits, and amino acids are the building units of proteins (*see Table 25.1*).

Methods of preparation

Oxidation
As carboxylic acids are the products of oxidation of aldehydes and ketones, they can be obtained by refluxing these substances with acidified sodium dichromate(VI) and then distilling to separate the constituents of the reaction mixture, e.g.

$$RCH \xrightarrow{[O]} RC.OH$$
$$\overset{\parallel}{O} \qquad\quad \overset{\parallel}{O}$$

Table 25.1 Some carboxylic acids of natural origin

Formula	Name	Common name	Source
HCOOH	Methanoic acid	Formic acid	Ants, nettles
CH_3COOH	Ethanoic acid	Acetic acid	Vinegar
COOH | COOH	Ethanedioic acid	Oxalic acid	Sorrel
CH(OH)COOH | CH_2COOH	2-Hydroxybutanedioic acid	Malic acid	Apples
CH(OH)COOH | CH(OH)COOH	2,3-Dihydroxybutanedioic acid	Tartaric acid	Grapes
$CH_3CH(OH)COOH$	2-Hydroxypropanoic acid	Lactic acid	Sour milk
CH_2COOH | C(OH)COOH | CH_2COOH	2-Hydroxypropane-1,2,3-tricarboxylic acid	Citric acid	Lemons
NH_2 | CH_2COOH	Aminoethanoic acid	Glycine	Meat extract

In view of the fact that aldehydes and ketones are themselves obtained by oxidation of primary and secondary alcohols, respectively, it is more usual to derive the acids directly from alcohols (or even, in some cases, from hydrocarbons) (cf. *Table 25.2*):

$$CH_3CH_2CH_2OH \xrightarrow{[O]} CH_3CH_2COOH$$
Propan-1-ol   Propanoic acid

Oxidation of aromatic hydrocarbons results in side chains being converted into carboxylic acid groups attached directly to the aromatic ring, as for example

$$C_6H_5.C_2H_5 \xrightarrow{[O]} C_6H_5.COOH$$
Ethylbenzene   Benzenecarboxylic (benzoic) acid

Table 25.2 Oxidation number of the carbon atom becoming the one in the carboxyl group

Compound type	RCH_3	RCH_2OH	$RCHO$	$RCOOH$	$[CO_2]$
Oxidation number	-4	-2	0	$+2$	$+4$

Hydrolysis of nitriles

Cyano derivatives (nitriles) are converted into acids by hydrolysis in the presence of acid or alkali

$$CH_3.CN \xrightarrow{H_2O} CH_3.COOH$$

Ethanonitrile   Ethanoic acid

or, in more detail as shown in *Figure 25.1*. Carbonyl compounds react with

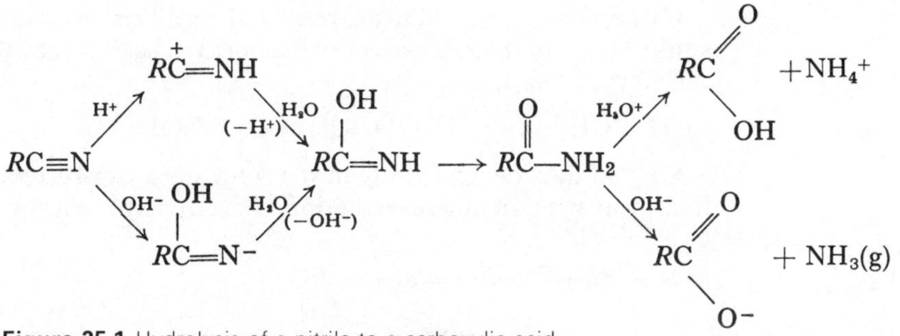

Figure 25.1 Hydrolysis of a nitrile to a carboxylic acid

the cyanide ion to give cyanohydrins (p. 505); hydrolysis of these compounds results in the formation of hydroxy-acids, e.g.,

$$CH_3.CH \xrightarrow{HCN} CH_3.CH.CN \xrightarrow{H_2O} CH_3.CH.CO.OH + NH_3(g)$$
$$\underset{O}{\parallel} \qquad\qquad \underset{OH}{|} \qquad\qquad\qquad \underset{OH}{|}$$

Ethanal   2-Hydroxypropanoic (lactic) acid

From natural products

Acids are often one of the products of the hydrolysis of natural products. When butter goes rancid, butanoic acid is one of the acids produced, whilst the turning sour of milk is caused by the appearance of 2-hydroxypropanoic acid.

Fats and oils are usually compounds of higher aliphatic 'fatty' acids and propane-1,2,3-triol. Boiling of these with sodium hydroxide solution converts them into the sodium salts of the acids and propane-1,2,3-triol (p. 478).

Proteins, on prolonged refluxing with acid or alkali, yield amino acids, hydrolysis occurring at the 'polypeptide' link (p. 558). The nature of proteins ensures that the resultant hydrolysates are complex mixtures, but separation can usually be effected by paper chromatography (p. 568).

$$...R—CONH—R'—CONH... \longrightarrow NH_2.R.CO.OH + NH_2.R'.CO.OH + ...$$
$$\quad\; HOH \qquad\quad HOH$$

From malonates and acetoacetates

Acids, especially of fairly complex structure, can be synthesized by using the esters of propanedioic (malonic) or 3-oxobutanoic (acetoacetic) acids (p. 553).

From carbon oxides

The oxides of carbon are sometimes used as a source of acids. For instance, carbon monoxide reacts with sodium alkoxide at elevated temperatures and high pressure:

$$RO^-Na^+ + CO \rightarrow RCO.O^-Na^+$$

and the free acid can be obtained from the salt by subsequent treatment with mineral acid.

Carbon monoxide and steam react with an alkene at elevated temperature and pressure in the presence of phosphoric(V) acid as catalyst to produce a carboxylic acid, e.g.,

$$RCH = CH_2 + CO + H_2O \rightarrow RCH_2.CH_2.COOH$$

Carbon dioxide, preferably in the solid state, can be reduced to acid by treatment with an ethereal solution of a Grignard reagent

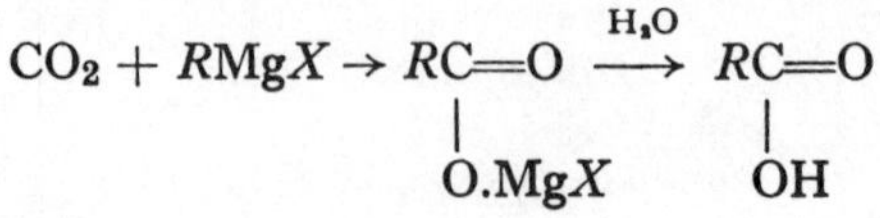

$$CO_2 + RMgX \rightarrow RC{=}O \xrightarrow{\;H_2O\;} RC{=}O$$
$$\qquad\qquad\qquad\;\; | \qquad\qquad\;\; |$$
$$\qquad\qquad\qquad O.MgX \qquad OH$$

Properties

Carboxylic acids are considerably less volatile than hydrocarbons of comparable relative molecular mass (*Table 25.3*).

Table 25.3 Volatility of hydrocarbons and related acids

Hydrocarbon	Relative mol. mass	B.p./°C	Acid	Relative mol. mass	B.p./°C
Propane	44	−44	Methanoic	46	102
Butane	58	−1	Ethanoic	60	118
Pentane	72	36	Propanoic	74	141

Calculation of the van't Hoff factor (p. 117) from measurements of, for example, freezing-point determinations of solutions in non-ionizing solvents indicates that hydrogen bonding, arising from the marked inductive effect of oxygen, results in association into double molecules, with relatively low volatility:

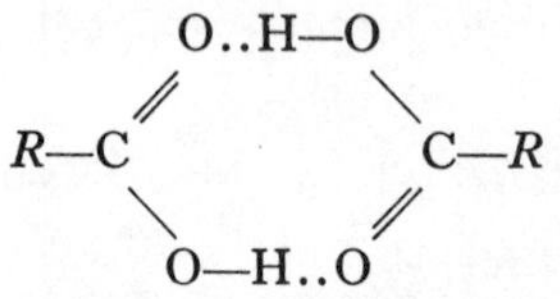

$$\begin{array}{ccc} & O..H{-}O & \\ & /\!/ \qquad\quad \backslash & \\ R{-}C & & C{-}R \\ & \backslash \qquad\quad /\!/ & \\ & O{-}H..O & \end{array}$$

Interaction with a polar solvent, e.g., water, causes the association to be replaced, at least partly, by ionization. Their aqueous solutions accordingly behave as acids

$$RCOOH + H_2O \rightleftharpoons RCO.O^- + H_3O^+$$

This reaction does not occur to a very great extent and they are therefore very weak acids in aqueous solution, particularly the higher homologues which are in any case only slightly soluble in water. It follows that aqueous solutions of carboxylic acid salts are extensively hydrolysed

$$RCO.O^- + H_2O \rightleftharpoons RCOOH + OH^-$$

For instance, if an aqueous solution of iron(III) ethanoate is boiled, a precipitate of basic salt is produced:

$$(CH_3.CO.O)_3Fe + 2H_2O \rightarrow (CH_3.CO.O)Fe(OH)_2(s) + 2CH_3.COOH$$

As the acidic strength depends on the extent of electron withdrawal from the relevant hydrogen atom, the dissociation constant will be increased by the introduction of electronegative groups into the molecule and decreased if hydrogen is replaced by a more electropositive group. Furthermore, the effect is most marked if the replacement is on the carbon atom next to the carboxyl group. For instance, methanoic acid is a stronger acid than ethanoic, where hydrogen has been replaced by an alkyl group of greater electron availability. Thereafter, as the series is ascended, the dissociation constant varies only slightly (see *Table 25.4*).

The acidity of the substituted carboxylic acids increases with the electronegativity of the element concerned. Fluorine has a greater effect than chlorine, chlorine than bromine, and bromine than iodine; the influence exerted increases with the number of halogen atoms present. Trichloroethanoic acid, for instance, has a dissociation constant comparable with those of mineral acids (*see Table 25.4*).

The benzene nucleus has a slight electron-donating effect relative to hydrogen because of conjugation between the carboxyl group and the ring leading to an electron drift from the ring, so that benzoic acid is weaker than methanoic acid. The introduction of electron-withdrawing groups into the benzene ring increases the acidity, however, especially when the substituents are in positions 2 and 4 (*Table 25.4*).

Dicarboxylic acids, by virtue of the increase in oxygen in the molecule, are stronger than the corresponding monobasic acids. Once again, the effect is less the further apart are the acid groups (*Table 25.4*).

The acidity of all carboxylic acids can be increased by using a solvent capable of accepting a proton (a protophilic solvent), the extent depending on the particular solvent employed. For example, the acid will dissociate more completely in a base like pyridine than in water (p. 173).

Table 25.4 Dissociation constants of acids

Acid	K_a/mol dm^{-3}	pK_a	Remarks
HCOOH	2.0×10^{-4}	3.7	
CH$_3$.COOH	1.6×10^{-5}	4.8	Acid weakened by electron-repelling effect of —CH$_3$; more distant alkyl groups have little effect
CH$_3$.CH$_2$.COOH	1.3×10^{-5}	4.9	
(CH$_3$)$_2$CH.COOH	1.3×10^{-5}	4.9	
FCH$_2$.COOH	2.0×10^{-3}	2.7	Halogen, with electron-attracting power, increases acidity, especially as the number of halogen atoms is increased
ClCH$_2$.COOH	1.6×10^{-3}	2.8	
BrCH$_2$.COOH	1.4×10^{-3}	2.9	
ICH$_2$.COOH	7.3×10^{-4}	3.1	
Cl$_2$CH.COOH	5.0×10^{-2}	1.3	
Cl$_3$C.COOH	1.2	−0.1	

⬡—COOH	6.3×10^{-5}	4.2	
⬡—COOH (NO$_2$)	6.5×10^{-3}	2.2	Electron-withdrawing effect of —NO$_2$ increases acidity, especially in position 2
⬡—COOH (O$_2$N)	3.5×10^{-4}	3.5	
O$_2$N—⬡—COOH	3.7×10^{-4}	3.4	

COOH / COOH	6.3×10^{-2}	1.2	adjacence of a second carboxyl group increases acidity; the effect declines with increasing distance between the —COOH groups
H$_2$C(COOH)(COOH)	1.6×10^{-3}	2.8	
⬡(COOH)(COOH)	1.0×10^{-3}	3.0	
⬡(COOH)(COOH)	3.2×10^{-4}	3.5	

$$RCOOH + H_2O \rightleftharpoons RCO.O^- + H_3O^+$$

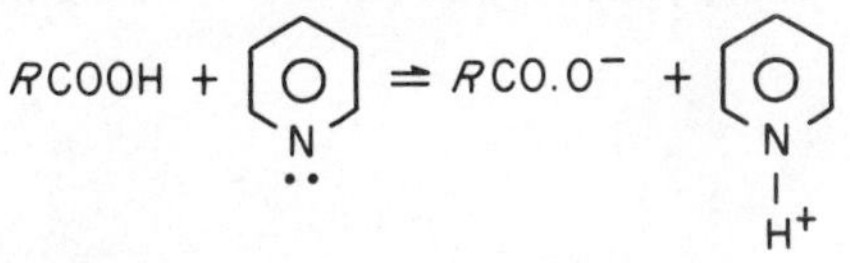

$$RCOOH + \left[\begin{array}{c} O \\ \underset{\cdot\cdot}{N} \end{array}\right] \rightleftharpoons RCO.O^- + \left[\begin{array}{c} O \\ \underset{\underset{H^+}{|}}{N} \end{array}\right]$$

Reactions

The structure of the carboxyl group is normally represented as

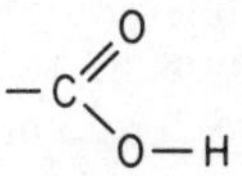

$$-C\underset{O-H}{\overset{O}{\diagup\!\!\!\backslash}}$$

but the acids and the derived anions show very few properties of the individual groups because of hydrogen bonding in the former and resonance in the latter

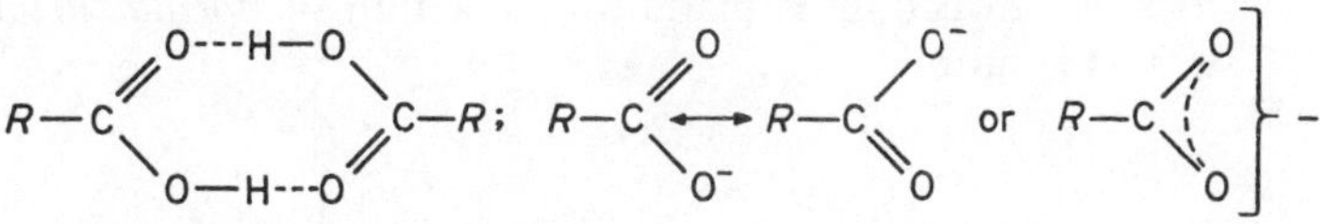

$$R-C\underset{O-H\cdots O}{\overset{O\cdots H-O}{\diagup\!\!\!\backslash}}C-R; \quad R-C\underset{O^-}{\overset{O}{\diagup\!\!\!\backslash}} \longleftrightarrow R-C\underset{O}{\overset{O^-}{\diagup\!\!\!\backslash}} \quad \text{or} \quad R-C\underset{O}{\overset{O}{\diagup\!\!\!\backslash}} \Big]^-$$

However, the carbonyl group can be reduced with lithium tetrahydridoaluminate, the final product being a primary alcohol

$$RCO.OH \xrightarrow{\text{LiAlH}_4} RCH_2.OH$$

The carbon of the carbonyl group also undergoes nucleophilic attack by the oxygen atom of alcohols, with the formation of esters, the reaction being catalysed by mineral acid (p. 480).

$$RCOOH + HOR' \rightleftharpoons RCO.OR' + H_2O$$

Chlorine progressively substitutes hydrogen in the aliphatic chain in a manner reminiscent of carbonyl compounds (p. 508), e.g.,

$$CH_3.COOH \xrightarrow[(-HCl)]{Cl_2} CH_2Cl.COOH \xrightarrow[(-HCl)]{Cl_2} CHCl_2.COOH \xrightarrow[(-HCl)]{Cl_2} CCl_3.COOH$$

Sulphur dichloride oxide and phosphorus chlorides attack acids in a similar way to alcohols (p. 438), giving acid chlorides

$$RCOOH \xrightarrow{SOCl_2} RCOCl$$

It follows from what has been said earlier that ionization of the carboxylic group will be favoured by treatment with alkali (i.e., salts will be formed):

$$RCOOH + OH^- \rightarrow RCO.O^- + H_2O$$

The resultant carboxylate ion possesses neither the carbonyl nor the hydroxyl group (see earlier) but has its own distinctive properties. For instance, it can be decarboxylated by treatment with alkali, e.g.,

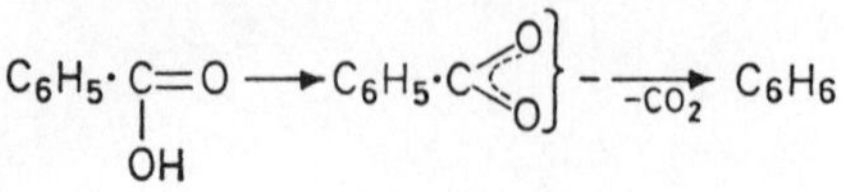

$$C_6H_5 \cdot C = O \longrightarrow C_6H_5 \cdot C\!\!\left<\begin{array}{c}O\\O\end{array}\right\} \xrightarrow[-CO_2]{} C_6H_6$$
$$\quad\ \ |$$
$$\quad\ \ OH$$

Stereoisomerism of carboxylic acids

Optical isomerism

If an α-hydrogen atom of propanoic acid is replaced by a hydroxyl group, then that carbon atom is attached to four different groups, i.e., it is asymmetric (p. 403) and the structure can exist in two mirror-image (enantiomorphic), optically active forms (*Figure 25.2*), and in a third inactive, or racemic, form which is an equimolecular mixture of the two active forms. This latter form of lactic acid is the variety that is present in sour milk. (+)-Lactic acid, rotating the plane of polarized light in a clockwise direction, occurs in muscle tissue and (−)-lactic acid, which rotates it in an anti-clockwise direction, is made by the action of *Bacillus acidi laevolactiti* on sucrose solutions.

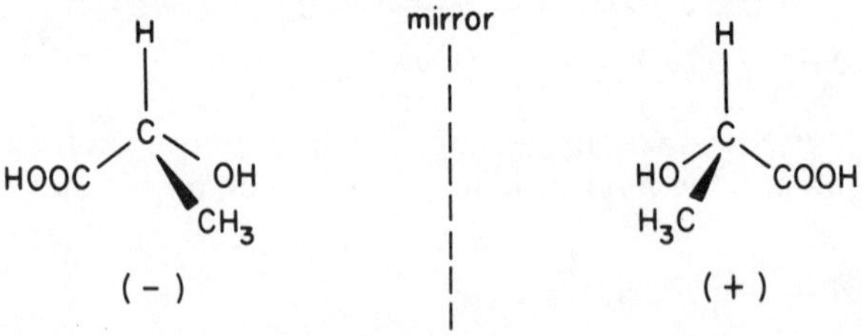

Figure 25.2 Enantiomorphs of 2-hydroxypropanoic (lactic) acid

A similar situation exists if the α-hydrogen of propanoic acid is replaced by the amino group. Thus 2-aminopropanoic acid (alanine) is optically active. The enantiomorph derived by hydrolysis of protein, like all the other amino acids containing asymmetric carbon and used in protein synthesis, is the (−)-form. As the enzymes used in the manufacture of natural products are themselves protein, and therefore optically active, it is to be expected that the products will be optically active, but why they should all be of the (−)-form is not clear.

Geometric isomerism

Unsaturated acids exhibit that form of stereoisomerism known as geometric isomerism (p. 402). The simplest examples are *cis*- and *trans*-butenedioic (maleic and fumaric) acids, respectively

HC.CO.OH   HO.OC.CH
‖   ‖
HC.CO.OH   HC.CO.OH
    cis   *trans*

Table 25.5 Properties of butenedioic acids

Property	cis	trans
Melting point/°C	130	287
Solubility/(g/100 g water)	79	0.7
Acid dissociation constants:		
$pK_a(1)$	1.92	3.02
$pK_a(2)$	6.22	4.38
Enthalpy of combustion/kJ mol^{-1}		

Unlike optical isomers, geometric isomers show several differences in chemical as well as physical properties. Indeed, some of these permit separation and identification of the isomers. For instance, cis-butenedioic acid is readily converted into the anhydride on heating; trans-butenedioic acid requires much stronger heating before an anhydride is formed, and then it gives cis-butenedioic anhydride. The dipole moment of trans-butenedioic acid is virtually zero.

$$
\begin{array}{ccccc}
\text{HC.CO.OH} & & \text{HC.C}=\text{O} & & \text{HC.CO.OH}\\
\| & \xrightarrow{\text{easy}} & \| \quad \diagdown{O} & \xleftarrow{\text{difficult}} & \|\\
\text{HC.CO.OH} & & \text{HC.C}=\text{O} & \text{HO.OC.CH} &
\end{array}
$$

cis-Butenedioic acid   cis-Butenedioic anhydride   trans-Butenedioic acid

Some of the differences between the two acids are summarized in *Table 25.5*.

Industrially important acids

Methanoic (formic) acid, H.COOH

The first member of an homologous series often has atypical properties. Methanoic acid is no exception. It contains the aldehyde group and is consequently a strong reducing agent, reducing Fehling's solution to copper:

$$Cu^{2+} + 2OH^- + HCOOH \rightarrow Cu(s) + CO_2(g) + 2H_2O$$

It is also easily oxidized to carbon dioxide by potassium manganate(VII)

$$HCOOH + [O] \rightarrow H_2O + CO_2(g)$$

Concentrated sulphuric acid dehydrates formic acid to carbon monoxide

$$HCOOH - H_2O \rightarrow CO(g)$$

It is a stronger acid than its homologues and is used in industry, e.g., dydstuffs manufacture, where a cheap and fairly strong acid is required; its fairly high volatility means that any residuum, e.g., on textiles, evaporates quickly. It is also used as a coagulant of rubber latex.

Methanoic acid is made on the large scale by passing carbon monoxide into a concentrated solution of sodium hydroxide at about 1.5 MPa pressure and ~200 °C temperature. Methanoic acid is liberated from this solution by adding dilute sulphuric acid (concentrated sulphuric acid would dehydrate it—see p. 535). Subsequent distillation gives an azeotrope rich in methanoic acid

$$CO + OH^- \rightarrow HCOO^- \xrightarrow{H^+} HCOOH$$

Ethanoic (acetic) acid, CH_3COOH

Ethanoic acid has been known for a long time as 'sour wine', the result of bacterial activity on ethanol, itself the result of the fermentation of carbohydrates. The process by which ethanol is converted into ethanoic acid is one of oxidation and takes place through the intermediate ethanal. It can be effected by oxidation of the vapour of the alcohol:

$$\underset{\text{Ethanol}}{CH_3.CH_2.OH} \xrightarrow[-2[H]]{[O]} \underset{\text{Ethanal}}{CH_3.CHO} \xrightarrow{[O]} \underset{\text{Ethanoic acid}}{CH_3.COOH}$$

It is also economically feasible to proceed from the intermediate, ethanal, because of the relatively low cost of this chemical. Recently, ethanoic acid, together with methanoic and propanoic acids, have been obtained directly from the light-oil fraction of crude petroleum by oxidation of the liquid at elevated temperatures; separation of the products is effected by fractional distillation.

Ethanoic acid is required commercially for conversion into esters (p. 549), the manufacture of ethanoic anhydride (p. 547), as a solvent and in the drug, dye and foodstuffs industries.

Several aryloxy derivatives of ethanoic acid are plant hormones, e.g.,

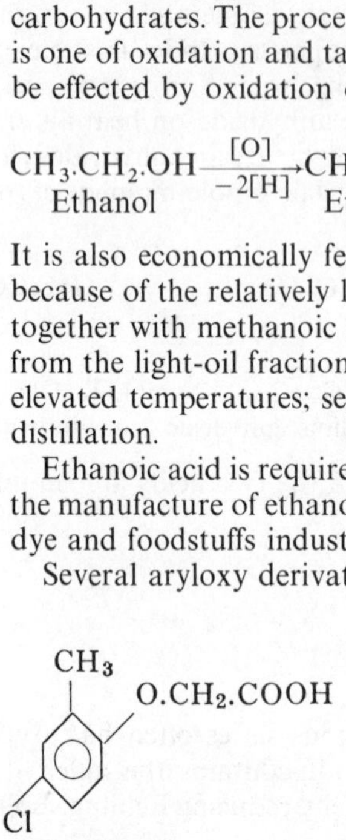

M.C.P.A. (4-chloro-2-methyl-phenoxyacetic acid)

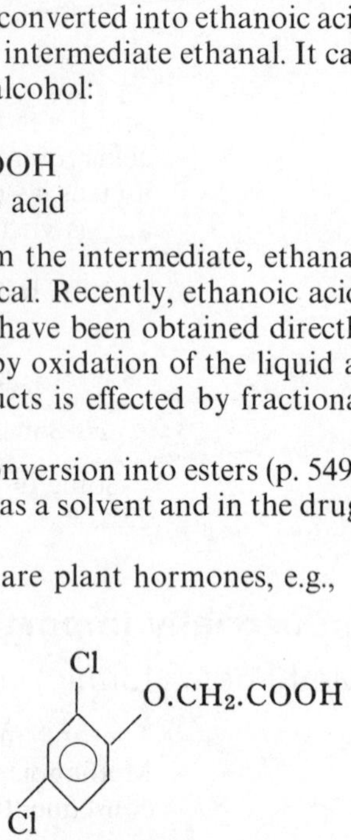

2,4-D.C.P.A. (2,4-dichloro-phenoxyacetic acid)

Some of these are fairly specific in their action, so that application of the hormone in sufficient quantities results in excessive, irregular growth and subsequent death of certain plants only. Selective weed-killing is now, within certain limits, a practical possibility.

Hexanedioic (adipic) acid

In recent years this has come into prominence as an intermediate in the manufacture of Nylon (p. 560). It is manufactured by oxidation of cyclo-hexanol, which is itself obtained from phenol, e.g.,

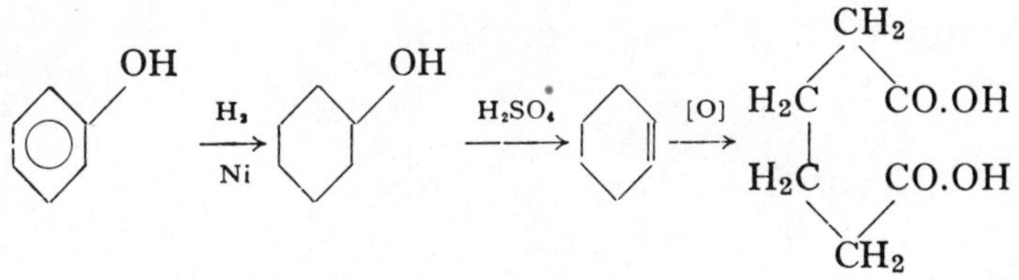

Ethanedioic (oxalic) acid

This is the parent member of the series of dicarboxylic acids, i.e., it is in a comparable position with methanoic acid. The two acids have several properties in common. Ethanedioic acid is dehydrated to carbon monoxide (and carbon dioxide) by concentrated sulphuric acid

$$\begin{array}{l} \text{COOH} \\ | \qquad -H_2O \rightarrow CO(g)+CO_2(g) \\ \text{COOH} \end{array}$$

It is also oxidized readily by acidified potassium manganate(VII) (particularly above 60 °C):

$$5(COOH)_2 + 2MnO_4^- + 6H^+ \rightarrow 8H_2O + 2Mn^{2+} + 10CO_2(g)$$

Furthermore, the two acids can be readily interconverted. Decarboxylation of ethanedioic acid, by heating its propane-1,2,3-triyl monoester, gives rise to methanoic acid:

$$\begin{array}{lll} \text{CH}_2.\text{O}.\text{CO}.\text{CO}.\text{OH} & & \text{CH}_2.\text{OH} \\ | & & | \\ \text{CH}.\text{OH} & + H_2O \longrightarrow & \text{CH}.\text{OH} \ + HCO.OH(g) + CO_2(g) \\ | & & | \\ \text{CH}_2.\text{OH} & & \text{CH}_2.\text{OH} \end{array}$$

Sodium methanoate on heating evolves hydrogen to give sodium ethanedioate. This is the basis of the large-scale production of the acid. The salt solution is treated with calcium chloride solution to cause precipitation of the calcium salt, from which the acid is liberated by addition of excess of dilute sulphuric acid:

$$\begin{array}{lll} \text{HCO.O}^- & \xrightarrow{\text{Heat}} & \text{CO.O}^- \\ + & & | \qquad + H_2(g) \\ \text{HCO.O}^- & & \text{CO.O}^- \end{array}$$

$$\begin{array}{lllll} \text{CO.O}^- & \xrightarrow{\text{Ca}^{2+}} & \text{CO.O}^- & \xrightarrow{H_2SO_4} & \text{CO.OH} \\ | & & | \qquad \text{Ca}^{2+}(s) & & | \qquad + CaSO_4(s) \\ \text{CO.O}^- & & \text{CO.O}^- & & \text{CO.OH} \end{array}$$

The acid is used in the dyestuffs industry and in the manufacture of inks. Some of its salts are used as photographic developers.

Benzenedicarboxylic (phthalic) acids

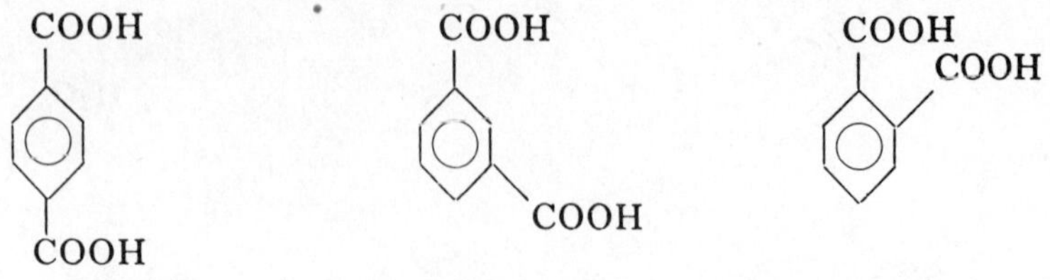

Benzene-1,4-dicarb-
oxylic (terephthalic) acid   Benzene-1,3-dicarb-
oxylic (isophthalic) acid   Benzene-1,2-dicarb-
oxylic (phthalic)
acid

Terephthalic acid is manufactured by the oxidation of 1,4-dimethylben-
zene (*p*-xylene) with nitric acid:

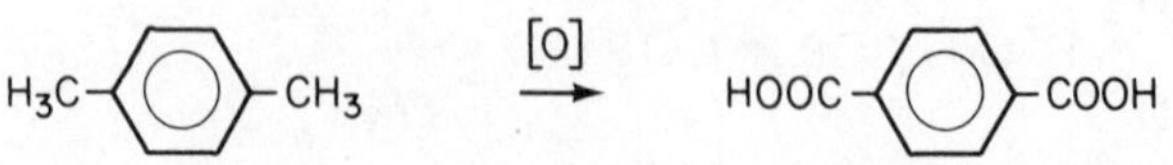

It is indirectly esterified with ethane-1,2-diol to form 'Terylene' (p. 555).

Phthalic acid is manufactured *via* the anhydride. Passing naphthalene
vapour and air over a hot vanadium pentoxide catalyst causes the reaction:

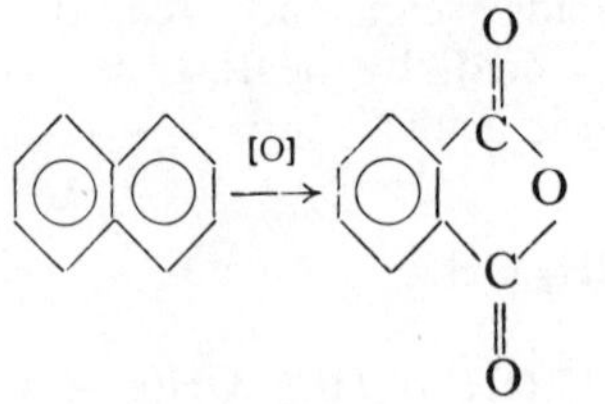

The anhydride is then boiled with sodium hydroxide solution to give the
sodium salt, from which the free acid is liberated by the addition of excess of
mineral acid

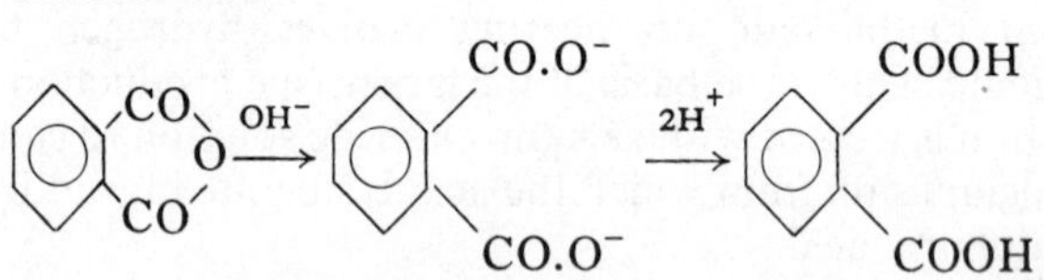

This acid is used for the manufacture of various esters, those with
monohydric alcohols being used as plasticizers, e.g. dibutyl phthalate, and
those with polyhydric alcohols, e.g. glycerol, as glyptal or alkyd resins
(p. 548).

Analysis of carboxylic acids

The carboxylic acid group is readily identified by the fact that it reacts, with
effervescence, with sodium carbonate solution. The 'anilide' is often prepared

as a derivative, first by converting the acid to the acid chloride and then reacting this with phenylamine (p. 471).

The elements present are estimated quantitatively in the usual way. Determination of the *true* relative molecular mass (association and dissociation often lead to erroneous results) permits evaluation of the molecular formula.

By adding excess silver nitrate solution to a neutral solution of an organic acid (prepared by boiling an excess of ammonium hydroxide with the carboxylic acid until free ammonia is no longer present), the silver salt is precipitated. The precipitate is washed, dried and weighed. By igniting the dry silver salt, a residue of silver is obtained; this, too, is weighed and the relative molecular mass of the acid calculated, providing the basicity has been separately determined. For a monobasic acid

$$RCOOH \rightarrow RCO.O^-NH_4^+ \rightarrow RCO.OAg \rightarrow Ag(s)$$
$$m\ g \qquad\qquad\qquad (m-1)+108\ g \quad 108\ g$$

Example. 0.830 g of a silver salt of a dibasic acid, on ignition, gave 0.540 g of silver. Hence, 2×108 g of silver would result from $(2 \times 108 \times 0.830)/0.540$ g of the silver salt, i.e., 332.0 g, which must be the relative molecular mass of the silver salt. The relative molecular mass of the acid is then $332 - (2 \times 108) + 2 = 118$.

Summary

Carboxylic acids are capable of both association and dissociation. Association into double molecules is the result of hydrogen bonding and reduces the tendency to acidity

$$
\begin{array}{ccc}
 & O \ldots H - O & \\
R-C & & C-R \\
 & O - H \ldots O &
\end{array}
$$

Dissociation into ions is encouraged by polar solvents such as water which are able to accept protons

$$R-C\!\!\begin{array}{c}{}^{\diagup O}\\{}_{\diagdown O-H}\end{array} \xrightarrow{H_2O} \ R-C\!\!\begin{array}{c}{}^{\diagup O}\\{}_{\diagdown O}\end{array}^{-} + \ H_3O^+$$

Both association and dissociation suppress the typical properties of the carbonyl and hydroxyl groups, so that carboxylic acids behave to a large extent as monofunctional compounds.

Some characteristic reactions of ethanoic acid are given in *Figure 25.3*.

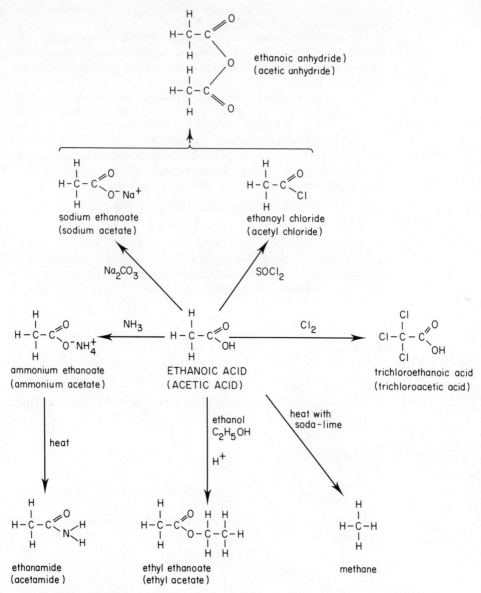

Figure 25.3 Characteristic reactions of ethanoic acid

Questions

(1) How might the basicity of a carboxylic acid be determined?

(2) How could you distinguish between the sodium salts of methanoic, ethanoic, ethanedioic, and benzenecarboxylic acids?

(3)   Give the structural formulae of isomers of molecular formula $C_4H_6O_2$. How would you attempt to distinguish between them?

(4)   Suggest methods of preparing the mononitrobenzoic acids. Account for the different acidities of the three acids.

(5)   What directive influence would you expect a carboxylic acid group to exert when attached to a benzene ring?

(6)   How might 2-benzoylbenzoic acid be prepared? The action of concentrated sulphuric acid on this compound produces a ring closure to give a quinone; suggest the structural changes involved and give the formula of the quinone produced.

(7)   A compound of molecular formula $C_9H_{10}O_3$ contains one carboxylic acid group, and although it does not react with bromine, it can be oxidized to a compound $C_9H_8O_3$, which imparts a coloration to Schiff's reagent. The original compound when strongly heated gives a substance of formula $C_9H_8O_2$, producing benzoic acid on oxidation. Suggest possible structures for the compound.

(8)   4.59 g of the silver salt of a carboxylic acid yields, on complete ignition, 3.24 g of silver; 1.15 g of the acid just neutralizes 250 cm^3 of 0.1 M sodium hydroxide solution. Name the acid.

(9)   Suggest ways in which *cis-* and *trans-*butenedioic acids can be synthesized from ethene.

(10)   Write an essay on the optical activity of natural products.

(11)   Explain the following facts about methanoic, ethanoic and benzene-carboxylic acids:

(*a*) Methanoic acid is the strongest acid of the three.
(*b*) Benzenecarboxylic acid is the most readily decarboxylated.
(*c*) Benzenecarboxylic acid is the least soluble in water.
(*d*) Methanoic acid is the only reducing agent of the three.

(12)   Write an account of the production of carboxylic acids by natural processes.

(13)   A 0.1 M solution of a monobasic organic acid in water had an osmotic pressure of 271 kPa at 27 °C. A 0.1 M solution of the same acid in benzene had an osmotic pressure of only 123 kPa at the same temperature. What can you deduce from this information?

(14)   Suggest an explanation for the fact that in the esterification of 2-aminopentane-1,5-dioic (glutamic) acid using ethanolic hydrochloric acid, the half-ester produced appears at the 1-position.

(15)   The hydrolysis of ethyl ethanoate using water enriched with $H_2{}^{18}O$ results in the accumulation of ^{18}O in the ethanoic acid and not in the alcohol. Suggest a reason for this observation and explain the technique used.

(16)   A substance A can be formed by any of the following methods: (a) heating 1-phenylethanol, (b) heating 2-phenylethanol, (c) heating 3-phenylprop-2-en-1-oic acid, (d) passing phenylethane over zinc oxide at $600\,°C$, or (e) treating ethene with benzene in the presence of aluminium chloride. Identify A, giving equations for the above reactions. Give two reactions of A.

(17)   Write the structural formulae for

(a) *cis*-Butenedioic acid (maleic acid).
(b) *trans*-Butenedioic acid (fumaric acid).
(a) (i)   What type of isomerism is shown by these two acids?
    (ii) What is the cause of this type of isomerism?
    (iii) Give two physical properties which are different for these two acids.
    (iv) Give one chemical reaction which is different for these two acids.
(b) The addition of hydrogen bromide separately to each of these two acids gives the same products, a pair of isomers, in each case.
    (i)   Write structural formulae which distinguish the two isomers.
    (ii) What type of isomerism is shown by these two isomers?
    (iii) What is the cause of this type of isomerism?

[A.E.B.]

26 Acid derivatives

Carboxyl compounds • Acid chlorides • Acid anhydrides • Esters • Amides • Peptides and proteins • Industrial aspects of amides

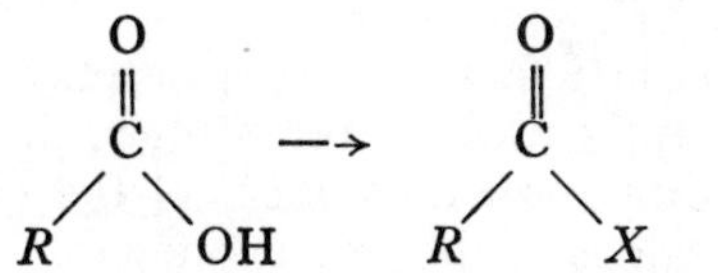

Acid Acid derivative

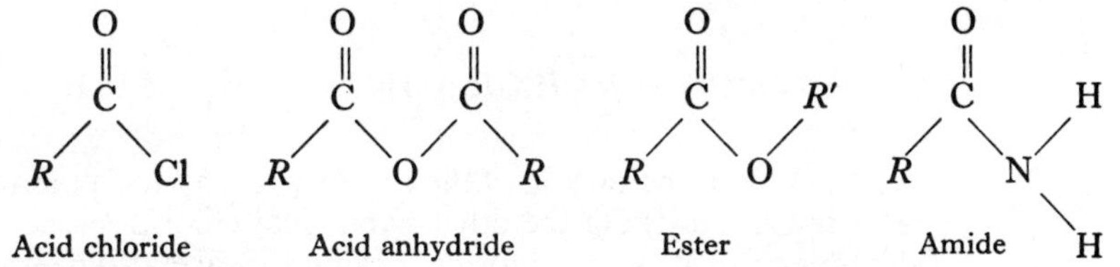

Acid chloride Acid anhydride Ester Amide

Carboxyl compounds

All carboxylic acid derivatives contain the carbonyl group, and it is interesting and helpful to consider how the characteristic properties of this group are modified by the substituent X. If X is hydroxyl, then the substance is a carboxylic acid and, as was seen in the previous Chapter, hydrogen bonding in the undissociated acid or resonance in the anion result in the virtual disappearance of carbonyl properties. Such is not the case with the acid derivatives; *nucleophilic addition at a carbon atom is still characteristic of these compounds*, although the readiness with which this reaction takes place can be affected by the electron-withdrawing capacity of the substituent X. In the case of acid chlorides (X = Cl), for instance, there is such a drift of electrons towards chlorine that the latter can even be ionized; e.g., it is claimed that ethanoyl (acetyl) chloride is ionized to the extent of 40 per cent. Consequently, the carbon atom is very polarized, and acid chlorides are easily the most reactive of the acid derivatives. On the other hand, the nitrogen atom of acid amides tends to share its lone pair with carbon, and this counteracts to some extent the polarization due to the carbonyl bond; amides are therefore the least reactive of acid derivatives:

$$R\overset{\overset{O}{\|}}{C}Cl \longleftrightarrow R\overset{\overset{O}{\|}}{C}^{+} + Cl^{-}$$

$$R\overset{\overset{O}{\|}}{C}\text{—}\ddot{N}H_2 \longleftrightarrow R\overset{\overset{O^{-}}{|}}{C}\text{=}\overset{+}{N}H_2$$

It does not follow that addition of a nucleophile to the carbon will result in any significant and complete change taking place. Much depends on the relative 'leaving capacity' of the groups attached to carbon in the intermediate formed:

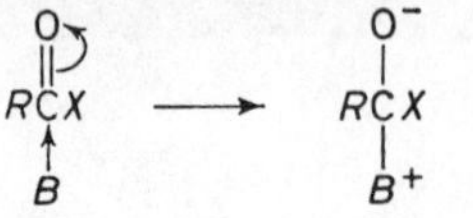

B = base (nucleophile)

Consider, for example, the reaction with the nucleophile *R*OH. Aldehydes react in the presence of hydrogen chloride as catalyst to give acetals, which represent the end product of the reaction and from which the original aldehyde is readily regenerated (p. 507)

$$2R\text{OH} + R'\overset{\overset{\displaystyle \text{O}}{\|}}{\text{C}}\text{H} \rightleftharpoons R'\overset{\overset{\displaystyle \text{O}R}{|}}{\text{C}}\text{H.O}R + \text{H}_2\text{O}$$

(Ketones, with the carbon of the carbonyl group less polarized, do not even react in this way.) On the other hand, acid chlorides react vigorously with alcohols to give esters, chlorine leaving the 'acetal-type' intermediate

$$R'\overset{\overset{\displaystyle \text{O}}{\|}}{\text{C}}\text{Cl} + R\text{OH} \rightarrow \left[R'\underset{\underset{\displaystyle R\overset{+}{\text{O}}\text{H}}{|}}{\overset{\overset{\displaystyle \text{Cl}}{|}}{\text{C}}}\text{.O}^- \right] \xrightarrow[-\text{H}^+]{-\text{Cl}^-} R'\overset{\overset{\displaystyle \text{O}}{\|}}{\text{C}}\text{.O}R$$

A similar reaction takes place with anhydrides, usually in the presence of a catalyst:

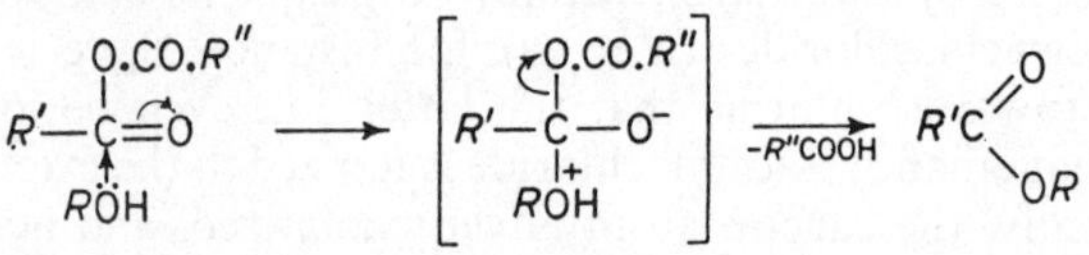

Nucleophilic attack by an alcohol even takes place upon an ester, as is shown by the fact that the alkyl group can be replaced by another, if the ester is left in contact with an alcohol:

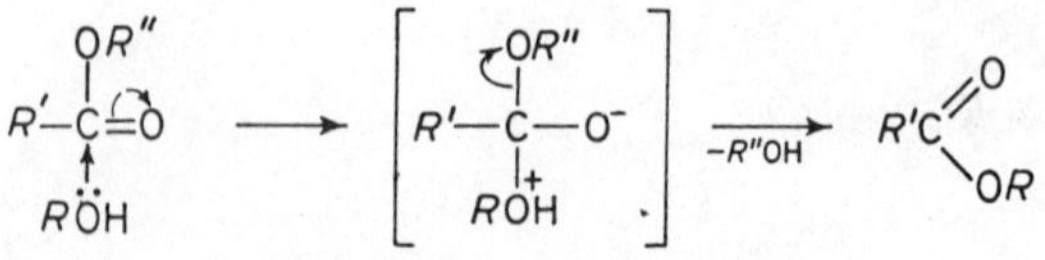

On the other hand, attack by alcohol upon an acid amide is negligible.

Table 26.1 Boiling points and melting points

		B.p./°C	M.p./°C		B.p./°C	M.p./°C
Acids	$CH_3.CO.OH$	118		$C_2H_3.CO.OH$	141	
Esters	$CH_3.CO.OCH_3$	56		$C_2H_5.CO.OCH_3$	79	
Chlorides	$CH_3.COCl$	52		$C_2H_5.COCl$	80	
Anhydrides	$(CH_3.CO)_2O$	140		$(C_2H_5.CO)_2O$	166	
Amides	$CH_3.CO.NH_2$		82	$C_2H_5.CO.NH_2$		77

Volatility

Both esters and acid chlorides are more volatile than acids of comparable relative molecular mass, because there is no prospect of hydrogen bonding and the substances exist as discrete molecules. Hydrogen bonding exists, though, with acid amides, and these are less volatile than the corresponding acids:

$$^\delta-O\ldots\ldots H^{\delta+}O^{\delta-}$$
$$\|\qquad\quad\;|\quad\|$$
$$\ldots HNH.CR\qquad NH.CR$$

The boiling or melting points of some common acids and their derivatives are shown in *Table 26.1.*

Acid chlorides, *RCOCl*

The systematic names of acid chlorides are formed from those of the parent acids by using the suffix *-yl* in place of the *-ic* of the acid, e.g.,

$C_2H_5.COOH$ $C_2H_5.COCl$
Propanoic acid   Propanoyl chloride

Because of their extreme reactivity, acid chlorides cannot be prepared from other acid derivatives; rather are they used for preparing the others, e.g.,

$RCOCl + R'OH \rightarrow RCO.OR' + HCl$

They can be obtained, however, by distilling the carboxylic acid with sulphur dichloride oxide or phosphorus chlorides; the former reagent is normally preferred because of the volatility of the by-products (and thus reduced contamination of the product itself), e.g.,

$CH_3.COOH + SOCl_2 \rightarrow CH_3.COCl + SO_2(g) + HCl(g)$

$C_6H_5.COOH + PCl_3 \rightarrow C_6H_5.COCl + H_3PO_3$

Acid chlorides are colourless, pungent liquids, fuming in moist air. Their great reactivity makes them important in synthetic work, e.g., ethanoyl (acetyl) chloride

$$CH_3.\overset{\overset{\displaystyle O}{\|}}{C}Cl$$

$$HO{-}H \longrightarrow CH_3.CO.OH \quad + \quad HCl$$

$$RO{-}H \longrightarrow CH_3.CO.OR \quad + \quad HCl$$

$$H_2N{-}H \longrightarrow CH_3.CO.NH_2 \quad + \quad HCl$$

$$RHN{-}H \longrightarrow CH_3.CO.NHR \quad + \quad HCl$$

$$RCOO^-\;Na^+ \longrightarrow CH_3.CO.O.OCR \; + \quad Na^+Cl^-$$

The mechanisms of these reactions have been indicated above.

The reaction between an acid chloride and alcohols or phenols is used as a means of characterizing the latter. The chloride normally used is 3,5-dinitrobenzoyl chloride, and the reaction is carried out in the presence of sodium hydroxide solution (Schotten–Baumann reaction), e.g.,

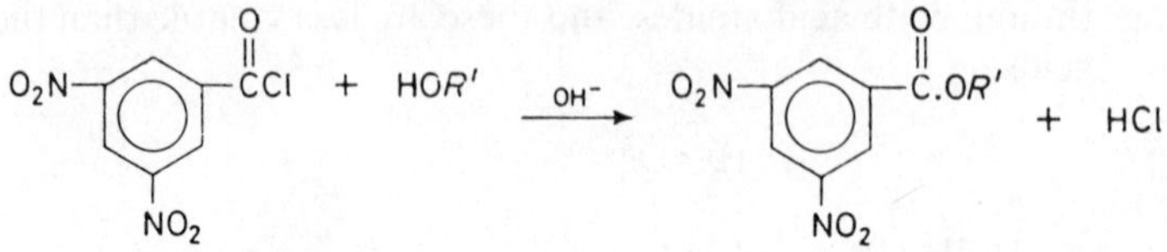

Amines can be converted into suitable derivatives in a similar way by treatment with benzoyl chloride in the presence of dilute sodium hydroxide solution

$$C_6H_5.COCl + H_2NR \rightarrow C_6H_5.CO.NHR + HCl$$

Anhydrous aluminium chloride encourages more complete ionization of acid chlorides by forming the complex $[AlCl_4]^-$ ion; the other ion formed, $[RCO]^+$, is a powerful electrophile and is capable of reacting with the benzene ring:

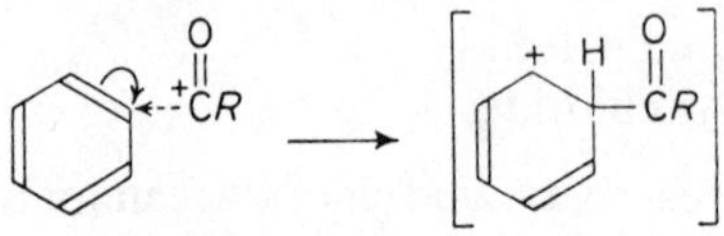

Expulsion of a proton from this intermediate results in the formation of a ketone (Friedel–Crafts reaction, see also p. 414):

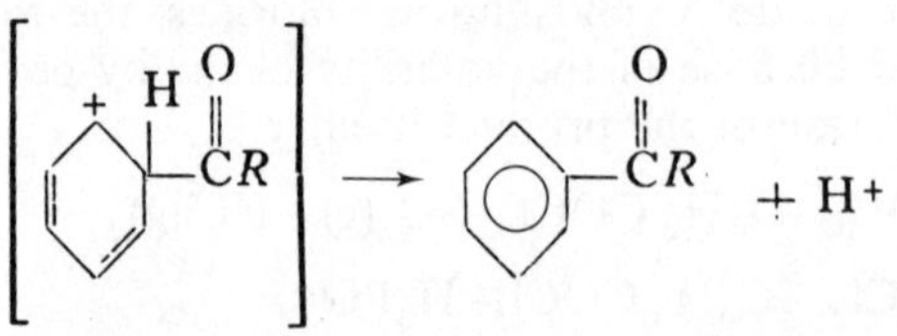

Acid chlorides are reduced by hydrogen in the presence of a palladium catalyst, first to aldehydes and then to alcohols [in the Rosenmund reaction

(p. 501), the palladium catalyst is 'poisoned' with barium sulphate, so that the reaction stops at the aldehyde stage]:

$$RCCl \xrightarrow{[2H]} RCH + HCl$$
(with O double-bonded above each of RCCl and RCH)

$$RCH \xrightarrow{[2H]} RCH_2.OH$$
(with O double-bonded above RCH)

Analysis

The chlorine in acid chlorides is so labile that hydrolysis rapidly takes place in the presence of aqueous silver nitrate; the resultant chloride ion then instantly reacts with the silver ion to give a white precipitate of silver chloride

$$RCOCl + H_2O \rightarrow RCOOH + H^+ + Cl^-$$

$$Ag^+ + Cl^- \rightarrow AgCl(s)$$

Suitable derivatives for characterization are prepared easily by the Schotten–Baumann reaction with phenylamine, e.g.,

$$CH_3.COCl + C_6H_5.NH_2 \rightarrow C_6H_5.NH.CO.CH_3$$

N.B. Methanoyl (formyl) chloride is unknown; attempts to prepare it always result in the production of carbon monoxide and hydrogen chloride:

$$HCOOH + SOCl_2 \rightarrow SO_2(g) + CO(g) + 2HCl$$

Acid anhydrides

$$RC.O.CR'$$
(with O double-bonded above each of RC and CR')

Acid anhydrides represent the removal of one molecule of water from two molecules of a monobasic acid; they are named from the parent acid by replacing 'acid' by 'anhydride'; e.g., $CH_3.COOH$, ethanoic (acetic) acid, gives rise to $CH_3.CO.O.CO.CH_3$, ethanoic (acetic) anhydride.

Dibasic acids often give *internal* anhydrides simply on distillation of the acid in an inert solvent, e.g.,

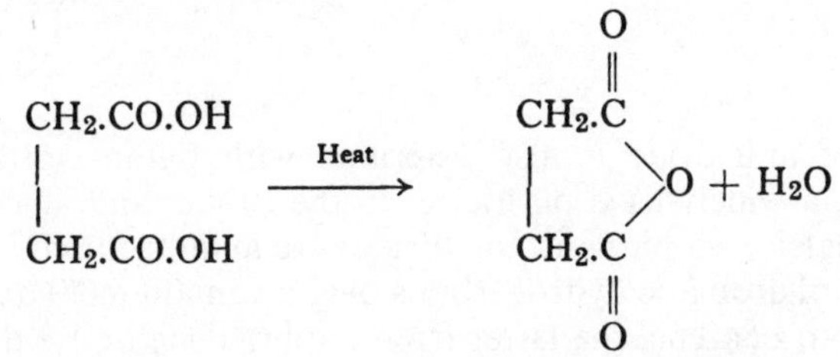

$$CH_2.CO.OH$$
$$|$$
$$CH_2.CO.OH$$
$$\xrightarrow{Heat}$$

Butanedioic acid
(succinic acid)

Butanedioic anhydride
(succinic anhydride)

Anhydrides of monobasic acids, however, usually require a more devious method of preparation, namely the distillation of an acid chloride with the anhydrous sodium salt of the acid. Here the active nucleophile is the carboxylate ion, e.g.,

$$\begin{array}{cc} \underset{\substack{\displaystyle\|\\ \text{CH}_3.\text{CCl}}}{\text{O}} & \underset{\substack{\displaystyle\|\quad\|\\ \text{CH}_3.\text{C.O.C.CH}_3}}{\text{O}\quad\text{O}} + \text{Cl}^- \\ \uparrow & \\ \text{CH}_3\text{COO}^- & \end{array}$$

Acid anhydrides are pungent, colourless liquids. They are not so violently reactive as acid chlorides and are sometimes preferred to them on this account for making amides and esters, e.g.,

$$\underset{\substack{\|\quad\|\\ \text{CH}_3.\text{C.O.C.CH}_3}}{\text{O}\quad\text{O}} + \text{NH}_3 \longrightarrow \underset{\substack{\|\\ \text{CH}_3.\text{C.NH}_2}}{\text{O}} + \text{CH}_3.\text{CO.OH}$$

$$\underset{\substack{\|\quad\|\\ \text{CH}_3.\text{C.O.C.CH}_3}}{\text{O}\quad\text{O}} + \text{C}_2\text{H}_5.\text{OH} \longrightarrow \underset{\substack{\|\\ \text{CH}_3.\text{C.O.C}_2\text{H}_5}}{\text{O}} + \text{CH}_3.\text{CO.OH}$$

Two very important anhydrides on the industrial front are *cis*-butenedioic (maleic) anhydride and benzene-1,2-dicarboxylic (phthalic) anhydride. The potential diacid functions present in these substances raise the possibility of condensation polymerization with polyfunctional alcohols to give polyesters; in fact, these are readily obtained. Known as 'alkyd resins' (*alcohol + acid*), they have many desirable commercial properties

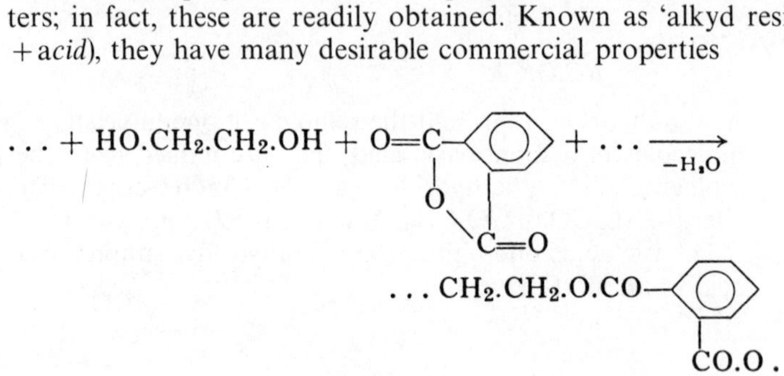

Phthalic anhydride is also esterified with butan-1-ol to give dibutyl phthalate, a widely-used plasticizer in the rubber and allied industries.

Both maleic and phthalic anhydrides are manufactured by vapour-phase oxidation of aromatic hydrocarbons over a vanadium(V) oxide catalyst, the former from benzene, the latter from naphthalene or 1,2-dimethylbenzene (*o*-xylene):

$$\text{benzene} \xrightarrow[V_2O_5]{O_2} \begin{array}{c} \overset{\displaystyle O}{\underset{\displaystyle \parallel}{CH.C}} \\ \underset{\displaystyle CH.C}{} \\ \underset{\displaystyle \parallel}{} \\ O \end{array}\!\!\!\! O \qquad \text{o-xylene} \xrightarrow[V_2O_5]{O_2} \begin{array}{c} \overset{\displaystyle O}{\underset{\displaystyle \parallel}{C}} \\ \\ \underset{\displaystyle \parallel}{C} \\ O \end{array}\!\!\!\! O$$

Esters

$$\overset{\displaystyle O}{\underset{\displaystyle \parallel}{RC.OR'}}$$

Esters are derived from acids by replacement of the acidic hydrogen by alkyl or aryl groups and are named accordingly, e.g., ethyl ethanoate, $CH_3.CO.OC_2H_5$, represents the replacement of the hydrogen atom in ethanoic acid, $CH_3.COOH$, by the ethyl radical.

The alkyl or aryl groups are usually provided by alcohols or phenols. Alcohols react with acids in a way which bears a superficial resemblance to the reaction of alkalis with acids but, on the other hand, the reaction velocity of *esterification* is fairly low and the reaction does not, in the absence of other substances, proceed to completion; in short, it is reversible and provides the classical example of chemical equilibrium (p. 141):

$$RCOOH + R'OH \rightleftharpoons RCO.OR' + H_2O$$

Addition of concentrated sulphuric acid to the reaction mixture, however, not only catalyses the reaction (i.e., accelerates its rate) but, by removing the water (as hydrates) as it is formed, shifts the equilibrium in favour of ester formation.

Protons provided by the mineral acid protonate the oxygen of the carbonyl group:

$$\overset{\displaystyle O:}{\underset{\displaystyle RC.OH}{\parallel}} \xrightarrow{H^+} \overset{\displaystyle OH^+}{\underset{\displaystyle RC.OH}{\parallel}}$$

The drift of electrons towards this oxygen atom is therefore increased, with the result that the carbon atom at the other end of the bond becomes more positively charged, i.e., more electrophilic, and attracts the oxygen atom of the alcohol. Eventually elimination of water gives the ester:

$$\overset{\displaystyle OH^+}{\underset{\underset{\displaystyle R'\overset{..}{O}H}{\uparrow}}{\underset{\displaystyle RC.OH}{\parallel}}} \longrightarrow \overset{\displaystyle OH}{\underset{\underset{\displaystyle R'OH^+}{}}{\underset{\displaystyle RC.OH}{|}}} \xrightarrow[-H^+]{-H_2O} \overset{\displaystyle O}{\underset{\displaystyle RC.OR'}{\parallel}}$$

Reaction of alcohols with acid chlorides and anhydrides is more pronounced than with the acid itself. Indeed, acid chlorides can be used to esterify phenols which, because of the modified properties of the hydroxy group when attached to an aromatic system (p. 479), do not react so readily with carboxylic acids, e.g.,

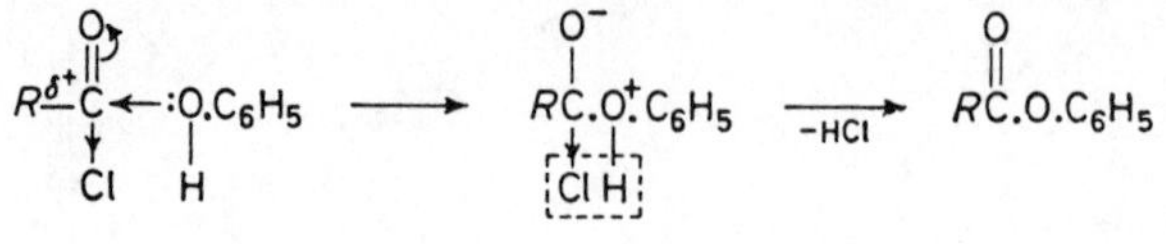

Properties

The lower esters are colourless, fragrant liquids, mobile and insoluble in water. Many of them are found in nature as the 'essential oils' of fruit, but by far the largest quantity of ester occurring naturally is in the form of fats and oils. These are, respectively, solid and liquid condensation products derived from the trihydric alcohol, propane-1,2,3-triol (glycerol), and monobasic acids which almost always contain an even number of carbon atoms, in accordance with their synthesis from acetyl coenzyme *A* units, $CH_3.CO—A$, formed from 2-oxopropanoic (pyruvic) acid during respiration (p. 520), e.g.

$$\begin{array}{l} CH_2.O.CO.CH_2.CH_2.CH_3 \\ | \\ CH.O.CO.CH_2.CH_2.CH_3 \\ | \\ CH_2.O.CO.CH_2.CH_2.CH_3 \end{array}$$

Propane-1,2,3-triyl tributanoate
(which occurs in butter)

The melting point of a fat depends to a large extent upon the state of saturation of the acid involved; acids containing double bonds give esters of lower melting point than do saturated acids and therefore tend to be present in the liquid oils. For example, stearin, a glyceride of the saturated octadecanoic (or stearic) acid, $CH_3(CH_2)_{16}COOH$ (rel. mol. mass 284), melts at 65 °C and is therefore solid at room temperature; olein, a glyceride of the unsaturated *cis*-octadec-9-enoic (or oleic) acid (rel. mol. mass 282) $CH_3(CH_2)_7CH{=}CH(CH_2)_7COOH$ melts at −6 °C and is therefore liquid at room temperature. Consequently, saturation by the addition of hydrogen across the double bonds provides a method for converting vegetable oils into fats. It is carried out by means of molecular hydrogen in the presence of a finely divided nickel catalyst (Sabatier–Senderens reduction):

$$\begin{array}{cc} H & H \\ | & | \\ —C{=} & C— \end{array} \xrightarrow[Ni]{H_2} \begin{array}{cc} H & H \\ | & | \\ —C— & C— \\ | & | \\ H & H \end{array}$$

This reaction is of fundamental importance in the conversion of materials such as whale oil and groundnut oil into margarine.

Saturation can also be effected by the addition of atmospheric oxygen to the double bond. Linseed oil, containing glycerides of the unsaturated acids, octadec-9,12-dienoic (linoleic) and octadec-9,12,15-trienoic (linolenic), becomes saturated in this way when exposed to the air and accordingly finds application as a 'drying oil' in paints.

Hydrolysis and ammonolysis

Esters can be converted into the corresponding acids and alcohols or phenols by hydrolysis, for example by refluxing the esters with strong alkali and then distilling. The distillate consists of the alcohol, whilst the acid remains as the non-volatile sodium or potassium salt in the residue from which it can be extracted by treatment with excess of mineral acid. If the esters used are oils or fats, then sodium or potassium salts of long-chain acids result and, because these are soaps, the process is known as *saponification*:

$$
\begin{array}{lll}
CH_2.O.COR & CH_2.OH & RCO.O^- \\
| & | & + \\
CH.O.COR' \; +3OH^- \; \longrightarrow \; CH.OH & + & R'CO.O^- \\
| & | & + \\
CH_2.O.COR'' & CH_2.OH & R''CO.O^- \\
\text{Glyceride} & \textbf{Glycerol} & \text{Soap anions}
\end{array}
$$

In the case of phenolic esters, the solution, after refluxing the ester with alkali, contains the alkali salts of both the acid and the phenol, and so no separation is effected by distillation. Acidification liberates the two acidic components:

$$C_6H_5.O.COR + 2OH^- \rightarrow C_6H_5.O^- + RCO.O^- + H_2O$$

$$C_6H_5.O^- + H^+ \rightarrow C_6H_5.OH$$

$$RCO.O^- + H^+ \rightarrow RCOOH$$

Advantage can now be taken of the marked difference in acidity between the two substances. If both are dissolved in ether and the ethereal phase is shaken with sodium carbonate, the carboxylic acid is the only component to react:

$$2RCOOH + CO_3^{2-} \rightarrow 2RCO.O^- + H_2O + CO_2(g)$$

As it does, it forms the hydrophilic salt which dissolves in the aqueous phase, from which the acid is reliberated by treatment with excess of mineral acid. Distillation of the ether phase leaves the phenol.

The mechanism of alkaline hydrolysis of an ester is based upon the addition of the nucleophilic hydroxide ion to the carbon of the carbonyl group and the subsequent departure of the alkoxy or phenoxy ion from the intermediate:

$$
\begin{array}{ccc}
O & O^- & O \\
\| & | & \| \\
RC.OR' \rightarrow & RC.OR' \rightarrow & RC.OH \; + \; {}^-OR' \rightarrow \; R{-}C{\Big\{}_{\!\!O}^{\!\!O}{\Big]}^- \; + \; R'OH \\
\uparrow & | & \\
\ddot{O}H^- & OH &
\end{array}
$$

Nitrogen compounds can act in a similar way to hydroxyl groups. *Concentrated ammonia solution converts esters into amides* [particularly in the case of ethanedioates (oxalates)] and hydroxylamine brings about conversion into hydroxamic acids, e.g.,

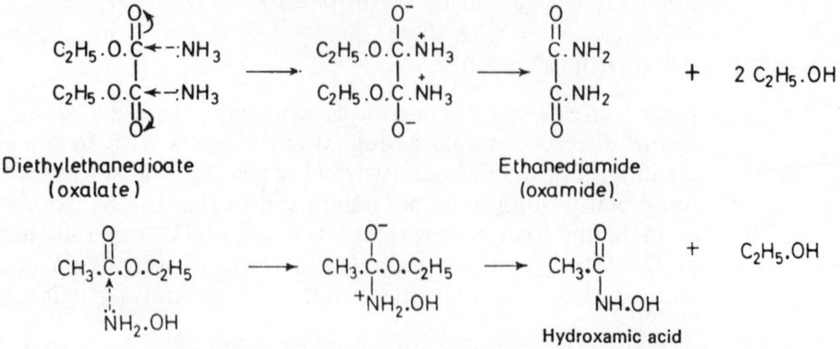

$$
\begin{array}{c}
\underset{\substack{\text{Diethylethanedioate}\\ \text{(oxalate)}}}{
\begin{array}{l}
\text{C}_2\text{H}_5.\text{O}.\overset{\overset{\displaystyle O}{\|}}{\text{C}}\!\!\leftarrow\!\!:\text{NH}_3\\
\text{C}_2\text{H}_5.\text{O}.\underset{\underset{\displaystyle O}{\|}}{\text{C}}\!\!\leftarrow\!\!:\text{NH}_3
\end{array}}
\longrightarrow
\begin{array}{l}
\text{C}_2\text{H}_5.\text{O}.\overset{\overset{\displaystyle O^-}{\mid}}{\underset{}{\text{C}}}.\overset{+}{\text{N}}\text{H}_3\\
\text{C}_2\text{H}_5.\text{O}.\underset{\underset{\displaystyle O^-}{\mid}}{\text{C}}.\overset{+}{\text{N}}\text{H}_3
\end{array}
\longrightarrow
\underset{\substack{\text{Ethanediamide}\\ \text{(oxamide)}}}{
\begin{array}{l}
\overset{\overset{\displaystyle O}{\|}}{\text{C}}.\text{NH}_2\\
\underset{\underset{\displaystyle O}{\|}}{\text{C}}.\text{NH}_2
\end{array}}
+\quad 2\,\text{C}_2\text{H}_5.\text{OH}
\end{array}
$$

$$
\underset{\underset{\displaystyle \text{N}\text{H}_2.\text{OH}}{\mid}}{\text{CH}_3.\overset{\overset{\displaystyle O}{\|}}{\text{C}}.\text{O}.\text{C}_2\text{H}_5}
\longrightarrow
\underset{\underset{\displaystyle {}^+\text{NH}_2.\text{OH}}{\mid}}{\text{CH}_3.\overset{\overset{\displaystyle O^-}{\mid}}{\text{C}}.\text{O}.\text{C}_2\text{H}_5}
\longrightarrow
\underset{\substack{\text{NH}.\text{OH}\\ \textit{Hydroxamic acid}}}{\text{CH}_3.\overset{\overset{\displaystyle O}{\|}}{\text{C}}}
+\quad \text{C}_2\text{H}_5.\text{OH}
$$

Hydroxamic acids give characteristic colours with iron(III) chloride, and the reaction therefore provides a useful test for esters, although acid chlorides also react in this manner.

Reduction

The carbonyl group in an ester can be reduced to an alcohol by the use of lithium tetrahydridoaluminate:

$$RCO.OR' + 4[H] \rightarrow RCH_2.OH + R'OH$$

Reduction with Grignard reagents and malonic and acetoacetic esters

Grignard reagents react with esters to form tertiary alcohols, as in *Figure 26.1*.

$$
\underset{}{\overset{\overset{\displaystyle O}{\|}}{R\text{C}.\text{O}R'}} + R''\text{Mg}X \xrightarrow{\text{H}_2\text{O}}
\underset{\underset{\displaystyle R''}{\mid}}{\overset{\overset{\displaystyle O\text{Mg}X}{\mid}}{R\text{C}\!-\!\text{O}R'}}
\longrightarrow \overset{\overset{\displaystyle O}{\|}}{R\text{C}R''} + \text{Mg(OH)}X + R'\text{OH}
$$

$$
\overset{\overset{\displaystyle O}{\|}}{R\text{C}R} + R''\text{Mg}X \longrightarrow
\underset{\underset{\displaystyle R}{\mid}}{\overset{\overset{\displaystyle O\text{Mg}X}{\mid}}{R\;\text{C}R''}}
\xrightarrow{\text{H}^+}
\underset{\underset{\displaystyle R}{\mid}}{\overset{\overset{\displaystyle \text{OH}}{\mid}}{R\;\text{C}R''}} + \text{Mg(OH)}X
$$

Figure 26.1 Grignard reagents and esters

Propanedioic (malonic) and 3-oxobutanoic (acetoacetic) esters
It has been pointed out that the presence of two oxygen atoms attached to the same carbon atom results in such a marked inductive effect that hydrogen attached to an oxygen is able to ionize:

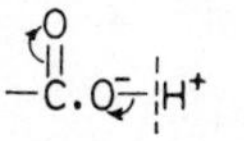

$$-\overset{\overset{\displaystyle O}{\|}}{C}.O^-\,H^+$$

In the case of 1,3-diketones, electron drift towards the oxygen renders the hydrogen attached to carbon-2 so labile that the ketone is in equilibrium with the *enol* form. This type of dynamic isomerism is known as *tautomerism*:

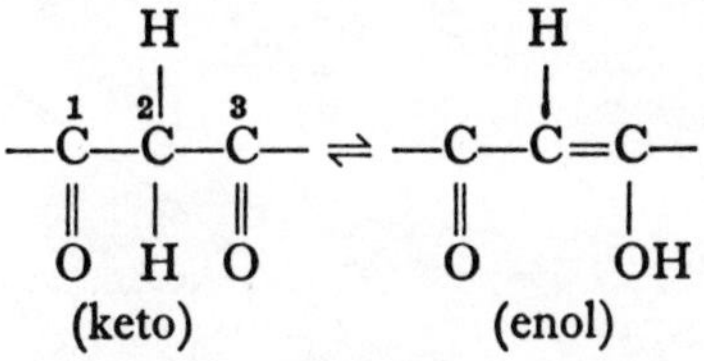

$$\underset{\underset{\displaystyle \text{(keto)}}{\overset{\displaystyle O\ \ H\ \ O}{\overset{\displaystyle \|\ \ \ |\ \ \ \|}{}}}}{-\overset{1}{C}-\overset{\overset{\displaystyle H}{|}}{\underset{2}{C}}-\overset{3}{C}-} \rightleftharpoons \underset{\underset{\displaystyle \text{(enol)}}{\overset{\displaystyle O\ \ \ \ \ OH}{\overset{\displaystyle \|\ \ \ \ \ \ \ |}{}}}}{-\overset{\overset{\displaystyle H}{|}}{C}-C=C-}$$

Diethyl propanedioate (*diethyl malonate*) is useful in synthesis because of this keto–enol isomerism. The hydrogen of the methene group may be replaced by sodium, and then a series of reactions such as those shown in *Figure 26.2* can be performed.

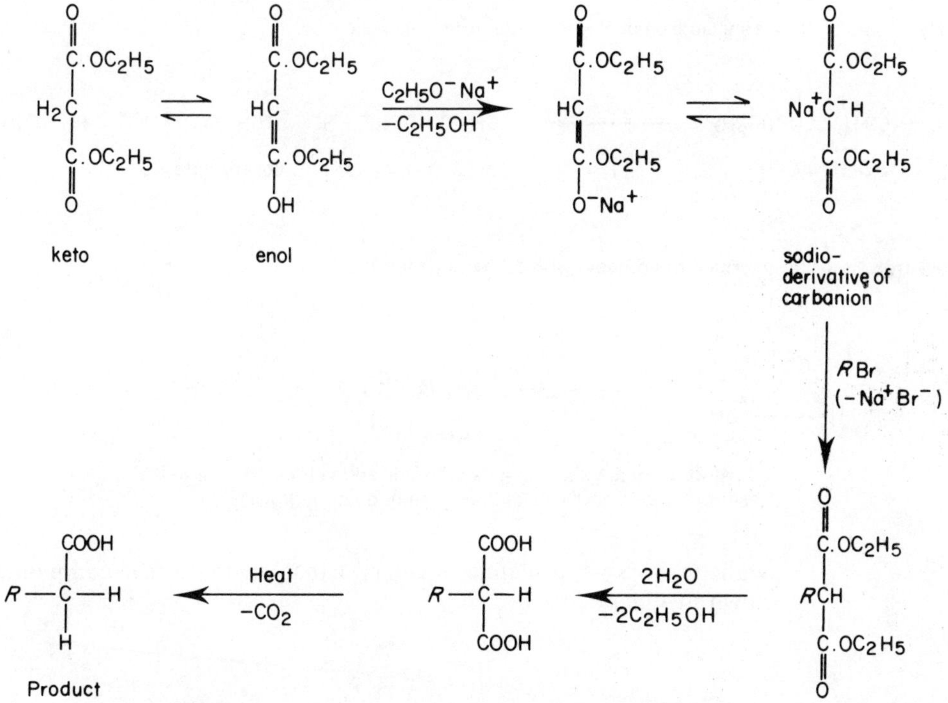

Figure 26.2 Synthesis of a carboxylic acid by the use of diethyl propanedioate and an alkyl bromide

Ethyl 3-oxobutanoate (*acetoacetic ester*) also exhibits tautomerism; in fact, the equilibrium mixture contains about 7 per cent of the enol form. Like diethyl malonate it is a very reactive compound and is used widely in

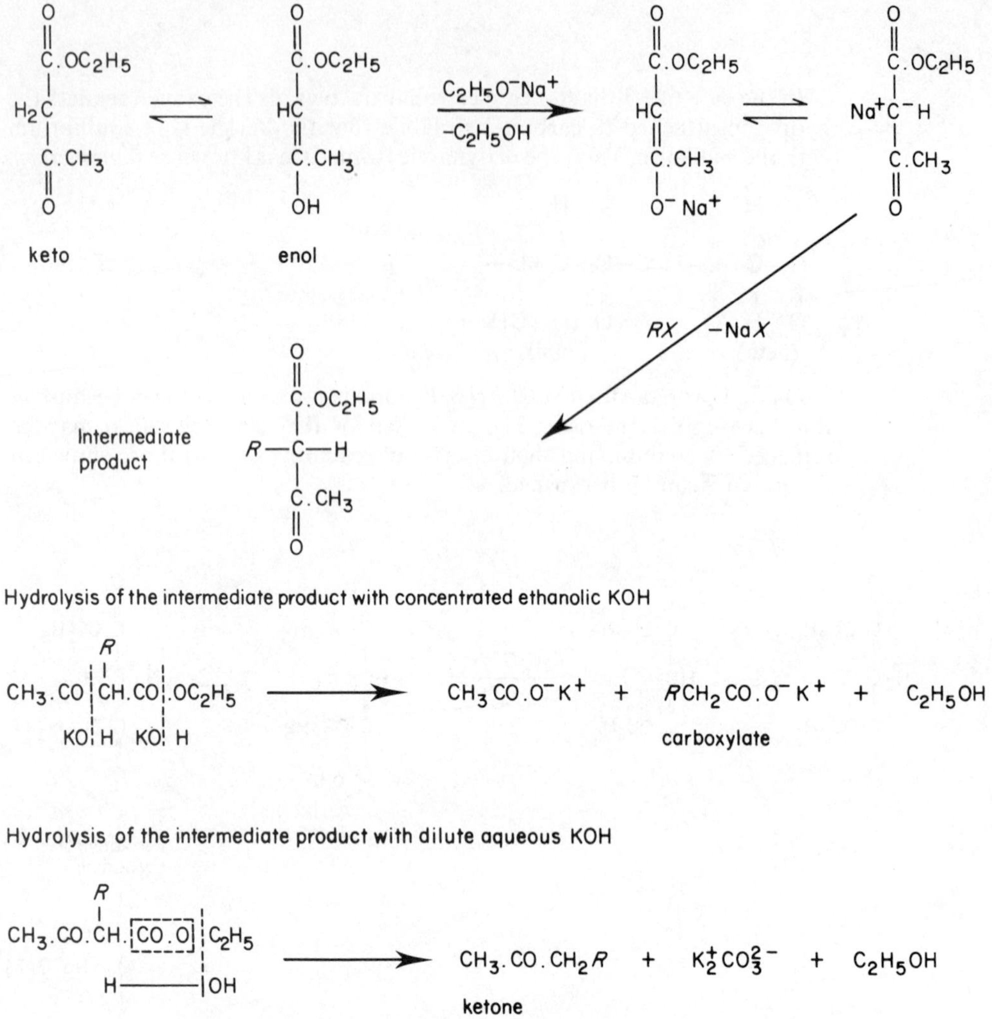

$$
\underset{\text{keto}}{
\begin{array}{c}
\overset{\displaystyle O}{\underset{\displaystyle \|}{}} \\
C.OC_2H_5 \\
| \\
H_2C \\
| \\
C.CH_3 \\
\| \\
O
\end{array}}
\rightleftharpoons
\underset{\text{enol}}{
\begin{array}{c}
\overset{\displaystyle O}{\underset{\displaystyle \|}{}} \\
C.OC_2H_5 \\
| \\
HC \\
\| \\
C.CH_3 \\
| \\
OH
\end{array}}
\xrightarrow[-C_2H_5OH]{C_2H_5O^-Na^+}
\begin{array}{c}
\overset{\displaystyle O}{\underset{\displaystyle \|}{}} \\
C.OC_2H_5 \\
| \\
HC \\
\| \\
C.CH_3 \\
| \\
O^-\,Na^+
\end{array}
\rightleftharpoons
\begin{array}{c}
\overset{\displaystyle O}{\underset{\displaystyle \|}{}} \\
C.OC_2H_5 \\
| \\
Na^+C^-H \\
| \\
C.CH_3 \\
\| \\
O
\end{array}
$$

$$RX \diagup -NaX$$

Intermediate product

$$
\begin{array}{c}
\overset{\displaystyle O}{\underset{\displaystyle \|}{}} \\
C.OC_2H_5 \\
| \\
R-C-H \\
| \\
C.CH_3 \\
\| \\
O
\end{array}
$$

Hydrolysis of the intermediate product with concentrated ethanolic KOH

$$
\underset{KO!H \quad KO!\,H}{CH_3.CO\,\lfloor CH.CO\rfloor.OC_2H_5}
\longrightarrow
CH_3\,CO.O^-K^+ \;+\; \underset{\text{carboxylate}}{RCH_2\,CO.O^-K^+} \;+\; C_2H_5OH
$$

Hydrolysis of the intermediate product with dilute aqueous KOH

$$
\underset{H \;\longrightarrow\; |OH}{CH_3.CO.CH.\lfloor CO.O\rfloor|C_2H_5}
\longrightarrow
\underset{\text{ketone}}{CH_3.CO.CH_2R} \;+\; K_2^+CO_3^{2-} \;+\; C_2H_5OH
$$

Figure 26.3 Synthesis of carboxylic acids and ketones by the use of ethyl 3-oxobutanoate and two alternative methods of hydrolysis

synthetic work, in a manner similar to that of the former compound, as shown in *Figure 26.3*.

Analysis

The presence of an ester is suggested by the hydroxamic acid test. The ester should then be hydrolysed and the acid and alcohol or phenol separated and identified.

Industrial applications

Esters are widely used as solvents, particularly for certain types of paints and varnishes. Since they provide the flavouring for many natural products, they can be used in synthetic essences.

Enormous quantities of oils and fats are hydrolysed and converted into soaps, sodium hydroxide producing hard, and potassium hydroxide soft, soaps.

In more recent years, considerable advances have been made in the manufacture of long-chain polyesters from difunctional acids and alcohols. For example, Terylene is obtained by polymerizing ethane-1,2-diol with dimethylbenzene-1,4-dicarboxylate (methyl terephthalate):

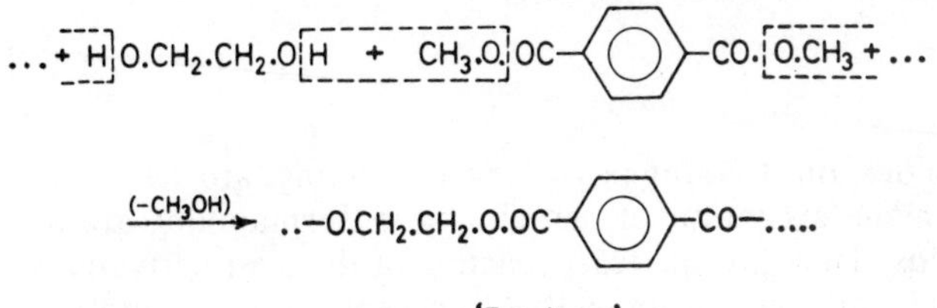

'Terylene'

Amides

$$\overset{O}{\underset{\|}{R C}}.NH_2$$

The formula of amides suggests that they can be prepared from ammonia, and this is indeed the case; acids or any of the derivatives mentioned earlier in this Chapter can be used as the other reactant. With acids, the first stage of the reaction is salt formation, followed by elimination of water on distillation

$$RCOOH + NH_3 \rightarrow RCO.O^-NH_4^+$$

$$RCO.O^-NH_4^+ \rightarrow RCO.NH_2 + H_2O$$

With acid derivatives, the initial stage is nucleophilic addition of ammonia to the carbon of the carbonyl group, followed by expulsion of halogen, acyl or alkoxy groups from chlorides, anhydrides or esters, respectively:

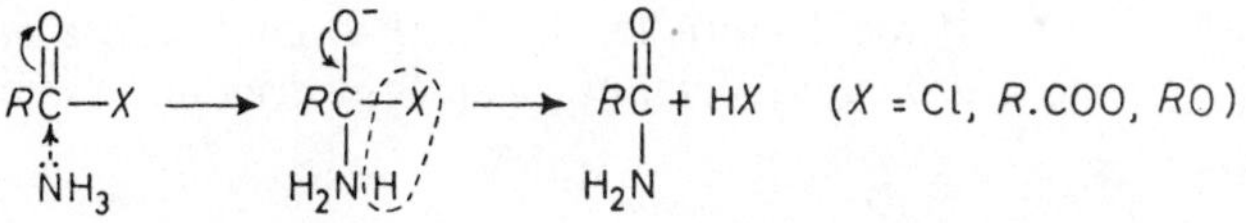

e.g.

$$CH_3.CO.OC_2H_5 + NH_3 \rightarrow CH_3.CO.NH_2 + C_2H_5.OH$$
Ethyl ethanoate   Ethanamide ·
(ethyl acetate)   (acetamide)

$$COCl_2 + 2NH_3 \rightarrow CO(NH_2)_2 + 2HCl$$
Carbonyl   Carbamide
chloride   (urea)

Substituted amides, i.e., with alkyl or aryl groups attached to the nitrogen, can be similarly prepared by using amines instead of ammonia, e.g.,

$$\underset{\substack{\text{Benzenecarbonyl}\\\text{chloride}}}{\overset{\overset{\displaystyle O}{\parallel}}{C_6H_5.CCl}} \quad + \quad \underset{\substack{\text{Phenylamine}}}{\overset{\overset{\displaystyle H}{\mid}}{HN.C_6H_5}} \quad \xrightarrow[-HCl]{} \quad \underset{\substack{N\text{-phenyl benzene-}\\\text{carboxamide (benzanilide)}}}{\overset{\overset{\displaystyle O}{\parallel}}{C_6H_5.C.NH.C_6H_5}}$$

A different method of preparation involves the hydrolysis of nitriles with either acid or alkali. The preliminary step is the addition of the hydrogen ion to the carbon, or the hydroxide ion to the nitrogen, of the nitrile bond (p. 529)

$$R - C \equiv N + H_2O \rightarrow RCONH_2$$

Properties

Amides, on account of hydrogen bonding, are relatively non-volatile, and with the exception of methanamide (formamide), are all white crystalline solids. They are the least reactive of the acid derivatives.

The nitrogen atom is able to accept a proton, and so amides have basic tendencies. They dissolve in acids to form salts:

$$RCO.NH_2 + H^+ \rightarrow RCO.NH_3^+$$

Amides liberate ammonia on boiling with alkali (unlike ammonium salts which liberate ammonia even with cold alkali):

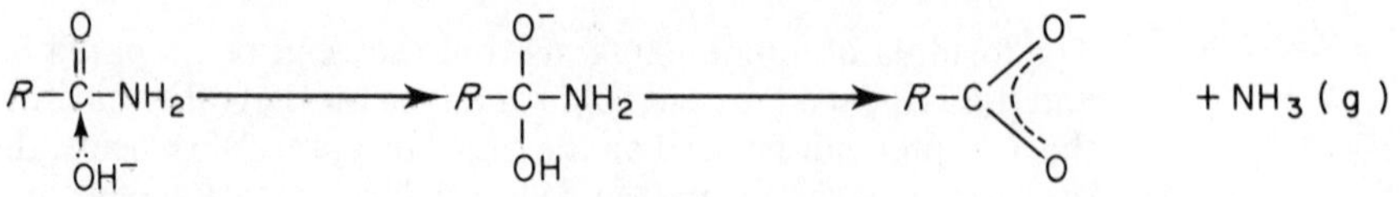

Distillation of amides with phosphorus(V) oxide results in dehydration to the nitrile:

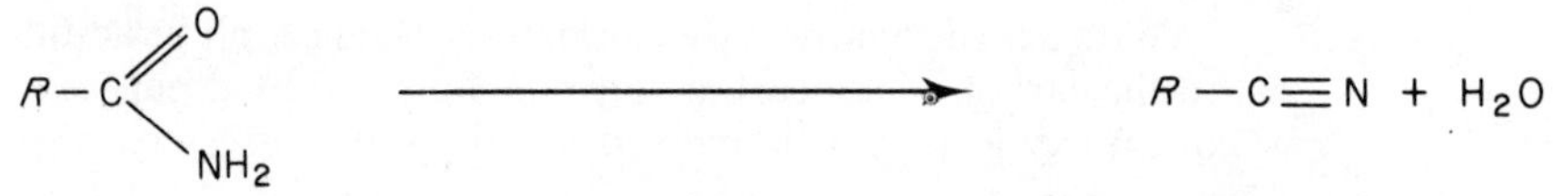

If amides are treated with a cold, dilute mixture of sodium nitrite and hydrochloric acid, nitrogen is evolved and the carboxylic acid is formed:

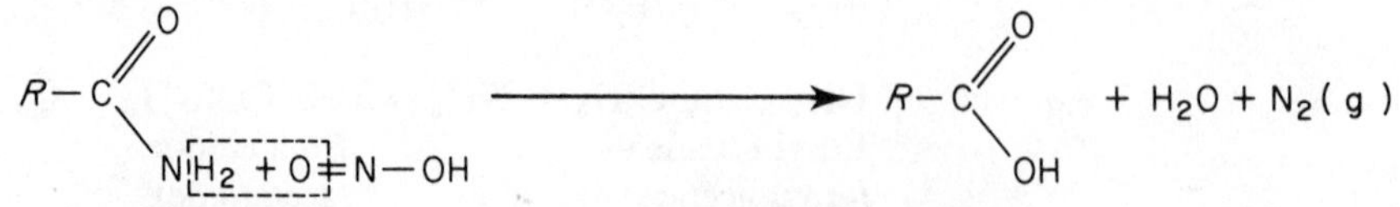

This reaction should be compared with those of amines with nitrous acid (p. 460).

Another important reaction of amides is the *Hofmann degradation* (p. 453), so called because it provides a means of removing a carbon atom from an

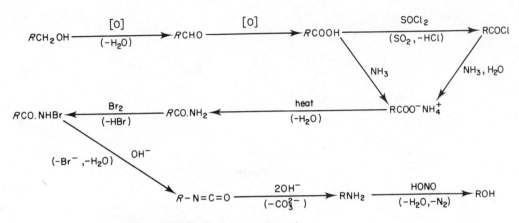

Figure 26.4 Descent of a homologous series

organic compound and, hence, of descending an homologous series, e.g., as shown in *Figure 26.4*.

Carbamide (urea), $O=C(NH_2)_2$ under similar treatment breaks down into carbonate and nitrogen:

$$CO(NH_2)_2 + 8OH^- \rightarrow CO_3^{2-} + 6H_2O + N_2 + 6e^-$$

$$3Br_2 + 6e^- \rightarrow 6Br^-$$

The nitrogen can be measured in a gas burette and affords a method of estimating urea quantitatively.

Sulphonamides

The amides so far described have been derived from carboxylic acids, but amides of sulphonic acids exist and are prepared by analogous methods, e.g.,

$$\underset{\text{benzene}}{C_6H_5.H} \xrightarrow[(-H_2O)]{HOSO_2Cl} \underset{\substack{\text{benzenesulphonyl} \\ \text{chloride}}}{C_6H_5SO_2Cl} \xrightarrow[(-HCl)]{NH_3} \underset{\text{benzenesulphonamide}}{C_6H_5.SO_2.NH_2}$$

Many sulphonamide derivatives are valuable drugs. e.g. 'M & B 693' is derived from 4-aminobenzenesulphonic (sulphanilic) acid as shown in *Figure 26.5*. These drugs bear a resemblance to 4-aminobenzoic acid, and it is believed that the invading bacteria are unable to distinguish between the latter, vital to their well-being, and the drug (or at least a breakdown product of it). Metabolic processes consequently go awry, with disastrous results for the bacteria:

$$H_2N-\!\!\!\left\langle \bigcirc \right\rangle\!\!\!-COOH$$
4-Aminobenzoic acid

$$H_2N-\!\!\!\left\langle \bigcirc \right\rangle\!\!\!-SO_2NH_2$$
Sulphanilamide

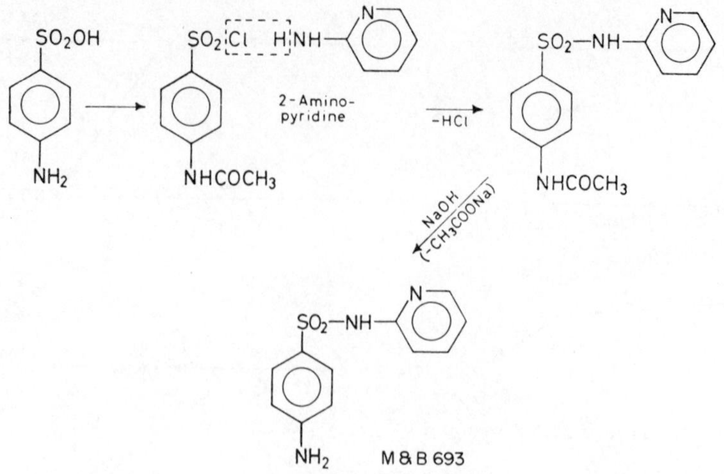

Figure 26.5 Preparation of a 'sulpha' drug, M&B 693

Peptides and proteins

Monobasic carboxylic acids and monoamines condense to form substituted amides

$$RCO.OH + R'NH_2 \rightarrow RCO.NHR' + H_2O$$

If amino acids are involved in condensation, then clearly the process can be perpetuated, and a long fibre results

$$\ldots + H_2N.R.CO.OH + H_2N.R.CO.OH + \ldots \longrightarrow \ldots NH.R.CO.NH.R.CO\ldots$$

The result of this polymerization, if carried to sufficient lengths, is a *protein* molecule, and the repeating —CO.NH— linkage, representing the residues of the acid and amine functions, is called the *peptide* link.

Proteins are vital to living processes, and their production comprises a fundamental part of the activity of the living cell. An impressive feature of the metabolism of the cell is the specificity of protein synthesis, that is to say, the manner in which a cell makes the correct protein. For example, a muscle cell makes muscle and not hair tissue from the amino acids available in the cell.

This specificity of protein synthesis is a consequence of the precise and peculiar arrangement of the bases in the relevant RNA molecules (p. 467). As this arrangement is itself a consequence of the formulation of the base in the parent DNA molecule from which the RNA is generated, it follows that, unless something goes amiss and a mutation occurs, the types of proteins synthesized in a particular cell will be directed by the types of DNA present.

It is believed that every one of the twenty or so amino acids involved has its own particular 'transfer' RNA. The amino acid is first of all activated by reaction with an enzyme and a molecule of adenosine triphosphate(V)(ATP). It is then transferred to its own RNA molecule by exchange of the adenine in

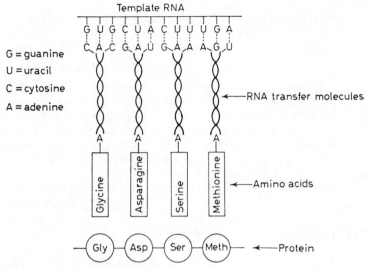

Figure 26.6 Protein synthesis

the ATP residue for adenine in the RNA. The RNA molecule, with its amino acid attached, links with the appropriate and complementary site of the 'template' RNA through, it is thought, three base units. As the process continues with other 'transfer' RNA molecules and their corresponding amino acids, a situation is arrived at where, because of the complementary nature of the 'template' and the much smaller 'transfer' RNA molecules, amino acids are brought together in the correct sequence; condensation then takes place with the formation of a *polypeptide*, which consequently leaves the RNA as a specific protein (*Figures 26.6 and 26.7*).

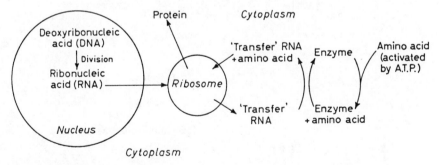

Figure 26.7 Cellular processes

Proteins can be reconverted into amino acids by hydrolysis, during digestive processes; often the degradation is only partial and produces molecules intermediate between the protein and the individual amino acids, and known as *peptones*, which are capable of build-up into new proteins. The amino acids themselves can be further degraded and de-aminated by respiratory processes (p. 520):

$$RCH.CO.OH \xrightarrow[-2H]{} RC.CO.OH \xrightarrow{H_2O} RC.CO.OH + NH_3$$

with NH_2 below the first, NH (‖) below the second, and O (‖) below the third, leading to **Pyruvic acid, etc.** and **Urea, etc.**

A qualitative test for proteins is the *biuret* test. Compounds containing the peptide link, —CO.NH—, give a purple colour on treatment with sodium hydroxide and a drop of copper(II) sulphate solution. Biuret itself is obtained by the action of heat on urea, two molecules condensing together by the elimination of one molecule of ammonia. Because it contains the peptide link, biuret gives a positive result with the above test:

$$H_2N.CO.NH|H + H_2N|CO.NH_2 \xrightarrow[-NH_3]{} H_2N.CO.NH.CO.NH_2$$
$$\text{Biuret}$$

Industrial aspects of amides

Nylon is undoubtedly the most common synthetic polyamide and is manufactured by the condensation polymerization of, for example, hexane-1,6-diamine with hexane-1,6-dioic acid (or the acid chloride):

$$...+ H_2N(CH_2)_6NH|_2 + HO|.OC(CH_2)_4CO.|OH +... \longrightarrow ...HN(CH_2)_6NH.CO(CH_2)_4CO...$$
$$\text{6,6-Nylon}$$

Both reactants are made from cyclohexane (which can itself be obtained from benzene or phenol, as shown in *Figure 26.8*.

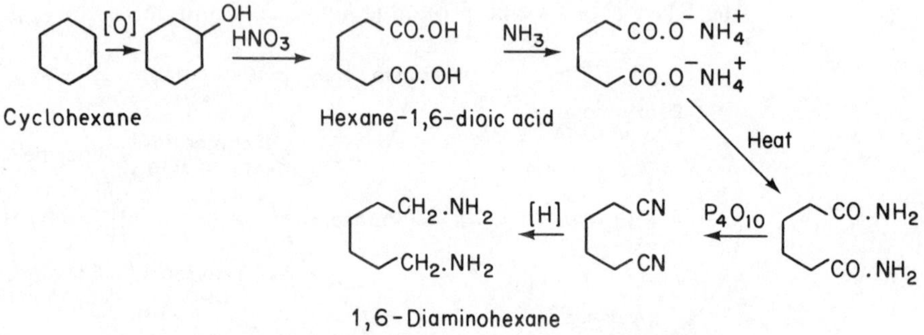

Figure 26.8 Production of components of nylon

Urea is utilized in the manufacture of resins. In the presence of alkali it reacts with methanal (formaldehyde);

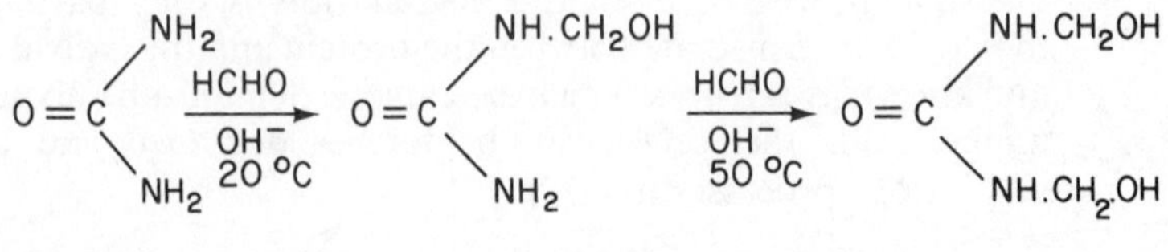

With a suitable catalyst, for example triethyl phosphate(V), this product polymerizes to give two- and three-dimensional structures based on the unit

$$\ldots O.CH_2.NH.CO.NH.CH_2 \ldots$$

Urea also condenses with carboxylic acids; if dibasic acids are involved, then cyclic derivatives called *ureides* are formed, e.g.,

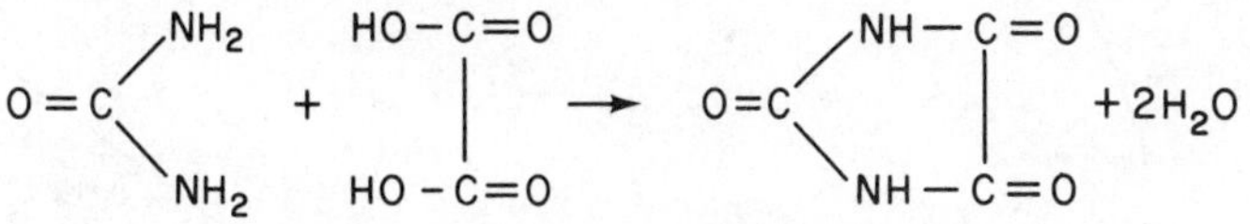

Propanedioic (malonic) acid gives rise to *barbiturates,* useful as sedatives and hypnotics; for example, phenobarbitone is the ureide of ethylphenylmalonic acid:

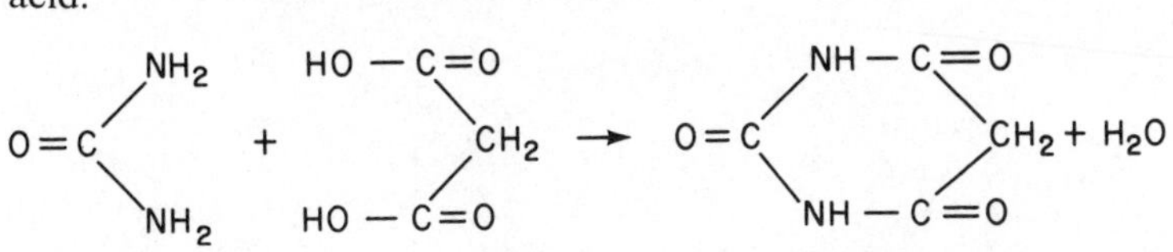

Barbituric acid

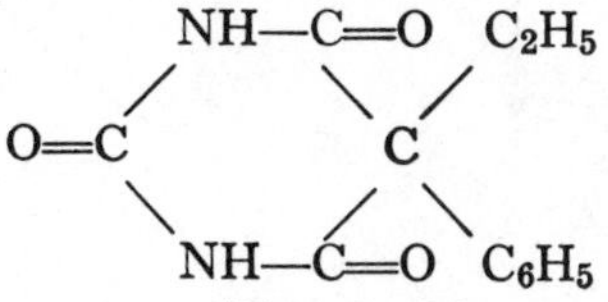

Phenobarbitone

Summary

The derivatives of carboxylic acids mentioned in this Chapter possess carbonyl groups which, because of polarization, are able to undergo nucleophilic attack at carbon by either nitrogen or oxygen. The readiness with which reaction ensues depends upon the ease with which substituent groups leave the intermediate formed:

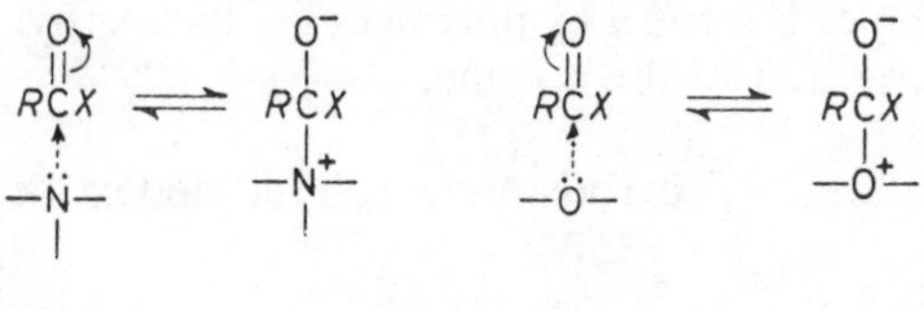

Acid chlorides are the most, and acid amides the least, reactive (*Figure 26.9*).

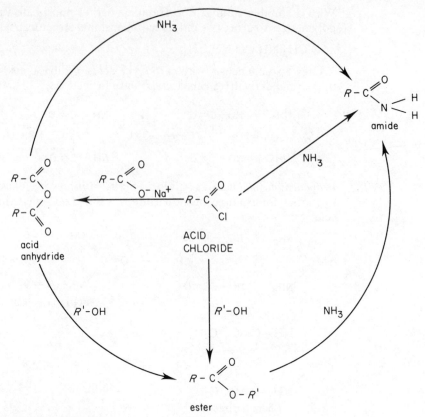

Figure 26.9 Reactions of an acid chloride

Questions

(1) Name the compounds and give details of the reactions indicated in the flow chart

$$CH_3.NH_2 \quad CH_3.CO.OH \quad CH_3.CO.OC_2H_5$$

$$CH_3CO.NH_2 \leftarrow CH_3.COCl \longrightarrow (CH_3.CO)_2O$$

$$[(CH_3)_4N]^+Br^- \quad CH_3.CN \quad CH_3.CHO$$

(2) How could 2-methylprop-2-enoic acid (methacrylic acid) be prepared from propanone? Perspex is formed by polymerizing the methyl ester of this acid; give the structural unit of the polymer.

(3) How might benzene-1,2-carboximide (phthalimide), be prepared from naphthalene?

(4)   The following steps are involved in the synthesis of a compound, starting from methylbenzene: (*a*) treatment with chlorosulphonic acid and separation of the *ortho* derivative; (*b*) oxidation of this with potassium manganate(VII); (*c*) reaction with ammonia and (*d*) heating of the product to eliminate water. Deduce the formula of the compound and give the experimental conditions for each step. What is the trivial name of the product?

(5)   A compound A, of molecular formula C_3H_5NO, on being boiled with dilute hydrochloric acid produced an acid B, $C_3H_6O_3$, which on oxidation gave a dibasic acid, C. C formed an acid anhydride on heating alone. Identify A, B and C and give equations for the reactions involved.

(6)   Describe how bromoethane might be prepared from 1-bromopropane.

(7)   A compound A, C_3H_5N, when treated with dilute acid yielded B, $C_3H_6O_2$. This compound, when treated with ammonia and heated gave C, C_3H_7ON. C when boiled with alkali liberated ammonia and a solid residue, D, remained. This was dissolved in water and acidified. Extraction with ether yielded a compound E which, upon treatment with lithium tetrahydridoaluminate, gave propan-1-ol. C when treated with bromine and alkali gave substance F which, with nitrous acid, yielded some ethanol. Deduce the nature of A–F and write equations for all the reactions.

(8)   Show, by means of structures, the possible reactions of linseed oil with oxygen.

(9)   Indicate how propan-1-ol can be converted into (*a*) ethanol (*b*) butan-1-ol.

(10)   Write an account of the industrial utilization of esters.

(11)   Write an essay on proteins.

(12)   4.0 g of commercial aspirin were heated on a water-bath for half an hour with 100 cm^3 of M NaOH. After cooling, it was made up to 500 cm^3 of solution and 25 cm^3 were titrated against 0.1M HCl, with phenol red as indicator: 31.2 cm^3 of acid were required to neutralize the excess of alkali. Say what you can about these reactions and about the condition of the original aspirin.

(13)   Devise a practical scheme of analysis to distinguish between solid samples of phenol, urea, phenylammonium chloride, ammonium ethanedioate, calcium ethanoate, sodium methanoate, sodium benzenecarboxylate, ethanamide, *N*-phenylethanamide, aminoethanoic acid and glucose.

(14)   The rate of conversion of ammonium cyanate into urea is proportional to the square of the concentration of the ammonium salt. This salt also hydrolyses in aqueous solution. Show that these observations could indicate either of the mechanisms:

$$NH_4^+ + CNO^- \rightarrow (NH_2)_2CO \leftarrow NH_3 + HCNO$$

Explain which mechanism you think is more probable and suggest further investigations which might help in deciding the 'true' mechanism.

(15) (a) Write the structural formula of each of the following derivatives of ethanoic acid (acetic acid): ethanoic anhydride (acetic anhydride); ethyl ethanoate (ethyl acetate); ethanoyl chloride (acetyl chloride); and ethanamide (acetamide).
   (b) Each of the derivatives in (a) can be hydrolysed.
      (i)   Explain what is meant by hydrolysis.
      (ii) Give the conditions for the hydrolysis of each derivative.
      (iii) Write the derivatives in order of ease of hydrolysis, putting the easiest first.
   (c) Stating your reasons, arrange the following in order of increasing acid strength (i.e., putting the least acidic first): ethanoic acid; monobromoethanoic acid; monochloroethanoic acid; monoiodoethanoic acid.

[A.E.B.]

(16) When a compound J ($C_{19}H_{18}O_4$) was refluxed with aqueous sodium hydroxide, it was quantitatively converted into a mixture of K and L. Compound K ($C_5H_{10}O_2$) formed as an oily liquid during the reaction, whereas L ($C_7H_6O_2$) was precipitated as a solid on addition of concentrated hydrochloric acid to the aqueous phase.

Ozonolysis of K produced two acidic compounds M ($C_3H_6O_3$) and N ($C_2H_4O_3$). Reaction of K with dichromate(VI) in dilute sulphuric acid yielded P ($C_5H_6O_4$) . Compound Q ($C_4H_4O_4$) was obtained when P was treated with chlorine in aqueous sodium hydroxide. On heating, Q readily formed a compound R ($C_4H_2O_3$).

Clearly outlining your reasoning, identify compounds J–R.

Hydrogenation of K produced S ($C_5H_{12}O_2$) which, on treatment with sulphuric acid, yielded T ($C_5H_{10}O$) amongst other products. Compound T did not decolorize a solution of bromine (in tetrachloromethane) nor did it react with metallic sodium.

Suggest a structure for T and for one of the other products which would be expected from the reaction of S with sulphuric acid.

[Cambridge 1978]

27 Modern experimental methods

Introduction • Isolation and purification • Analysis • Mass
spectroscopy • Electromagnetic spectroscopy • Magnetic and
electric measurements • X-Ray analysis • Polarography

Introduction

The investigation of an unknown substance roughly comprises the following
steps: isolation, purification and analysis. The ultimate aim is then the
identification of the different atomic groups and their orientation in the
molecule, so giving a complete *structural formula*. However, first the atomic
ratios of the different atoms must be determined to provide the *empirical
formula*, from which, knowing the relative molecular mass, the *molecular
formula* can be deduced.

Example. A substance, of relative molecular mass 60, contains 40.0 per cent
carbon, 6.7 per cent hydrogen and 53.3 per cent oxygen. Dividing by the
relevant relative atomic masses gives the atomic ratios as 40.0/12 for carbon,
6.7/1 for hydrogen and 53.3/16 for oxygen. Changing these values to the
smallest whole numbers gives the empirical formula as $C_1H_2O_1$. But as the
relative molecular mass is 60, the molecular formula must be twice the
empirical formula, i.e., $C_2H_4O_2$, and the possible structural formulae are

$$\underset{CH_3.COH}{\overset{\overset{\displaystyle O}{\|}}{}} \quad and \quad \underset{HCOCH_3}{\overset{\overset{\displaystyle O}{\|}}{}}$$

For complex molecules, particularly naturally occurring organic compounds, independent methods of synthesis are often explored completely to
confirm their structures. Further elucidation involves a detailed examination
of the physical properties of the compound and the mechanisms by which it
takes part in chemical reactions, with a possible view to the commercial use
of the substance.

Few reactions, particularly in organic chemistry, give simple, readily
separable products, and this factor, together with the yield obtainable, limits
the usefulness of many reactions. However, with the improvement in physical
techniques, many complex mixtures can now be separated. For example, the
separation of the amines resulting from the reaction between organic halides
and ammonia under pressure (p. 442) now relies on the precise control which
can be applied in fractional distillation, whereas the alternative has long been
a lengthy chemical separation.

The modern use of *vacuum lines*, which consist of closed systems of reaction
vessels suitably connected, so that the addition of reagents, their reaction
together and the products formed can be maintained under controlled

conditions, has allowed the separation of many new compounds which exist only under special circumstances, e.g., in inert atmospheres at low temperatures. The instability of many of these substances causes considerable difficulty in determining their molecular formulae, let alone their structures.

Similarly, *flash photolysis* and the use of *shock waves* result in the transient formation of new species. In the former method, high energies of suitable frequencies are supplied for a very short time (e.g., a few microseconds), whilst in the latter, the reaction vessel is divided by a partition which can be readily pierced so that, by building up a pressure on one side, a shock wave can pass along the rest of the tube when the membrane is ruptured; the shock wave produces a high temperature by adiabatic compression. By these methods, species of short life can be observed spectroscopically (p. 572).

The increasing demand for high purity, required in, e.g., semiconductor work, has meant that the classical methods of analysis can no longer be used. Instead, more sensitive methods, based on atomic and molecular properties (e.g., spectral analysis) have been developed. The use of these methods in turn requires high purity of the primary standards, and this may be achieved by, for example, chromatographic separation. In the majority of cases the criterion of purity is decided spectroscopically, reliance on the old-established methods of sharp melting and boiling points generally being impossible because of the instability of the substances at the high temperature necessary.

Methods of determining relative molecular masses have already been dealt with in Chapters 4 and 5. Besides these methods there are those based on mass spectroscopy (p. 570), measurements of diffusion in solutions and the rates of sedimentation under centrifugal forces, the latter being used exclusively for colloidal systems.

A certain fraction of the molecules of a compound can be 'labelled' by making one of the elements radioactive or by incorporating a different isotope into the molecule. Such methods find application in the investigation of reaction mechanisms, e.g., the sequence of reactions involving radioactive sulphur, S^*,

$$S^* + SO_3^{2-} \xrightarrow{\text{boil}} S^*SO_3^{2-} \xrightarrow{2H^+} S^*(s) + H_2O + SO_2(g)$$

shows the non-equivalence of the sulphur atoms in the thiosulphate ion since, if they were in similar positions, the radioactivity of the sulphur precipitated would be halved, whilst in fact all the radioactivity is recovered in the sulphur liberated.

Isolation and purification

The methods available are:

Filtration

This can be used when the particle size is too large for the substance to pass through the pores of a filter medium. It can also be carried out under reduced pressure as well as at any suitable temperature, particularly if use is made of the range of filter media of controlled porosity and inertness which are available.

Fractional crystallization

This can be used to separate solids of differing solubilities in a particular solvent; precise temperature control is necessary, and usually several recrystallizations are required to obtain satisfactory separation. Also, for this process to be suitable, conditions must be such that one of the substances will crystallize from the mother liquor before the solution becomes saturated with respect to the other solutes.

Centrifugation

Centrifuging is an alternative to filtration, and by using high-speed (ultra-) centrifuges, values of up to 250 000 times the acceleration due to gravity may be obtained, allowing the separation of substances in the colloidal state.

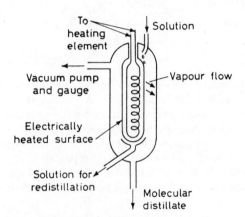

Figure 27.1 Molecular distillation

Distillation

Steam and fractional distillations have been discussed already (pp. 105–109). Distillation under reduced pressure clearly allows a lowering of the normal boiling point and is therefore useful for substances which tend to decompose on heating. In *molecular distillation*, the mean free path (the average distance travelled by a gas molecule between successive collisions) of the vapour molecules produced by reducing the pressure to less than 0.13 Pa (10^{-3} mmHg), is increased to a few centimetres, allowing molecules from a liquid film to travel straight to an adjacent cool surface where condensation occurs (*Figure 27.1*); using molecular distillation, liquid mixtures may be fractionated at more than 100 kelvins below their normal boiling points.

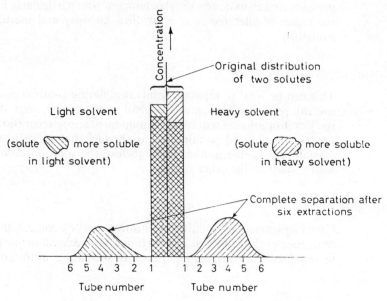

Figure 27.2 Solvent extraction

Solvent extraction

The distribution of solutes between two immiscible solvents (see p. 104 for simple partition) has been made into a semicontinuous process—*countercurrent distribution*. After each extraction, the mixtures are allowed to separate, and the lighter fraction is transferred to another extraction tube containing the heavier solvent, whilst more of the lighter solvent is added to the residual heavy medium. The extraction is repeated in this manner many times, and as it proceeds from tube to tube, separation of the solutes occurs. Typical distribution curves of the solutes between the solvents are shown in *Figure 27.2*.

Chromatography

Chromatographic separation represents a continuous extraction process using one mobile solvent with either (*a*) a solid absorbent replacing the second solvent (*adsorption chromatography*) or (*b*) the second solvent acting as a stationary phase because of its adsorption on to a solid. In *paper chromatography*, for instance, the moisture adsorbed on the cellulose fibres acts as the immobile second solvent. The mobile solvent may also be a gas (*gas chromatography*) (*Figure 27.3*), and a further possibility is that, if the solutes carry any charge, then application of an electric potential across the immobile phase may aid separation (*continuous electrophoresis*). To a first approximation, the stationary phase takes up the substance of highest adsorbability or solubility to the exclusion of the other solutes; in any case, complete separation can be effected by drawing a suitable *eluting agent* through the column of the adsorbent. In gas chromatography, the vapour of

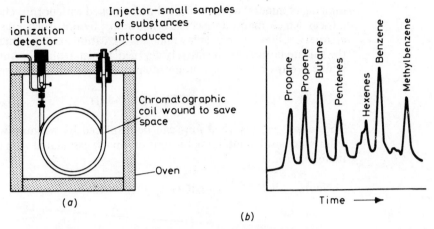

Figure 27.3 Gas chromatography: (a) the basic unit (b) a gas chromatogram of a mixture of hydrocarbons

the substance least adsorbed on the stationary phase is swept along by the carrier gas (e.g., argon or nitrogen), reaches the end of the adsorption column first and can be detected, for example, by observations based on the changes in the thermal conductivity of the gas mixtures. The extraction of the separated solutes on a *chromatogram* depends on first identifying their positions; for solutes which are not coloured, reliance is placed either on their response to ultraviolet light or to reagents added after drying the chromatogram (*Figure 27.4*).

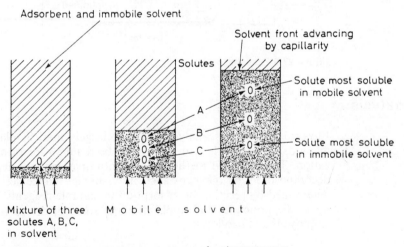

Figure 27.4 Stages in the development of a chromatogram

Ion exchange

This finds widespread application, particularly for the selective concentration of certain ionic species. The simplest application is in the deionizing of mineral water, using columns packed with porous charged crystal lattices which maintain electrical neutrality by adsorbed ions of opposite charge. Zeolites and most clay minerals are cation exchange materials, since the silicate lattices are negatively charged, whilst artificial resins containing amine groups behave as anion exchange substances in acid solution, e.g.,

$$-N(CH_3)_3{}^+OH^- \xrightarrow{Cl^-} -N(CH_3)_3{}^+Cl^- + OH^-$$

The selectively adsorbed ions can be displaced by washing with a strong solution of a salt containing the ions originally associated with the exchange material, e.g.,

$$-N(CH_3)_3{}^+Cl^- \xrightarrow{OH^-} -N(CH_3)_3{}^+OH^- + Cl^-$$

Zone refining

This is now much used for obtaining elements in a very pure state. It consists of extruding the impure substance slowly through a moving high-temperature region. The impurities, provided their solubility increases with temperature, concentrate in the hotter part of the specimen, so that after several cycles most of the contaminants will have been swept along to one end of the sample (*Figure 27.5*).

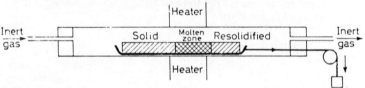

Figure 27.5 Zone refining

Analysis

Physical methods of analysis are now largely used in place of chemical methods, although the latter may still be required for primary standardization. The chemical methods of quantitative analysis depend upon the decomposition of the substance into simpler molecules which can be determined by reaction or adsorption in suitable reagents (for example p. 431). The principles and applications of some of the physical techniques employed are as follows:

Mass spectroscopy

This is a technique in which ions of the substance are produced by the use of electron bombardment at low pressure, e.g., $Ar + e^- \rightarrow Ar^+ + 2e^-$. The ions are focused by electric and magnetic fields on to a photographic plate or other form of detector (*Figure 27.6*). The deflection of the ions in the fields,

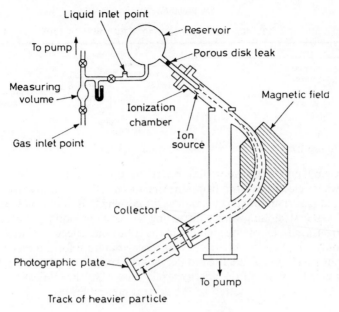

Figure 27.6 The mass spectrograph

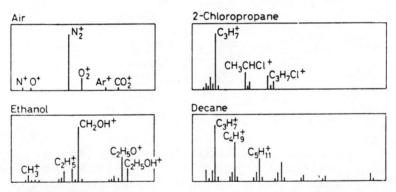

Figure 27.7 Mass spectra

and hence their position on the plate, depends upon their mass and charge, and this affords a method of identifying the species present; for example, ethanol gives rise to 24 fragments, with the following ions predominating (*Figure 27.7*):

$$CH_3^+ \qquad C_2H_5^+ \qquad CH_2OH^+ \qquad C_2H_5O^+ \qquad C_2H_5OH_2^+$$

Mass spectrometry is now the accepted method for obtaining accurate values of relative atomic and molecular masses.

Electromagnetic spectroscopy

All substances are sensitive to some part of the electromagnetic spectrum (*Figure 27.8*), although comparatively few exhibit colour, i.e., respond to the

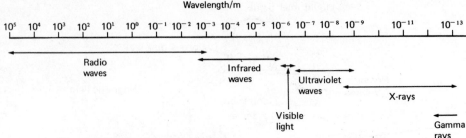

Figure 27.8 The electromagnetic spectrum

visible part of the spectrum. Energy is absorbed in (*a*) *electronic transitions*, where electrons jump from one orbit to another, and involves larger energy changes than (*b*) *vibrational transitions* which, in molecules, are represented by the stretching and contracting of covalent bonds, whilst (*c*) *rotational transitions*, involving changes in the rotational energies of molecules, require only small energy changes. The energy diagram for a molecule can then be shown (*Figure 27.9*). Fortunately for the spectroscopist, changes cannot take place between all the energy levels shown, so that the spectral lines can in many cases be allotted to the correct energy levels.

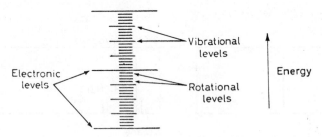

Figure 27.9 Molecular energy levels

Because of the relation $\Delta E = h\nu$ (where ΔE = the energy change, h = Planck's constant and ν = frequency of the spectral line) among the common spectral ranges, those of the infrared involve the lowest frequencies, and the resulting spectra arise from the lowest energy transitions, i.e., from molecular vibration and rotation. For a diatomic group, the stretching vibration between the atoms is related to the bond strength, and spectral lines result from the vibrational frequency of the bond, as energy of the appropriate frequency is absorbed. Thus compounds containing this group produce a characteristic frequency, which is modified by the nature of the adjacent bonding. Rotational spectra are governed by the moments of inertia of the system, so that information regarding the mass and interatomic distances is obtained from these spectra. In general, infrared spectra provide 'fingerprints' of the substances, permitting the identification of many groups (*Figure 27.10*).

The visible and ultraviolet spectra are more complicated, because electronic transitions are also possible. For substances called *fluorescers*, i.e., certain compounds containing loosely bound π electrons in delocalized

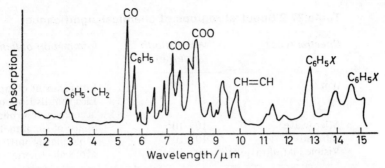

Figure 27.10 Infrared spectrum of ethyl 3-phenylpropenoate (cinnamate)

systems, absorption of ultraviolet radiations results in the emission of visible light, of lower frequency and energy, since the excited electrons do not return directly to their original energy states.

As indicated above, all these transitions are governed by quantum rules which explain the characteristic line and band structures in the spectra. The spectrum of a group may undergo a *shift* in frequency when combined in different molecules. This is particularly the case when conjugated double bonds, i.e., extended π clouds, are possible, and it results in a shift to lower frequencies so that, by introducing appropriate groups into a molecule originally having an ultraviolet spectrum, a spectrum in the visible range results: the new substance exhibits colour. Groups which produce this effect are called *chromophores* (*Table 27.1*).

Table 27.1 Chromophoric groups

Carbonyl	$>C=O$	Nitroso	$-N=O$
Nitro	$-N\overset{O}{\underset{O}{\diagup\!\!\diagdown}}$	Azo	$-N=N-$
Azoxy	$-N=N\overset{\diagup}{\underset{O}{\diagdown}}$	Azoamino	$-N=N-N\overset{\diagup}{\underset{H}{\diagdown}}$

The different types of spectroscopy are

Emission spectroscopy

In this, energy is supplied to excite electrons to higher electronic states. When these electrons return to their lower orbits, characteristic radiations are emitted. The excitations may be accomplished by applying an electric discharge to the substance, resolving the spectrum using a prism or diffraction grating and finally recording the result photographically. In another form of apparatus (the *flame photometer*), a solution of the

Table 27.2 Spectral regions of chemical significance

Spectral type	Wavelength range/nm	Information obtained
n.m.r.	$10^{11}-10^9$	Environment of atom
e.s.r.	$10^{-9}-10^{-6}$	Electron distribution (and occupancy of orbitals)
Microwave	10^5-10^5	Shape of small molecules
Infrared	10^5-10^3	Molecular geometry
Visible and ultraviolet	10^4-10^2	Molecular energy levels

substances is drawn into a gas flame under a constant air pressure and the radiations are passed through appropriate colour filters before measurement with a photoelectric cell. The flame photometer is used for the determination of sodium, potassium, calcium, strontium and barium ions in solution.

Absorption spectroscopy

Incident radiation is absorbed by molecules, provided it is of the correct frequency to satisfy the quantum rules. The absorption bands produced may give quantitative results on the number and types of bonds present (*Table 27.2*). The standard types of absorption spectroscopy include the use of ultraviolet, visible and infrared spectrophotometers and also of measurements made with frequencies less than those encountered in the infrared range. Such frequencies can produce only rotational transitions and inversions (e.g., *see* ammonia, p. 280) and are studied by employing radio techniques (*microwave spectroscopy*).

Spin resonance spectroscopy

As each extra-nuclear electron possesses spin (p. 21), so each nucleon also has spin. These spins are quantized and because both electrons and protons are charged, their spins give rise to magnetic moments just as do orbiting electrons. Under the influence of a suitable magnetic field, alignment of the spins either with or against the field occurs, producing two distinct energy levels. At suitable radio frequencies and high magnetic fields, adsorption of energy results, as nuclei jump from one energy level to another. This effect is called *nuclear magnetic resonance* (n.m.r.).

Nuclei capable of exhibiting n.m.r. are those which possess an odd number of protons, e.g., ^1_1H, $^{14}_7N$, and $^{19}_9F$. The method is generally used in identifying the structural units containing hydrogen nuclei, since the effective field at a nucleus depends not only on the applied field, but also on the shielding effect of the extranuclear electrons; for example, separate proton resonance lines are obtained for the hydrogen nuclei found in the $—CH_3$, $>CH_2$ and OH groups present in ethanol (*Figure 27.11*).

Magnetic and electric measurements

The pairing of electrons results in a substance possessing *diamagnetism* (i.e., it experiences a repulsion when placed in a magnetic field), whilst substances

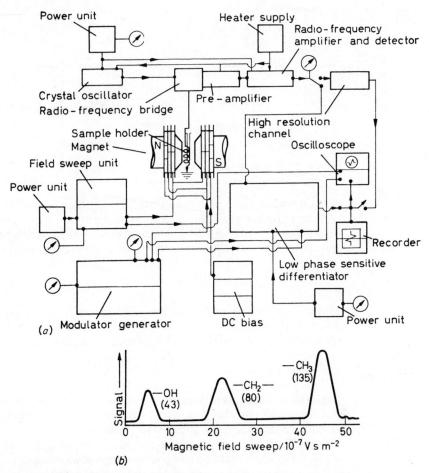

Figure 27.11 Nuclear magnetic resonance: (a) layout of equipment; (b) n.m.r. spectrum for ethanol

containing unpaired electrons exhibit *paramagnetism* (i.e., are attracted by a magnetic field) because of the increase in the number of lines of force passing through the specimen (*Figure 15.4*, p. 352). The latter phenomenon leads to the concept of *paramagnetic susceptibility*, the value of which is given by $\mu_M/3kT$, where μ_M = the magnetic moment, k = the Boltzmann constant and T = the absolute temperature (diamagnetic susceptibility is independent of the actual temperature). The importance of paramagnetism in indicating the number of unpaired electrons has been discussed in Chapter 15; determinations are made by finding the force resulting from the application of a powerful magnetic field to a long sample of the substance suspended in a glass tube from one arm of a balance (Gouy balance).

Dipole moments
Dipole moments have been used for distinguishing between geometric isomers (e.g. p. 535); their values are found by measuring the capacitances of a

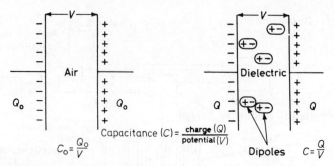

Figure 27.12 The meaning of relative permittivity: $\varepsilon_r = C/C_0 = Q/Q_0$

capacitor when filled with the substance and when empty. The ratio of the two capacitances gives the *relative permittivity* (*dielectric constant*) of the substance (*Figure 27.12*). The relation between the relative permittivity, ε_r, and the permanent *dipole moment*, p_e, of a polar compound is given by

$$\frac{\varepsilon_r - 1}{\varepsilon_r + 2} \cdot \frac{M_r}{\rho} = \frac{4\pi}{3} L(\alpha_0 + p_e^2/3kT)$$

where M_r = relative molecular mass, ρ = density, L = Avogadro constant and α_0 = the distortion polarization arising from temporary displacement of the electrons and nuclei caused by the applied electrostatic field.

X-Ray analysis

Since the wavelengths of X-rays are comparable with the atomic diameters of most elements and therefore to the distance between the lattice planes in a crystal, reflections from these layers can produce interference patterns (*Figure 27.13*). If the difference in path length between the reflected rays (i.e., $DEC - ABC$) is equal to an integral number n of wavelengths of the radiation λ, then a maximum occurs in the diffraction pattern. If d is the distance between lattice planes, then

$$n\lambda = 2d\sin\theta$$

Electrons can be similarly diffracted by reflection at, say, a nickel surface, demonstrating their wave nature (*see* Davisson and Germer, p. 20). In

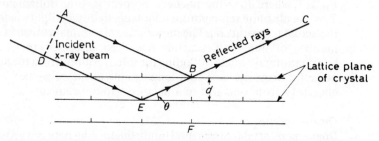

Figure 27.13 Bragg reflection produced by|diffraction of X-rays in a crystal

practice, monochromatic X-rays are used and, in order to produce a sufficient number of diffraction spots on the photographic film, either a small crystal is rotated at the centre of a camera, moving laterally to and fro, or a stationary sample of powder (containing randomly orientated crystallites) is employed in conjunction with a stationary camera. In both cases, some lattice layers are bound to be in a position to produce diffraction patterns, but in the latter the powder photograph obtained is too complex for ready determination of the crystal structure, so that this method is employed only for comparing substances (cf. infrared 'fingerprints').

X-Ray diffraction is used initially to obtain the relative molecular mass from a determination of the cell dimensions of a crystal of the compound under investigation. The mass of substance in a unit cell of volume V is ρV, where ρ = density, and by multiplying by the Avogadro constant, the relative molecular mass, M_r, is found

$$M_r = \frac{L\rho V}{n}$$

where n is here the number of molecules in a unit cell. In practice, the information derived from the photographs is often sufficient also to predetermine the number n.

The final use of X-ray analysis in determining the absolute structure of a compound is time-consuming and difficult, particularly if no previous information, such as to the number and types of functional groups present, is available. The chief difficulty lies in determining the phase differences between the diffracted X-rays, so that the reflecting planes of the crystal can be correctly labelled. Sometimes a model is built, and the results expected from such a structure are compared with the actual diffraction pattern, giving further information from which the model can be refined, so that eventually the correct structure may be obtained.

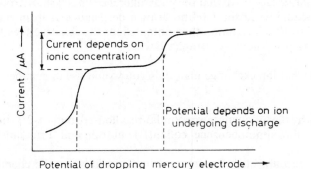

Figure 27.14
Polarogram for two dischargeable ions

Polarography

A certain potential is required to oxidize or reduce any particular ion, and when this value is reached, a surge of current results (*Figure 27.14*). For consistent results, the measuring electrode must not become contaminated with the products of the electrode reactions. This can be achieved by using a

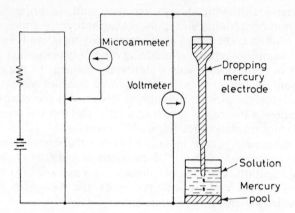

Figure 27.15 Basic circuit of the polarograph

dropping mercury cathode (for redox reactions involving cations), the potential of which can be controlled, and a large pool of mercury as the anode (*Figure 27.15*). Polarographic techniques have revealed the existence of many ionic species previously unknown, besides furnishing standard methods of quantitative analysis at low concentrations (about 10^{-5} mol dm^{-3}).

Questions

(1) Suggest how you could ascertain the molecular and structural formulae of sulphuric acid, phosphorus(V) chloride and ethanoic acid.

(2) From *Figure 27.13* prove that the path difference on reflection from adjacent planes is equal to $2d \sin \theta$. (Hint: draw a perpendicular from B on to DE, produced to F). The relation then obtained for the constructive interference of the X-rays is known as Bragg's law.

(3) Describe what ion exchange materials you would use in demineralizing tap water.

(4) Suggest why the following compounds are coloured: butane-2,3-dione, prop-2-en-1-al, 4-azoaminobenzene, cobalt(II) chloride and 'Prussian blue'.

(5) Write an account of the various methods available to the chemist for the purification of organic compounds.

(6) Two sixth-form students obtained the following results for the gas chromatography of a sample of petrol.

   One of the resistances of a Wheatstone bridge was placed in the stream of gas emerging from the adsorbent, its resistance being affected by its environment. The bridge was adjusted for zero deflection before introduction of the sample of petrol: the results in *Table 27.3* were obtained [Rosser, W. E., *School Science Review*, **49**, 167, 180 (1967)]:

Table 27.3

Deflection of galvanometer	1	2	2.5	6	2.5	22	1.5
Persistence of deflection/s	31	40	152	30	30	170	60

Say what you can about the composition of the petrol sample from these results.

(7) Write an essay on spectroscopic methods in analysis.

(8) The *X*-ray spectra of the elements are caused by the loss of inner shell electrons, e.g., from the *K* shell, followed by electrons from an outer shell falling in to fill the vacancies. On this basis explain why the frequencies for the *K* series are greater than that for the *L* series for elements of the same atomic number.

(9) A plot of the potential energy for two atoms forming a chemical bond against their internuclear distance is given in *Figure 27.16*. The horizontal lines represent the energies of the possible vibrational states of the atom pair.

Figure 27.16

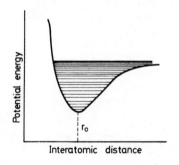

(*a*) Describe carefully the change in the potential energy of a pair of interacting atoms as their distance of separation is reduced. What is the significance of r_0?

(*b*) What physical explanation might there be for the steep rise in potential energy to the left of r_0?

(*c*) The horizontal lines merge into a continuum as the energy rises. Where does this continuum start and what is its significance?

(*d*) The band spectrum of a diatomic molecule converges to a limit at 270 nm. Explain what information can be obtained from this value.

(*e*) Superimpose on the graph a curve for an electronically excited state of the molecule and draw a comparison with the energy level diagram of *Figure 27.9*.

(10) *Figure 27.17* shows a simple arrangement for determining ionization energies.

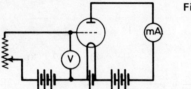

Figure 27.17

If the valve is filled with argon, describe the shape of the graph obtained by plotting the current produced as the applied voltage is increased and show how the ionization energy of argon can then be found.

(11) Radioactive indicator or tracer methods are based on the isotopy of radioactive atoms with stable elements. Using examples discuss this statement.

28 Chemistry and society

The image of chemical industry

It is difficult to imagine any area of modern Western life lying completely outside the influence of Chemistry. Chemical energy under control provides a major part of the basis of what might be called 'civilized life'; out of control it threatens the very heart of that civilized way of life. Fertilizers, herbicides and pesticides, judiciously applied, result in food crops that would have astounded our forefathers, whilst chemical additives enable food to be preserved for long periods of time in edible condition. Drugs help fight disease and pain, and in some cases even affect our 'spiritual' state of mind. The list could be extended. Of course, in one sense chemistry has always been with us—what we are talking about here is the deliberate application of chemical knowledge to affect man's condition. And to begin to understand the social implications, we need to explore the various relationships: between producer and consumer, between those employed at different levels in chemical manufacture, and between chemical industry and society-at-large. This exploration is the chief purpose of this final chapter.

We can clearly relate to objects, events and people through first-hand or secondary experience. For most of us, it is the latter which provides our main knowledge and informs our attitudes towards chemical industry. And the mass media represent overwhelmingly the chief agencies for these processes. To illustrate their potency we will take just one case, that of *dioxin*, a name that has rapidly entered the public consciousness as thalidomide did twenty years earlier.

The dioxin story

There are two points to make at the start: the story is almost certainly not finished, nor is it complete in the sense of the documentation having been thoroughly researched. It is merely a fragmented tale built up from what can probably be called average exposure to the media.

Dioxin is formed as an unwanted by-product of the manufacture of 2,4,5-trichlorophenol, itself produced from the chlorination of benzene, followed by partial hydrolysis:

1,2,4,5-tetra–
chlorobenzene

Sodium salt of
2,4,5-trichlorophenol

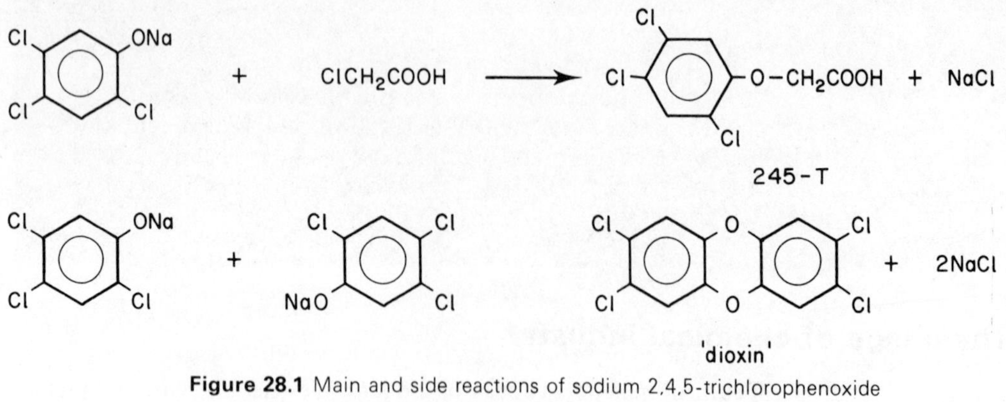

Figure 28.1 Main and side reactions of sodium 2,4,5-trichlorophenoxide

The sodium trichlorophenoxide is used, among other things, to manufacture 2,4,5-trichlorophenoxyethanoic acid (2,4,5-trichloroacetic acid), by condensation with chloroethanoic acid. 2,4,5-Trichlorophenoxyethanoic acid (245-T) is a very effective herbicide, mimicking the effects of a plant hormone and killing dicotyledonous plants. However, a small amount of a side reaction can occur, especially if the temperature rises above 200 °C during the partial hydrolysis, producing 2,3,7,8-tetrachlorodibenzo-1,4-dioxin ('dioxin'), which has turned out to be extremely poisonous to animals and humans (*Figure 28.1*).

245-T soon became a popular herbicide, and in 1962 the American military high command, frustrated by the successes of communist guerillas in South Vietnam, decided to use it, mixed with another herbicide, as Agent Orange, to defoliate the dense forests of Vietnam and so deny the enemy its natural cover. Shortly afterwards several Vietnamese civilians showed signs of poisoning. Much later, in the seventies, this suggestion of poisonous side-effects was taken up by American war veterans from Vietnam, who claimed that chloracne, birth defects and cancer were the direct result of previous contact with Agent Orange.

On July 10, 1976, there was an explosion at the Icmesa factory in Seveso, northern Italy (owned by the multi-national Hoffman–La Roche), accompanied by the eruption of a mysterious white cloud. The next day, animals and plants began to die and subsequently there were about six hundred cases of chloracne, as well as an increase in nervous disorders, miscarriages and cancer[1].

The Summer, 1977, edition of 'Science for People' (representing the British Society for Social Responsibility in Science) carried a lengthy article under the title 'SEVESO: NO ACCIDENT' describing the sequence of events in Seveso after the explosion. In its analysis, it included attacks on both the industrial company and the state authorities:

After use of 245-T in Vietnam, there is now the risk that the Seveso affair could lead to the development of a new chemical weapon in dioxin. This could explain the secret analysis made by SETAF in Vicenze using samples taken from the area by non-uniformed NATO military experts... The true face of Hoffman–La Roche is revealed: an

American company giving over production of chemical weapons to another company to maintain its own credibility. We should not forget, moreover, that the president of Hoffman–La Roche, E. Bobst, was one of the most important advisers to President Nixon, especially during the period in which the USA developed its aggressive policies towards the Vietnamese.

What has happened is that the Lombardy regional authority has reproduced within itself all the inefficiencies (and other faults) of central government. All the investigations and research have been conducted in accordance with the interests of the Christian Democrats. Their aim was to minimize any risks and calm the fears of the population, using science or other methods.

... Three different theories could follow from the statement that pollution is always the consequence of a particular economic structure. The first theory is that we must accept the inevitable side effects of pollution and disease if we want progress... The second theory is that we can use technology itself to solve the problems... The third theory is that the destruction of our environment and harm to our health is the inevitable consequence of the system of production that seeks to maximize profits... IN THE CASE OF SEVESO, THE STATE HAS RESPONDED THROUGH ITS INSTITUTIONS IN TERMS OF THE FIRST TWO THEORIES.

On 6 November 1979 the *Daily Mirror* newspaper launched a 'special' enquiry by devoting its entire front page to the following message:

In keeping with its image as the supporter and representative of organized labour and the 'common man' the *Daily Mirror* kept the pot boiling for a few months by, for instance, giving front page reference on March 13 1980 to the call by farm-workers on the Government to outlaw 245-T, under the headline:

The front page also carried the photograph of a small girl with the caption 'Tragic. Full Story—page 7'. Page 7 was largely devoted to this story, with another photograph of the child as well as of her mother, under the headline:

The article began:

> The mother of a doomed child pleaded yesterday for a ban on the controversial weedkiller 245-T. Three-year-old Kerry Hogben was born with a large hole in her heart and other defects.
>
> Her mother, Irene, believes there is a link with 245-T... she came into contact with 245-T when she worked for a firm of chemical distributors before her daughter's birth.

It is apparent then, that at the time of writing this (April 8, 1980) the *Daily Mirror* was conducting a vigorous campaign to persuade the British government to ban the use of 245-T.

On March 17, *Horizon* on BBC 2, presented 'Portrait of a Poison', in which the real culprit, dioxin, was exposed to a 'popular scientific' analysis, as befits the nature of this television programme. Apart from considering its formation as a condensation product of 2,4,5-trichlorophenol, it examined the research that had been carried out into the clinical effects of dioxin and concluded that it was the most dangerous chemical known. There was reference to accidents that had occurred in factories producing 2,4,5-trichlorophenol and to the greater incidence of miscarriages in zones in the USA which had been sprayed with 245-T as a weedkiller. The point was also made that it was technically feasible to remove the dioxin from 245-T but that the process was expensive; it also showed that some commercial preparations containing 245-T for public sale contained less dioxin than could be detected by the extremely sensitive techniques of gas chromatography and mass spectroscopy (Chap. 27). The tone generally was informative rather than polemical. Not so the poem by Roger Woddis in *Radio Times* for that week:

All things bright and beautiful;
All creatures great and small
Thrive on toxic chemicals
And seldom die at all.

> Give thanks to Agent Orange
> That bears a lovely name;
> Once treated with dioxin,
> You'll never feel the same.

All well-tried defoliants,
All poisons that they spray
Make our bodies healthier
And guard our DNA

> From Vietnam to Seveso
> Their wonders are still sung;
> They made the women fertile
> And fortified the young.

One professional response to this situation took the form of the following letter to the *Guardian* newspaper of July 22, 1980:

245-T: another side to those one-sided media reports

Sir, — Environmentalists and union spokesmen sometimes talk about the "powerful chemical lobby" ensuring a good image for the pesticide industry. But just think about it amid all the barrage of accusation and innuendo against the weedkiller 245-T and the way pesticides are cleared for use in Britain; when did you last read a word for the defence?

It has always been a principle of British justice that both sides should be heard, but the public has been left with a distinctly one-sided story from TV, the Daily Mirror, and various trade unions.

But there *is* another side to tne story. Nobody has pointed out that Britain has the finest safety record for pesticides in the world; not one single farm fatality in recent years and, out of a total of 4000 farm accidents in 1979, only seven were linked with sprays. Against this background of u.1matched success the TUC General Council is demanding 'improvements,' which simply means worker participation and collective bargaining with the distinguished independent experts who are at present responsible for advising Government.

But what if the unions are wrong? There is a lot at stake; this country has dragged itself out of the mire of 19th-century, unrelieved pain and untimely deaths by the use of chemical technology. In 1900, a person's life expectancy was 44; today it is 75. Pesticides have played and continue to play a vital part.

The effect of not spraying tropical crops would of course be disastrous, and the resulting famine would be the greatest disaster the world has ever known.

The benefits of pesticides cannot excuse irresponsibility, and the chemicals in today's world need careful surveillance. The need is for skilled and impartial watchdogs who have no political, commercial or emotional axe to grind.

Watchdogs yes, but not witch-hunts. We don't need witch-hunters who claim that the 245-T used in Britain has been *proved* to harm people, which is not so. We don't need witch-hunters who claim that Seveso was a pesticide factory, which it wasn't. And most of all we don't need witch-hunters who publicly insult the ability and integrity of all scientists and experts who do not happen to agree with them.
(Dr) D. G. Hessayon
Chairman, British Agrochemicals Association, London SE1.

What conclusions can we draw from these extracts from various separate sources? They reveal the 'snowballing' effect of newsworthy items (there was also an article in *Nature* in March, 1980). They also remind us very forcefully that there is no neutrality in reporting: the language alone reveals a particular perspective or prejudice.

The dioxin story also underlines the pressures commercial firms are under to market their products as rapidly and as cheaply as possible and of the potency of a link between industry and the armed forces (the military–

industrial complex). So there will always be a need for vigilance and legislation to protect the health of chemical workers, consumers and the public-at-large.

Such accounts as those summarized above not only inform our attitudes towards chemical industry but reveal something of the complexity of the possible relationships between chemistry and society as well as the potential hazards to both consumer and industrial workers. Decisions might be taken within the board-room or between a team of research chemists which have the most profound effects on social customs. In the case of the contraceptive pill, for example, there appears to have been a far-reaching effect on the attitude of Western women towards sex. It might be claimed, for instance, that chemistry has liberated woman. And yet the decision to carry out research into a female, rather than a male, pill is arguably sexist and chauvinist and might be seen as an example of the continuing exploitation of women by men. Which is another way of saying that scientific research is a male-dominated activity. On the other hand, since it is the woman rather than the man who suffers most from an unwanted pregnancy, it could be argued that it is to woman's advantage to have contraception under her control.

Ideology

The influence of Chemistry on society is far from being one-way. Society itself controls who shall be trained in chemical research and what form that training shall take. Whilst the route into 'pure' academic research is a good honours degree at a University, chemists taking up 'applied' research in industry may well have attended courses at a Polytechnic or Technical College and obtained various certificates. According to *Statistics of Science and Technology*, graduates after training were located as in *Table 28.1*, where *basic research* is customarily regarded as that carried out with no thought to practical utilization, *applied research* is concerned with solving problems with clear practical implications, and *development* is even more consumer-oriented. The classic example of development is the scaling-up of a process hitherto confined to small-scale laboratory operations into a fully-fledged industrial procedure (*Figure 28.3*). But these distinctions are to some extent political rather than scientific, reflecting divisions in our society. Thus the academic, enjoying high status, carries out research at the frontiers defined by his own reference group of fellow-intellectuals. His identity as a member of this group is confirmed through a ritualistic 'exchange of gifts'; that is by the

Table 28.1 Placement of chemistry graduates after training in one year, and type of research undertaken

Placement	%	Research	%
Industry	60	Basic	12.6
Universities, schools etc.	25	Applied	26.0
Government departments	15	Development	61.4

Figure 28.3 Cartoon: 'The bench-scale results were so good we by-passed the pilot plant'
[Reproduced by permission from *The Modern Inorganic Chemicals Industry*, Ed. R. Thompson. Proceedings of the Inorganic Chemicals Group of the Industrial Division of the Chemical Society.]

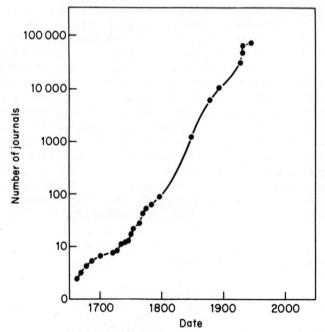

Figure 28.4 Total number of scientific journals being published at various dates

publication of research papers in learned journals, without thought of financial reward[2]. The extent to which this activity has developed over the years is shown in *Figure 28.4* and *28.5* which interestingly reveals an increase in the tendency to produce joint papers, reflecting research by teams rather than by individuals working alone, often self-supporting.

It is a frequent complaint in the United Kingdom that our low level of industrial performance is at least partly attributable to the fact that too many of our most capable scientists work in the academic field rather than in industry. This is not surprising in view of the hierarchical status embedded in the notions of pure, applied and developmental outlined above. The irony lies in the fact that industry plays its own considerable part in perpetuating these divisions in society. The argument being advanced here is that the problem is *ideological* and that we need to explore the nature of the dominant ideology before we can begin to understand the nature of the relationships between chemists of differing status, industry and other sectors of society. This dominant ideology of the Western world has the following distinctive features:

(a) It is *capitalistic*; that is, it is wedded to the notion of accumulating capital and generating profit.

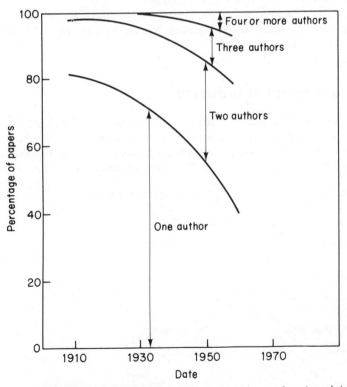

Figure 28.5 Incidence of multiple authorship as a function of date [Reproduced by permission from Price, *Little Science, Big Science*]

(b) It is *competitive*; insofar as there is cooperation, it is within one team competing with another[3].

(c) It is *exploitative* in its use of resources, whether material or human. Thus, human beings are seen as providers of labour and as consumers. The system is at its most efficient when, on the one hand, it strikes the hardest bargain with its labour supply and, on the other, persuades as many as possible to consume its product

Central to this ideology is the concept of *growth*, with markets being indefinitely expanded and output constantly increased. Associated with it are the ideas of the division of labour and mass production, bringing with them the alienation of the worker from the product of his labour. It is an ideology which in the last century has fuelled a remarkable increase in industrial output, and in the standard of living (as measured by the capacity to consume that industrial output)[4].

That the above may be a simplistic analysis is suggested by the fact that many Western countries operate a mixed economy, with a more-or-less uneasy balance between private and public sectors, that is, between private enterprise and nationalized concerns. On the other hand, both these sectors may be imbued with the capitalistic spirit: certainly the ongoing conflict between employers and Trades Unions is common to both private and public sectors, and takes the same shape.

This, then, would appear to be the over-riding ethos within which the professional chemist has to operate. Let us see how he is located within chemical industry.

The structure of chemical industry

The concept of 'capital' is particularly pertinent to the chemical industry, since it is largely capital-intensive; that is to say, the emphasis is upon sophisticated and expensive equipment to take advantage of the continuous nature of chemical processes. It is not unusual for an operative in the

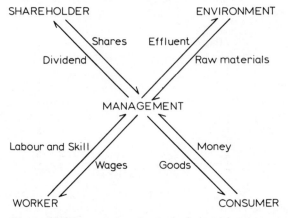

Figure 28.6 Interested parties to the industrial system

chemical industry to be in control of plant worth several hundreds of thousands of pounds. Another way of looking at this is in terms of the low level of labour demand and the relatively slight impact the chemical industry has upon the employment situation.

When this aspect of capital intensity is considered together with such factors as rapid obsolescence of chemical plant, then the need to be able to raise capital readily is appreciated; devices such as take-overs and mergers are often resorted to. The chemical industry illustrates the tendency of giant corporations to emerge, crossing frontiers and redefining national sovereignty, so that, for example, a decision taken in the board-room of a multinational corporation can nullify or outweigh decisions taken by the government of the country concerned[5]. The response of, for example, the British Labour Party to this growing concentration of power is to accept it as a *fait accompli*:

> The greatest single problem of modern democracy is how to ensure that the handful of men who control these concentrations of power can be responsive and responsible to the nation (*Signpost for the Sixties*).

This increasing concentration of power is reinforced by the fact that more and more shares in industry are being held by large financial institutions (banks, insurance and pension funds), so that investment *policy* is in relatively fewer hands (*Figures 28.7* and *28.8*).

The typical structure of an industrial organisation is hierarchical and pyramidal. That is to say, a small number of individuals constitute the controlling unit at the top, with lines of communication passing down vertically to increasing numbers below. The way in which chemists fit into this structure is illustrated in *Figure 28.9*. It will be seen from this that the chemists' contributions to the industry are categorized as Production, Research and Development. The former area includes the analysis of raw materials and end-products, to ensure satisfactory quality control. It also includes control of the actual chemical processes within the factory. For example, in the case of a firm manufacturing plastics, it may be necessary for the works chemist to carry out appropriate tests at certain times in order to ascertain the precise stage of polymerization.

If industry were completely static, in the sense that there was no search for novelty or improvement, the above would presumably constitute the sum total of the chemists' contribution. But for a firm to survive in a competitive situation, it must produce at the lowest cost, and be on the look-out for something different with which to woo the customer. It is therefore essential that there be improvements to existing processes and the discovery of new ones. Some of the chief problems confronting the Research and Developments units in these respects are:

(a) *The availability and cost of reactants.* As the supply of raw materials becomes weaker, so the price will rise and the firm may be forced to seek alternative routes and reactants.

(b) *The yield of reaction.* Whilst a favourable equilibrium constant is preferred, operating costs may be so high that a high rate constant

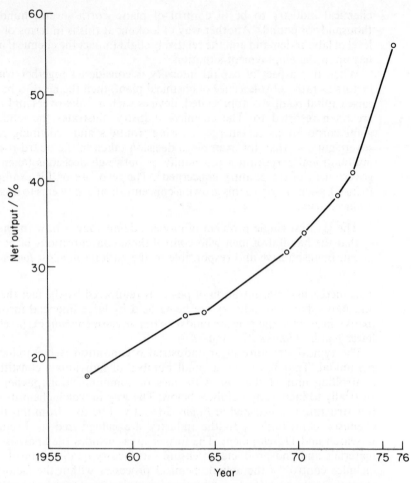

Figure 28.7 Ordinary shares held by financial institutions in Britain

overrides it. It may be worth a firm's devoting considerable resources towards research into new catalytic processes and the discovery of fresh catalysts.

(c) *The production of by-products.* One way of reducing the costs of a chemical operation is to produce by-products of commercial value. This, too, points towards the need for fundamental research into the nature of chemical reactions.

It was pointed out earlier that the division of research into basic, applied and developmental was difficult to justify on scientific or logical grounds. In fact, industry generally makes an equally suspect division between 'Research' and 'Development' but in reality, for the purposes of investment and policy decisions, regards them as a whole. Nevertheless, it is the Research Division which customarily possesses the initiative. Once it has made a breakthrough, for example, in synthesizing a new material, it becomes more the concern of

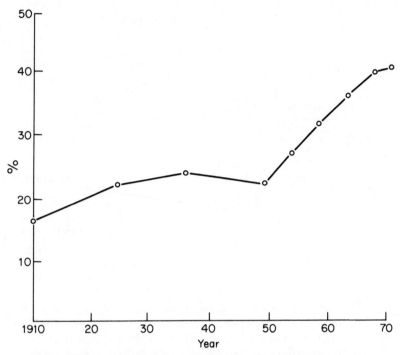

Figure 28.8 Percentage of total due to hundred largest British manufacturing enterprises

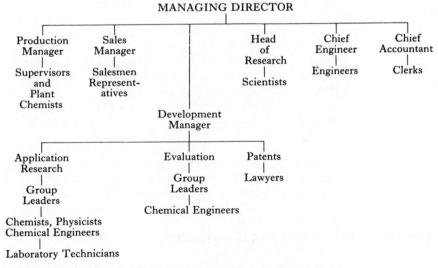

Figure 28.9 The top of the industrial tree, with detail of one of its branches

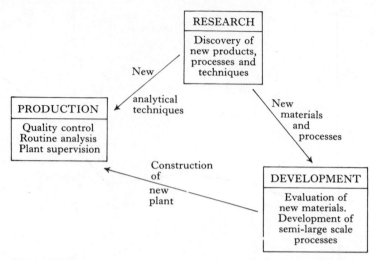

Figure 28.10 Possible links between production, research and development

Development in establishing the feasibility of the process for large-scale production in economically sound terms (*see Figure 28.10*). Because of the large capital required to construct modern chemical plant, a pilot (semi-technical) plant is usually set up as intermediate and this is then evaluated to see how far performance matches promise. If the indications at this stage are sufficiently promising, then the full-scale plant is constructed, but not before being protected by patent, so that any rival firm will be compelled to pay certain rights before being able to emulate the pioneer.

Efficient use of capital resources demands that, as far as possible, production is continuous, so that expensive plant is kept working at full capacity around the clock. Where appropriate, processes are 'programmed' so that they can be operated automatically by computer. There is, of course, little point in automating a plant if the data being processed is too crude to enable the programme to function efficiently—working hand in hand with increased automation, therefore, is improved instrumentation. The information thus made available passes to the control unit, which sends appropriate instructions to both the store and arithmetic unit: relevant data is supplied to the latter unit from the store and the subsequent results fed back to the plant, so that appropriate action can be taken: automatically, if the plant is fully automated (*Figure 28.11*). Automation can thus be seen as a means of reducing the need to employ large numbers of personnel and of reducing the menial and repetitive nature of the job.

The location of chemical industry

The industrialist is primarily concerned with bringing together in to the most efficient combination, capital, raw materials, energy, a suitable labour force and the customer. This is clearly, to some extent, a geographical matter and

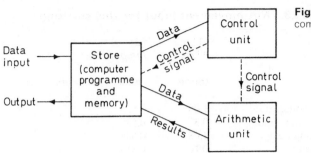

Figure 28.11 A computer circuit

involves siting the factory in the best place for optimum utilization of the above factors, and hence the maximization of profits.

Another way of putting this is to say that for efficient running of a factory, transport costs must be kept down to a minimum. In terms of raw material, this means siting the factory near the source of raw material if:

(a) The raw material is bulky.
(b) There is a considerable amount of waste or loss in mass during manufacture.
(c) Freight charges are high.
(d) The value of raw material is low compared with the final product.

A similar position exists for the supply of energy; that is, if the process is particularly energy-intensive, it will tend to be located near the energy source. Alternatively, the energy supply can be taken to the industry: for example one of the chief considerations in damming the Zambesi River to produce Lake Kariba was in order to provide hydroelectric power for the copper industries of Northern Rhodesia (now Zambia).

The availability of cheap energy can, in fact, be a critical factor for the chemical industry which, as *Table 28.2* shows, is generally very energy-intensive.

Table 28.2 Cost of energy as percentage value added to cost of manufactured products (*from U.S. Annual Survey of Manufactures*, 1962)

Industry	%	Industry	%
All industries	3.5	Cement manufacture	22.6
Lime production	30.0	Blast furnaces and steel mills	12.2
Aluminium extraction	24.7	Glass manufacture	8.7

P. Haggett (*Location Analysis in Human Geography*, 1965 quoted in Open University Social Science Course Reader *Understanding Society*) analyses the cost (in terms of 'net movement input') for the transport of raw material, fuel and final product for zinc smelting (*Table 28.3*).

The situation, based on the above calculations, can be illustrated graphically for a market up to 160 km from the source of raw materials. It is clear from this simplified picture (*Figure 28.12*) that location of the smelter at source is desirable.

Table 28.3 Net movement input for zinc smelting

Input	(1) Mass	(2) Freight rate	(1) × (2) Net distance input
	tonne	(dollars per tonne–km)	
Zinc concentrate	1.00	0.625	0.625
Reduction coal	0.37	0.688	0.255
Heating coal	1.08	0.688	0.743
Fireclay	0.10	0.313	0.031
		Total input	1.654
Output			
Slab zinc	0.54	1.313	0.709

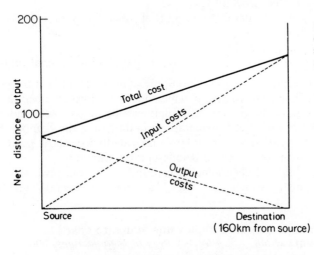

Figure 28.12 Transport costs for location of zinc smelter between source and destination of 160 km

The above considerations ignore the *labour* factor. Yet an adequate supply of suitably skilled and experienced personnel can be an overriding factor in locating an industry. Often a new industry is sited near to similar, but declining industries because of the availability of redundant workers with suitable expertise. New industries also tend to be sited in industrial areas for the same reason that there will probably be a seam of trained labour to be tapped. On the other hand, growing automation in the chemical industry means that there is correspondingly less reliance upon a pool of labour (*Table 28.4*).

So far this is a simplistic and incomplete discussion of the principles of industrial location, since it ignores the political dimension. Government policy may override the above considerations in order to achieve a politically desirable goal. For instance, in the UK, governments have provided financial incentives to firms to move into areas with declining industries and growing unemployment (the so-called 'development areas'). Private industry may also

Table 28.4 Labour costs as percentage value of products, USA (*from U.S. Annual Survey of Manufactures*)

Industry	%
All manufacturing	24
Steel rolling/finishing	22
Paint industries	17
Basic chemicals	16
Oil refining	6

take what are essentially political decisions to take advantage of relatively cheap labour in underdeveloped countries, to employ women rather than men, or to take on only non-unionized labour.

Problems facing modern chemical industry

Since a capitalist economy is dependent upon the generation of profit (because only then will it attract financial support away from its competitors) it can be assumed, in the absence of evidence to the contrary, that it is inextricably tied to the ideas of *growth* and *obsolescence*. The likely outcome of this is short-sighted use of finite, capital resources and pollution of the environment with waste and used products. Let us look at the problem of resources first.

The problem of finite resources

One of the chief resources of the chemical industry is *energy*. It is a resource in the sense that it fuels the chemical industry and *Table 28.3* shows that certain chemical processes are extremely hungry for energy. It is also a resource in that certain processes, notably of combustion, are markedly exothermic and can therefore be used for generating energy. The fuel which has dominated Western industry in the past few decades is petroleum. Our lives have become so dependent upon it that it illustrates forcibly the interrelationships between politics and the economy. That aspect will be taken up later. What is of relevance here is the fact that petroleum is a fossil fuel and hence is irreplaceable (at least in the time scale which we are considering). Although the development of the oil industry is a tribute to the vigour and ingenuity of modern industrial chemistry (*see Figure 19.4* for a simplified flow-sheet of the industry) it is nevertheless the case that a capitalistic economy, responding to market forces, has quickly used up a large proportion of our capital reserves of organic compounds, mainly for the purpose of exothermic oxidation to such plentiful substances as carbon dioxide and water (*Figure 28.13*). There is no incentive to look for alternative sources of energy whilst profit margins remain adequate, although *Figure 28.14* shows that known reserves of petroleum are far less than one year's solar radiation. The fact that the

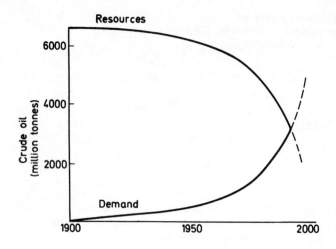

Figure 28.13 Exponential demand for oil, and diminishing resources

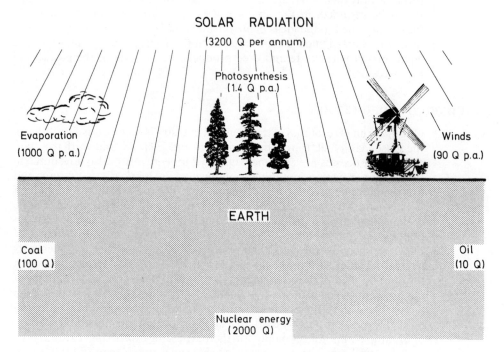

Figure 28.14 Available sources of energy $(1Q = 3 \times 10^{21} J = 8 \times 10^{14}$ kW-hr)

overwhelming source of energy in the Western world is in the form of fossil fuel like coal and oil is a reflection of the unattractiveness of harnessing alternative forms like wind and solar energy to the gigantic corporations which are increasingly becoming the norm.

Yet, in principle, energy is available on a current account, as well as a capital one. This option, however, is not available with other resources of the chemical industry. As a glance at *Figure 28.15* will show, at present exponential growth rates, all known reserves of most metals will have been exhausted within a few generations. Even for the abundant metal, iron, accessible and exploitable reserves are now within a life-span. Of course, it is naive to extrapolate in this fashion towards disaster. Relative scarcity will result in higher prices, and it will become profitable to recycle raw materials. The question to ask, however, is 'Can industry be persuaded to conserve raw materials before being compelled by economic forces?'

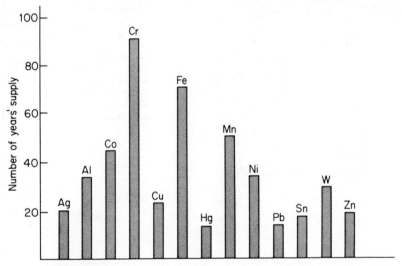

Figure 28.15 Limits of known reserves of metals at present exponential rates of consumption

Problem of pollution

The problem of pollution is closely linked with that of conservation of resources. If raw materials are used sparingly, if used materials are *recycled* to the fullest extent, if the required chemical reaction is maximized to eliminate side-reactions, and if by-products are utilized or absorbed, then pollution will be minimized (*Figure 28.16*). But pollution can take many forms. For example, the environment can be damaged by the disposal of waste (which it is unprofitable for a firm to process further) or through the use of the manufactured products themselves. In the case of the latter, a striking example is the use of chemical products to effect biological control. Not only can these so upset the ecosystem that the pest flourishes because of blanket decimation of its predators; but insecticides can be built up and concentrated

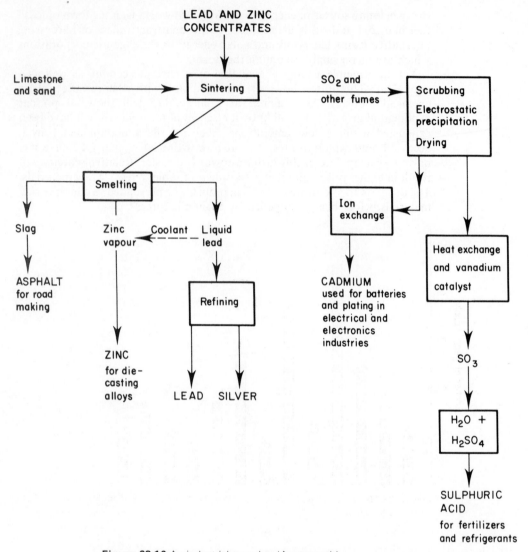

Figure 28.16 An industrial complex (Avonmouth)

through food-chains, so that they create havoc in the biosphere, particularly among those animals at the end of the food-chain. Some of these outcomes are the result of ignorance or narrow specialization (for example, chemists with little biological understanding) but they may also be an inevitable result of firms cutting corners in order to reach the market-place first or to reduce costs of production.

Clearly, one way forward is through education, by developing ecological awareness and becoming familiar with, for instance, the four laws of ecology as formulated by Barry Commoner[5]:

Everything is connected to everything else
Nature knows best
Everything must go somewhere
There is no such thing as a free meal.

But in the end, the crucial action may have to come from government, in the form of tax-relief or subsidies for effective waste-control, and penalties for carelessness or indifference. However, not only might it prove extremely difficult to establish the source of pollution; there are also the problems of establishing the mode of action of the pollutant and laying down a safety limit. Lead pollution illustrates these aspects of mechanism, threshold and source nicely. Although lead has been recognized as a toxic metal for a long time, it is only recently that it has been shown to affect the functioning of the enzyme ALA-dehydrostase in the blood and the brain, explaining why symptoms of lead poisoning include anaemia and loss of coordination, and suggesting a possible cause of some educational difficulties with inner city children. Whilst the generally-recognized danger limit is 80 micrograms of lead per 100 cm^3 of blood, there are some authorities who argue for a considerable downward revision of this figure. With regard to the source, it may be lead paints, lead smelting or the lead additives in petrol. What *can* be relied upon in this rather confused area is the initial response of the manufacturers to the suggestion that they are in any way implicated. A direct and unambiguous link of cause and effect is difficult to establish.

The problem of the Third World

Pollution is not distributed evenly in either geographical or social terms. The rich can easily avoid the worst excesses of industrial pollution. It is the labouring class who by and large have to bear the brunt of the decisions taken by the powerful. So much is obvious. What is more interesting is that with the growth of multinational corporations, there is a tendency for the underdeveloped countries (the 'Third World') to become the dustbins of industry. And such is their economic condition that they are likely to embrace what might be seen by them as a first move in to the industrial age.

Poor countries eager to acquire smokestacks and the jobs which they bring, are likely to view their unused or underused waste absorptive capacity as a resource to be exploited in international economic competition, much like mineral reserves or fertile farmland. The response of investors to pollution differentials among countries in some ways parallels that to wage differentials. In effect, firms are beginning to locate pollution-intensive phases of their operations in countries with low pollution levels, much as they have located labour-intensive aspects of their operations in low-wage countries, most prominently Mexico, Taiwan, Hong Kong, Singapore and South Korea in the past decade[7].

Some Third World countries have rich reserves of raw materials (in that sense we need to distinguish between the various Third World countries, since their economic outlooks can differ dramatically) and these have provided valuable revenue as exports to the chemical organizations of the First and Second Worlds (i.e., the Western world and the Communist bloc countries). They have also helped sustain the huge differences in standard of living because of the wealth accruing to the industrial nations through development of the raw material. The response of the underdeveloped countries to this form of exploitation is taking different forms:

(a) Exporting countries are beginning to bargain collectively, as witness the OPEC (Organization of Oil Exporting Countries) negotiations of recent years.
(b) The supply of raw materials is being restricted or banned, as the exporting countries prefer long-term internal development to short-term gain.
(c) Raw materials are being at least partially processed before exporting.

The ultimate problem concerns the nature of the relationship between the developed and underdeveloped parts of the world; whether it is to be one of mutual exploitation, or of cooperation, and whether Third World countries will demand entry as full members of the industrial nations or seek other alternatives.

The problem of unemployment

Reference has already been made to the capital-intensive nature of chemical industry, with sophisticated equipment and automated processes. It does comparatively little, then, to solve the employment problems of the modern world. It might be argued that, provided sufficient wealth is generated, all is well and that we can turn our attention to shorter working weeks and increased leisure. To some extent this is a matter of values; if one believes that human beings achieve fulfilment through meaningful work, then these alternatives are unacceptable. If, on the other hand, work is seen as something intrinsically undesirable, the less of it around the better.

The situation is complicated, however, by the rapid growth in world population (to some extent the result of the efficiency of modern science and technology and their application in medicine) which, allied to the problems outlined above, puts a large question mark against the capacity of the industrial nations to deliver the goods. To alleviate this problem E. F. Schumacher ('*Small is beautiful*') advocated *intermediate technology*[5], whereby instead of maximizing output per man, as in conventional technology, the work opportunities for the unemployed would be maximized. The aim of intermediate technology would be to provide workshops in the areas where people already live, using local material and simple techniques. These workshops would be intermediate in cost between traditional and modern workplaces and would be labour-intensive. This idea of intermediate technology may seem at first sight highly attractive. What is missing, though, is an analysis or description of *relationships* between man and man in the work-place, or between man and nature. It is implicitly ideology-free and can therefore be accommodated within the prevailing mode of production, perpetuating some of the shortcomings outlined above (*see* next section).

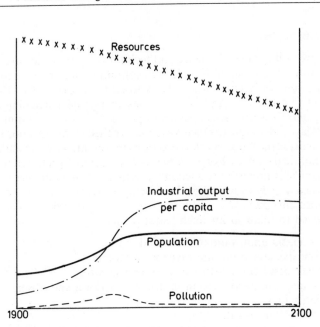

Figure 28.17 Stabilized world model

An interesting, if much-criticized, attempt to put the house of capitalism into some sort of order is that by the Club of Rome, a group of industrial leaders, scientists and economists. Concerned at the exponential growth of population, pollution and use of finite resources, it commissioned the 'Limits of Growth' project at the Massachusetts Institute of Technology. The outcome was a computer model capable of assessing the interactive effects of changes in the variables upon each other. For example, *Figure 28.17* shows the situation expected to arise if:

(a) Population were stabilized (birth-rate equal to death-rate) by 1975.
(b) Industrial capital ('material-crunching' capital) were stabilised in 1990 by setting the investment rate at the same level as depreciation rate. Capital would instead be diverted towards servicing, recycling and pollution control etc.
(c) Resource consumption per unit of industrial output were reduced to one-fourth in 1975.
(d) Pollution generation per unit of industrial output were reduced to one-fourth in 1975.

That such action is in the realms of fantasy is indicated by the fact that the deadline for three of the four variables concerned, 1975, is past and nothing has been remotely attempted on a global scale. Modern governments sell themselves to their electors on a policy of economic growth and improvement in the 'standard' of living. There appears to be a complete impasse. As the U.S. government's Science adviser commented, with reference to the Club of Rome, 'Technology assessment is fine as long as it doesn't mean technology arrestment'.

Alternatives in life styles

The prevailing view of science and technology is that they are 'neutral'; that is, knowledge is acquired without contamination with ideology. Values, politics etc. only enter when this knowledge is applied to the solution of 'worldly' problems. And yet decisions to pursue particular research in a particular direction are dependent upon certain values and in this sense *are* political or ideological. Once we recognize this, we are in a position to expose and analyse the various value-positions represented by particular ideological and political stances. For example, if cooperation is preferred to competition, if equality is preferred to inequality, if integration is preferred to alienation etc. then a technology might be developed which is a *genuine* alternative to the present pervasive technology. Such a *counter-technology* would be expected to incorporate such features as:

Emphasis upon human skills and crafts.
Low specialization and division of labour.
Weak distinction between work and leisure.
Democratic participation in decision-making and absence of hierarchies.
Low pollution rate.
Low energy consumption.
Emphasis upon use of renewable, rather than finite resources, etc.

A profound difficulty facing any such *pluralistic* society, with alternative life-styles and technologies persisting side by side, is that of mutual tolerance and accommodation. This difficulty is quickly revealed by the example of a group of people wishing to live collectively and simply in a 'back-to-nature' style; unless they already own land or wealth, they find it very difficult to get started. We are back to the old problem of power and ownership. And so, as the gap between rich and poor increases and as the problems recognized by the Club of Rome become more insistent, the world polarizes into crude ideological divisions, each bent on conquering the rest.

Decision-taking

The nature of decision-taking procedures in chemical industry depends on the structures within it. Typically, the board of directors at the apex of the organizational pyramid take fundamental decisions on the firm's policy. The extent of the advice they take and the amount of publicity given to the public on the decision and the discussions preceding it is a matter for the board[9]. The rationale underpinning this structure is *elitist* or *meritocratic*; crudely, people are either born into positions of power or they work their way to the top. Justification for this can be provided on genetic or environmental grounds; either leaders are born, not made, or they reveal their potential by making the most of their opportunities on their way up. Of course, they will need to be expert in their field but it goes without saying that considerable expertise will have been acquired in the course of their previous experiences, either through the educational system or on the job[10].

So much for private industry. In the public arena, the situation can be remarkably similar. Decisions are taken by 'experts' who can be relied upon to act in the interests of the general public. (The 'public' are here assumed to be largely ignorant of scientific matters but share a common outlook; that is, they all see the virtues of the system-as-it-is and are not therefore expected to question it.)

What does all this mean for scientific research, particularly in the field of chemistry? Since a private company is in the business of generating profits, even in the short-term, research can be expected to be pragmatic and instrumental in character, with a clear expectation of pay-off. The public sector is not so clearly defined. Funds can be provided for research in universities, in government departments and the nationalized industries. Since research can nowadays be a very costly matter and there is competition for funds, there is naturally an acute dilemma over the criteria to use in preferring one research proposal to another. Various suggestions have been made to resolve this dilemma. For example, Alvin Weinberg, of the Oak Ridge National Laboratories, Tennessee, categorized appropriate criteria as internal and external:

> *Internal*: the quality of the scientists involved in the research programme;
> the ripeness of the research for exploitation.
> *External*: whether the likely advances are important for research in other fields;
> the potential technological value;
> whether the research is likely to serve desirable social ends.

The character of such criteria serves to illustrate the complexity of the problem and the possible arbitrariness of the eventual decision. The total amount of publicly supported research will depend on other factors, too, such as the financial health of the nation and the political complexion of the government (a Labour government can always be relied upon to give relatively stronger support to the public sector than a Conservative one, even though in absolute terms, the support might be less).

Openness

A plea for openness in the field of scientific research can rest on at least two supports. In the first place, science is characterized by the public nature of its knowledge, with experiments available for replication and the conclusions drawn from them open to criticism. In the second place, since the application of scientific research has clear social implications, then society at large has a right to know about it, and perhaps partake in decisions whether or not to utilize the knowledge available. But such pleas run into difficulties very quickly. In the case of private investment, it can be argued that 'private' is the operative word and that industrial research would lose its point if its outcomes were to be available to competitors. A second area of difficulty concerns that 'public' research conducted on military grounds. Clearly such research must be ruthlessly concealed from a potential enemy, and in the United Kingdom we have the Official Secrets Acts to enforce such concealment.

In some cases, the military and commercial interests overlap. For example, one product of the 'peaceful' utilization of nuclear energy is plutonium, the basic ingredient of some nuclear weapons. It is perhaps this uneasy relationship between the interests of the industrial and the military which elicits most public disquiet. Such disquiet that the British government was compelled to hold a public enquiry into the demand by British Nuclear Fuels Limited that they be allowed to reprocess spent fuel at their nuclear plant in Cumbria (the Windscale Enquiry)[11]. Mr Justice Parker, who chaired the enquiry, was able to shape the enquiry along the lines of three questions:

(a) Should oxide fuel from United Kingdom reactors be reprocessed within the UK at all?
(b) Should reprocessing be at Windscale?
(c) Should the plant be double the size required for UK spent fuel and used to reprocess foreign fuel as a commercial operation?

The enquiry lasted a hundred days and considered both the applicants' and objectors' cases with regard to

(a) Nuclear weapons proliferation.
(b) Energy conservation.
(c) Risks from terrorism.
(d) Balance of payments.
(e) Prospects for employment.

Lord Parker then came down firmly in favour of BNFL's application, while calling for improved security and a tightening-up of radiological protection standards.

What is of particular relevance about this enquiry was that the terms of reference excluded a discussion of the desirability of a nuclear energy programme in the first place. Since this was taken for granted, arguments concerning

(a) Cost, availability and security of raw material.
(b) Safety of operation.
(c) Flexibility of scale.
(d) Effects on employment.
(e) Environmental impact.
(f) Return on capital investment.
(g) Net energy yield.

particularly in relation to other sources of energy were irrelevant.

A continuing issue, so far as 'openness' is concerned, therefore, is the extent and nature of public discussion[12]. In other words, who decides what issues are available for public discussion and the grounds on which the debate shall take place?

A case-study: plastics division, Ciba-Geigy (UK) Limited, Duxford, Cambridge*

To illustrate some of the generalizations made in this chapter and to put them into a more realistic perspective, we are going to trace the development over the last five decades of what is now Britain's largest manufacturer of synthetic resin adhesives and a member of the CIBA-GEIGY organization. We wish to thank the management of Plastics Division, CIBA-GEIGY (UK) Limited, Duxford, for their help and co-operation in this enquiry.

Historical

The forerunner of the present factory was Aero Research Limited, incorporated on April 7th, 1934. Both the siting and founding of this company were the results of the interests of Dr N. A. de Bruyne, F.R.S. At the time, he was a young don of Trinity College, Cambridge, interested both in flying and the use of plastics in aircraft construction. He needed land that was cheap and able to house not only a workshop but a hangar for his own aircraft, and not too remote from Cambridge. It was these considerations which led to the purchase of a large field on the outskirts of the small, agricultural village of Duxford.

Dr de Bruyne's research into new structural materials for aircraft quickly led to a research contract from de Havilland and the development of 'Gordon Aerolite' with a higher strength/weight ratio than any material known at the time (as well as honeycomb and sandwich structures which anticipated some of the present products of the Bonded Structures division.) Work on wooden aircraft structures led logically to resins for bonding the wood, rather than replacing it. 'Aerolite', a urea-formaldehyde glue, was the product resulting from this new orientation. Its resistance to heat, water and moulds represented a considerable advance over wood glues currently available and in April, 1937, it received official approval for the glueing of wooden aircraft structures. The simultaneous development of GB hardeners meant that resin and hardener could be applied separately, extending the commercial possibilities into areas such as the *furniture industry*. Sales were approaching £300 a year, with a production of about 1 tonne a month when, in September 1939, war was declared.

War meant an end to the furniture component of the market, but Aero Research Limited was able to survive by repairing war-damaged aircraft and by making laminated plastic tail planes. And not merely survive: it was still able to carry out research and anticipate future developments. One of these concerned the glueing of metals, to obviate the need to weld or rivet. Work on this problem led to the production in 1942 of 'Redux', a phenol–formaldehyde resin which remains one of the strongest adhesives known to man.

* In this industrial section, names and terminology commonly used in industry are being retained where appropriate in preference to the latest nomenclature used elsewhere in this book.

Figure 28.18 The original urea–formaldehyde plant

The ending of the war in 1945 meant the opening up once again of commercial markets. Increased interest in the production of reconstituted wooden products such as chipboard virtually guaranteed the economic viability of urea-formaldehyde resins. The urgent problem now was the

Figure 28.19 The present plant floodlit for the 'coming of age' celebration 21 years later in 1955

provision of sufficient capital to enable a new factory to be built to meet the likely rise in demand. In 1946, negotiations were started with the CIBA Company of Basle, leading to the acquisition of a controlling interest in Aero Research Limited and an eventual change of name to CIBA (A.R.L.) Limited. The capital was now available to build a new process plant for the production of the urea-formaldehyde 'Aerolite' (*Figure 28.18*). Also available to the Duxford factory were CIBA products, such as the 'Araldite' epoxy resins. These epoxy resins combine toughness with resistance to water, chemicals and excellent electrical insulation, as well as unique adhesion and non-shrink solidification. In 1958, a new process plant (*Figure 28.19*) was constructed for the production of 'Araldite' epoxy resins. Steady growth since then has led to an annual production of 85 000 tonnes of resin, compared with a pre-war figure of about 10 tonnes p.a. The merger of CIBA Ltd with J. R. Geigy S.A. (October 20, 1970) means that, whereas the entire staff of Aero Research Limited during the war was not much more than a dozen, the Duxford factory is now part of an organization of 65 000 employees[13].

Chemical

Urea-formaldehyde Resins (see *Figure 28.20*)
Oxidation of methanol Formaldehyde (methanal) is produced from me-

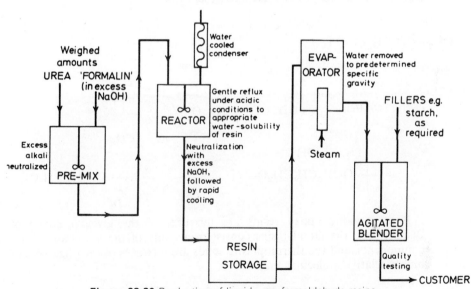

Figure 28.20 Production of liquid urea–formaldehyde resins

thanol by vapour phase dehydrogenation at 600 °C in the presence of a silver catalyst. The reaction may be basically represented as:

$$CH_3OH \xrightarrow[-2H]{\substack{Ag \\ 600-650\,°C}} HCHO$$

Reaction with urea Since urea is 4-functional and formaldehyde bifunctional there are many possibilities for reaction and the chemistry is very complex. Among the variables affecting the particular course of reaction are pH, temperature and the relative quantities of reactants.

Initial products include the methylolureas, formed by addition at the carbonyl bond of the formaldehyde:

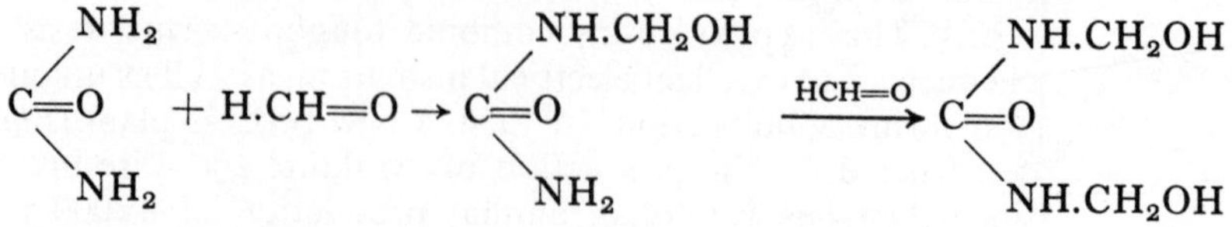

$$
\begin{array}{c}
NH_2 \\
| \\
C{=}O \\
| \\
NH_2
\end{array}
+ H.CH{=}O \rightarrow
\begin{array}{c}
NH.CH_2OH \\
| \\
C{=}O \\
| \\
NH_2
\end{array}
\xrightarrow{HCH{=}O}
\begin{array}{c}
NH.CH_2OH \\
| \\
C{=}O \\
| \\
NH.CH_2OH
\end{array}
$$

Subsequent polymerization involves to some extent the formation of methylene bridges between adjacent methylol and amino-groups, e.g.:

$$H_2N.CO.NH.CH_2OH + H_2N.CO.NH.CH_2OH$$

$$\xrightarrow[-H_2O]{} H_2N.CO.NH.CH_2.NH.CO.NH.CH_2OH$$

Epoxy resins (see Figure 28.21)
Epichlorhydrin reacts with 'Bisphenol-A' in alkaline solution, with the elimination of HCl, to give glycidyl ethers:

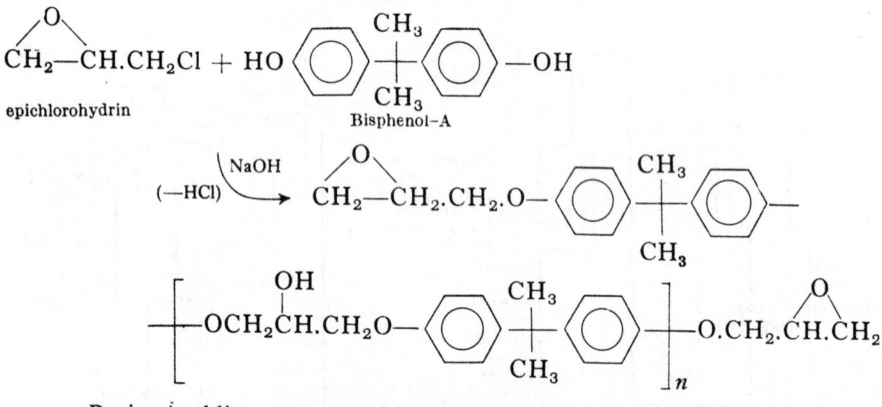

epichlorohydrin   Bisphenol-A

Basic Åraldite epoxy resins are mixtures of the glycidyl ethers of Bisphenol-A; the liquid resins consist principally of the $n = 0$ and $n = 1$ components, and the thermoplastic solids are mixtures of the oligomers of higher relative molecular mass.

Technical

Practitioners of chemistry will need little reminding that there is a world of technical detail and know-how between the representation of a chemical reaction in the manner illustrated above, and the production of well-characterized reactants in efficient quantity.

To give the student of chemistry a flavour of this technical world of

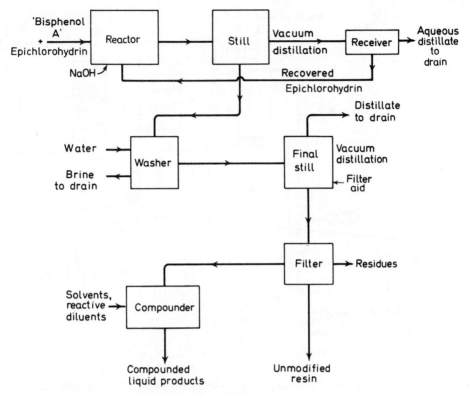

Figure 28.21 Production of liquid Araldites

chemical industry, we will now follow in simplified and summarized form, the actual procedure used at Duxford for the conversion of methanol to formaldehyde (*see Figures 28.22* and *28.23*).

A homogeneous mixture of 64 per cent methanol (stored in mild steel tanks) and 36 per cent water is pumped through a stainless steel heat exchanger to the top bubble cap tray of a vertical stainless steel evaporator. The evaporator has a vertical calandria to which dried steam is supplied, while above the calandria a measured mass of air is filtered and supplied to a circular distribution manifold. The sump of the evaporator contains a weak methanol solution boiling at about 90 °C, and equilibrium conditions are maintained by means of an evaporator level controller adjusting the methanol/water feed to give a constant evaporator level for a set steam/air flow rate.

Above the evaporator is a stainless steel superheater in which the feed vaporizing in the evaporator is heated to 95–100 °C, before passing through a copper gauze flame trap into the convertor chamber. Here the gases come into contact with an electrolytically refined silver crystal catalyst laid to a depth of nearly half an inch on fine-mesh copper gauze. The catalyst is first heated by nichrome toaster elements to initiate the reaction, which begins at about 450 °C. When the conversion of methanol to formaldehyde has begun, the heaters are switched off and the temperature raised to 600 °C by

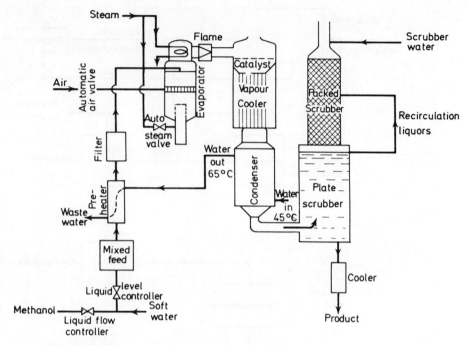

Figure 28.22 Schematic flow sheet of formaldehyde plant

increasing the air flow thereby increasing the oxygen supplied to the reaction.

The hot gases, consisting of formaldehyde, methanol, water vapour and air, leave the catalyst holder and pass into a vertical stainless steel vapour cooler where by heat exchange with condensate they generate steam for the evaporator, at the same time being cooled to 150 °C. The gases are further cooled to 50 °C by passing through a vertical stainless steel condenser. Condensed liquors and uncondensed gases pass from the condenser into the bottom of a stainless steel plate scrubber. Equilibrium conditions are maintained in the scrubber by recirculating liquors collecting in the top tray of the plate scrubber to a point half way up the packed scrubber above it. Condensed and absorbed gases run from the bottom of the plate scrubber into two aluminium 1900 gallon tanks, where they are blended with water and methanol to give a product of constant quality. Final storage is in two 110 000 gallon lagged mild-steel resin-lined tanks, kept at 40 °C by thermostatically controlled steam heating coils.

The plant is run by one man on each shift who regulates the process from a control room and tests and standardizes the product. The automatic control of all major operations is maintained from a central instrument panel. The rate of steam flow to the evaporator is set by a flow controller operating a valve in the main steam line, and a liquid level controller regulates the feed flow to the evaporator, keeping a steady level for a given evaporation rate. A further ratio controller makes sure that the evaporator feed is of a constant methanol/water ratio, whilst a butterfly valve operated by an air flow controller is adjusted from the control panel to give a constant flow of air

Figure 28.23 Part of the formaldehyde plant showing, from left to right, the evaporator, vapour cooler and condenser, and scrubber. This should be compared with the flow sheet of Figure 28.22. (Photograph courtesy of CIBA-GEIGY (U.K.) Ltd.)

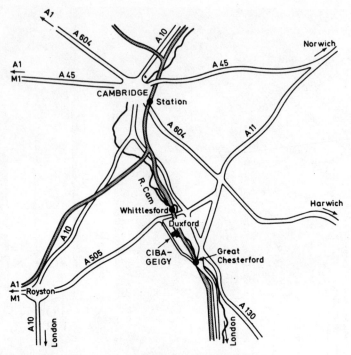

Figure 28.24 Road and rail network around Duxford

from the blowers. The temperature of the catalyst bed is recorded by a three-point recorder from three chromel-alumel thermocouples. Should the catalyst temperature rise to 700 °C, the recorder sounds a warning hooter, whilst the plant is automatically closed down if there is a further rise to 750 °C. This is just one of several safety precautions taken in the factory.

Socio-economic

As was indicated in the historical section, the Duxford factory is now part of the CIBA-GEIGY group, with total assets of over 3500 million Swiss francs. Not only does this allow more imaginative and longer-term investment, but absorption into the fields of pharmaceuticals, dyestuffs, plastics, agrochemicals and photographic products increases the intellectual and technical resources as well. In an area as capital and research-intensive as the chemical industry, these are both considerable assets and the net result is that both capital investment and sales value per employee are high at the Duxford factory, even compared with the rest of the chemical industry which, it will be remembered from points made earlier in this chapter, compares very favourably with other industrial sectors.

It may be wondered how far a site chosen originally for personal reasons measures up to the demands accompanying the expansion that has taken place since the war. Reference to the map (*Figure 28.24*) will reveal that the Duxford factory has ready access to roads to London, the Midlands, the East

Coast and hence Europe. The main Cambridge–London railway runs close
to the site boundary, while the River Cam forms part of the boundary itself.
(There is no likelihood of the latter ever providing transport links with the
outside, but it is at least a ready recipient for purified effluent—see later). The
factory has access to the railway network *via* sidings at Great Chesterford
and Whittlesford, about 3 km up-track towards London. Finally, there is the
added bonus of water existing in quantity in the chalk beneath the site, from
which it is pumped at the rate of up to 3 million gallons (15×10^6 dm^3) per
day. It can be seen, then, that the site has several advantages perhaps not
immediately apparent to Dr de Bruyne when he bought it 50 years ago. The
very fact that the area of about 100 acres (40 hectare) is more than is needed
for industrial development means that, apart from sports fields for cricket,
football, hockey and tennis, there is land at present available for the
cultivation of cereal crops, to maintain an open, rural outlook to the factory.

Despite all this, the classical economist might still have had reservations
about the value of the Duxford site in terms of the availability of labour. The
development of the petrol and oil engine has, however, had an impact that
would have been difficult to foresee even half a century ago. Increasing
mechanization of the land has led to a surplus of labour which the factory has
been able to tap. The availability of the motor-car, on the other hand, means
that employees are able to travel swiftly and conveniently from outlying
areas so that Cambridge itself, with all its amenities, is almost literally just
around the corner.

Environmental

Perhaps owners of a site as rural as this have added responsibilities to the
community in the way of preservation. It is quickly apparent, on visiting the
factory that there has been a keen eye for the aesthetic in designing and siting
the various buildings; not only is there functional elegance and cleanness of
line, but there is a scale about it which ensures that nearby housing estates
will not be swamped and dominated. When the landscaping and gardening
projects are also considered, it is readily apparent that a different philosophy[14]
is at work from that which characterized the Industrial Revolution of a
couple of centuries ago (*see Figures 28.25* and *28.26*).

What is more relevant than the aesthetic environment, however, is the
chemical. We will end this study, therefore, by looking at the way in which the
effluent from the factory is treated (*see Figure 28.27*).

The bactericidal nature of the contaminated waste water generated by the
factory poses problems in terms of effluent treatment. Fortunately it has been
found that these problems can be solved by using a bacteria-rich medium
assisted by the addition of domestic sewage as an additive and submitting the
mixed waste to biological oxidation. This, in fact, is the main principle
involved in the treatment of effluent. It remained to put this principle into
effective practice by installing appropriate equipment; this was developed
during 25 years to keep pace with expanding production resulting in the
outlay of a quarter of a million pounds on plant large and efficient enough to
serve a population of 40 000.

The first stage of the treatment involves the pumping of liquid waste into a

Figure 28.25 The factory at Duxford photographed from the air and showing its rural setting

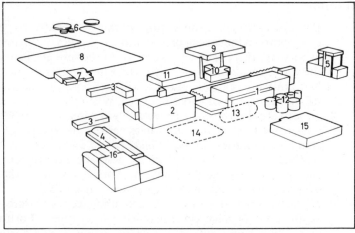

1 = Urea formaldehyde 'aerolite' plant
2 = 'Araldite' epoxy resin plant
3 = Research buildings
4 = Technical development laboratories
5 = Spray drier
6 = Effluent plant
7 = Dining hall
8 = Sports field
9 = Bonded structures
10 = Boiler house
11 = Engineering services
12 = Methanol & formalin tanks
13 = Liquid resin loading bay
14 = Tank farm
15 = Chemical stores
16 = Multi-purpose plant

Figure 28.26 Plan of the factory, showing the buildings and plant photographed in Figure 28.25

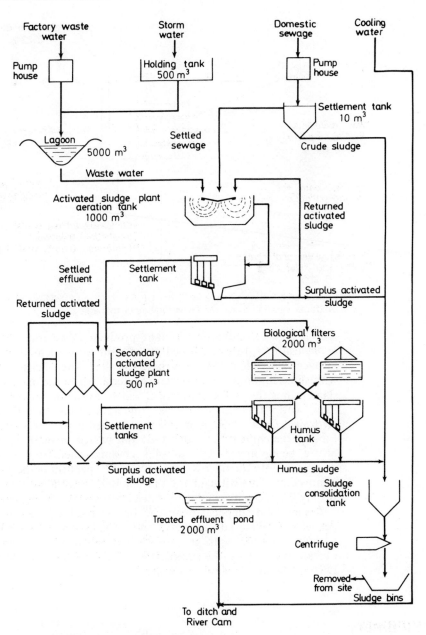

Figure 28.27 Effluent treatment flow sheet

lagoon with a capacity of a million gallons (5×10^6 dm³). This capacity represents the output of four days and so there is the opportunity at this stage of some 'evening-out' of concentrations. From here, waste water is pumped to the primary *activated sludge plant*, where there is intensive aeration and turbulence. Purification takes place here by means of a bacteriological floc known as activated sludge which is separated from the treated water and

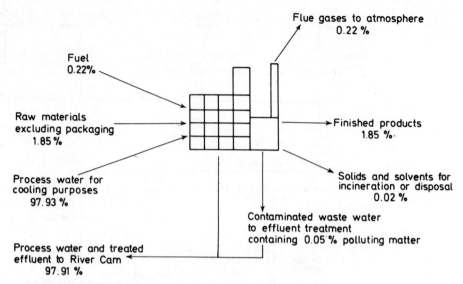

Figure 28.28 Mass balance for CIBA-GEIGY (UK) Ltd., Plastics Division, Duxford, with figures expressed as total input tonnage

recycled. A second type of plant producing bacterial oxidation is the *biological filter*, through which the effluent subsequently passes; the blast-furnace slag which makes up these so-called 'filter' beds becomes covered with slime containing appropriate bacteria, whilst the open structure of the slag ensures that sufficient oxygen will be available for the bacteria to do their work. Effluent which has undergone this treatment passes into *settlement tanks*, where humus sludge formed by bacterial action separates. Finally, the effluents are combined and passed through a balancing pond before being mixed with surplus cooling-water from the factory and discharged into the river. Tests show that 99 per cent of the pollution load is removed by this treatment and that the BOD (biochemical oxygen demand) is only 2 mg dm^{-3}, similar to that of many clean British rivers.

A successful chemical plant can be regarded as the site of interaction between raw material and energy under conditions discovered and provided by human ingenuity, to give the desired products in maximum yield. The overall situation can be summarized in a mass balance sheet; that for the Duxford factory is shown in *Figure 28.28*.

Summary

This chapter represented an attempt to present a brief account of the interaction of chemistry and society. Society can arrange for chemistry to be taught:

(*a*) So that it enables people to have more direct control over their own lives.
(*b*) So that students can enter the profession of chemistry or work in chemical industry at different levels.

(c) So that people can understand and take part in discussions on chemical matters affecting their future.

These possible aims of a chemical education are not mutually exclusive; how they interact and the emphasis each receives will depend, among other factors, on the nature of society and the ideology(ies) underpinning it. In a totalitarian dictatorship, one can imagine chemical education being used as a means of providing the appropriate labour force on the one hand, and, on the other, producing compliant citizens who obey without question. In a democratic society, a chemical education could be expected to produce free-thinking autonomous individuals, bringing a critical perspective to bear, and acting morally and responsibly. Just how complex the situation really is can be seen from how a capitalist economy in an apparently democratic society can produce totalitarian features in that society; such as the allocation of people to roles of apparently permanent subservience, the use of chemistry in the cause of exploitation, the denial of access to decision-making procedures, the location of power in the hands of a few individuals. Such are some of the contradictions that we experience today in the interactions between science, the economy and the individual.

Questions

(1) This Chapter has taken a generalised view of capitalism. What variations of capitalism might exist which repudiate or weaken the arguments advanced?

(2) Consider the factors involved in the location of an aluminium plant.

(3) Consider in the light of this Chapter the implication of *Figures 28.29* and *28.30*.

(4) How far do you think it essential that a scientist is a morally responsible person (a) in the choice of a research project (b) in applying his knowledge in the field of industry? Emphasize any particular difficulties a scientist may face in these areas.

(5) Work out a recycling programme for some industrial material.

(6) Discuss methods currently available for the disposal of waste in gas, liquid and solid phases. Suggest how these methods may be improved so as to approach genuine waste control.

(7) Indicate possible reasons for the present rate of increase of 0.2 per cent carbon dioxide in the atmosphere per annum. Suggest some possible long-term effects of the trend, if it continues, and suggest what mechanisms of Nature tend to control this trend.

(8) The case-study described in this chapter follows the development of a small business run by the founder-innovator to a large business which is now part

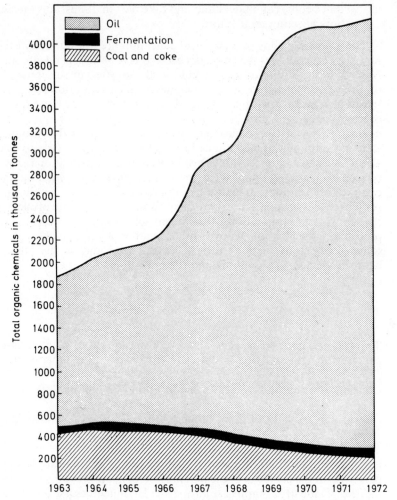

Figure 28.29 Sources of production of organic chemicals in the
United Kingdom, 1963–1972

of a gigantic multinational corporation. In what ways do you think
relationships may have changed during this period?

(9)   'Many of the products of modern technology are used to solve problems
which are the result of the industrial system in the first place'. Discuss with
regard to drugs.

(10)   The total amount of solar radiation reaching the biosphere is 1.4×10^3 W
m^{-2} min^{-1}.

(a) If the area of the Earth's biosphere facing the sun at any one time is $1.45
\times 10^{14}$ m^2 what is the total solar radiation intercepted by the biosphere?

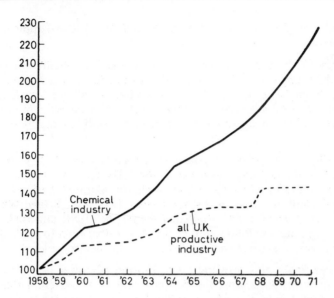

Figure 28.30 Chemicals production, with 1958 = 100, seasonally adjusted

(b) The mean distance of the Earth from the Sun is 1.50×10^3 km. Calculate the solar radiation which would reach the surface of a hypothetical planet 7.50×10^7 km from the Sun.

(c) 1 watt = 14·3 cal min^{-1} and 33 per cent of the radiation incident upon the upper layers of the biosphere is absorbed by them. Calculate the energy level at the Earth's surface in terms of cal cm^{-2} min^{-1}.

(d) If the velocity of light is 3×10^8 m s^{-1}, how long will it take to travel from the Sun to the earth?

(e) The total energy utilized in all the photosynthetic processes in the biosphere is estimated to be 4×10^{13} W. What percentage is this of the total incoming radiation at the Earth's surface?

[London 1975]

Notes

(1) Shortly after the accident, in August 1976, the scientific weekly *New Scientist* published an article by its specialist on industrial safety matters Lawrence McGinty, giving a detailed account of the accident, its consequences, the chemistry concerned, and other related incidents. So far as the world scientific community was concerned, therefore, within six weeks of the accident considerable information about 245-T, dioxin, and the related bactericide, hexachlorophane, was fully available, making later talk of secret processes' less acceptable.

(2) Of course there are rewards in terms of recognition and promotion. The notion of *disinterested* knowledge and the pursuit of 'knowledge for its own sake' takes little account of the motives for people's actions. For this and

other reasons, the divisions between, for example, basic and applied research are dubious.

(3) Whilst this element of competition is most apparent in commercial research, it might be claimed that the academic researcher is not as pure or open as he might like to think but is also contaminated by the system. See, for example, *The Double Helix* by James Watson for an account (admittedly retrospective and image-building) of 'pure' research in a highly competitive field.

(4) Almost the entire development of science and technology, and also of medicine and pharmacy, has occurred in the last few centuries, spearheaded in those countries which embraced capitalism and free market forces. It is perhaps significant that in the years since 1917 little, if any, scientific or technological novelty has originated within the USSR and their industry tends to imitate and follow that of the West rather than to lead. Discovery and innovation are not characteristic of controlled societies.

(5) Interestingly, this growing concentration of power is a phenomenon confined to Britain among the Western nations. It is accompanied by the lowest growth rate. *See The Guardian* Financial Extra for Jan 31 1977 by Kevin Page: 'Is it time to break up the Giants?'

(6) *See* Commoner, B. *The Closing Circle; confronting the Environmental Crisis.*

(7) *See Guardian Extra* for Dec 5 1973: 'Grubbing for Earth's Crust' by Lester R. Brown.

(8) *See* Schumacher, E. F. *Small is Beautiful.* (London, Blond and Briggs, 1973).

(9) The chairman of the board presents his annual report at the shareholders' meeting but this is rarely an occasion for fundamental criticism.

(10) This expertise will naturally reflect the prevailing ideology. Experts are seldom able to 'go native' in alien cultures. Witness, for example, the reaction of Harley Street to Chinese medicine.

(11) *The Windscale Enquiry.* Report by Hon. Mr Justice Parker Jan 26 1978. Vol 1 (HMSO).

(12) *See*, for example, *Undercurrents*, number 9, 'Nuclear Power, the Devil we don't Know'.

(13) It is doubtless significant that the Duxford factory had to clear this study with Head Office before publication was allowed.

(14) A different philosophy perhaps. But it could be argued that the central ideology was unchanged; that by providing various amenities the firm is helping the worker identify with the organization.

Appendix 1

Constants and conversion factors

Relative atomic masses

Relative atomic masses were originally based on a chemical standard—the natural mixture of the isotopes of hydrogen (i.e. $H = 1.0000$). Since the introduction of accurate mass spectroscopy, physical standards have been adopted; until recently, the most abundant isotope of oxygen has been used as the standard (i.e. $^{16}O = 16.0000$), but now the value $^{12}C = 12.0000$ is employed.

The SI Units

SI (Système International) units have been used in this book where possible, but difficulty is met with in dealing with molar concentrations previously expressed in mol(e)/litre. Concentration, according to SI nomenclature, should be molal, i.e. $mol\ kg^{-1}$, although little difference will exist between $mol\ dm^{-3}$ and $mol\ kg^{-1}$ for dilute aqueous solutions and the symbol M for such solutions has been retained.

The SI fails to recognize g-molecule, g-equivalent, etc; instead mole is used for such quantities, where one mole is the amount of substance which contains as many elementary particles as there are atoms in 0.012 kilogram of carbon-12. The elementary unit must be specified and may be an atom, molecule, ion, radical electron, photon, etc., or a specified group of such entities. Thus one mole of $\frac{1}{2}H_2SO_4$ is the same as one equivalent of H_2SO_4 and one mole of $SO_4{}^{2-}$ is used in place of one g-ion of $SO_4{}^{2-}$.

Prefixes for SI units

Fraction	Prefix	Symbol	Multiple	Prefix	Symbol
10^{-1}	deci	d			
10^{-2}	centi	c			
10^{-3}	milli	m	10^3	kilo	k
10^{-6}	micro	μ	10^6	mega	M
10^{-9}	nano	n	10^9	giga	G
10^{-12}	pico	p	10^{12}	tera	T
10^{-15}	femto	f			
10^{-18}	atto	a			

Definitions of units

Ampere. The ampere is that constant current which, if maintained in two straight parallel conductors of infinite length, of negligible circular cross section, and placed 1 metre apart in a vacuum, would produce between these conductors a force equal to 2×10^{-7} newton per metre of length. Symbol A

Newton. The newton is that force which, when applied to a body of mass 1 kilogram, gives it an acceleration of one metre per second per second. Symbol $N = kg\ m\ s^{-2}$

Joule. The joule is the unit of energy and is the work done when the point of application of a force of 1 N is displaced through a distance of 1 metre in the direction of the force. Symbol $J = kg\ m^2\ s^{-2}$

Coulomb. The coulomb is the unit of charge and is the quantity of electricity transported in 1 second by a current of 1 ampere. Symbol $C = A\ s$

Kelvin. The units of kelvin and Celsius temperature interval are identical. A temperature expressed in degrees Celsius (°C) is equal to the temperature expressed in kelvin less 273.15, i.e., $K = °C + 273.15$. (For practical purposes, $K = °C + 273$)

Ohm. The unit of electrical resistance is the ohm and is the resistance between two points of a conductor when a constant potential difference of 1 volt, applied between these points, produces in the conductor a current of 1 ampere, provided that the conductor itself is not a source of any electromotive force. Symbol $\Omega = V\ A^{-1}$. (Electrical conductance, the reciprocal of electrical resistance, is sometimes measured in siemens, $S = \Omega^{-1}$)

Volt. The volt is the unit of electrical potential and is the difference of potential between two points across which a current of 1 ampere flows, when the power dissipated between these points is equal to 1 watt. Symbol $V = W\ A^{-1}$

Watt. The unit of power is the watt and is equal to 1 joule per second. Symbol $W = J\ s^{-1}$

Pascal. A pressure of $1\ N\ m^{-2}$ is referred to as one pascal (Pa)

Metre. The metre is the unit of length, equal to 1 650 763.73 wavelengths in vacuum of the radiation corresponding to the transition between the levels $2p_{10}$ and $5d_5$ of the krypton-86 atom. Symbol m

Kilogram. The unit of mass is the kilogram and is equal to the mass of the international prototype made of platinum–iridium kept at Sevres. Symbol kg

Second. The second is the duration of 9 192 631 770 periods of the radiation corresponding to the transition between the two hyperfine levels of the ground state of the caesium-133 atom. Symbol s

Conversion factors

Ångström unit, $Å = 10^{-10}\ m = 10^{-1}\ nm = 10^2\ pm$

Micron unit, $\mu = 10^{-6}\ m = 1\ \mu m$

Calorie, $cal = 4.184\ J$

Litre, $l = 1\ dm^3$

Atmosphere, $atm = 760\ mmHg = 760\ torr = 101\ 325\ N\ m^{-2} \approx 10^5\ Pa$

Logarithm, $\log_e = \ln = 2.303 \times \log_{10}$

Appendix 2

Table of Relative Atomic Masses and Numbers
(By courtesy of the International Union of Pure and Applied Chemistry)

	Symbol	Atomic number	Relative atomic mass		Symbol	Atomic number	Relative atomic mass
Actinium	Ac	89		Mendelevium	Md	101	
Aluminium	Al	13	26.98	Mercury	Hg	80	200.61
Americium	Am	95		Molybdenum	Mo	42	95.95
Antimony	Sb	51	121.76	Neodymium	Nd	60	144.27
Argon	Ar	18	39.944	Neon	Ne	10	20.183
Arsenic	As	33	74.91	Neptunium	Np	93	
Astatine	At	85		Nickel	Ni	28	58.71
Barium	Ba	56	137.36	Niobium	Nb	41	92.91
Berkelium	Bk	97		Nitrogen	N	7	14.008
Beryllium	Be	4	9.013	Nobelium	No	102	
Bismuth	Bi	83	209.00	Osmium	Os	76	190.2
Boron	B	5	10.82	Oxygen	O	8	16
Bromine	Br	35	79.916	Palladium	Pd	46	106.4
Cadmium	Cd	48	112.41	Phosphorus	P	15	30.975
Calcium	Ca	20	40.08	Platinum	Pt	78	195.09
Californium	Cf	98		Plutonium	Pu	94	
Carbon	C	6	12.011	Polonium	Po	84	
Cerium	Ce	58	140.13	Potassium	K	19	39.100
Caesium	Cs	55	132.91	Praseodymium	Pr	59	140.92
Chlorine	Cl	17	35.457	Promethium	Pm	61	
Chromium	Cr	24	52.01	Protactinium	Pa	91	
Cobalt	Co	27		Radium	Ra	88	
Copper	Cu	29	63.54	Radon	Rn	86	
Curium	Cm	96		Rhenium	Re	75	186.22
Dysprosium	Dy	66	162.51	Rhodium	Rh	45	102.91
Einsteinium	Es	99		Rubidium	Rb	37	85.48
Erbium	Er	68	167.27	Ruthenium	Ru	44	101.1
Europium	Eu	63	152.0	Samarium	Sm	62	150.35
Fermium	Fm	00		Scandium	Sc	21	44.96
Fluorine	F	9	19.00	Selenium	Se	34	78.96
Francium	Fr	87		Silicon	Si	14	28.09
Gadolinium	Gd	64	157.26	Silver	Ag	47	108.880
Gallium	Ga	31	69.72	Sodium	Na	11	22.991
Germanium	Ge	32	72.60	Strontium	Sr	38	87.63
Gold	Au	79	197.0	Sulphur	S	16	32.066†
Hafnium	Hf	72	178.50	Tantalum	Ta	73	180.95
Helium	He	2	4.003	Technetium	Tc	43	
Holmium	Ho	67	164.94	Tellurium	Te	52	127.61
Hydrogen	H	1	1.0080	Terbium	Tb	65	158.93
Indium	In	49	114.82	Thallium	Tl	81	204.39
Iodine	I	53	126.91	Thorium	Th	90	232.05
Iridium	Ir	77	192.2	Thulium	Tm	69	168.94
Iron	Fe	26	55.85	Tin	Sn	50	118.70
Krypton	Kr	36		Titanium	Ti	22	47.90
Kurchatovium		104		Tungsten	W	74	183.86
Lanthanum	La	57	138.92	Uranium	U	92	238.07
Lawrencium	Lw	103		Vanadium	V	23	50.95
Lead	Pb	82	207.21	Xenon	Xe	54	131.30
Lithium	Li	3	6.940	Ytterbium	Yb	70	173.04
Lutetium	Lu	71	174.99	Yttrium	Y	39	88.92
Magnesium	Mg	12	24.32	Zinc	Zn	30	65.38
Manganese	Mn	25	54.94	Zirconium	Zr	40	91.22

* The basis for relative atomic masses is $^{12}C = 12$.

† Because of natural variations in the relative abundance of the isotopes of sulphur the relative atomic mass of this element has a range of ± 0.003.

Appendix 3

Summary of definitions

Absolute temperature. This is recorded in K (kelvin): K $= 273.15 +$ °C, since at -273.15 °C, i.e., 0 K, all molecular motion ceases.

Acid. A proton donor (Brønsted–Lowry) or an electron-pair acceptor (Lewis acid). In aqueous solution, an acid, HA first donates a proton to water, giving H_3O^+, the oxonium ion. BF_3 is a Lewis acid as it can accept the lone pair on ammonia:

$$F_3B \leftarrow :NH_3$$

Addition reaction. A general reaction of unsaturated compounds producing saturation; addition takes place in stages, not by direct addition across double bond.

Alkali. A soluble base producing OH^- ions.

Anode. The electrode to which anions (negatively charged atoms or groups of atoms) travel.

Aqueous solution. Chiefly or entirely water as the solvent.

Avogadro's constant. The number of molecules in one mole (6.023×10^{23} mol^{-1}).

Avogadro's law. Equal volumes of all gases at the same temperature and pressure contain the same number of molecules. This was the breakthrough in determining relative molecular masses.

Base. A proton acceptor. The oxide of a metal is a base because of the reaction: $O^{2-} + 2H_3O^+ \rightarrow 3H_2O$. Some metal oxides are *amphoteric*, that is, capable of reacting with an acid or an alkali, e.g.,

$$ZnO + H_2O + 2OH^- \rightarrow [Zn(OH)_4]^{2-}$$

Organic bases are amines—substituted ammonia, e.g., $:N(CH_3)_3$ (compare with $:NH_3$); the lone pair on the N is capable of accepting H^+ giving, e.g., NH_4^+, $[NH(CH_3)_3]^+$

Boyle's law. At constant temperature, the volume of a fixed mass of gas is inversely proportional to the pressure.

Buffer solution. One which resists any change in pH when treated with small amounts of acid or alkali. Might consist of a weak acid HA and its sodium salt NaA; the ionization of the weak acid is suppressed by its sodium salt and is therefore able to neutralize added alkali or reverse its ionization when further acid is added, through reaction with anions of the salt.

$$HA + OH^- \rightarrow H_2O + A^-$$

$$A^- + H^+ \rightarrow HA$$

They may also consist of weak bases with their chlorides, e.g., ammonia and ammonium chloride:

$$NH_3 + H^+ \rightarrow NH_4^+$$

$$NH_4^+ + OH^- \rightarrow NH_3 + H_2O$$

Catalyst. A substance which increases the rate of a reaction but can be recovered at the end of the reaction.

Cathode. The electrode to which cations (positively charged atoms or groups of atoms) travel.

Charles' law. At constant pressure, the volume of a fixed mass of gas is directly proportional to the absolute temperature. (It was the discovery that *all* gases expanded as if they would have zero volume at about $-273\,°C$ that originally gave rise to the idea of an absolute temperature scale.)

Colligative property. One which depends only on the number of particles, not on their type, e.g., osmotic pressure.

Colloidal solution. A suspension of particles sufficiently small to allow the effect of gravity to be balanced by their Brownian motion, i.e., their random motion caused by bombardment from solvent molecules.

Conductance. For a solution, conductance is equal to the reciprocal of its electrical resistance.

Complex salt. One containing at least one complex ion in which the central atom has associated with it a number of ligands which sufficiently modify the properties of the central atom to cause it not to behave as a simple ion, e.g., $[Fe(CN)_6]^{3-}$ fails to give the characteristic properties of the simple Fe^{3+} ion.

Critical point. The transition point where the liquid and gaseous states merge into each other; characterized by a critical pressure and temperature.

Dalton's law of partial pressures. The total pressure of a mixture of gases is equal to the sum of their partial pressures. The partial pressure of a gas is the pressure which it would exert if it alone occupied the total space.

Degree of ionization (dissociation). That fraction of one mole of a weak electrolyte which ionizes; unlike dissociation constant it varies with concentration. The degree of dissociation (α), molarity (C) and dissociation constant K are related by Ostwald's dilution law:

$$K = \frac{\alpha^2 C}{(1 - \alpha)}$$

Diamagnetism. See paramagnetism.

Diffusion. See osmosis.

Dipole moment. The product of the internuclear distance and the charge separation in a heteroatomic bond; traditionally measured in debye (D): $1\,D = 3.336 \times 10^{-30}$ C m.

Disproportionation. A redox reaction in which the same substance acts as its own reducing and oxidizing agent.

Distribution law. The ratio of the concentrations of a substance dissolved in two immiscible liquids in equilibrium with each other is a constant at a fixed temperature.

Dulong and Petit's law. The product of the specific heat capacity of a solid element and its relative atomic mass is approximately a constant equal to 26 $J K^{-1} mol^{-1}$.

Electrode potential. The difference in electrical potential established between an electrode and the solution in which it is placed. If the solution is molar with respect to the ions forming the electrode then the *standard electrode potential* may be measured provided that the system is in equilibrium and can be coupled to another half-cell.

Electrolysis. The decomposition of an electrolyte by the passage of an electric current.

Electrolyte. A dissolved or fused substance which conducts electricity by means of mobile ions, with chemical changes at the electrodes. If the substance is completely ionized it is a *strong electrolyte*; partly ionized it is a *weak electrolyte*.

Electron affinity. The enthalpy change occurring when electron capture by a gaseous atom or ion takes place. Process is exothermic for single capture by a neutral atom.

Electronegativity. A measure of the electron-attracting power of an atom *within a molecule*.

Electrolytic conductivity. The conductance of a solution contained between electrodes of unit area and separated by unit distance.

Electrophilic reagent. One which attacks high electron density regions in other molecules, e.g., NO_2^+ reacting with benzene.

Enantiotropy. See monotropy.

Energy barrier. The *activation energy* required to bring reacting species together into an *activation complex* or *transition state*, from which new products form or the reactants are regenerated depending on the reversible nature of the overall reaction.

Enthalpy change. The change in heat content for a reaction; exothermic if heat is evolved, ΔH negative; endothermic if heat is absorbed, ΔH positive.

Entropy. A measure of disorder in a system; an increase in entropy allows for a greater redistribution of energy and is the natural tendency for the universe.

Equilibrium law. At equilibrium the ratio of the concentrations of the products of a reversible reaction raised to the power of their combining ratios divided by the product of the concentrations of the reactants raised to their combining ratios is a constant; the constant is the *equilibrium constant*, K_c or K_p, depending on the concentrations expressed in mole dm^{-3} or as partial pressures.

Faraday's laws of electrolysis. (1) The amount of substance liberated is proportional to the quantity of electricity (coulombs) passed. (2) To liberate one mole of an *n*-valent ion required nF coulombs, where $1 F = 96\,500$ C. (This charge is called the Faraday, 96.5 kC mol^{-1}).

First law of thermodynamics. Energy can neither be created or destroyed, but only changed from one form to another; it is the *Law of conservation of energy*.

Free energy change. The maximum useful work which a system can do (ΔG negative for a spontaneous reaction).

Functional group. Generally the more reactive part of a molecule, depending on the reagent attacking it.

Gay-Lussac's law of combining volumes. When gases react together they do so in simple ratio by volume and if the products are also gaseous, their volumes will also be in simple ratio, provided that temperature and pressure remain constant.

Graham's law of diffusion. The rate of diffusion of a gas is inversely proportional to the square root of the density. Because of Avogadro's law we can also say that, for two gases at the same temperature and pressure, their rates of diffusion are inversely proportional to the square roots of their relative molecular masses.

Half cell. Consists of an electrode dipping into an electrolyte and to measure its potential difference requires coupling to another half-cell; single electrode potentials cannot be measured.

Henry's law. The mass of gas dissolved by a fixed mass of liquid is directly proportional to the gas pressure at constant temperature (does not apply to gases such as ammonia which react with water).

Hess' law of constant heat summation. The total enthalpy change for a set of reactions is independent of the route taken.

Heisenberg's Uncertainty Principle. The product of the errors in measuring simultaneously the position and momentum of an electron equals approximately the value of Planck's constant; i.e., the position and momentum of a small particle cannot both be precisely known at the same time.

Heteroatomic (polar) bond. A chemical bond formed between two atoms of different electronegativities; gives rise to a bond dipole moment.

Homoatomic bond. A chemical bond formed between two identical atoms.

Homologous series. Adjacent members differ by CH_2; all possess the same functional group(s).

Hund's rule. This requires that electrons should exert their maximum multiplicity whenever energy requirements permit, i.e., electrons in the same set of orbitals will, if possible, be unpaired.

Hybridization of atomic orbitals. This permits their mixing to produce new orbitals, e.g., one s and three p orbitals produce four sp^3 hybrid orbitals.

Ion. An atom or group of atoms which have lost or gained one or more electrons; necessarily present in *ionic (electrovalent)* compounds as opposed to *covalent* compounds.

Ionic product of water. The product of the ionic concentrations of hydrogen and hydroxide ions in solution; this product must equal 1×10^{-14} mol^2 dm^{-6} at 25 °C in any aqueous solution.

Ionization energy or potential. The energy required to remove an electron from a gaseous atom or ion.

Isotopes. Atoms having the same number of protons (therefore belonging to the same element) but different number of neutrons.

Kohlrausch's law of independent migration of ions. At infinite dilution the molar conductivity is the sum of the separate ionic conductivities comprising the electrolytic solution.

Law of conservation of mass. Matter can neither be created nor destroyed in a chemical reaction.

Law of constant composition. No matter how a pure substance is made it will always have the same composition.

Law of mass action. The rate of a chemical reaction is proportional to the product of the concentrations of reactant species raised to low powers (0, 1 or 2). Nowadays referred to as the *rate law*. Originally propounded by Guldberg and Waage.

Law of multiple proportions. If two elements form more than one compound, then the several masses of A which separately combine with a fixed mass of the second element B, are in simple ratio.

Law of octaves. A restricted and much earlier form of the *periodic law*, due to Newlands.

Law of partial pressures. *See* Dalton's law.

Le Chatelier's principle. If a constraint is applied to a system in equilibrium, then the system adjusts itself so as to try to overcome the effect of the constraint.

Ligand. An atom, group of atoms, or ion capable of forming a bond with a central atom or ion generally by donating a lone pair of electrons, e.g., :CN^- in $[Fe(CN)_6]^{3-}$.

Markownikoff's rule. Addition of HX to an unsaturated bond takes place so that the H attaches itself to the carbon already carrying the greater number of hydrogen atoms.

Mass number. The total number of nucleons (neutrons and protons) in an atom; it therefore approximates to the relative atomic mass.

Molality. The concentration of a solution in moles of solute per kg of solvent.

Molarity. The concentration of a solution in moles of solute per dm^3 of solution.

Molar conductivity. The electrolytic conductivity per unit concentration.

Molecular depression (cryoscopic) constant. The theoretical lowering of freezing point of 1 kg solvent by the addition of one mole of solute.

Molecular elevation (ebullioscopic) constant. The theoretical elevation of the boiling point of 1 kg of a solvent by the addition of 1 mole of solute.

Mole. The amount of substance which contains as many elementary particles as there are atoms in 0·012 kg of carbon-12.

Mol fraction. The ratio of the mole of a substance to the total moles in a mixture.

Monotropy. This is displayed by those substances which exist in more than one polymorphic form and in which one of these forms is always metastable with respect to the others, e.g., 'white' phosphorus is the more energetic form than 'red' phosphorus no matter what conditions prevail. *Enantiotropy* is the case where each form is stable over different temperature intervals, e.g., α- and β-sulphur.

Moseley's law. The square root of the frequency of the characteristic X-rays emitted from a metal is proportional to the atomic number of the element. (This law was the first demonstration that atomic number had a physical meaning beyond being the ordinal number in the list of relative atomic masses.)

Nucleophilic reagent. One which attacks low electron density regions in other molecules, e.g., OH^-, :NH_2 (electron pair seeking positive centre).

Orbit. The path taken by a particle.

Orbital. The spatial envelope in which there is a, e.g., 00% chance of finding a particular electron.

Order of reaction. The sum of the powers to which the concentrations must be raised in order to obtain an expression for the rate law.

Osmosis. The passage of solvent from low concentration to high through a semipermeable membrane. *Diffusion* is the spreading out of the solute from high concentration to low, i.e., osmosis is negative diffusion.

Osmotic pressure. Pressure required to prevent osmosis occurring; if applied pressure is greater than osmotic pressure, *reverse osmosis* occurs.

Paramagnetism. This is shown by atoms with unpaired electrons; molecules are then pulled into a region of higher magnetic field; opposite effect for *diamagnetic* substances, where paired electrons only are present.

Pauli exclusion principle. No more than two electrons in an atom can have the same *quantum numbers*, i.e., n, l and m can have the same values for two electrons, but spin quantum numbers must then differ, i.e., $s = +\frac{1}{2}$ or $-\frac{1}{2}$.

Periodic law. The properties of the elements are periodic functions of their atomic numbers (Mendeleev).

pH. The negative logarithm to the base 10 of the hydrogen ion concentration in mol dm^{-3}; $pH = -\log_{10}[H_3O^+]$. pK_a and pK_b are similarly defined for the acid dissociation constant, K_a and the base dissociation constant, K_b, where these are the equilibrium constants for the ionizations:

$$HA \rightleftharpoons H^+ + A^- \quad \text{and} \quad MOH \rightleftharpoons M^+ + OH^-$$

$$K_a = \frac{[H^+][A^-]}{[HA]} \qquad K_b = \frac{[M^+][OH^-]}{[MOH]}$$

Pi (π) bond. The bond formed by 'sideways' overlap of atomic orbitals at right angles to the axis joining combined atoms.

Photon. A quantum of light.

Polar bond. See heteroatomic bond.

Polarized light. Light whose waves are vibrating in one plane only.

Raoult's law. The relative lowering of vapour presure of a solvent caused by a solute is *equal to* the mol fraction of the solute.

Rate law. See law of mass action.

Redox reaction. A reaction in which one substance donates electrons (is oxidized) and another substance accepts these electrons (is reduced).

Relative atomic mass. The number of times one atom of an element is greater in mass than one-twelfth of an atom of carbon-12.

Relative molecular mass. The number of times one molecule of a substance is greater in mass than one-twelfth of an atom of carbon-12.

Resonance energy. The stabilization energy of a compound caused by existence of more than one classical electronic structure for the molecule.

Salt. A salt is formed by replacing the hydrogen of an acid by a metal or other cation as in a neutralization reaction.

Saturated compound. One in which only σ (sigma) bonds are present; no double or triple bonds. Such compounds undergo substitution reactions.

Second law of thermodynamics. For any spontaneous process, the total entropy change is positive.

Sidgwick–Powell theory. This predicts the shape of molecules by accounting for all the bonding and non-bonding electron pairs.

Sigma (σ) bond. The bond formed by endwise overlap of atomic orbitals along the axis of two combined atoms.

Sequestering agent. A reagent which forms stable complexes (*chelates*) with cations.

Solubility product. The product of the ionic concentrations of a sparingly soluble salt raised to the powers of their combining ratio.

Substitution reaction. One in which there must be a leaving group before reaction can be complete.

Transition temperature. The temperature at which two modifications of the same substance are in equilibrium.

van der Waals forces. Weak interatomic forces caused by dipole interactions and polarizations.

van't Hoff factor. The correcting factor in the general gas equation, $pV = inRT$, to allow for the dissociation of substances, or, for colligative properties,

$$i = \frac{\text{effect observed}}{\text{effect calculated in absence of dissociation}}$$

Appendix 4

For further reading

Readers interested in pursuing certain topics beyond the scope of this book should find the following useful:

ATKINS, W. A., *Physical Chemistry*, Oxford University Press, London (1979)

BROWN, G. I., *A New Guide to Modern Valency Theory*, Longman Group, London (1980)

COTTON, F. A. and WILKINSON, G., *Basic Inorganic Chemistry*, John Wiley and Sons, London (1976)

EDELMAN, J. and CHAPMAN, J. M., *Basic Biochemistry*, Heinemann Educational Books, London (1978)

MOORE, J. W. and MOORE, E. A., *Environmental Chemistry*, Academic Press, London (1978)

TEDDER, J. M., NECHVATAL, A. and JUBB, A. H., *Basic Organic Chemistry*, Part 1, John Wiley and Sons, London (1980)

Index